Microbial Mats

Structure, Development and Environmental Significance

NATO ASI Series

Advanced Science Institutes Series

A series presenting the results of activities sponsored by the NATO Science Committee, which aims at the dissemination of advanced scientific and technological knowledge, with a view to strengthening links between scientific communities.

The Series is published by an international board of publishers in conjunction with the NATO Scientific Affairs Division

A Life Sciences B Physics	Plenum Publishing Corporation London and New York
C Mathematical and Physical Sciences D Behavioural and Social Sciences E Applied Sciences	Kluwer Academic Publishers Dordrecht, Boston and London
F Computer and Systems Sciences G Ecological Sciences H Cell Biology I Global Environmental Change	Springer-Verlag Berlin Heidelberg New York London Paris Tokyo Hong Kong Barcelona Budapest

NATO-PCO DATABASE

The electronic index to the NATO ASI Series provides full bibliographical references (with keywords and/or abstracts) to more than 30000 contributions from international scientists published in all sections of the NATO ASI Series. Access to the NATO-PCO DATABASE compiled by the NATO Publication Coordination Office is possible in two ways:

- via online FILE 128 (NATO-PCO DATABASE) hosted by ESRIN, Via Galileo Galilei, I-00044 Frascati, Italy.
- via CD-ROM "NATO Science & Technology Disk" with user-friendly retrieval software in English, French and German (© WTV GmbH and DATAWARE Technologies Inc. 1992).

The CD-ROM can be ordered through any member of the Board of Publishers or through NATO-PCO, Overijse, Belgium.

Series G: Ecological Sciences, Vol. 35

Microbial Mats

Structure, Development and Environmental Significance

Edited by

Lucas J. Stal

Laboratory for Microbiology
University of Amsterdam
1018 WS Amsterdam
The Netherlands

Pierre Caumette

Laboratoire d'Oceanographie
Biologique
Université de Bordeaux I
F-33120 Arcachon, France

Springer-Verlag
Berlin Heidelberg New York London Paris Tokyo
Hong Kong Barcelona Budapest
Published in cooperation with NATO Scientific Affairs Division

Proceedings of the NATO Advanced Research Workshop on Structure, Development and Environmental Significance of Microbial Mats, held in Arcachon, France, September 27–October 1, 1993

ISBN-13: 978-3-642-78993-9 e-ISBN-13: 978-3-642-78991-5
DOI: 10.1007/978-3-642-78991-5

CIP data applied for

Softcover reprint of the hardcover 1st edition 1994

Typesetting: Camera ready by authors
SPIN 10122391 31/3130 - 5 4 3 2 1 0 - Printed on acid-free paper

Preface

Microbial mats develop in a wide variety of different environments. They are known from intertidal coastal sediments, coral reefs, hypersaline ponds and lakes and marine salterns, thermal springs, Antarctic lakes, dry and hot deserts. Virtually all these microbial mats are formed by phototrophic microorganisms but also mats of the chemotrophic *Beggiatoa* spp. and more rarely of fungi are known. The majority of microbial mats are formed by the oxygenic prokaryotic cyanobacteria but mats of anoxygenic phototrophic organisms or eukaryotic algae are also known. Of particular interest are chemotrophic microbial mats that are found in deep-sea hydrothermal vents and in certain thermomineral sulfurous caves.

It is difficult to give a good definition of a microbial mat. Most researchers refer to laminated microbial ecosystems which are characterized by the cohesive, viscous layers they produce and that may vary in thickness from a few millimetres to 1 metre. In general, the biologically active layer is limited to the upper few mm or cm. However, other benthic communities of (phototrophic) microorganisms that do not form tough coherent structures, have also been referred to as microbial mats or as biofilms. The unifying property of microbial mats and biofilms is the densily packed biomass in which steep and fluctuating gradients of physical and chemical factors are present. Microbial mats are increasingly viewed as important model systems for the study of ecological relationships and biogeochemical cycles and may even serve as models for global circulation. Microbial mats exchange matter with both geo- and atmosphere and due to the high biomass content these systems may contribute to, and, perhaps, regulate global processes.

The investigation of microbial mats requires multidisciplinary studies and close cooperation between geologists, sedimentologists, chemists, biochemists, microbiologists, biogeochemists, taxonomists, physiologists and molecular biologists. Thus, the aim of this international meeting on microbial mats was to bring together specialists of different disciplines and stimulate exchanges and discussions about the structure, development and environmental significance of microbial mats. We hope that this book will contribute to and stimulate microbial mat research and encourage more scientists to study these systems.

This meeting on microbial mats has been preceded by 2 other workshops on the same topic. The first workshop on microbial mats was held in Woods Hole (USA) in 1982 (Cohen et al. 1984) followed by a second meeting in Eilat (Israel) in 1987 (Cohen and Rosenberg 1989). Whereas the Woods Hole meeting focussed on microbial mats as recent analogues of fossil stromatolites, the Eilat meeting was devoted to ecophysiology of microbial mat communities. These meetings were very successful and greatly stimulated research on microbial mats. Considerable progress has been made in microbial mat research since the Eilat workshop and this justified the organization of a new workshop. The 3rd meeting on microbial mats took place from September 27 to October 1, 1993 in the charming city of Arcachon, on the south-west coast of France.

The idea to have this third meeting on microbial mats developed during the 6th International Symposium on Microbial Ecology (ISME), held in September 1992 in Barcelona, Spain. During this symposium an evening round table discussion was organized by the editors of this book. One of the conclusions of this round table discussion was, that although our knowledge on microbial mats has increased tremendously during the last decade, very little is known about their structure and development. Other aspects that were discussed, included the possible applicability of microbial mats and the question whether environments in which microbial mats develop should be protected. It was stated that the natural environments of microbial mats are disappearing rapidly. This is in part due to increased anthropogenic activities in coastal zones, which are environments pre-eminently suited for microbial mats. Other environments such as alkaline inland lakes and sulfur hot springs disappear because of human intervention as well as by other natural causes. Recently, we have become aware of the enormous genetic potential of microbial mats and biofilms that may be used e.g. for degrading xenobiotics (Characklis and Wilderer 1989).

We begin to realize that microbial mats are more than interesting objects for study by a few microbiologists, biogeochemists and geologists. For instance, microbial mats developed on top of the oil that was released into the environment during the Gulf war and it was suggested that these mats are active in the degradation of the oil (Sorkhoh et al. 1992). Increasing interest also exists for the possibility to use natural or constructed microbial mats for degradation of xenobiotics and for site remediation (Bender et al. 1993; Bender and Phillips 1993; Phillips et al. 1993). Another possible application may be to prevent erosion of coastal zones that are currently increasingly endangered due to an increasing population that utilize these areas. Growth and metabolism of microorganisms, notably cyanobacterial mats and diatom biofilms, are significantly limiting erodibility of coastal sediments. The general conclusion of these discussions was that the investigation of applied aspects of microbial mats requires more information on their structure and development (Stal and Caumette 1993).

One year after the 6th ISME, the Arcachon meeting on 'Structure, Development and Environmental Significance of Microbial Mats' was a fact. The meeting brought together scientists working on microbial mats and related fields. The workshop was attended by 47 scientists from 9 different countries.

The study of the structure of microbial mats is facilitated by a suite of new or improved methods. Methodology is one of the most serious constraints for microbial mat research. Since the electrochemical microsensors have been introduced into microbial mat research an enormous progress has been made in the unraveling of the structure of this ecosystem in which steep physical and chemical gradients are present at μm scale. The electrochemical microsensors have been improved considerably and are now applied as standard tool for microbial mat research by many groups. Other microsensors are currently developed such as for instance gas diffusion microprobes. In the (near) future, optrodes - sensors with optical signal transduction - may become available which may further widen experimental possibilities. The construction of fiber optic microprobes with the possibility of scalar light measurements has for the first time revealed detailed information about the light climate in microbial mats.

Nucleic acid-based probes are now available and are used to elucidate the structure of the community and its microbial diversity. New microscopical techniques, often combined with fluorescent probes and digital imaging are now available to study the structure of microbial mats and the behaviour of individual species.

Important new observations were made with respect to interrelations of species in microbial mats. An increasing number of researchers have observed that microbial mats can no longer be regarded as vertically stratified populations of microorganisms (Margulis et al. 1980). Most of the microorganisms are present in the top layer of the mat. For instance, high numbers of the sulfate-reducing bacteria can be found in the immediate vicinity of the cyanobacteria. Until recently, sulfate-reducing bacteria were considered to be obligately anaerobic organisms. Today, a suite of different sulfate-reducing species is known which differ with respect to their behaviour towards oxygen. Some species even respire oxygen. Others may even reduce sulfate under oxic conditions. It seems that cyanobacteria are the motor of sulfate reduction since they produce glycolate as a product of photorespiration during the day and excrete low-molecular fermentation products during the night. Several new species of microorganisms involved in the sulfur cycle have been discovered. Worth mentioning are organisms that disproportionate thiosulfate, sulfite and elemental sulfur. Their importance in microbial mats has now to be established. This also applies to the discovery of the anoxygenic iron-oxidizing phototrophic bacteria. Not much is known about their occurrence in microbial mats but considering the importance of iron in marine microbial mats these organisms can be expected to be present. All these discoveries will further increase our knowledge on biodiversity in microbial mats.

We would like to thank the NATO Science Committee for the generous support without which the organization of this Advanced Research Workshop would have been impossible. Additional support was obtained from Sanofi Elf Bio Recherches and the Universities of Amsterdam and Bordeaux. We are particularly grateful to the City of Arcachon for their hospitality and for the congress facilities they offered. Yehuda Cohen, David Des Marais and Bo Barker Jørgensen are acknowledged as members of the organizing committee for their invaluable advise for the composition of the scientific programme of this workshop. We are most indebted to Simone Behrens, Danielle Moreno and Françoise Truong for the phantastic help during the meeting and to Hans Reith for his expert assistance in the editing of this volume. Their efforts are deeply appreciated.

Lucas J. Stal and Pierre Caumette,
February 1994.

References

Bender J, Murray R, Phillips P (1993) Microbial mat degradation of chlordane. *In situ* and on-site bioreclamation. The Second International Symposium Proceedings, San Diego, Apr 5-7, 1993

Bender J, Phillips P (1993) Implementation of microbial mats for bioremediation. *In situ* and on-site bioreclamation. The Second International Symposium Proceedings, San Diego, Apr 5-7, 1993

Characklis WG, Wilderer PA (eds) (1989) Structure and function of biofilms. Dahlem Workshop Reports 46 Wiley, Chichester

Cohen Y, Castenholz RW, Halvorson HO (1984) Microbial Mats: Stromatolites. Alan R Liss Inc, New York, 498 pp

Cohen Y, Rosenberg E (1989) Microbial Mats: Physiological ecology of benthic microbial communities. ASM, Washington, 494 pp

Margulis l, Barghoorn ES, Ashendorf D, Banerjee S, Chase D, Francis S, Giovannoni S, Stolz JF (1980) The microbial community in the layered sediments at Laguna Figueroa, Baja California, Mexico: does it have Precambria analogues? Precambrian Res 11:93-123

Phillips P, Bender J, Word J, Niyogi D, Denovan B (1993) Mineralization of naphtalene, phenanthrene, chrysene and hexadecane with a constructed silage-microbial mat. *In situ* and on-site bioreclamation. The Second International Symposium Proceedings, San Diego, Apr 5-7, 1993

Sorkhoh N, Al-Hasan R, Radwan S, Höpner T (1992) Self-cleaning of the Gulf. Nature 359:109

Stal LJ, Caumette P (1993) Microbial mats: Future research developments. In: Guerrero R, Pedrós-Alió C (eds) Trends in Microbial Ecology, SSM, Barcelona, pp 147-152

Table of contents

Opening lecture

I. Colonization and initial processes in mat formation

II. New methods in microbial mat research

IV. Diel and spatial variations of physico-chemical parameters and processes in microbial mats

V. New physiological groups of organisms in microbial mats

VI. Bioremediability and biogeochemical cycles (panel discussion)

General considerations

Participants

Brad M. BEBOUT	Max-Planck Institute for Marine Microbiology, Fahrenheitstraße 1, D-28359 Bremen, GERMANY
Donald E. CANFIELD	Max Planck Institute for Marine Microbiology, Department of Biogeochemistry, Fahrenheitstraße 1, D-28359 Bremen, GERMANY
Richard W. CASTENHOLZ	Biology Department, University of Oregon, Eugene, Oregon 97403, USA
Pierre CAUMETTE	Institut Universitaire de Biologie Marine, Station Biologique d'Arcachon, Université de Bordeaux I, 2, rue du Professeur Joylet, F-33120 Arcachon, FRANCE
Yehuda COHEN	The Moshe Shilo Center for Marine Biogeochemistry, The Institute of Life Sciences, The Hebrew University of Jerusalem, Jerusalem 91904, ISRAEL
Ralf CONRAD	Max-Planck Institute for Terrestrial Microbiology, Karl-von-Frisch-Straße, D-35043 Marburg, GERMANY
Heribert CYPIONKA	ICBM, University of Oldenburg, P.O. Box 2503, D-2900 Oldenburg, GERMANY
Alan W. DECHO	US Geological Survey, Water Res. Div., MS 465, Menlo park, CA 94025, USA
David J. DES MARAIS	NASA Ames Research Center, Mail Stop 239-4, Moffet Field, CA 94035, USA
Ben DE WINDER	Laboratory for Microbiology, University of Amsterdam, Nieuwe Achtergracht 127, NL-1018 WS Amsterdam, THE NETHERLANDS
Rutger DE WIT	Institut Universitaire de Biologie Marine, Station Biologique d'Arcachon, Université de Bordeaux I, 2, rue du Professeur Joylet, F-33120 Arcachon, FRANCE

Isabel ESTEVE	Unit of Microbiology, Department Genetics and Microbiology, Autonomous University of Barcelona, E-08193 Bellaterra (Barcelona), SPAIN
Jack D. FARMER	NASA Ames Research Center, Mail Stop 239-4, Moffet Field, CA 94035, USA
Ferran GARCIA-PICHEL	Max Planck Institute for Marine Microbiology, Department of Biogeochemistry, Fahrenheitstraße 1, D-28359 Bremen, GERMANY
Maria Carmen GUERRERO	Department of Ecology, Faculty of Sciences, Universidad Autonoma de Madrid, Edificio Biologicas, Canto Blanco, E-28049 Madrid, SPAIN
Rodney A. HERBERT	Department of Biological Sciences, University of Dundee, Perth Road, DD1 4HN Dundee, UNITED KINGDOM
Mariona HERNÁNDEZ-MARINÉ	Botany Department, Faculty of Pharmacy, University of Barcelona, Avda. Juan XXIII s/n, E-08028 Barcelona, SPAIN
Bas W. IBELINGS	Department of Botany, University of Bristol, Bristol BS8 1UG , UNITED KINGDOM
Bo Barker JØRGENSEN	Max Planck Institute for Marine Microbiology, Department of Biogeochemistry, Fahrenheitstraße 1, D-28359 Bremen, GERMANY
Aaron KAPLAN	Department of Botany, The Hebrew University of Jerusalem, Jerusalem 91904, ISRAEL
Michael KÜHL	Max Planck Institute for Marine Microbiology, Department of Biogeochemistry, Fahrenheitstraße 1, D-28359 Bremen, GERMANY
Carsten LASSEN	Institute of Genetics and Ecology, University of Aarhus, Ny Munkegade, Building 550, DK 8000 Aarhus, DENMARK
Roy MOEZELAAR	Laboratory for Microbiology, University of Amsterdam, Nieuwe Achtergracht 127, NL-1018 WS Amsterdam, THE NETHERLANDS

Gerard MUYZER	Max Planck Institute for Marine Microbiology, Department of Biogeochemistry, Fahrenheitstraße 1, D-28359 Bremen, GERMANY
Thomas R. NEU	University of Groningen, Laboratory for Material Technica, Antonius Deusinglaan 1, NL-9713 AV Groningen, THE NETHERLANDS
Lars-Peter NIELSEN	Institute of Genetics and Ecology, University of Aarhus, Ny Munkegade, Building 550, DK 8000 Aarhus, DENMARK
Aharon OREN	The Moshe Shilo Center for Marine Biogeochemistry, The Institute of Life Sciences, The Hebrew University of Jerusalem, Jerusalem 91904, ISRAEL
Hans W. PAERL	Institute of Marine Sciences, University of North Carolina, Chapel Hill, 3407 Arendell Street, Morehead City, North Carolina 28557, USA
David M. PATERSON	Gatty Marine Laboratory, University of St-Andrews, KY16 8LB Fife, UNITED KINGDOM
Lee PRUFERT-BEBOUT	Max-Planck Institute for Marine Microbiology, Fahrenheitstraße 1, D-28359 Bremen, GERMANY
Niels B. RAMSING	Max-Planck-Institut für Marine Mikrobiologie, Fahrenheitstraße 1, D-28359 Bremen, GERMANY
Niels-Peter REVSBECH	Institute of Genetics and Ecology, University of Aarhus, Ny Munkegade, Building 550, DK 8000 Aarhus, DENMARK
James Bruno RISATTI	Illinois Geological Survey, Microbial Geochemistry Laboratory, 615 E. Peabody Drive, Champaign, IL 61820, USA
Serban M. SARBU	Department of Biological Sciences, 821A Rieveschl Hall (ML 006), University of Cincinnati, Cincinnati, Ohio 45221-0006, USA
Bart SCHAUB	Department of Microbiology, University of Groningen, P.O. Box 14, NL 9750 Haren, THE NETHERLANDS

David A. STAHL	University of Illinois, 2001 S Lincoln Ave, Urbana, IL 61501, USA
Lucas J. STAL	Laboratory for Microbiology, University of Amsterdam, Nieuwe Achtergracht 127 , NL-1018 WS Amsterdam, THE NETHERLANDS
John F. STOLZ	Department of Biological Sciences, Duquesne University, Mellon Hall, Pittsburgh, Pennsylvania, 15282-1502, USA
Frank P. VAN DEN ENDE	Department of Microbiology, University of Groningen, P.O. Box 14, NL 9750 Haren, THE NETHERLANDS
Hans VAN GEMERDEN	Department of Microbiology, University of Groningen, P.O. Box 14, NL 9750 Haren, THE NETHERLANDS
Marlies VILLBRANDT	Laboratory for Microbiology, University of Amsterdam, Nieuwe Achtergracht 127 , NL-1018 WS Amsterdam, THE NETHERLANDS
Pieter T. VISSCHER	U.S. Geological Survey, 345 Middlefield Road, Menlo Park, CA 94025, USA
David M. WARD	Department of Microbiology, Montana State University, Bozeman, MT 59717-0352, USA
David T. WELSH	Department of Biological Sciences, University of Dundee, Perth Road, DD1 4HN Dundee, UNITED KINGDOM
Friedrich WIDDEL	Max Planck Institute for Marine Microbiology, Department of Biogeochemistry, Fahrenheitstraße 1, D-28359 Bremen, GERMANY
Julian WIMPENNY	School of Pure and Applied Biology, University of Wales College of Cardiff , Cardiff CF1 3TL, UNITED KINGDOM
George A. ZAVARZIN	Institute of Microbiology, Russian Academy of Sciences, Moscow University, Prospect 60 Letija Oktjabria 7, Moscow, 117811, RUSSIA

Opening lecture

Microbial mat research: The recent past and new perspectives

Richard W. Castenholz

Department of Biology, University of Oregon, Eugene, Oregon 97403, USA

Scientists of diverse interests, and even laymen, are beginning to understand what is meant by a "microbial mat". But this knowledge is still superficial even for the specialist, both figuratively and literally. Most of what we know of microbial mats up to this point is still relegated to the superficial upper layer or layers which are usually composed mainly of photosynthetic prokaryotes. It is these that are responsible for the accretion and for the conversion of atmospheric and aqueous CO_2 into reduced energy-rich compounds on which the rest of the community is dependent. It is also these superficial layers that define most of the Precambrian remnants of microbial mats, the lithified mats or stromatolites which also contain the microfossils that are often recognizable as morphotypes of cyanobacteria. The laminae or underzones of even living mats are still "black boxes" for the most part, with little known of the actual microbial makeup, the matrix of exopolymers, and of the species interactions and metabolic activitites. Although the microbial mat community may be regarded as an ecosystem characterized by primarily a molecular flow of carbon and chemical energy (since grazing invertebrates and even smaller phagotrophs are generally rare or absent), most mat communities are now known to be of considerable taxonomic and biochemical complexity, and a great diversity of communities exists.

It will become evident during the course of this workshop that considerable advances have been made in several areas since the last of these workshops in 1987, but most of these cannot be discussed in this introduction. The present review will concentrate mainly on phototrophic organisms and photosynthetic aspects of microbial mats, avoiding in part the mire of the "black boxes".

Similarity and diversity among modern microbial mats

Although differences and similarities of topology, fabric, and species composition among microbial mats have been known for many years, in fact only in the last few years have we been shown how remarkably alike some of these appear over a great geographic and climatic expanse. The similarities in mat morphology are most striking; even the cyanobacterial species composition appears surprisingly similar, although the comparison of species using only morphological criteria for identification says little about genetic relatedness (e.g. Giani et al.

NATO ASI Series, Vol. G 35
Microbial Mats
Edited by Lucas J. Stal and Pierre Caumette

1989; Mir et al. 1991; Zhang and Hoffman 1992; Golubic 1992; Bauld et al. 1992; Ward et al. 1992a; Des Marais et al. 1992a; Guerrero and de Wit 1992). The similarities apply more to marine, hypersaline and inland saline mats, but to only a few types of hot spring mats when compared on a global basis. This categorical difference between marine/saline mats and hot spring mats should be expected, since geothermal waters differ dramatically with respect to chemistry. Although it has recently been established that the species composition of a single hot spring mat includes several different genotypes with sofar indistinguishable phenotypes, it is not known whether many or all of these genotypes occur over a restricted or broad geographical range. However, the typical *Synechococcus / Chloroflexus* mat from > 70°C to 45 or 50°C known for many years only in western North America has recently been well described in China (Yun 1986). Nevertheless, it still appears certain that such high temperature mats dominated by *Synechococcus* do not occur in Iceland, Europe, and New Zealand even though a wide variety of acceptable chemical niches are available in the temperature range of 64-73° C.

In addition to recent discoveries of mat communities in new habitats and over a great geographic range (Pierson 1992a), there have been fairly recent descriptions of new, more extreme, non-marine types which differ greatly from each other, usually as a result of differences in the combinations of temperature, salinity, pH, and sulfide concentrations (see Castenholz 1988; Ward et al. 1992a). These include mats where primary production is entirely or significantly due to anoxygenic photoautotrophic bacteria, such as some forms of *Chloroflexus, Chlorobium,* or purple sulfur bacteria (see Ward et al. 1989, 1992a; Castenholz et al. 1992), and in some cases due to extensive mats of non-photosynthetic, sulfide-oxidizing autotrophs such as thermophilic *Thermothrix* or to *Beggiatoa* (Nelson et al. 1989). For obvious reasons grazing invertebrates are absent or rare in these habitats, but in most of these cases even cyanobacteria are absent, usually due to intolerance of high sulfide levels at higher temperatures or to low pH or darkness.

One of the most interesting discoveries during the last 6 years has been the revelation that cyanobacterial microbial mats (including laminated types) dominate the benthic communities of thousands of ponds and small lakes in south Antarctica (i.e. south of 75°) (Vincent 1988; Vincent et al. 1993). The variation in the morphology of these mats mimicks those of hypersaline types in temperate latitudes and those of many hot springs (Fig. 1). The reason that this is surprising is that cyanobacteria, in general, are known for having temperature optima much above the 0-8°C experienced in the Antarctic ponds during mid-summer. And it appears that at least some of the Antarctic forms too have much higher temperature optima for maximum growth rates, presumably growing at extremely slow rates at their *in situ* temperatures (Castenholz and Schneider 1993; Seaburg et al. 1981). The dominance of benthic cyanobacteria in these cold ponds may be a result primarily of the lack of grazing invertebrates above the level of rotifers and the ability to withstand freezing and/or osmotic stress better than most eukaryotic algae. The freshwater ponds freeze solidly in winter and those that are more saline (e.g. ca. 10mS cm^{-1}) concentrate a salt brine (> 150 mS cm^{-1}) in their bottom contact zone during the winter freeze which lasts for about 9 months (Schmidt et al. 1991).

Fig. 1. Cyanobacteria-dominated mats in Antarctic ponds. Left: "Ice Ridge Pond" with mat in foreground. Right: Section of laminated, *Phormidium* -dominated, "Conophyton"- like mat from "Fogg-Horne Pond". Scale on right is in millimeters. Both ponds are on the ablation moraine of the McMurdo Ice Shelf west of Bratina Island.

Tropical scleractinian corals are seldom recognized as a substrate for cyanobacterial/ microbial mats. Some species of these corals become lethally "infected" with "black band" or "red band" diseases which, in reality, are multispecies microbial mats which spread over the living surfaces of certain coral species, and are dominated by motile, oscillatorian cyanobacteria, particularly one species in the case of black band (Rützler and Santavy 1983; Richardson 1992).

Another example of perennial cyanobacterial mats in unexpected habitats is in an ultraoligotrophic lake (TDS < 7 mg l^{-1}) in the Cascade Mountains of Oregon (USA) where slow-growing stromatolitic shelf mats develop, possibly because there is insufficient calcium to support an aquatic gastropod population (Castenholz, unpublished observations).

Part of the significance of these and other findings is that they help to confirm the theory that microbial mats (other than ephemeral biofilms) do not develop and persist in the presence of efficient grazers (i.e. benthic invertebrates such as gastropods) (see Farmer 1992). The inverse correlation between grazers and the development of mats in geothermal, hypersaline, and frequently desiccated environments has long been known (Javor and Castenholz 1981). However, mats persistent enough to develop laminae sometimes do develop in marine intertidal habitats with normal salinity and regular wettings, but often these are habitats where sulfide concentrations reach high enough levels to discourage most invertebrate grazers. Such mat communities (sulfureta) have been studied recently in the Orkneys (see van Gemerden et al. 1989). Also, extensive mats dominated by typical cyanobacteria (e.g. *Microcoleus*) have developed in the Persian Gulf specifically in association with widespread intertidal crude oil deposits where grazing invertebrates had been exterminated by the pollutants (Sorkhoh et al. 1992). It follows then, with even greater reality, that the advent of "efficient" grazing invertebrates ended the "Age of Stromatolites" at or near the end of the Proterozoic, although competition with ever increasing populations of eucaryotic macroalgae was possibly an important factor as well.

One of the most endless, but interesting and challenging, problems is the determination of species composition within even a single microbial mat. Few prokaryotic microorganisms are distinctive enough in their morphology or pigmentation to be recognized with light microscopy. Some exceptions are filamentous or otherwise complex cyanobacteria and eukaryotic algae (e.g. diatoms) and a few other photosynthetic microorganisms, which in some cases can be categorized with respect to family or larger group by pigment characteristics using microspectrophotometry or epifluorescence microscopy. A number of other microscopic techniques have also been developed or improved upon in recent years, such as improved imaging by computer enhancement and by confocal laser microscopy (Caldwell et al. 1992). Numerous innovations in staining technology have also come into use. Although some of the larger more dominant components of microbial mats may constitute more than 90% of the biomass, they probably represent only a small percentage ($\lesssim$ 5 %) of the species present. Nor does the biomass of any one component reflect its contribution in terms of metabolic activity. Turnover may be very rapid in some of the minor biomass contributors.

Many major microorganisms of microbial mat communities are in culture, but it is commonly estimated that these represent only a small percentage (perhaps < 5%) of the autochthonous, active species. Several of the remaining species could probably be cultured without much difficulty, but there have been insufficient efforts so far. Many, if not most, may be unculturable without innovative or radically different methods. In lieu of this, other techniques for elucidating the species diversity within mats and other microbial ecosystems have been advanced over the last few years. Great efforts have been made in identifying microbial components of mats and other microbial communities with respect to characteristic nucleotide sequences of a conservative element, namely several hundred to over a thousand nucleotides of the 16S subunit of ribosomal RNA (see Ward et al. 1992b). The usefulness of r-RNA sequences (or probes that complement them) is in distinguishing different families (or higher categories), genera, and even species from each other phylogenetically. However, the various methods used do not give a name to the organism unless the sequence is essentially identical to that of a reference species that is or has been in culture. Since many types of microorganisms that cover an extensive array of physiological diversity are now in culture, it is often (or usually) possible to relate an unknown sequence taken from the natural community at least to a category. For example, a fairly large data base for 16S r-RNA sequences is developing for cyanobacteria (see Wilmotte and Golubic 1991; Wilmotte et al. 1992). However, within the realm of most groups of heterotrophic bacteria, even closely related species can be quite different physiologically. Thus, phylogenetic relatedness (based on r-RNA sequences) may not tell us what the unknown organism is doing or capable of doing in the mat community. There are partial exceptions: e.g. most cyanobacteria have a similar photoautotrophic mode of life, yet some are capable of nitrogen fixation, nitrate reduction, chemoheterotrophy, or anoxygenic photosynthesis, and some are not. However, even species specific probes (with quantification) may give an idea of potential metabolic activity but not of the actual activity. Thus, activity must be measured separately, perhaps with electrochemical microprobes or by enzyme based genetic probes.

The extensive review of molecular methods of assessing microbial population diversity (Ward et al. 1992b) also discusses the applications and pitfalls of the various methods. In addition, Muyzer et al. (1993) have developed a new, more rapid method of separating PCR-amplified 16S rDNA fragments by electrophoresis in polyacrylamide gels containing a linearly increasing gradient of denaturants.

Nevertheless, species component identification is a long and tedious task. In studies of a relatively simple mat at 50-55°C in a single hot spring, sequence data indicate several species of unicellular cyanobacteria and *Chloroflexus*-like organisms, none of which are identical to those in culture from the same or similar hot springs (Weller et al. 1991,1992; Ward et al. 1992b).

At present the most common method of establishing cultures of representative microorganisms of various communities is the elective enrichment method with subsequent clonal isolation. Nevertheless, it has long been realized that this may often favor minor, often opportunistic, species in the population. Whenever possible the natural sample collected should be homogenized and severely diluted to near extinction with several replicates before enrichments and isolations are made, so that it is more likely that an isolate represents a common component of the community.

The use of membrane and other cellular components, such as fatty acids and lipids (Vestal and White 1989; Hedrick et al. 1991; Shiea et al. 1991; Tunlid and White 1992; Thompson et al. 1993; Zeng et al. 1992a,b), hopanoids (Buchholz et al. 1993), and polyamines (Hamana and Matsuzaki 1992) as biomarkers has recently increased to a significant extent. Specific signatures of various components may now be used to identify a large number of bacteria and other microorganisms, although this also requires a data bank of known organisms.

The Proterozoic biosphere

The revelation of the taxonomic and biochemical diversity within each modern microbial mat has also had great impact on the interpretation of ancient lithified mats, the stromatolites. Of importance is the realization that even the primary producers of modern microbial mats may differ, not simply in terms of genera or species of cyanobacteria, but also in some specialized environments, in terms of major categories of primary producers. In these environments anoxygenic photoautotrophic bacteria may be the sole primary producers (e.g. *Chloroflexus sp.* -Castenholz 1973; Giovannoni et al. 1987; *Chlorobium tepidum* -Castenholz et al. 1990; Wahlund et al. 1991; *Chromatium tepidum* -Madigan 1986; also see Ward et al. 1989, 1992a). The presence of such mats today, although rare, nullifies the dogma that all early Precambrian stromatolites had accreted through the photosynthetic activity of oxygenic (O_2-producing) cyanobacteria or their oxygenic ancestors. In addition, the recent discovery that ferrous-iron oxidizing, anoxygenic, anaerobic photoautotrophs exist today (Widdel et al. 1993) lends credence to the hypothesis that oxidized iron-rich sediments, such as the "banded iron formations" of the Proterozoic, could have been deposited in a completely anoxic environment.

B. K. Pierson pointed out (Pierson 1992b) that little is yet known about the fabric of diverse microbial mats and little emphasis has been placed on studies of diagenesis. Although the study of the internal fabric of mats may be boring to most microbiologists, and appropriate stages of diagenesis are hard to find, detailed studies of these aspects along with those of external mat morphology are needed if convincing homologies between modern and ancient stromatolites are to be made.

Enormous progress has been made in the last few years in the description and interpretation of sedimentary geology, biogeochemistry, ancient atmospheres, oceans, micropaleontology, and paleobiogeography of the Proterozoic, the principal era of microbial mat predominance on Earth and, thus, the "age of stromatolites". This is best evidenced by the huge compilation of chapters in Schopf and Klein (eds.), The Proterozoic Biosphere, a Multidisciplinary Study (1992). The recent, most convincing evidence of Archean microfossils that are cyanobacteria-like, at about 3.5 Gyr (Schopf 1993), will have a sizeable influence on the proposed evolutionary timescale for life on Earth, particularly on that for phototrophs. In the future, chemical signatures from Precambrian and Phanerozoic sediments will be of even greater importance in relating ancient and modern microbial mats (see DesMarais et al. 1992b).

The microenvironment within modern microbial mats

Light attenuation and microbial responses

The efforts to measure the total irradiance (i.e. scalar irradiance) experienced by microorganisms at submillimeter depths within compact mats and sediments have been considerable in the last few years, and improvements in equipment, including those of spectral radiometry, have been applied to detailed measurements of oxygenic photosynthesis vs irradiance, pigment/organism identification, and action spectra of oxygenic photosynthesis (Lassen et al. 1992a,b; Ploug et al. 1993; Pierson et al. 1990). Important findings include the discoveries that scalar irradiance levels at shallow depths in a mat (e.g. < 0.1-0.3 mm) could be considerably higher than the incident irradiance at the mat surface (Lassen et al. 1992b), further confirmation that maximum photosynthetic "efficiency" occurred below the surface layer (at least under high intensity), and that cyanobacteria appeared more efficient at lower intensities than did diatoms, due in part to preferential absorption of penetrating wavelengths (Ploug et al. 1993). Of potential importance to mat studies are the findings that some phototrophic bacteria (i.e. green sulfur bacteria) have the ability to grow, albeit slowly, at irradiance levels far below those formerly thought possible (e.g. Overmann et al. 1992). These findings may modify present concepts of photosynthetic compensation depths in mats where light gradients are often extremely steep.

The dynamic aspects of the species assemblages comprising microbial mats have attracted many more investigations in the last few years. Microbial populations of mats are not necessarily sedentary, and several studies have shown vertical migration patterns of oscillatorian cyanobacteria that are very sensitive to environmental stimuli such as light (Richardson and Castenholz 1987; Castenholz et al. 1991; Jørgensen et al. 1992; Garcia-Pichel et al. (submitted); Lassen et al. 1992b; Ploug et al. 1993). Not only are oscillatorian cyanobacteria involved, but also unicellular *Synechococcus* (cyanobacterium) in some hot spring mats (M. Teiser and R.W. Castenholz, unpublished), *Chloroflexus, Heliothrix* and relatives in some hot spring mats (B.K. Pierson and R.W. Castenholz, unpublished), *Chromatium tepidum* (purple sulfur bacterium) in hot spring mats (Jue 1990), and *Beggiatoa* spp. in both marine and warm spring mats (see Garcia-Pichel et al., submitted; and Fig. 2). For the phototrophs this generally means surfacing under low incident irradiance and in darkness, and migration downwards (1-2 mm or less) into the mat under high incident light. At least some of these responses are photophobic reactions or negative or positive phototaxis, and in at least one case, UV-A radiation is the most effective spectral region in preventing upward movement under high solar irradiance (see Garcia-Pichel and Castenholz, this workshop). To what extent UV radiation irreversibly inhibits motility or photomovements in nature is unknown (see Donkor and Häder 1991; Donkor et al. 1993).

In almost all cases, the migrating population of cyanobacteria is either plainly seen at the surface of the mat as a dark cover (usually rich in phycobilins as well as chlorophyll) or is absent (when below the surface) exposing a lighter colored mat matrix visible at the surface (see Fig. 2). In one case, however, it was possible to determine the exact vertical position of the migrating band during a diel period through the use of multiple micro-coring, and also establish that the top of the band not only avoided high, photoinhibitory intensities,but also positioned itself at a depth within the mat that allowed saturated (maximal) photosynthetic rates (Garcia-Pichel et al., submitted). As in other cases of vertical migratory behavior, the motile cyanobacteria retained a high content of light-harvesting pigments, an obvious advantage under variable incident light intensities.

One of the most dramatic vertical migrations within mats is that of *Chromatium* in some hot spring mats (Jue 1990). Swarms of swimming cells ascend to the surface at night and descend again at dawn (Fig. 2). The ascending cells contain elemental sulfur which accumulated during the day (by photosynthetic sulfide oxidation) and apparently consume much of the sulfur during the night, using it to power dark chemolithotrophy with O_2 as the oxidant (Jue and

Castenholz, unpublished data). There is precedence for this type of dark metabolism in other purple sulfur bacteria, particular in marine mats (see de Wit and van Gemerden 1990).

Another revelation of the last few years has been the demonstration that photoprotective pigments are extremely abundant and effective in the cyanobacteria that comprise the uppermost layer of many microbial mats. Carotenoids, of course, have been known as photoprotective pigments for many years, but the sheath pigment, scytonemin, provides a powerful, but passive, shielding protection, especially against UV-A (320-400 nm) and violet-blue wavelengths, but also against UV-B (280-320 nm) (Garcia-Pichel and Castenholz 1991; Garcia-Pichel et al. 1992). This stable pigment allows water-stressed, salt-stressed, or nutrient-limited cyanobacteria to tolerate high light exposure even when in a dormant or stationary phase and unable to rapidly repair or resynthesize damaged or destroyed cell components (Fig. 3) (See Garcia-Pichel and Castenholz, this workshop).

Fig. 2. Evidence for vertical migration in microbial mats. Left: Hunter's Hot Spring mat at about 45-50° C. Open-ended rings placed on mat 2 hr earlier at mid-day. Left-most ring had been covered with aluminum foil and now shows swarming *Chromatium* . Right ring was left open and still retains top mat of motile *Oscillatoria cf. terebriformis* which descends in darkness, as on left. Right: Covers removed from sandy mat at Great Sippiwissett Marsh near Woods Hole, Mass.,USA. after 1.5 hr of coverage near mid-day. The dark areas are ascended populations of *Oscillatoria cf. margaritifera.* The darkest patch resulted with about 5% of the incident radiation, the next darkest with about 12%; three others (less dark) were under opaque covers.

The dynamic chemical and metabolic environment within microbial mats

The chemical microenvironment of mats, although usually changing in a predictable, diel manner, goes through an enormous range of concentration changes brought on by the combined metabolic activities of the microorganisms. Some compounds, such as H_2, which are produced in large quantities in many anoxic habitats, are probably used as rapidly as produced, and seldom accumulate. Essentially all of the activities can be traced back to the photosynthetic components which provide reduced carbon for their own growth and accretion, but either through excretion, leakage, or death provide the micro-or macro-molecules of reduced carbon necessary to fuel the various chemoheterotrophs within the mat, or in some cases merely provide the carbon skeleton for photoheterotrophs, such as some *Chloroflexus* types, purple non-sulfur bacteria, or heliobacteria. Little is yet known of the actual compounds from photoautotrophs on which the rest of the mat metabolism depends, nor of what routes are taken in the molecular food web. Certain photoexcretion products, such as glycolate, have been identified in hot spring mats (Bateson and Ward 1988), along with a few expected fermentation products (Anderson et al. 1987). A recent study of the extracellular low molecular weight organic compounds released by thermophilic, mat-forming *Synechococcus* (cyanobacterium) in culture has revealed a great assortment of products, more than would be expected intuitively (Teiser and Castenholz, unpublished data). By HPLC about 20 compounds were identified that comprised over 90% of the total released. Many of the compounds could be used by *Chloroflexus* , the photoheterotroph constantly associated with the cyanobacterium in nature.

The secondary waste product, O_2, from the usually oxygenic photosynthesizers of the uppermost layers, influences greatly what goes on below because of daytime diffusion and nighttime exclusion, and within the the compact photosynthetic layers as well (such as rates of nitrogen fixation). O_2, of course, is predictably periodic except in the polar summers. However, there, as well as elsewhere, intensities and photosynthetic output depend on cloud cover, sun height (i.e. time of day, latitude, and season). There has been much work recently demonstrating that non-heterocystous cyanobacteria fix nitrogen, but do so at the highest rates during darkness when O_2 is low (e.g. Fay 1992; Bebout et al. 1993; Villbrandt et al. 1990).

Primary chemical factors are also important, i.e. if the waters bathing the mats carry high or low combined nitrogen; high, low, or no sulfide; high or low sulfate, and inummerable other nutrients and metabolic reductants or oxidants used in aerobic or anaerobic processes. These variables determine, for example, whether the species composition is skewed towards nitrogen fixers (particularly in the phototrophs), whether sulfide-dependent anoxygenic photosynthesizers (obligate or facultative) predominate (as in many hot springs), and with respect to sulfate, whether biogenic sulfidogenesis predominates over methanogenesis within the mats.

There is still much to know about the control of metabolic processes within microbial mats, how species interact, and about feedback mechanisms. Even with the development of new and improved microsensors, one is still measuring the changes in content of interstitial water and not probing the activity of single species. What lies outside of the cells represents the net output and consumption of many organisms. Thus, great emphasis needs to be placed on the ever increasing knowledge of metabolic diversity of microorganisms in culture that were isolated from mats or at least are likely to be capable of participating in mat environments. New advances in electrochemical microsensors have certainly helped in the study of internal processes in mats and in understanding the early stages of mat formation through biofilms. Since the last workshop there have been significant improvements in oxygen, sulfide, and pH microelectrodes, and many of the controls and readouts have been automated, digitized, and computerized. Also new types, such as the oxygen/nitrous oxide microsensor (see Dalsgaard and Revsbech 1992), ammonium, nitrate, and CO_2 microsensors (see Revsbech, this workshop) have been developed. Although not yet in a comparable micro scale, a gas diffusion probe for measuring H_2, CO, or CH_4 can now be used to gauge various microbial activities in sediments or mats (Krämer and Conrad 1993).

The study of dynamics of diel activity in mats has often overlooked the cases in which nominal dormancy exists, where photosynthetic activity is slight even in daylight hours. Photosynthetic rates per unit chlorophyll in the cold Antarctic freshwater mats appears to be very low, as does accretion rate (Vincent et al. 1994). In warmer environments, mats in the spray or steam of the edges of hot springs also appear to exhibit very slow growth and accretion rates (S. Brenowitz and Castenholz, unpublished observations). Many intertidal cyanobacterial mats of temperate or warmer regions persist because of the absence of grazers due to long-term desiccation and/or severe osmotic stress (e.g. in sabkhas). Mats that remain uncovered by water for several days or longer are common. Many mats experience daily increases in salinity of over 100% as hypersaline seawater evaporates in shallow depressions on the mats during tidal ebb. Nutrient stress too, in some mats, perhaps because of water stagnation and crowding of the microorganisms, may often lead to stationary phase or dormancy being the most common conditon. In other cases it seems that photosynthetic rates may be high due to nutrient cycling within the mat, even though there is a serious allochthonous nutrient stress (see Canfield and Des Marais, this workshop).

An extensive review of these complex subjects is outside the scope of this introduction, but a few contributions related to the "black box" of the undermat need to be cited. For example, sulfate-reducing activity has now been measured under aerobic conditions and is even stimulated by photosynthetic activity at the mat surface (Canfield and DesMarais 1991; Fründ and Cohen 1992; Marschall et al. 1993). The presence and utilization of thiosulfate, polysulfides, and other sulfur moieties in mats has finally been documented (Jørgensen and Bak 1991; Visscher et al. 1990). The number of substrates that diverse sulfate- or sulfur-reducing organisms can use has been expanded greatly (Widdel 1988; Visscher et al. 1992), and many newly described types are likely to participate in intermediate or terminal mineralization processes within mats.

In addition, it needs to be stressed that the use of laboratory cultures of many types of mat inhabitants in controlled experiments is still an essential part of eventually understanding the various metabolic processes occurring in mats.

Fig. 3. Evidence of scytonemin-rich sheathed cyanobacteria *in situ*. Left: hot spring stream in Midway Geyser Basin, Yellowstone National Park. Central portion of stream is bright orange due to carotenoid-rich cyanobacteria; dark borders of stream (lower temperature) composed of scytonemin-containing cyanobacteria (mainly *Calothrix* sp.). Right: cold-water pond in upper Victoria Valley, Antarctica containing benthic cover of *Pleurocapsa* -like cyanobacterium rich in scytonemin. Water depth < 1 m., temperature ca. 4° C.

Future directions

The direction of future research on modern microbial mats, at least, will depend on the results of this workshop, what present paths are productive and interesting, and to what extent some attempts are made at integrating the rapidly expanding knowledge.

There are some areas that I consider important that seem to be lacking or slighted at present. Actual determinations of global microbial mat coverage and productivity and the contributions that mats of a variety of types make to global cycling of carbon, sulfur, nitrogen, etc. should be attempted. These must be more than merely extrapolating "prime time" primary production rates from a few mats. The expanded use of remote imaging from appropriate satellites could be quite helpful, at least for the intertidal and very shallow mat environments.

As mentioned earlier, it would be interesting to determine to what extent the microorganisms of mats are under stress, and thus not photosynthesizing and metabolizing to their full potential. There are many stresses, such as high and low light; high and low temperature; high and low pH; high salinity; desiccation; high concentrations of poisons such as sulfide and toxic metals; and stress due to low nutrients. All are common as predictable or erratic conditions for many microbial mats. Thus, to what extent do microbial mats represent "life and death in stationary phase"?

Another related question of interest is: are microbial mat ecosystems stochastic environments or are most stable and predictable environments? A corollary question is: what is the rate of evolution of the microbial components? Does the extremely compact nature of most mats increase the rate of genetic exchange, since so many microbes are in constant, close association?

To even begin to answer some of these questions, a very few, selected mats should be studied in more exhaustive detail and in an interdisciplinary manner. A good start has been made with certain modern mats in the salinas of Guerrero Negro, Mexico, but it would also be a good idea to select one or more sites that are located very close to a permanent laboratory, if this is possible.

In conclusion, it seems that some direct use of microbial mats for human benefit may be found. The natural importance of biofilms and mats in binding unconsolidated sediments and thus slowing erosion could be more emphasized, and perhaps some manipulations such as introduction of inoculum would be possible. The inadvertant experiment in the Persian Gulf should also direct future research into the possible use of microbial mat turf in the bioremediation of oil polluted beaches and intertidal flats.

References

Anderson K, Tayne TA, Ward DM (1987) Formation and fate of fermentation products in hot spring cyanobacterial mats. Appl Environ Microbiol 53: 2343-2352

Bateson MM, Ward DM (1988) Photoexcretion and fate of glycolate in a hot spring cyanobacterial mat. Appl Environ Microbiol 54: 1738-1743

Bauld J, D'Amelio E, Farmer JD (1992) Modern microbial mats. In: Schopf JW, Klein C (eds) The Proterozoic biosphere. Cambridge Univ. Press, New York, pp 261-269

Bebout BM, Fitzpatrick MW, Paerl HW (1993) Identification of the sources of energy for nitrogen fixation and physiological characterization of nitrogen-fixing members of a marine microbial mat community. Appl Environ Microbiol 59: 1495-1503

Buchholz B, Laczko E, Pfennig N, Rohmer M, Neunlist S (1993) Hopanoids of a recent sediment from Lake Constance as eutrophication markers. FEMS Microbiol Ecol 102: 217-223

Caldwell DE, Korber DR, Lawrence JR (1992) Confocal laser microscopy and computer image analysis in microbial ecology. In: Marshall KC (ed) Advances in Microbial Ecology. vol 12, Plenum, New York, pp 1-67

Canfield DE, Des Marais DJ (1991) Aerobic sulfate reduction in microbial mats. Science 251: 1471-1473

Castenholz RW (1973) The possible use of sulfide by the filamentous phototrophic bacteria of hot springs. Limnol Oceanogr 18: 863-876

Castenholz RW (1988) The green sulfur and nonsulfur bacteria of hot springs. In: Olson JM, Ormerod JG, Amesz J, Stackebrandt E, Trüper HG (eds) Green photosynthetic bacteria. Plenum, New York, pp 243-255

Castenholz RW, Schneider AJ (1993) Cynaobacterial dominance at high and low temperatures: optimal conditions or precarious existence? In: Guerrero R, Pedrós-Alió (eds) Trends in microbial ecology. Spanish Society for Microbiology, Barcelona, pp 19-24

Castenholz RW, Bauld J, Jørgensen BB (1990) Anoxygenic microbial mats of hot springs: thermophilic *Chlorobium* sp. FEMS Microbiol Ecol 74: 325-336

Castenholz RW, Jørgensen BB, D'Amelio E, Bauld J (1991) Photosynthetic and behavioral versatility of the cyanobacterium *Oscillatoria boryana* in a sulfide-rich microbial mat. FEMS Microbiol Ecol 86: 43-58

Castenholz RW, Bauld J, Pierson BK (1992) Photosynthetic activity in modern microbial mat-building communities. In: Schopf JW, Klein C (eds) The Proterozoic biosphere. Cambridge Univ. Press, New York, pp 279-285

Dalsgaard T, Revsbech NP (1992) Regulating factors of denitrification in trickling filter biofilms as measured with the oxygen/nitrous oxide microsensor. FEMS Microbiol Ecol 101: 151-164

De Wit R, van Gemerden H (1990) Growth of the phototrophic purple sulfur bacterium *Thiocapsa roseopersicina* under oxic/anoxic regimens in the light. FEMS Microbiol Ecol 73: 69-76

Des Marais DJ, Bauld J, Palmisano AC, Summons R, Ward DM (1992b) The biogeochemistry of carbon in modern microbial mats. In: Schopf JW, Klein C (eds) The Proterozoic biosphere. Cambridge Univ. Press, New York, pp 299-308

Des Marais DJ, D'Amelio E, Farmer JD, Jørgensen BB, Palmisano AC, Pierson BK (1992a) Case study of a modern microbial mat-building community: the submerged cyanobacterial mats of Guerrero Negro, Baja California Sur, Mexico. In: Schopf JW, Klein C (eds) The Proterozoic biosphere. Cambridge Univ. Press, New York, pp 325-333

Donkor VA, Amewowor DHAK, Häder D-P (1993) Effects of tropical solar radiation on the motility of filamentous cyanobacteria. FEMS Microbiol Ecol 12: 143-148

Donkor V, Häder D-P (1991) Effects of solar and ultraviolet radiation on motility, photomovement and pigmentation in filamentous, gliding cyanobacteria. FEMS Microbiol Ecol 86: 159-168

Farmer J (1992) Grazing and bioturbation in modern microbial mats. In: Schopf JW, Klein C (eds) The Proterozoic biosphere. Cambridge Univ. Press, New York, pp 295-297

Fay P (1992) Oxygen relations of nitrogen fixation in cyanobacteria. Microbiol Rev 56: 340-373

Fründ C, Cohen Y (1992) Diurnal cycles of sulfate reduction under oxic conditions in cyanobacterial mats. Appl Environ Microbiol 58: 70-77

Garcia-Pichel F, Castenholz RW (1992) Characterization and biological implications of scytonemin, a cyanobacterial sheath pigment. J Phycol 27: 395-409

Garcia-Pichel F, Sherry ND, Castenholz RW (1992) Evidence for an ultraviolet sunscreen role of the extracellular pigment scytonemin in the terrestrial cyanobacterium *Chlorogloeopsis* sp. Photochem Photobiol 56: 17-23

Giani D, Seeler J, Giani L, Krumbein WE (1989) Microbial mats and physicochemistry in a saltern in the Bretagne (France) and in a laboratory scale saltern model. FEMS Microbiol Ecol 62: 151-162

Giovannoni SJ, Revsbech NP, Ward DM, Castenholz RW (1987) Obligately phototrophic *Chloroflexus*: primary production in anaerobic hot spring microbial mats. Arch Microbiol 147: 80-89

Golubic S (1992) Microbial mats of Abu Dhabi. In: Margulis L, Olendzenski L (eds) Environmental evolution. MIT Press, Cambridge, pp 103-130

Guerrero MC, de Wit R (1992) Microbial mats in the inland saline lakes of Spain. Limnetica 8: 197-204

Hamana K, Matsuzaki S (1992) Polyamines as a chemotaxonomic marker in bacterial systematics. Crit Rev Microbiol 18: 261-283

Hedrick DB, Nivens DE, Stafford C, White DC (1991) Rapid differentiation of archaebacteria from eubacteria by diffuse reflectance Fourier-transform IR spectroscopic analysis of lipid preparations. J Microbiol Methods 13: 67-73

Javor BJ, Castenholz RW (1981) Laminated microbial mats, Laguna Guerrero Negro, Mexico. Geomicrobiol J 2: 237-273

Jørgensen BB, Bak F (1991) Pathways and microbiology of thiosulfate transformations and sulfate reduction in a marine sediment (Kattegat, Denmark) Appl Environ Microbiol 57: 847-856

Jørgensen BB, Castenholz RW, Pierson BK (1992) The microenvironment within modern microbial mats. In: Schopf JW, Klein C (eds) The Proterozoic biosphere. Cambridge Univ. Press, New York, pp 271-278

Jue CS (1990) The effect of aerobic environments on *Chromatium* cf. *tepidum*, a thermophilic purple sulfur bacterium. Master's thesis University of Oregon, Eugene

Krämer H, Conrad R (1993) Measurement of dissolved H_2 concentrations in methanogenic environments with a gas diffusion probe. FEMS Microbiol Ecol 12: 149-158

Lassen C, Ploug H, Jørgensen BB (1992a) A fiber optic scalar irradiance microsensor: application for spectral light measurements in sediments. FEMS Microbiol Ecol 86: 247-254

Lassen C, Ploug H, Jørgensen BB (1992b) Microalgal photosynthesis and spectral scalar irradiance in coastal marine sediments of Limfjorden, Denmark. Limnol Oceanogr 37: 760-772

Madigan MT (1986) *Chromatium tepidum* sp. nov., a thermophilic photosynthetic bacterium of the family Chromatiaceae. Int J Syst Bacteriol 36: 222-227

Marschall C, Frenzel P, Cypionka H (1993) Influence of oxygen on sulfate reduction and growth of sulfate-reducing bacteria. Arch Microbiol 159: 168-173

Mir J, Martínez-Alonso M, Esteve I, Guerrero R (1991) Vertical stratification and microbial assemblage of a microbial mat in the Ebro Delta (Spain). FEMS Microbiol Ecol 86: 59-68

Muyzer G, De Waal EC, Uitterlinden AG (1993) Profiling of complex microbial populations by denaturing gradient gel electrophoresis analysis of polymerase chain reaction-amplified genes coding for 16S rRNA. Appl Environ Microbiol 59: 695-700

Nelson DC, Wirsen CO, Jannasch HW (1989) Characterization of large, autotrophic *Beggiatoa* spp. abundant at hydrothermal vents of the Guaymas Basin. Appl Environ Microbiol 55: 2909-2917

Overmann J, Cypionka H, Pfennig N (1992) An extremely low-light-adapted phototrophic sulfur bacterium from the Black Sea. Limnol Oceanogr 37: 150-155

Pierson BK (1992a) Introduction. In: Schopf JW, Klein C (eds) The Proterozoic biosphere. Cambridge Univ. Press, New York, pp 247-251

Pierson BK (1992b) Unsolved problems and future research directions. In: Schopf JW, Klein C (eds) The Proterozoic biosphere. Cambridge Univ. Press, New York, pp 339-340

Pierson BK, Sands VM, Frederick JL (1990) Spectral irradiance and distribution of pigments in a highly layered marine microbial mat. Appl Environ Microbiol 56: 2327-2340

Ploug H, Lassen C, Jørgensen BB (1993) Action spectra of microalgal photosynthesis and depth distribution of spectral scalar irradiance in a coastal marine sediment of Limfjorden, Denmark. FEMS Microbiol Ecol 102: 261-270

Richardson LL, Castenholz RW (1987) Diel vertical movements of the cyanobacterium *Oscillatoria terebriformis* in a sulfide-rich hot spring microbial mat. Appl Environ Microbiol 53: 2142-2150

Richardson LL (1992) Red band disease: a new cyanobacterial infestation of corals. Proceedings of the 10th Annual Academy of Underwater Sciences, Diving for Science 1992, pp 153-160

Rützler K, Santavy DL (1983) The black band disease of Atlantic reef corals. I. Description of the cyanophyte pathogen. Marine Ecol 4: 301-319

Schmidt S, Moskal W, De Mora SJ, Howard-Williams C, Vincent WF (1991) Limnological properties of Antarctic ponds during winter freezing. Antarctic Sci 3: 379-388

Schopf JW (1993) Microfossils of the early Archean apex chert: new evidence of the antiquity of life. Science 260: 640-646

Seaburg KG, Parker BC, Wharton RA Jr, Simmons GM Jr (1981) Temperature-growth responses of algal isolates from antarctic oases. J Phycol 17: 353-360

Shiea J, Brassell SC, Ward DM (1991) Comparative analysis of extractable lipids in hot spring microbial mats and their component photosynthetic bacteria. Org Geochem 17: 309-319

Sorkhoh N, Al-Hasan R, Radwan S, Höpner T (1992) Self-cleaning of the Gulf. Nature 359: 109

Thompson IP, Bailey MJ, Ellis RJ, Purdy KJ (1993) Subgrouping of bacterial populations by cellular fatty acid composition. FEMS Microbiol Ecol 102: 75-84

Tunlid A, White D (1992) Biochemical analysis of biomass, community structure, nutritional status, and metabolic activity of microbial communities in soil. Soil Biochem 7: 229-262

van Gemerden H, Tughan CS, De Wit R, Herbert RA (1989) Laminated microbial ecosystems on sheltered beaches in Scapa Flow, Orkney Islands. FEMS Microbiol Ecol 62: 87-102

Vestal JR, White DC (1989) Lipid analysis in microbial ecology: quantitative approaches to the study of microbial communities. BioScience 39: 535-541

Villbrandt M, Stal LJ, Krumbein WE (1990) Interactions between nitrogen fixation and oxygenic photosynthesis in a marine cyanobacterial mat. FEMS Microbiol Ecol 74: 59-71

Vincent WF (1998) Microbial Ecosystems of Antarctica. Cambridge Univ. Press New York, 304 pp.

Vincent WF, Castenholz RW, Downes MT, Howard-Williams C (1994) Antarctic cyanobacteria: light, nutrients and photosynthesis in the microbial mat environment. J Phycol 30: (in press)

Vincent WF, Downes MT, Castenholz RW, Howard-Williams C (1993) Community structure and pigment organisation of cyanobacteria-dominated microbial mats in Antarctica. Eur J Phycol 28: (in press)

Visscher PT, Nijburg JW, van Gemerden H (1990) Polysulfide utilization by *Thiocapsa roseopersicina*. Arch Microbiol 155: 75-81

Visscher PT, Prins RA, van Gemerden H (1992) Rates of sulfate reduction and thiosulfate consumption in a marine microbial mat. FEMS Microbiol Ecol 86: 283-293

Wahlund TM, Woese CR, Castenholz RW, Madigan MT (1991) A thermophilic green sulfur bacterium from New Zealand hot springs, *Chlorobium tepidum* sp. nov. Arch Micrbiol 156: 81-90

Ward DM, Bateson MM, Weller R, Ruff-Roberts AL (1992b) Ribosomal RNA analysis of microorganisms as they occur in nature. In: Marshall KC (ed) Advances in Microbial Ecology, vol 12, Plenum, New York, pp 219-286

Ward DM, Bauld J, Castenholz RW, Pierson BK (1992a) Modern phototrophic microbial mats: anoxygenic, intermittently oxygenic/anoxygenic, thermal, eukaryotic, and terrestrial. In: Schopf JW, Klein C (eds) The Proterozoic biosphere. Cambridge Univ. Press, New York, pp 309-324

Ward DM, Weller R, Shiea J, Castenholz RW, Cohen Y (1989) Hot spring microbial mats: anoxygenic and oxygenic mats of possible evolutionary significance. In: Cohen Y, Rosenberg E (eds) Microbial Mats, Am. Soc. Microbiol., Wash. DC, pp 3-15

Weller R, Bateson MM, Heimbuch BK, Kopczynski ED, Ward DM (1992) Uncultivated cyanobacteria, *Chloroflexus*-like inhabitants, and spirochete-like inhabitants of a hot spring microbial mat. Appl Environ Microbiol 58: 3964-3969

Weller R, Weller JW, Ward DM (1991) 16S rRNA sequences of uncultivated hot spring cyanobacterial mat inhabitants retrieved as randomly primed cDNA. Appl Environ Microbiol 57: 1146-1151

Widdel F (1988) Microbiology and ecology of sulfate- and sulfur-reducing bacteria. In: Zehnder AJB (ed) Biology of anaerobic microorganisms. J. Wiley & Sons, New York, pp 469-585

Widdel F, Schnell S, Heising S, Ehrenreich A, Assmus B, Schink B (1993) Ferrous iron oxidation by anoxygenic phototrophic bacteria. Nature 362: 834-836

Wilmotte A, Golubic S (1992) Morphological and genetic criteria in the taxonomy of Cyanophyta/Cyanobacteria. Arch. Hydrobiol. Suppl. 92, Algological Stud 64: 1-24

Wilmotte A, Turner S, Van de Peer Y, Pace N (1992) Taxonomic study of marine oscillatoriacean strains (cyanobacteria) with narrow trichomes. II. Nucleotide sequence analysis of the 16S ribosomal RNA. J Phycol 28: 828-838

Yun Z (1986) Thermophilic microorganisms in the hot springs of Tengchong geothermal area, West Yunnan, China. Geothermics 15: 347-358

Zeng YB, Ward DM, Brassell S, Eglinton G (1992a) Biogeochemistry of hot spring environments. 2. Lipid composition of Yellowstone (Wyoming, USA) cyanobacterial and *Chloroflexus* mats. Chem Geol 95: 327-345

Zeng YB, Ward DM, Brassell S, Eglinton G (1992b) Biogeochemistry of hot spring environments. 3. Apolar and polar lipids in the biologically active layers of a cyanobacterial mat. Chem Geol 95: 347-360

Zhang, Y, Hoffmann, L (1992) Blue-green algal mats of the salinas in San-ya, Hai-nan Island (China): structure, taxonomic composition, and implications for the interpretation of Precambrian Stromatolites. Precambrian Res 56: 275-290

I. Colonization and initial processes in mat formation

Microbial mats in coastal environments

Lucas J. Stal

Laboratory for Microbiology, University of Amsterdam,
Nieuwe Achtergracht 127, NL-1018 WS Amsterdam, The Netherlands

Introduction

Microbial mats develop in a wide variety of environments such as hot springs, hypersaline ponds, dry and hot deserts, alkaline lakes and coastal intertidal sediments (Cohen et al. 1984; Cohen and Rosenberg 1989). A noticeable similarity of these environments is their relative low species diversity. Particularly, multicellular organisms are excluded from such environments and it has been conceived that the absence or limited activity of grazing organisms is an important prerequisite for the development of microbial mats (Walter 1976). At first sight such conditions seem not to apply to coastal environments. However, intertidal sediments are alternatingly inundated and exposed to the atmosphere and this will cause fluctuations of water content, salinity and temperature. Fluctuations of environmental conditions are more extreme in the higher parts of the intertidal flats where microbial mats develop particularly well (Stal et al. 1985; Pierson et al. 1987).

Non-cohesive sandy sediments are often inhabited by filamentous cyanobacteria which form a dense net-work of entangled trichomes in which sand particles are trapped. The cyanobacterial mat is often covered by a thin layer of sand and sometimes considerable numbers of diatoms can be found at the sediment surface. These non-cohesive sediments consist of fine grained sand (63-200 μm) but may be mixed with varying amounts of silt (2-63 μm) and clay (< 2 μm). Cohesive sediments consist mainly of silt and clay particles. Cohesive intertidal sediments inhabit predominantly diatoms as the phototrophic component (Admiraal et al. 1982). Diatoms are moving up and down in the sediment according to the tidal cycle (Pinckney and Zingmark 1991) and they do not produce a tough microbial mat; they rather form biofilms (Paterson 1989). Non-cohesive sediments are very dynamic and also the small size of the sediment particles allows only very limited light penetration. Apparently, such sediments are not suitable for colonization by cyanobacteria, but the rapidly moving diatoms seem to be adapted to thrive in these dynamic environments.

Growth and metabolism of microorganisms in coastal sediments have been conceived to influence sediment transport and coastal morphodynamics (Paterson 1989; Yallop et al. 1994). Colonization of sediment particles and the excretion of extracellular polymeric substances (EPS) are significantly effecting sediment erodibility (Dade et al. 1990). Filamentous cyanobacteria may form a network in which sediment particles are trapped are contribute to the consolidation of the sediment. EPS forms a matrix in which the microbial

NATO ASI Series, Vol. G 35
Microbial Mats
Edited by Lucas J. Stal and Pierre Caumette

community is embedded (Decho 1990; Decho this volume). In non-cohesive sediments it seems that EPS which is excreted by diatoms plays an important role in sediment consolidation. These polymers have the capacity of binding to clay and silt particles thereby fixing the sediment and decreasing its erodibility (Decho this volume; Neu this volume).

Development and structure of intertidal microbial mats

The development of a microbial mat on sandy intertidal sediments is usually initiated by cyanobacteria. In most cases such environments are low in organic matter (however see van Gemerden et al. 1989) and photoautotrophy seems to be the most appropriate metabolism in this environment. There may be several reasons for the formation of microbial mats by cyanobacteria. For instance, many cyanobacteria are potential nitrogen fixers. Coastal marine environments are often depleted in combined nitrogen and the fixation of dinitrogen has been shown to be an important process in mat development (Stal 1993; Stal et al. this volume).

Photosynthetic CO_2 fixation enriches the sediment with organic matter which forms the basis of a complex microbial ecosystem. The organic matter produced by the cyanobacteria becomes available to other microorganisms in a number of different ways. (i) *excretion of glycolate as a result of photorespiration* (Renström and Bergman 1989). This is because in microbial mats oxygen accumulates in the light while CO_2 is depleted. Such conditions promote photorespiration. (ii) *dark fermentation.* During the dark oxygen is consumed rapidly and the sediment becomes anoxic. Under such conditions cyanobacteria start to ferment storage carbohydrate and excrete fermentation products (Moezelaar and Stal, this volume). (iii) *excretion of extracellular polysaccharides (EPS)* (see below). (iv) *cell death and lysis.*

The organic matter is decomposed by a variety of microorganisms. Because the microbial mat is anoxic below the photic zone and during the night anoxic conditions may be present throughout the mat, anaerobic sulfate reduction is the dominant process. It has been shown that glycolate is an important substrate for some sulfate reducing bacteria (Bateson and Ward 1988; Fründ and Cohen 1992). Fermentation products serve also as substrates for sulfate reducing bacteria (Anderson et al. 1987; Fründ and Cohen 1992) . The sulfide which is produced by sulfate-reducing bacteria is oxidized by different microorganisms. Under anoxic conditions anoxygenic phototrophic bacteria utilize sulfide as electron donor, while colorless sulfur bacteria oxidize sulfide under oxic conditions. Cyanobacteria may produce sulfide by reducing elemental sulfur when they experience anoxic conditions in the dark (Stal 1991). Under anoxic conditions in the light they are capable of oxidizing sulfide. The different groups of microorganisms show strong interrelationships (van Gemerden 1993; this volume).

Extracellular polysaccharides and sediment stabilisation

Cyanobacteria and diatoms excrete extracellular polymeric substances (EPS) (Decho 1990). The most important component of EPS is polysaccharide. This EPS formes a matrix in which the organisms are embedded. Several important functions have been ascribed to EPS. (Decho 1990; this volume) (i) it may protect organisms against grazing (ii) it may aid the organism to attach to a surface (iii) it may bind essential or toxic metals (iv) it may protect against desiccation. Very little is known about the controlling factors of EPS production and excretion. It may be a result of unbalanced growth, for instance when cyanobacteria are growing under a nitrogen limitation, or may play a role in motility of diatoms. It is conceived that EPS in marine sediments contribute to its stabilisation due to the capacity of polysaccharides to bind to particles (Dade et al. 1990; Paterson 1989; this volume). In table 1 the results of measurements along a transect of a sandy intertidal flat are shown. The critical erosion stress (the force that has to be applied to erode the sediment surface) seems to be positively correlated with the amount of colloidal carbohydrate and lesser with chlorophyll *a* content. A negative correlation was observed with sediment water content. The interactions are complex. It is well known that microbial mats develop better at locations that are less often inundated. On the other hand, the level of hydration of polysaccharides also determines their binding characteristics.

Table 1. Transect along a sandy intertidal flat. Station D was close to the low water border line, station A was close to the high water border line.

Station	Chl *a* (mg g^{-1})	H_2O (%)	Colloidal carbohydrate (mg g^{-1})	Critical erosion stress (N m^{-2})
A	4.2	18.5	1.1	17.1
B	3.4	17.6	0.9	4.38
C	2.0	20.0	0.5	>1.4
D	3.4	27.6	0.3	>1.4

Data from Krumbein et al. 1993a

In another series of measurements a comparison was made between a diatom dominated and a cyanobacteria dominated sediment (Yallop et al. 1994). In this case it was evident that the diatom community produced the highest amount of carbohydrate. This was not only true for the total amount of carbohydrate but also for colloidal carbohydrate which is assumed to be an important factor for sediment stabilisation. Notwithstanding the relative low content of colloidal carbohydrate the cyanobacterial mat possessed the highest stabilisation coefficient (table 2) .

Table 2. Biochemical parameters of 2 types of intertidal sediments colonized by either cyanobacteria or by diatoms and the stability of their sediment surfaces as measured against a non-colonized control.

Mat type	Chl *a*	Protein	Carbohydrate (total)	Carbohydrate (colloidal)	Stabilisation coefficient
			$\mu g\ g^{-1}$ sediment		$N\ m^{-2}$
Cyanobacterial mat	13.2	4900	2600	278	>21.7
Diatom biofilm	19.2	9600	10900	1460	2.9

Data from Yallop et al. (1994)

Table 3. Composition of the extracellular polysaccharide (EPS), capsular polysaccharide (CPS) and rest fraction of the mat-forming cyanobacterium *M. chthonoplastes* and the biofilm forming diatom *Navicula* sp.

Microcoleus chthonoplastes				
Fraction	EPS	CPS	Rest	Total
Protein	0	49	51	100
Hexose	31	50	19	100
Uronic acids	0	81	19	100
Navicula sp.				
Fraction	EPS	CPS	Rest	Total
Protein	49	16	35	100
Hexose	39	36	25	100
Uronic acids	100	0	0	100

Data from de Winder and Stal (see: Krumbein et al. 1993b)

In table 3 a comparison is made between the composition of extracellular polysaccharide (EPS) and capsular polysaccharide (CPS) fractions of the mat-forming cyanocbaterium *M. chthonoplastes* and the biofilm forming diatom *Navicula* sp. It was noted that the EPS of *M. chthonoplastes* is composed exclusively of glucose (table 4), whereas the EPS of *Navicula* sp. is much more complex and all uronic acids of this organism were found in this fraction. In *M. chthonoplastes* the uronic acids were found in the CPS fraction. The CPS fraction of the cyanobacterium showed a comparable complexity as the EPS of the diatom. However, the sugar composition of these fractions showed marked differences (table 4). *M. chthonoplastes* usually produced only small amounts of EPS and most of the polysaccharides were found as CPS. In the diatom *Navicula* sp. this was not the case and this organism produced only small amount of CPS while most of the polysaccharide was excreted as EPS. A notable difference of these 2 types of polysaccharides was the uronic acid content which was relatively low in the diatom. However, it must be considered that the amounts of EPS produced by the diatom community were much higher than by cyanobacteria (table 2). Also the fact that the cyanobacteria produce uronic acids only as a constituent of CPS may have important consequences for their action as a glue for sediment particles. It seems likely that the acid groups in the CPS of *M. chthonoplastes* serve among other things as binding places for iron (see below).

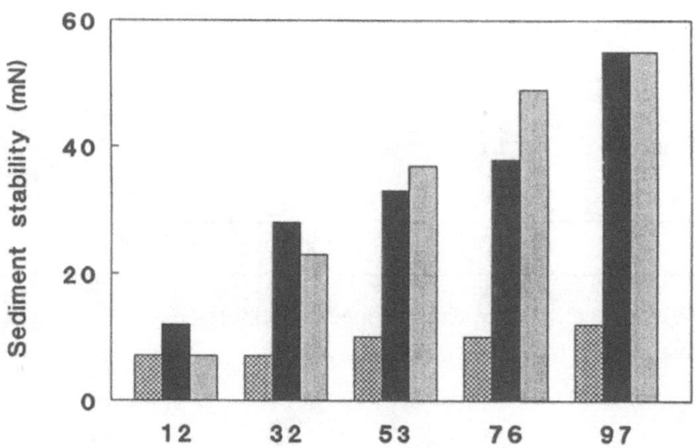

Fig. 1. Effect of growth of 3 different species of benthic marine cyanobacteria on the erodibility of a fine sandy sediment. Cross-hatched bar: *Merismopedia*; Solid bar: *Microcoleus chthonoplastes*; Dotted bar: *Phormidium* sp. (Data from H. Riege see Krumbein et al. 1993b).

The greater stability of cyanobacterial mats compared to diatom biofilms can not be explained by the production of EPS and by its uronic acid composition alone. It seems that the filamentous cyanobacteria form a dense network of entangled trichomes in which sediment particles are trapped. This network would also result in a physically stable sediment surface. It is evident that diatoms are not capable of forming such structures. Also unicellular cyanobacteria that sometimes are important in intertidal sediments are not capable of forming such structures. This was clearly shown in laboratory models (Fig. 1) (data from H. Riege, see Krumbein et al 1993b).

Table 4. Comparison of sugar composition of EPS and CPS of the cyanobacterium *Microcoleus chthonoplastes* and the diatom *Navicula* sp.

	Microcoleus chthonoplastes		*Navicula* sp.	
	EPS	CPS	EPS	CPS
arabinose	0	3	2	0
galactose	0	3	18	5
glucose	100	35	29	81
mannose	0	19	15	4
rhamnose	0	10	11	9
xylose	0	3	18	1
galacturonic acid	0	25	7	0
N-acetylglucosamin	0	0	0	0
total	100	100	100	100

Data from de Winder and Stal (see Krumbein et al. 1993b)

Function of iron

Well-developed coastal microbial mats usually have an extended black horizon of precipitated iron sulfide (FeS) below the photic zone. Very characteristic is also a rusty layer of precipitated iron hydroxides at the lower part of the cyanobacterial mat but always above the layer of the anoxygenic phototrophic purple sulfur bacteria. Although sea water contains only low amounts of iron (3 $\mu g\ l^{-1}$) this compound seems to accumulate in the sediment. This might explain that generally no free sulfide can be detected in coastal intertidal sediments. Sulfide will readily react with iron and precipitate as FeS. This sulfide is also termed acid-volatile sulfide (AVS). Part of sulfide may be deposited as pyrite (FeS_2). Apart from the chemical reaction of sulfide with iron, the concentration of free sulfide is also kept low due to the activity of anoxygenic phototrophic bacteria and of the colorless sulfur bacteria. The reactions of sulfide with iron are as follows (Ehrlich 1990):

$$2Fe^{3+} + H_2S \longrightarrow 2Fe^{2+} + 2H^+ + S^0 \qquad 1)$$

$$Fe^{2+} + H_2S \longrightarrow FeS + 2H^+ \qquad 2)$$

$$2Fe^{3+} + 2H_2S \longrightarrow FeS_2 + Fe^{2+} + 4H^+ \qquad 3)$$

The reduction of ferric iron by sulfide thus results in the formation of elemental sulfur. Microbial mats often contain a high amount of elemental sulfur (Table 5). Although elemental sulfur may be produced biologically by the oxidation of sulfide (e.g. by anoxygenic phototrophic purple sulfur bacteria), the chemical production of elemental sulfur should not be underestimated.

Ferrous iron in microbial mats is continuously oxidized. Oxidation of ferrous iron may be chemically or biologically. Ferrous iron reacts rapidly with oxygen according to the following equation:

$$Fe^{2+} + H^+ + \tfrac{1}{4}O_2 \longrightarrow Fe^{3+} + \tfrac{1}{2}H_2O \qquad 4)$$

The aerobic biological oxidation of iron (i.e. by *Thiobacillus ferrooxidans*) is probably not important in microbial mats. Under oxic conditions at pH above 5, ferrous iron autooxidizes rapidly. In marine microbial mats the pH rarely drops below 7. However, anaerobic oxidation of ferrous iron has been shown to be catalized by anoxygenic phototrophic bacteria that utilize it as electron donor (Widdel et al. 1993; this volume) according to the following reaction:

$$4\,FeCO_3 + 7\,H_2O \longrightarrow [CH_2O] + 4\,Fe(OH)_3 + 3\,CO_2 \qquad 5)$$

Therefore, oxygen produced by cyanobacterial photosynthesis as well as iron-dependent anoxygenic photosynthesis may be responsible for a continuous oxidation of iron. The ecological importance of iron-dependent anoxygenic photosynthesis has now to be determined but it seems possible that it forms the layer of iron hydroxydes which is often found below the cyanobacteria in coastal microbial mats.

Table 5 shows that free sulfide ion is below the limit of detection in a North Sea microbial mat. Sulfide is present as iron sulfide (FeS). Free iron is still present and it can be expected that any sulfide produced by sulfur or sulfate reduction will be precipitated as FeS. Considerable amounts of elemental sulfur can be found which may originate from biological sulfide oxidation or from the chemical reaction of ferric iron with sulfide (reaction 1).

Table 5. Elemental sulfur, free sulfide ion, iron and iron sulfide (acid volatile sulfide) in different layers of a microbial mat from a North Sea intertidal sediment.

Sediment layer	S^0	S^{2-}	Fe^{2+}/Fe^{3+}	FeS
	mg g^{-1} sediment			
cyanobacterial mat	2.88	0	0.130	0.032
black FeS layer	12.55	0	0.116	3.960
Oxidized layer	6.40	0	0.267	0.036

adapted from Stal (1991)

Microcoleus chthonoplastes is often found as the dominant organism in microbial mats. This cyanobacterium is capable of accumulating iron in its polysaccharide sheath (Camacho and Stal, manuscript in preparation). Table 6 shows how cell-associated iron depends on iron concentration in the growth medium. The optimum iron concentration was 100 μM, above which specific cell accumulation decreased. Concentrations of more than 200 μM Fe appeared to be toxic.

Table 6. Accumulation of Fe by *M. chthonoplastes*

Concentration of Fe (μM)	Cell-associated Fe (mg mg^{-1} Chl *a*)
9	0.5
50	2.2
100	17.6
150	13.1
200	11.5
250	6.0

M. chthonoplastes not only accumulated iron but was also capable of reducing ferric iron. In Fig. 2 it is shown that Fe accumulation and reduction was catalized by living cells only (data from Camacho and Stal, manuscript in preparation).

Iron accumulation by a mat-forming cyanobacterium such as *M. chthonoplastes* may be beneficial for the organism. Two possible adavantages were tested. (i) The accumulation of iron may protect the organism against deleterious concentrations of sulfide. (ii) ferrous iron may scavenge oxygen and keep its concentration in the vincinity of the organism low with that increasing efficiency of photosynthetic CO_2 fixation (low photorespiratory activity). When *M. chthonoplastes* was preincubated in medium with elevated iron and subsequently exposed to sulfide under anoxic conditions in the dark, the latter disappeared totally from the medium. Also no FeS was found to be associated with the trichomes. Presumably, the sulfide reacted with ferric iron to elemental sulfur (reaction 1). In contrast, cells that did not receive elevated iron did not remove sulfide from the medium (table 7).

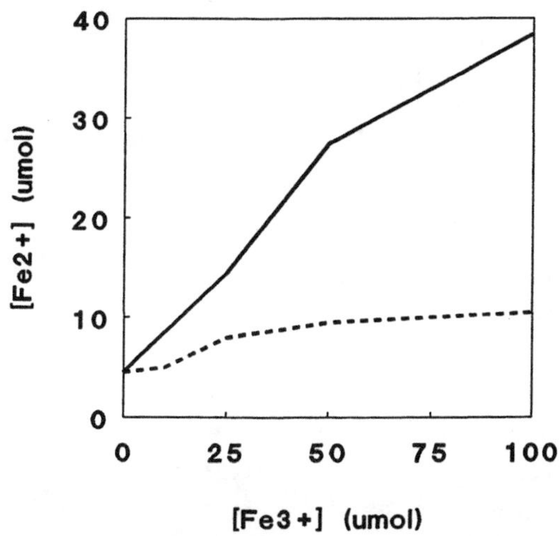

Fig. 2. Accumulation and reduction of iron by the cyanobacterium *Microcoleus chthonoplastes*. Ferric iron was reduced and ferrous iron accumulated by living cells only (solid line). As a control heat- or Hg_2Cl killed cells were used (dotted line).

Table 7. The effect of sulfide addition to suspensions of *M. chthonoplastes* preincubated with or without elevated amount of iron[1].

	Low Fe medium	High Fe medium
	$\mu mol\ l^{-1}\ S^{2-}$	
Sulfide addition	300	300
reovery from the medium	280	0
recovery from the cells	40	3

[1]Cells were preincubated for 24 h with 250 $\mu M\ Fe^{3+}$ in standard ASN 3 medium (Rippka et al. 1979) (high Fe medium) or in regular ASN 3 medium (low Fe medium). The cells were exposed to sulfide for 3 h in the dark under anoxic conditions.

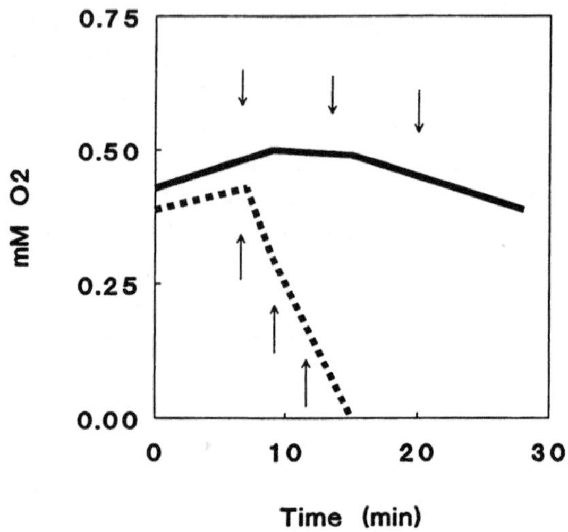

Fig. 3. Effect of 1 mM additions of sulfide (arrows) on oxygen exchange in *Microcoleus chthonoplastes*. Solid line represents a culture grown with low iron; dotted line a culture grown with high iron.

When *M. chthonoplastes* was exposed to 1 mM of sulfide, photosynthetic oxygen evolution ceased. Successive additions of sulfide resulted in oxygen uptake, probably as a result of the chemical reaction of sulfide with oxygen (Fig. 3). However, in cultures of cells that had accumulated high amounts of Fe, oxygen disappeared instantaneous after the addition of 1 mM of sulfide. Since neither sulfide alone nor ferric iron would show this oxygen uptake it seems that ferrous iron produced by sulfide mediated reduction of ferric iron was responsible for oxygen uptake.

The accumulation of Fe by *M. chthonoplastes* is probably mediated by the extracellular polysaccharide sheath. The EPS of *M. chthonoplastes* contains a considerable amount of uronic acids. These acid groups are conceived to be involved in the interaction with EPS of other organisms or with sediment particles. These groups also interact with metals such as iron. It has been observed that when *M. chthonoplastes* was grown with elevated amount of iron, the culture grew homogenously, whereas cultures grown in standard medium often aggregated and showed wall-growth (R. Moezelaar pers. com.). Therefore iron could decrease the capacity of EPS to bind to sediment particles. EPS seems to be capable of reducing ferric iron, and its capacity to keep a certain amount of iron in the reduced state, even in the presence of oxygen, seems to depend on its continuous synthesis since only living cells show iron reduction.

Acknowledgements. This work was supported by the Commission of the European Communities, DG XII, contract MAST-0027-C.

References

Admiraal W, Peletier WH, Zomer H (1982) Observations and experiments on the population dynamics of epipelic diatoms from an estuarine mudflat. Est Coast Shelf Sci 14:471-487

Anderson KL, Tayne TA, Ward DM (1987) Formation and fate of fermentation products in hot spring cyanobacterial mats. Appl Environ Microbiol 53:2343-2352

Bateson MM, Ward DM (1988) Photoexcretion and fate of glycolate in a hot spring cyanobacterial mat. Appl Environ Microbiol 54:1738-1743

Cohen Y, Rosenberg E (1989) Microbial Mats. Physiological ecology of benthic microbial communities. ASM, Washington.

Cohen Y, Castenholz RW, Halvorson HO (1984) Microbial Mats. Stromatolites. Alan R. Liss, New York.

Dade WB, Davis JD, Nichols PD, Nowell ARM, Thistle D, Trexler MB, White DC (1990) Effects of bacterial exopolymer adhesion on the entrainment of sand. Geomicrobiol J 8:1-16

Decho AW (1990) Microbial exopolymer secretions in ocean environments: their role(s) in food webs and marine processes. Oceanogr Mar Biol Ann Rev 28:73-153

Ehrlich H (1990) Geomicrobiology. Marcel Dekker, New York and Basel

Krumbein WE, Paterson DM, Stal LJ, Wipperman T (1993a) Microbially mediated processes in tide-influenced deposits and their importance in stabilization and diagenesis of sediments. In: Barthel KG, Bohle-Carbonell M, Fragakis C, Weydert M (eds) Mast days and euromar market. ECSC-EEC-EAEC, Brussels and Luxembourg, pp.242-259

Krumbein WE, Paterson D, Stal LJ, Wipperman T, de Winder B, Parsiegla K, Villbrandt M, Yallop M (1993b) Microbially mediated processes in tide influenced deposits and their importance in stabilisation and diagenesis of sediments. unpublished report EEC- MAST

Nicholson JAM, Stolz JF, Pierson BK (1987) Structure of a microbial mat at Great Sippewissett Marsh, Cape Cod, Massachusetts. FEMS Microbiol Ecol 45:343-364

Paterson DM (1989) Short-term changes in the erodibility of intertidal cohesive sediments related to the migratory behaviour of epipelic diatoms. Limnol Oceanogr 34:223-234

Pinckney J, Zingmark RG (1991) Effects of tidal stage and sun angles on intertidal benthic microalgal productivity. Mar Ecol Progr Series 76:81-89

Renström E, Bergman B (1989) Glycolate metabolism in cyanobacteria. I. Glycolate excretion and phosphoglycolate phosphatase activity. Physiol Plant 75:137-143

Rippka R, Deruelles J, Waterbury JB, Herdman M, Stanier RY (1979) Generic assignments strain histories and properties of pure cultures of cyanobacteria. J Gen Microbiol 111:1-61

Stal LJ (1991) The sulfur metabolism of mat-building cyanobacteria in anoxic marine sediments. Kieler Meeresforsch 8:152-157

Stal LJ (1993) Nitrogen fixation in microbial mats. In: Guerrero R, Pedrós-Alió (eds) Trends in Microbial Ecology, SSM, Barcelona, pp 141-146

Stal LJ, van Gemerden H, Krumbein WE (1985) Structure and development of a benthic marine microbial mat. FEMS Microbiol Ecol 31:111-125

van Gemerden H (1993) Microbial mats, a joint venture. Mar Geol 113:3-25

van Gemerden H, Tughan CS, de Wit, R, Herbert RA (1989) Laminated microbial ecosystems on sheltered beaches in Scapa Flow, Orkney Islands. FEMS Microbiol Ecol 62:87-102

Walter MR (1976) Stromatolites. Developments in Sedimentology 20. Elsevier, Amsterdam

Widdel F, Schnell S, Heising S, Ehrenreich A, Assmus B, Schink B (1993) Ferrous iron oxidation by anoxygenic phototrophic bacteria. Nature 362:834-836

Yallop ML, de Winder B, Paterson DM, Stal LJ (1994) Comparative structure, primary production and biogenic stabilisation of cohesive and non-cohesive marine sediments inhabited by microphytobenthos. Est Coast Shelf Sci (in press)

Species diversity in hot spring microbial mats as revealed by both molecular and enrichment culture approaches – relationship between biodiversity and community structure

David M. Ward, Michael J. Ferris, Stephen C. Nold, Mary M. Bateson, Eric D. Kopczynski and Alyson L. Ruff-Roberts

Department of Microbiology, Montana State University, Bozeman, Montana 59717 USA

We have conducted long-term studies of microbial mats of hot springs as model systems for investigating composition and structure of natural microbial communities and as modern analogs of stromatolites (Ward et al. 1984, 1987, 1989b). As recently as the last symposium on microbial mats our knowledge of the biodiversity within these communities was based solely on microbial species cultivated from such mat systems. The view has changed dramatically in the intervening years, because of the application of culture-independent techniques to recover and probe ribosomal RNAs (rRNAs) as biomarkers representing individual mat community members (Ward et al. 1992; Weller et al. 1992 and references cited therein), and more recently because of the renewed attempt to cultivate more relevant species. Here, we summarize what is currently known. The evidence suggests that the molecular methods we are using (Ward et al. 1992) may still not permit a complete understanding of the true complexity of the community. However, the approach does provide insight into understanding the basis behind this large biodiversity and into how more relevant species can be cultivated.

We have mainly tried to understand the composition of the Octopus Spring cyanobacterial mat. Table 1 compares what is known of species which are likely to be present based on those previously cultivated from this or similar geothermal habitats, those recognized by recovering small subunit (SSU) rRNA sequences, and those obtained in recent cultivation attempts. We have organized the table according to phylogenetic type rather than physiologic type (as we have done in previous reviews) because the SSU rRNA sequence information reveals the former, but not always the latter. This organization also provides a more evolutionarily relevant view of biodiversity. Our use of the term "species" is based on sufficient phenotypic difference to have justified published descriptions of the cultivated microorganisms, or on SSU rRNA sequence difference exceeding 3 to 5%. We have observed differences as small as 3-5% for only two pairs of SSU rRNAs (type-A vs. type-B and *Thermus* sp. E2 vs. *Th. aquaticus*); available sequences are usually much more dissimilar. Where comparisons can be made this degree of difference in SSU rRNA sequence exceeds that observed for isolates which have been described on the basis of phenotypic information as different species (e.g. *Thermus* spp., Bateson et al. 1990) or even genera (e.g., Gram-positive isolates, Bateson et al. 1989). It is possible that individual species might possess more than one SSU rRNA gene with different

NATO ASI Series, Vol. G 35
Microbial Mats
Edited by Lucas J. Stal and Pierre Caumette

Table 1. Species known or presumed to inhabit the **Octopus Spring cyanobacterial mat** community whose **SSU rRNA** sequences are known.

Basis for Recognition as Community Member					
Previously Cultivated	SSU rRNA recovered[1]	Recently Cultivated[1]	SSU rRNA Probed[2]	Highest Proven Abundance[3]	Pertinent References
I. Cyanobacteria					
Synechococcus sp. Y-7c-s[4]		*Syn.* sp. C1	+	10^1/mL(C,P)	1,2
	type-A		+	10^6/mL (P)	3,4
	type-B		+	10^6/mL (P)	3,4
	type-I		0		4
	type-J		+		4
	type-P	*Syn.* sp. B10		10^{10}/mL (C)	2,5
		Syn. sp. C9		10^9/mL (C)	2
II. Green nonsulfur bacteria and relatives					
Chloroflexus aurantiacus Y-400-fl			+		6
Thermomicrobium roseum[5]					7
	type-C		+		3,4
	OS-V-L-20				4
III. *Thermus/Deinococcus* group					
Thermus aquaticus YT-1		*Th. aq.* L2		10^3/mL (C)	8,9
Thermus sp. OS-Ramaley-4		*Th.* sp. E2		10^3/mL (C)	9,10
		Th. ruber N		10^6/mL (C)	9
IV. Planctomyces					
Isosphaera pallida IS1B[6]					11
	type-L[11]				3
V. Gram-positive bacteria					
Thermoanaerobium brockii HTD4[7]					12
Thermobacteroides acetoethylicus HTB2/W				10^7/mL (C)	13,14
Thermoanaerobacter ethanolicus JW200					15
Clostridium thermohydrosulfuricum 39E				$<10^3$/mL(C)	14,16
Clostridium thermosulfurogenes 4B				$<10^3$/mL(C)	17
Clostridium thermoautotrophicum JW701					18
VI. Thermodesulfobacterium					
Thermodesulfobacterium commune YSRA[8]					19
VII. Archaea					
Methanobacterium thermoautotrophicum Δ H[9]				10^7/mL (C)	20,21
VII. Spirochetes					
	type-H				3,4
	OS-V-L-7				4
	type-K				4
IX. Proteobacteria					
	OS-V-L-28				5
	type-G[11]				3
	type-O[11]				22
X. Green sulfur bacteria					
	type-E				5
	type-M[11]				3
	OS-III-9[11]				22
XI. Uncertain affiliation					
	type-D				3
	type-F				3
	type-N				3
	OP-I-2				5

[1] from 50° to 55°C regions of the **Octopus Spring** mat

[2] from 49° to 70°C regions of the **Octopus Spring** mat (Ruff-Roberts et al., submitted); blanks indicate that no probe experiments have been conducted.

[3] from either cultivation (C) or hybridization probing (P) (i.e., presence of **SSU rRNA** of that type) from the extincting dilution in **enrichment culture** ; previously reported high abundance for *Syn. lividus* in the **Octopus Spring** mat (Ward et al., 1987) should be disregarded, as this was based on direct counts of all autofluorescing rod-shaped cyanobacteria

[4] cultivated from Clearwater Springs, but presumed to inhabit **Octopus Spring** because of the presence of organisms with

similar morphology (Doemel and Brock 1977; Ward et al. 1987)
[5] cultivated from Toadstool Spring, but presumed to inhabit Octopus Spring because distinctive diol lipids are abundant in mat (Zeng et al. 1992a and b).
[6] cultivated from Kah-nee-tah Spring, but presumed to inhabit Octopus Spring because of the presence of organisms with similar morphology (Doemel and Brock 1977)
[7] cultivated from Washburn Spring, but Octopus Spring strains are known (Zeikus et la. 1979)
[8] cultivated from Ink Pot Spring, but Octopus Spring strains are known (Zeikus et al. 1983)
[9] cultivated from sewage sludge, but Octopus Spring strains are known (Zeikus et al. 1980; Sandbeck and Ward 1982; Wiegel, personal communication)
[10] only after enrichment culture with 10^{-1}-diluted mat inoculum
[11] based on Ribosomal Database Project analysis version 3.0 (personal communication, C. Woese, Univ. of Illinois)

References: 1, Kallas and Castenholz 1982; 2, Ferris and Ward unpublished; 3, Ward et al. 1990; 4, Weller et al. 1992; 5, Kopczynski and Ward unpublished; 6, Pierson and Castenholz 1974; 7, Jackson et al. 1973; 8, Brock and Freeze 1969; 9, Nold and Ward unpublished; 10, Ramaley personal communication; 11, Giovannoni,et al. 1987; 12, Zeikus et al. 1979; 13, Ben-Bassat and Zeikus 1981; 14, Zeikus et al. 1980; 15, Wiegel and Ljungdahl 1981; 16, Wiegel et al. 1979; 17, Schinck and Zeikus 1983; 18, Wiegel et al. 1981; 19, Zeikus et al. 1983; 20, Zeikus and Wolfe 1972; 21, Sandbeck and Ward 1982; 22, Ward et al. 1992.

sequences, so that the molecular method could overestimate species diversity (see Ward et al. 1992), but this is known to occur rarely, especially among bacteria. It seems more likely that by using such a conservative genetic marker (Pace et al. 1986; Ward et al. 1992) we are underestimating biodiversity.

Sixteen species are known on the basis of previous culture work, including two *Thermoleophilum* species, *T. album* and *T. minutum* (Zarilla and Perry 1984, 1986; Merkel et al. 1978), not included in Table 1 because their SSU rRNA sequences have not yet been determined. So far, twenty one new species have been suggested by SSU rRNA analysis. At face value the molecular approach seems to have slightly more than doubled the number of known species. However, one of the most striking features of Table 1 is that none of the SSU rRNAs of previously cultivated species have so far been recovered. This means that the views of community composition provided by previous cultivation efforts and by retrieval of SSU rRNA sequences are quite independent. This is further substantiated by the fact that the SSU rRNA approach revealed three new phylogenetic types of microorganisms (spirochetes, Proteobacteria, and green sulfur bacteria), whereas it has so far failed to detect phylogenetic types commonly recovered by cultivation (*Thermus*, Gram-positive bacteria, *Thermodesulfobacterium*, *Methanobacterium*). Thus, the biodiversity of the community is likely to be very poorly described by the collection of previously cultivated species.

In general the previously cultivated species provide a more limited view of biodiversity in the 50° to 55°C Octopus Spring mat than that provided by recovery of SSU rRNAs. The best example is that five SSU rRNA sequence types (A, B, I, J, P) were recovered, none of which was that of the single cultivated species *Synechococcus* sp. Y-7c-s. This strain was considered representative of thermophilic *Synechococcus* isolates, which, on the basis of nearly identical G + C ratios were all considered to be strains of a single species (Waterbury and Rippka 1989). We and others have in the past referred to this species as *Syn. lividus*. Because this now seems to be an oversimplification and because of the general confusion in *Synechococcus* taxonomy (Waterbury and Rippka 1989) we choose for the moment to refer to only the genus and strain designation. We have used hybridization probes to investigate the presence of *Synechococcus* sp. Y-7c-s in the 50° Octopus Spring mat, as well as in enrichment cultures inoculated with 10-fold serially diluted mat homogenates. Although *Synechococcus* sp. Y-7c-s SSU rRNA was not directly detected in mat samples or in the extincting dilution of the enrichment series, it was detected in an enrichment culture inoculated with 10^{-1}-diluted mat homogenate (Ferris et al. in preparation), implying its presence in the mat at low population density. We have recently been able to cultivate from a 10^{-1}-diluted homogenized mat sample a rapidly growing *Synechococcus* strain (C1), whose SSU rRNA sequence is nearly identical (99.3% similar) to that of strain Y-7c-s. New *Synechococcus* isolates from 10^{-9}- and 10^{-10}-diluted inocula grow much more slowly in culture, and have SSU rRNAs which were detected by cloning (strain B10 is type-P) or which are completely new (strain C9). The cultivation of a numerically relevant *Synechococcus* sp. (C9) whose SSU rRNA was not recovered by cloning suggests either a shortcoming of the molecular method (i.e., bias against recovery of this SSU rRNA) or that there is greater spatial or temporal heterogeneity of *Synechococcus* species distribution in the mat than we have previously assumed (i.e., samples for cloning and cultivation were not collected on the same dates and in exactly the same locations). The previous cultivation of *Synechococcus* sp. Y-7c-s combined with the simple rod-shaped morphology of cyanobacteria inhabiting the Octopus Spring 50° to 55°C mat misled we and others to think that this single cyanobacterium constructed the mat. In reality at least seven different cyanobacteria, apparently all with this same morphology, are present. In addition to belonging to one phylogenetically coherent group, these cyanobacteria presumably also all carry out oxygenic photosynthetic metabolism, and thus could be considered to belong to a single functionally defined guild, as defined by Odum (1983) to be a group of species that have a similar or comparable roles in the community.

Other phylogenetic groups (which might also be functionally defined guilds) appear to also have multiple species representation. Two *Chloroflexus*-like SSU rRNAs (type-C and OS-V-L-20) have been recovered and hybridization probe studies show that *C. aurantiacus* Y-400-fl is also present in high-temperature mat regions (Figure 1B), so that three *Chloroflexus*-like species are known. Recent dilution-based culture work has revealed that at least three *Thermus* species exist in the mat (Nold and Ward, unpublished). Three spirochete-like SSU rRNAs have also been recovered from the mat. The degree of biodiversity within each of these groups is large. The difference among the eight cyanobacterial SSU rRNAs is equivalent to about half of the difference known to exist in a large sampling of cyanobacteria (Weller et al. 1992). *Chloroflexus*-like SSU rRNAs are comparably diverse (Weller et al. 1992). The three *Thermus* species span the known breadth of difference in SSU rRNA sequences among species of this genus (Nold and Ward, unpublished). Two of the three spirochete-like SSU rRNAs belong to different major subdivisions of this phylogenetic group; the third is only loosely affiliated with the group (Weller et al. 1992).

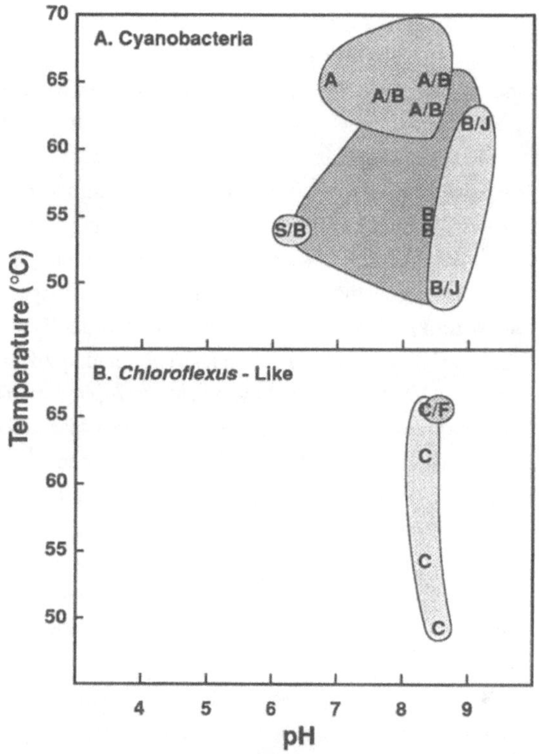

Fig. 1. Distribution of (A) cyanobacterial (type-A, -B, -J and *Synechococcus* sp. Y-7c-s (S)) and (B) *Chloroflexus*-like (type-C and *C. aurantiacus* Y-400-fl (F)) SSU rRNAs in hot spring microbial mats of varying temperature and pH, based on hybridization probing (Ruff-Roberts et al. submitted).

What is the basis for biodiversity within phylogenetic groups within the community? We believe that our results to date suggest two possible explanations. First, the species pool within the community is likely to be greater than the subset of species which are active under the environmental conditions of the moment (i.e., the active autochthonous community members). Allochthonous microorganisms which are dispersed to the community from some other habitat, but which may not be competitive in locally adverse conditions, may nevertheless survive and thus be capable of growth when the environment changes (including in enrichment culture). Zymogenous microorganisms, native to the community but waiting for more opportune conditions, may exist at low population density and behave similarly. As Baas-Becking (1934) suggested, "everything is everywhere : but the milieu selects...in nature and in the laboratory". Although nature may be more severe in controlling the survival of allochthonous and zymogenous species, there are undoubtedly some of these types of organisms present in all microbial communities. *Synechococcus* sp. Y-7c-s may, for example, be allochthonous or zymogenous to mats in alkaline hot springs. Hybridization probe studies suggest that its SSU rRNA is more abundant in a mat in a slightly acid spring (Figure 1A). Its possible preference for such habitats is indicated by the fact that its SSU rRNA decreases when samples from the acid mat are shifted to higher pH, and rises when more alkaline mat samples are shifted to lower pH (Ruff-Roberts et al. submitted). In the alkaline 50° to 55°C Octopus Spring mat *Synechococcus* sp. Y-7c-s seems to be in low population density, but in enrichment culture it is capable of rapidly overgrowing much more relevant *Synechococcus* species.

A second explanation for biodiversity within phylogenetic groups, especially relevant for active community members, is that the community may have substructure(s) involving specialization of populations within guilds to specific environmental features. The best known example is of genetically stable populations of *Synechococcus* adapted to different temperatures (Figure 2A and B). Castenholz (1973) showed how the overlapping distributions of such populations could result in redundancy of cyanobacteria at any given spatial location within the community (Figure 2A). For instance, at 50° to 55°C (vertical lines in Figure 2A) four populations are predicted to exist with those best adapted most abundant. Without knowing the genetic relationships among these populations they were interpreted as strains of the single species *Syn. lividus* (Castenholz 1981). We have observed that some of the Octopus Spring cyanobacterial SSU rRNA sequence types exhibit a distribution and behavior characteristic of temperature adaptation. Type-A, -B and -J SSU rRNAs are found principally at high, moderate and low temperatures, respectively (Figure 1A); shift from low to high temperature causes population restructuring in favor of the SSU rRNA which is normally distributed to the high temperature region (i.e., type-A increases, while type-B and -J decrease, Figure 3). Thus, temperature adaptation may explain some of the biodiversity among cyanobacterial SSU rRNA sequence types we have discovered. However, the similarity among these SSU rRNAs is much lower than one would expect among strains of a single species (Weller et al. 1992). Substructuring based on adaptation of different populations to different temperatures may help to provide homeostasis within the community, since at a given location within the mat there are flow-related and seasonal differences in temperature (Brock 1967).

Temperature adaptation may be a general phenomenon among phylogenetic groups (which might also constitute functional guilds) within the mat community. This behavior was reported by Bauld and Brock (1973) for organisms with a *Chloroflexus*-like metabolism (Figure 2C). Hybridization probe studies suggest that *C. aurantiacus* Y-400-fl is found at (Figure 1B) and is adapted to high temperature (Figure 3). The other *Chloroflexus*-like SSU rRNA sequences might represent populations adapted to lower temperatures. Similarly, the different *Thermus* species we have recently cultivated might represent temperature-adapted populations. *Th. ruber*, the most abundant of the three isolates from the 50°C mat is also the species reported to have the lowest temperature optimum (50° to 55°C, Loginova et al. 1975). *Thermus* sp. OS-Ramaley-4 was 1000-fold less abundant in the same sample and was only observed in enrichments at 70°C. Differential distributions of *Thermus* spp. optimized to different temperatures has been previously reported (Ramaley and Bitzinger 1975).

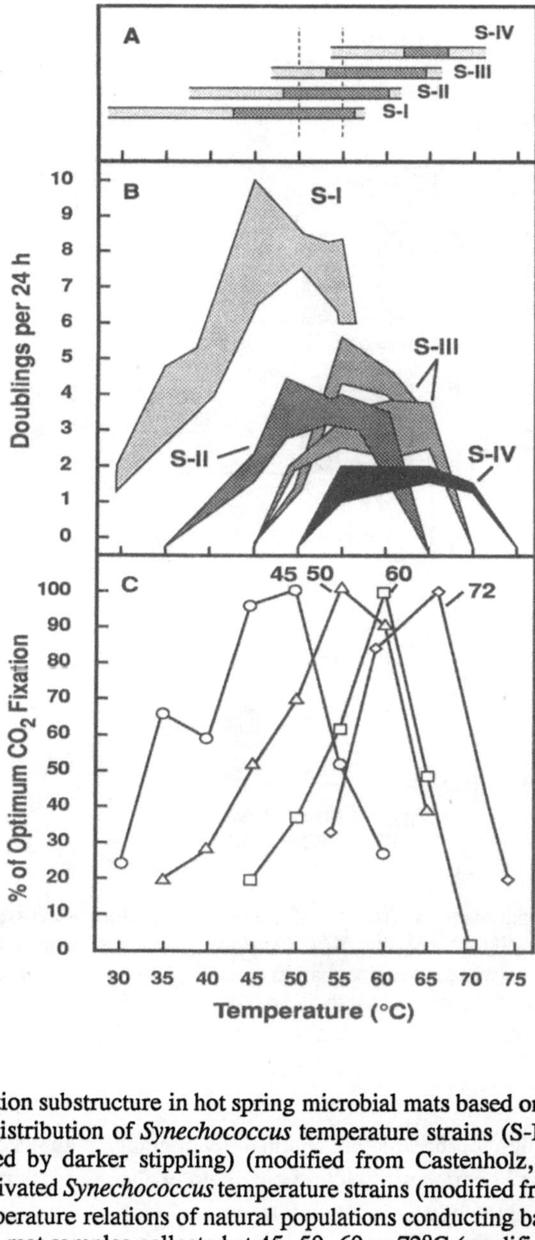

Fig. 2. Population substructure in hot spring microbial mats based on temperature adaptation. (A) Proposed distribution of *Synechococcus* temperature strains (S-I, -II, -III, -IV; optimum range designated by darker stippling) (modified from Castenholz, 1973), (B) temperature relations of cultivated *Synechococcus* temperature strains (modified from Peary and Castenholz 1964), (C) temperature relations of natural populations conducting bacterial photosynthesis in Octopus Spring mat samples collected at 45, 50, 60 or 72°C (modified from Bauld and Brock 1973).

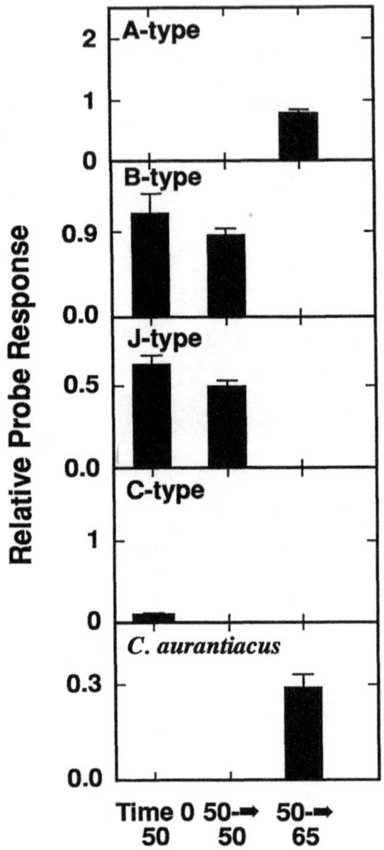

Fig. 3. Distribution of cyanobacterial (type-A, -B, and -J) and *Chloroflexus*-like (type-C and *C. aurantiacus* Y-400-fl) SSU rRNAs in a 50°C Octopus Spring mat sample before (Time 0, 50) and after shifting for one week *in situ* incubation at 65°C (50-->65). An *in situ* control (50-->50) was also included. Relative response is based on specific normalized by universal hybridization probe responses (Ruff-Roberts et al., submitted). Bars indicate standard error (n=3).

A lack of appreciation for the structure within the community, combined with the strong selection which can occur in enrichment culture may help explain the general incongruence between culture and SSU rRNA collections. High-temperature-adapted populations would likely contaminate and survive in cooler downstream regions of mat (Figure 2A). They might be considered to be zymogenous members of the cooler mat regions. Enrichment cultures inoculated with undiluted low-temperature mat samples, but incubated at high temperature, would favor the high-temperature population. The case of *Thermus* provides an example. Although *Th. ruber* was 1000-fold more predominant in the 50°C mat, *Thermus* sp. OS-

Ramaley-4 was selected during incubation at 70°C. Careful examination of the cultivation strategies which led to most of the previous isolates listed in Table 1 shows that these organisms were enriched (usually in broth medium) from mat samples (usually undiluted) which either came from high-temperature regions (where high-temperature-adapted populations should predominate), or from low-temperature regions followed by incubations at high temperature (favoring recovery of high-temperature-adapted populations). The SSU rRNA sequences, on the other hand, have come directly from the 50° to 55°C mat region.

There are other possible explanation for the incongruence between culture and SSU rRNA sequence collections. One is that the composition of some enrichment media may not accurately mimic the natural environment. For instance, *T. commune*, a sulfate-reducing bacterium, was enriched with lactate as energy source (Zeikus et al. 1983), but environmental measurements suggest that H_2 is the primary energy source driving sulfate reduction in the mat (Ward et al. 1984). The failure to recover SSU rRNAs of some community members might also be a consequence of community structure. Most of the cultivated species whose SSU rRNAs have not yet been detected belong to physiologic groups (fermentative, sulfate reducing, methanogenic) that should occupy the highest trophic levels in the community. It is reasonable to expect that the population densities of such community members should be lower and thus the SSU rRNAs less frequently detected. Empirical evidence in favor of this hypothesis comes from lipid biochemical marker analysis (Ward et al. 1989a; Zeng et al. 1992 a and b).

In summary, the SSU rRNA biomarker approach reveals that the species cultivated from the Octopus Spring cyanobacterial mat provide only a very limited view of the composition of the community. It is comforting that probe studies confirmed the presence of both cultivated species we have so far investigated. However the relevance of these species to the community is either minimal (*Synechococcus* sp. Y-7c-s) or is restricted to a specific part of the community (*C. aurantiacus* Y-400-fl). Furthermore, cultivation has grossly underestimated the biodiversity within phylogenetic groups. Probing of SSU rRNA enables distributional and behavioral analysis, which provides insight into possible adaptations which might explain such biodiversity, and which in turn leads to new cultivation approaches that are yielding numerically relevant new species. Our observations to date force us to expand our concept of mat community composition and structure. We must accommodate two general concepts. First, the community may contain members which are inactive and thus likely to be at low population density, but which are capable of rapidly responding to changing environmental conditions. A dramatic example of such environmental change is enrichment culture which, because of currently unknown differences between laboratory and natural environments, can select strongly for minor community members (e.g., *Synechococcus* sp. Y-7c-s from 50°C Octopus Spring mat). Second, among the active community members, specific phylogenetic groups of microorganisms (e.g., cyanobacteria, green nonsulfur bacteria, spirochetes, *Thermus*) may contain multiple populations which are adapted to specific environmental conditions. The populations within each group might constitute a functional guild within the community, and might thus provide a mechanism at the level of different populations which imparts homeostasis of function to dampen the effects of environmental variation on a biochemical process within the mat community. Adaptations of microbial processes in hot springs, have been previously interpreted as physiological optimizations of single populations (Brock 1967; Madigan and Brock 1977). One example may be that temperature-adapted populations could stabilize community functions against flow-related or seasonal variation in temperature. Adaptations of cyanobacterial populations to pH and light are also suggested by our hybridization probe studies (Ruff-Roberts et al. submitted). Populations of one thermophilic cyanobacterium, *Plectonema notatum*, differentially adapted to light intensity are know to exist (Sheridan 1976, 1979). This might provide a basis for additional biodiversity within photosynthetic guilds of the community.

Acknowledgements. This work was supported by grants from the U.S. National Science Foundation (BSR-9209677) and the U.S. National Aeronautics and Space Administration (NAGW-2764). We thank the U.S. National Park Service for permission to conduct research in Yellowstone National Park. We especially thank Dr. J.G. Kuenen for locating and translating the quote from Baas Becking (1934).

References

Baas Becking LGM (1934) Geobiologie of inleiding tot de milieukunde. W.P. van Stockum and Zoon N.V., Den Haag, The Netherlands

Bateson MM, Thibault K, Ward DM (1990) Comparative analysis of partial 16S ribosomal RNA sequences of *Thermus* species. System Appl Microbiol 13:8-13

Bateson MM, Wiegel J, Ward DM (1989) Comparative analysis of 16S ribosomal RNA sequences of thermophilic fermentative bacteria isolated from hot spring cyanobacterial mats. System Appl Microbiol 12:1-7

Bauld J , Brock TD (1973) Ecological studies of *Chloroflexis*, a gliding photosynthetic bacterium. Arch Mikrobiol 92: 267-284

Ben-Bassat A, Zeikus JG (1981) *Thermobacteroides acetoethylicus* gen. nov. and spec. nov., a new chemoorganotrophic anaerobic, thermophilic bacterium. Arch Microbiol 128:365-370

Brock TD (1967) Micro-organisms adapted to high temperatures. Nature 214:882-885

Brock TD, Freeze H (1969) *Thermus aquaticus* gen. n. and sp. n., a nonsporulating extreme thermophile. J Bacteriol 98:289-297

Castenholz RW (1973) Ecology of blue-green algae in hot springs. In: Carr NG, Whitton BA (eds) The biology of blue-green algae. Blackwell, Oxford/London, pp 379-414

Castenholz RW (1981) Isolation and cultivation of thermophilic cyanobacteria. In: Starr MP, Stolp H, Truper HG, Balows A, Schlegel HG (eds) The Prokaryotes. Springer-Verlag, Berlin/Heidelberg/New York pp 236-246

Doemel WN, Brock TD (1977) Structure, growth, and decomposition of laminated algal-bacterial mats in alkaline hot springs. Appl Environ Microbiol 34:433-452

Giovannoni SJ, Schabtach E, Castenholz RW (1987) *Isosphaera pallida*, gen. and comb. nov., a gliding, budding eubacterium from hot springs. Arch Microbiol 147:276-284

Jackson TJ, Ramaley RF, Meinschein WG (1973) *Thermomicrobium*, a new genus of extremely thermophilic bacteria. Int J System Bacteriol 23:28-36

Kallas T, Castenholz RW (1982) Internal pH and ATP-ADP pools in the cyanobacterium *Synechococcus* sp. during exposure to growth-inhibiting low pH. J Bacteriol 149:229-236

Loginova LG, Egorova LA (1975) An obligately thermophilic bacterium *Thermus ruber* from hot springs in Kamchatka. Microbiology 44:593-597

Madigan MT, Brock TD (1977). Adaptation by hot spring phototrophs to reduced light intensities. Arch Microbiol 113:111-120

Merkel GJ, Underwood WH, Perry JJ (1978) Isolation of thermophilic bacteria capable of growth solely in long-chain hydrocarbons. FEMS Microbiol Lett 3:81-83

Odum EP (1983) Basic ecology. Saunders College Publishing, Fort Worth, Texas

Pace NR, Stahl DA, Lane DJ, Olsen GJ (1986) The analysis of natural microbial populations by ribosomal RNA sequences. Adv Microbial Ecol 9:1-55

Peary JA, Castenholz RW (1964) Temperature strains of a thermophilic blue-green alga. Nature 202:720-721

Pierson BK , Castenholz RW (1974) A phototrophic gliding filamentous bacterium of hot springs, *Chloroflexus aurantiacus*, gen. and sp. nov. Arch Microbiol 100:5-24

Ramaley RF, Bitzinger K (1975) Types and distribution of obligate thermophilic bacteria in man-made and natural thermal gradients. Appl Microbiol 30:152-155

Ruff-Roberts AL, Kuenen JG, Ward DM Distribution of cultivated and uncultivated cyanobacteria and *Chloroflexus*-like bacteria in hot spring microbial mats (submitted)

Sandbeck KA, Ward DM (1982) Temperature adaptations in the terminal processes of anaerobic decomposition of Yellowstone and Icelandic hot spring microbial mats. Appl Environ Microbiol 44: 844-851

Schink B, Zeikus JG (1983) *Clostridium thermosulfurogenes* sp. nov., a new thermophile that produces elemental sulphur from thiosulphate. J Gen Microbiol 129:1149-1158

Sheridan RP (1976) Sun-shade ecotypes of a bluegreen alga in a hot spring. J Phycol 12:279-285.

Sheridan RP (1979) Seasonal variation in sun-shade ecotypes of *Plectonema notatum* (Cyanophyta). J Phycol 15:223-226

Ward DM, Beck E, Revsbech NP, Sandbeck KA, Winfrey MR (1984) Decomposition of hot spring microbial mats. In: Cohen Y, Castenholz RW, Halvorson HO (eds) Microbial mats: Stromatolites. A.R. Liss, Inc., New York pp 191-214

Ward DM, Tayne TA, Anderson KL, Bateson MM (1987) Community structure and interactions among community members in hot spring cyanobacterial mats. Symp Soc Gen Microbiol 41:179-210

Ward, DM, Shiea J, Zeng YB, Dobson G, Brassell S, Eglinton G (1989a) Lipid biochemical markers and the composition of microbial mats. In: Cohen Y, Rosenburg E (eds) Microbial Mats: Physiological ecology of benthic microbial communities. Am Soc Microbiol, Wash DC pp 439-454

Ward DM, Weller R, Shiea J, Castenholz RW, CohenY (1989b) Hot spring microbial mats: anoxygenic and oxygenic mats of possible evolutionary significance. In: Cohen Y, Rosenburg E (eds) Microbial Mats: Physiological ecology of benthic microbial communities. Am Soc Microbiol, Wash DC pp 3-15

Ward DM, Weller R, Bateson MM (1990) 16S rRNA sequences reveal numerous uncultured microorganisms in a natural community. Nature 345: 63-65

Ward DM, Bateson MM, Weller R, Ruff AL (1992) Ribosomal RNA analysis of microorganisms as they occur in nature. Adv Microbial Ecology 12:219-286

Waterbury, JB, Rippka R (1989) Subsection I. Order *Chroococcales*. In: Staley JT, Bryant MP, Pfennig N, Holt JG (eds) Bergey's Manual of Systematic Bacteriology, vol. 3. Williams and Wilkins, Baltimore, MD, pp 1728-1746

Weller R, Bateson MM, Heimbuch BK, Kopczinski ED, Ward DM (1992) Uncultivated cyanobacteria, *Chloroflexus*-like inhabitants and spirochete-like inhabitants of a hot spring microbial mat. Appl Environ Microbiol 58:3964-3969

Wiegel J, Ljungdahl LG (1981) *Thermoanaerobium ethanolicus* gen. nov., spec. nov., a new, extreme thermophilic, anaerobic bacterium. Arch Microbiol 128:343-348

Wiegel J, Ljungdahl LG, Rawson JR (1979) Isolation from soil and properties of the extreme thermophile *Clostridium thermohydrosulfuricum*. J Bacteriol 139:800-810

Wiegel J, Braun M, Gottschalk G (1981) *Clostridium thermoautotrophicum* species novum, a thermophile producing acetate from molecular hydrogen and carbon dioxide. Curr Microbiol 5:255-260

Zarilla KA, Perry JJ (1984) *Thermoleophilum album* gen. nov. and sp. nov., a bacterium obligate for thermophily and n-alkane substrates. Arch Microbiol 137:286-290

Zarilla KA, Perry JJ (1986) Deoxyribonucleic acid homology and other comparisons among obligately thermophilic hydrocarbonoclastic bacteria, with a proposal for *Thermoleophilum minutum* sp. nov. Int J System Bacteriol 36:13-16

Zeikus JG, Wolfe RS (1972) *Methanobacterium thermoautotrophicum* sp. n., an anaerobic, autotrophic, extreme thermophile. J Bacteriol 109:707-713

Zeikus JG, Hegge PW, Anderson MA (1979) *Thermoanaerobium brockii* gen. nov. and sp. nov., a new chemoorganotrophic, caldoactive, anaerobic bacterium. Arch Microbiol 122:41-48

Zeikus JG, Ben-Bassat A, Hegge PW (1980) Microbiology of methanogenesis in thermal, volcanic environments. J Bacteriol 143:432-440

Zeikus, JG, Dawson MA, Thompson TE, Ingvorsen K, Hatchikian EC (1983) Microbial ecology of volcanic sulphidogenesis: isolation and characterization of *Thermodesulfobacterium commune* gen. nov. and sp. nov. 129:1159-1169

Zeng YB, Ward DM, Brassell S, Eglinton G (1992a) Biogeochemistry of hot spring environments. 2. Lipid compositions of Yellowstone (Wyoming, USA) cyanobacterial and *Chloroflexus* mats. Chem Geol 95:327-345

Zeng YB, Ward DM, Brassell S, Eglinton G (1992b) Biogeochemistry of hot spring environments. 3. Apolar and polar lipids in the biologically active layers of a cyanobacterial mat. Chem Geol 95:347-360

Microbial mats in a thermomineral sulfurous cave

Serban M. Sarbu, Luminita Vlasceanu, Radu Popa[1], Peter Sheridan, Brian K. Kinkle, Thomas C. Kane

Department of Biological Sciences (ML-6), University of Cincinnati, Cincinnati, OH 45221-0006, USA

Introduction

The thermomineral sulfurous waters at Mangalia in southeastern Dobrogea, Romania, have been known and used as spa facilities for well over 2,000 years (Feru and Capotà 1991). Hydrogeological studies performed during the last 60 years (Macovei 1912; Ciocîrdel and Protopopescu-Pache 1955; Moissiu 1968; Feru and Capotà 1991) identified a deep captive sulfurous aquifer located in Barremian-Jurassic limestones, extending 15 km to the North and 50 km to the South of Mangalia. In the Mangalia region, a system of geological faults allows the deep water to ascend toward the surface and mix with the Sarmatian oxygenated waters (Lascu et al. 1993). The biological investigation of the subsurface ecosystems associated with the sulfurous waters at Mangalia commenced in the late eighties, after the discovery of Movile Cave and its unique subterranean chemoautotrophically based ecosystem (Sarbu, 1990).

The construction in 1986 of an artificial shaft has provided investigators access to a system of natural cave passages located at a depth of 20 meters, partly flooded by hydrogen sulfide-rich thermal waters (Lascu 1989). Despite the extreme isolation of this cave system from the surface (Lascu et al. 1993), the large population sizes of the invertebrate species and the abundance of predators in the trophic structure suggest a large energy base for the ecosystem (Sarbu and Popa 1992). Geophysical investigations showed that numerous other subsurface voids are present in the region, indicating that Movile Cave is part of an extensive network of fissures and cave passages associated with the sulfurous waters at Mangalia. This hypothesis is also supported by the recent discovery (Lascu et al. 1993) of representatives of the Movile Cave fauna in old hand-dug wells in the town of Mangalia, at a distance of several kilometers from the cave.

Isolated air pockets ('airbells') in the submerged passages of Movile Cave, contain microbial mats on the walls and floating on the water surface (Sarbu 1990). Thin brown microbial mats cover the walls of the submerged cave passages, and of the deep and dark sections of surface sulfurous springs located around Mangalia.

---

[1] "E. Racoviţa" Speleological Institute, Frumoasa str 31, Bucharest, Romania

NATO ASI Series, Vol. G 35
Microbial Mats
Edited by Lucas J. Stal and Pierre Caumette

Aerobic microbial mats

Both the floating microbial mats and the mats covering the emergent limestone walls occur only in the remote airbells of the cave (Figure 1.A) where the atmosphere is depleted in O_2 (7-10%), highly enriched in CO_2 (2.5-3%) and contains small amounts of CH_4 (0.5-1%). Experimental manipulation in microcosms containing a controlled atmosphere showed that floating mats do not develop when O_2 exceeds 15%. Thus, microbial mats are absent from the water surface of the lake and from the limestone walls of the Lake Room (Figure 1.B) where the O_2 reaches 19-20%. It is believed that the low O_2 levels in the airbell atmosphere inhibit the heterotrophic activity within the mat, while in the Lake Room the heterotrophic breakdown of the mat exceeds the rate of chemoautotrophic mat formation, and thus the microbial mats are absent.

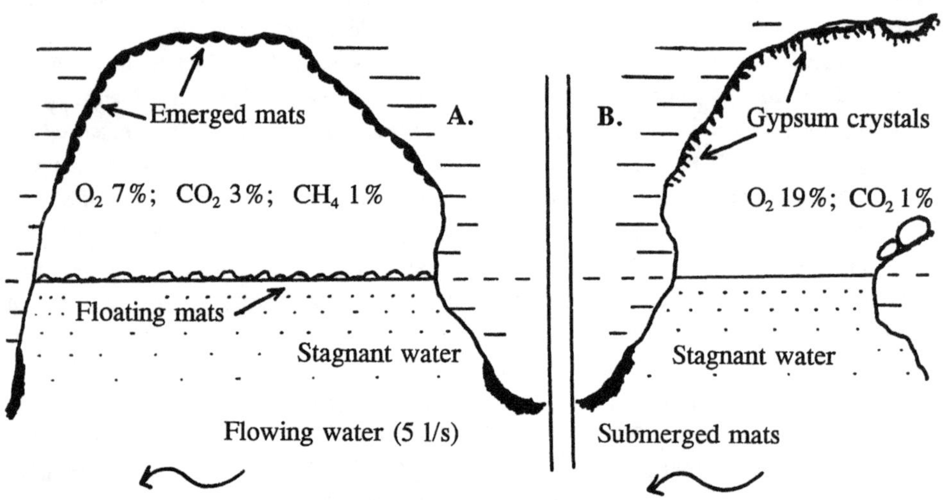

Fig. 1. Cross-section of the lower level of Movile Cave: A. Airbell; B. Lake-Room.

From the moment of their discovery it was speculated that the microbial mats contain sulfide-oxidizing chemolithoautotrophic microorganisms that take advantage of the sulfide-oxygen interface and are responsible for synthesizing sufficient organic matter to provide a food base for the cave's rich invertebrate community.

Floating microbial mats in Movile Cave are located at the interface between the H_2S-rich cave water and the oxygenated cave atmosphere. They are kept afloat by gas bubbles trapped beneath the mat that may reach volumes of 100 cm^3. If the mat is disturbed, the bubbles are lost and sections of the mat sink to the bottom of the cave passage. Scanning electron micrographs of floating mat show an extensive interwoven mycelial structure with closely associated bacterial rods and cocci (Photo 1). Preliminary investigations by P. Febbroriello indicate that the dominant fungal component of the mat appear to be Oomycetes with either terminal or mid-hyphal oogonia. Other species of fungi present in the mats are: *Gliocladium* sp., *Penicillium* sp., *Trichoderma* sp., and *Plasmopora* sp. A profile view scanning electron micrograph shows that the bulk of the mycelia appears to be dead biomass (Photo 2.A). The bacterial component appears to be actively growing due to the absence of fatty acid stress markers such as 16:1w7t/16:1w7c and 18:1w7t/18:1w7c.

Mats covering the exposed cave walls are present in the airbells. Gaseous H_2S can be smelled, but it could not be detected with Draeger tubes in the atmosphere of undisturbed airbells. Following disturbance of the water surface, H_2S concentrations of 0.5-1 ppm were measured. The high solubility of H_2S in water causes its dissolution into the water film covering the cave walls, hence the low levels of H_2S in the atmosphere of the airbell. This allows microbial mats to develop at the interface between H_2S rich water film and the oxygenated cave atmosphere. Sulfuric acid resulting as byproduct of sulfide oxidation is responsible for the low pH values of 3.7 to 4.2 measured on the surface of the wall mats as compared with pH values of 7.5 to 8 for water flowing through limestone. Beneath the mat, the carbonate walls are corroded by the sulfuric acid, leading to the enlargement of the cave passages by sulfuric-acid speleogenesis first described from caves in the USA by Egemeier (1981).

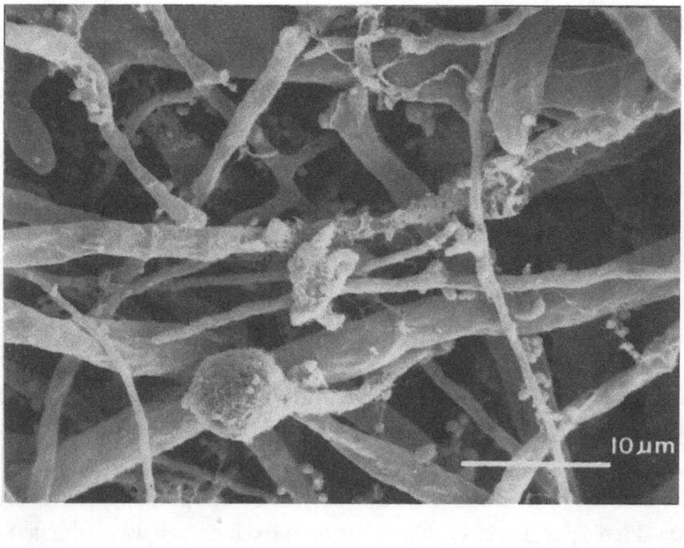

Photo 1. Floating mat consisting of mycelium associated with bacterial rods and cocci.

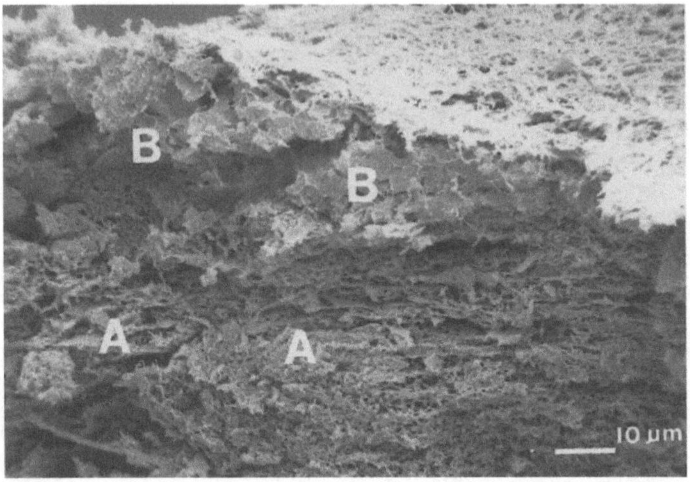

Photo 2. Cross-section through floating mat. A. Stratified layers of dead mycelium; B. Elemental sulfur.

Anaerobic submerged microbial mats

Submerged microbial mats were first discovered on the walls of the submerged cave passages in water depths exceeding 0.5 m. Recent field studies also identified slowly growing microbial mats on limestone walls in total darkness, deep within several of the sulfurous springs and in the deeper sections of some hand-dug wells in the town of Mangalia. Oxygen concentrations drop under 0.03 mg l^{-1} in these waters at depths exceeding 5 cm, while H_2S concentrations range between 10 and 43 mg^{-1}. Redox potential values range between -260 and -351 mV. SEM observations showed that the submerged mats are composed of morphologically similar fungal mycelia with little bacterial colonization.

Chemoautotrophic carbon fixation within the microbial mats

Our preliminary studies show that chemolithoautotrophic sulfide-oxidizing bacteria are present in the floating microbial mats (Sarbu et al. 1991). Samples of floating mat were incubated with [^{14}C]bicarbonate, followed by chloroform-methanol lipid extraction (Bligh and Dyer 1959). The microbial biomass was determined by analysis of the phospholipid component of the extracted lipids (Findlay et al. 1989). The amount of fixed carbon was estimated by measuring the radioactivity of the microbial lipids extracted at the end of the incubation time (McKinley et al. 1982). Increase in the radioactivity of the microbial lipids during a time course incubation indicated uptake of radiolabeled bicarbonate, suggesting chemoautotrophic carbon fixation.

Support for the hypothesis of chemoautotrophy has also been obtained from an assay for ribulose-1,5-bisphosphate carboxylase/oxygenase (RuBisCO). Incubation of an homogenate made from the mat with the substrate ribulose-1,5-bisphosphate resulted in an increase in acid-stable reaction products compared with the controls (Tabita et al. 1978). This assay indicates that RuBisCO may be present and active in the mats. A thick layer of elemental sulfur was identified by X-ray dispersion analysis under the surface of the floating mats (Photo 2.B). Sulfur is a byproduct of sulfide oxidation and accumulates in the mat at the sulfide-oxygen interface.

Ecological significance of the Movile Cave microbial mats

Numerous collembola graze on the surface of the mats, while underneath the mat there are rich populations of ciliates, nematodes, oligochaetes, rotifers, copepods, and amphipods, all feeding upon the mat microbiota. These data are all consistent with the hypothesis that the energy base of the Movile Cave community is chemoautotrophic production by microbes using H_2S as energy source. This indicates that Movile is unlike other cave communities, which rely on allochthonous energy sources, but rather has more similarities with chemoautotrophic systems such as those found near deep-sea vents. Movile Cave may, therefore, provide an accessible and simplified model for understanding these unusual chemoautotrophic systems.

Acknowledgements

The research presented here was supported by the National Geographic Society (grant 4639-91 to T.C. Kane), the Cave Research Foundation, the National Speleological Society, the Explorers Club, the Department of Biological Sciences of the University of Cincinnati, and the "Emil Racoviţa" Speleological Institute. The authors would like to thank Rick Olson for his help with the SEM work.

References

Bligh EG, Dyer WJ (1959) A rapid method of total lipid extraction and purification. Can J Biochem Physiol 37:911-917

Ciocîrdel R, Protopopescu-Pache E (1955) Consideraţii hidrogeologice asupra Dobrogei. Studii tehnice şi economice. Seria E 3 hidrogeologie pp 1-49

Egemeier SJ (1981) Cavern development by thermal waters. NSS Bull 43:31-51

Feru MU, Capotà A (1991) Les eaux thermominérales karstiques de la zone de Mangalia (Roumanie). Theor Appl Karstol 4:143-157

Findlay RH, King GM, Watling L (1989) Efficacy of phospholipid analysis in determining microbial biomass in sediments. Appl Environ Microbiol. 55:2888-2893

Lascu C (1989) Paleogeographical and hydrogeological hypothesis regarding the origin of a peculiar cave fauna. Mics speol Rom 1:13-18

Lascu C, Popa R, Sarbu SM, Vlasceanu L, Prodan S (1993) La grotte de Movilé: une faune hors du temps. La Recherche 258:1092-1098
Macovei G (1912) Cîteva observaţiuni asupra hidrogeologiei subterane în Dobrogea de Sud. Dari de Seama Instit Geol Rom III. pp 123.
McKinley VL, Federle TW, Vestal JR (1982) Effects of hydrocarbons on plant litter microbiota of an Arctic lake. Appl Environ Microbiol. 43:129-135
Moissiu C (1968) Consideraţii hidrochimice privind apele freatice din Dobrogea de Sud. Hidrotehnica Gospodarirea Apelor Meteorologia. 13:417-420
Sarbu S (1990) The unusual fauna of a cave with thermomineral waters containing H_2S, from southern Dobrogea, Romania. Memoires de Biospeologie. XVII:191-195
Sarbu SM, Popa R (1992) A unique chemoautotrophically based cave ecosystem. In: Camacho A (ed) The Natural History of Biospeleology. Luis Arguero Publ. Madrid pp 637-666
Sarbu SM, Popa R, Goliat I (1991) Chemoautotrophic production in a thermomineral sulfurous cave. Trav Inst Speol XXX:59-61
Tabita FR, Caruso P, Whitman W (1978) Facile assay of enzymes unique to the Calvin cycle in intact cells with special references to ribulose 1,5-bisphosphate carboxylase. Anal Biochem 84:462-472

Establishment of phototrophic purple sulphur bacteria in microbial mat systems

Rodney A. Herbert and Dave T. Welsh

Department of Biological Sciences, University of Dundee,
Dundee DD1 4HN, UK

Introduction

Laminated microbial mat ecosystems, dominated by phototrophic microorganisms are widely distributed in geothermal, hypersaline and intertidal marine environments (Castenholz 1984; Cohen 1984; Jørgensen and Cohen 1977; Stal et al. 1985; Van Gemerden et al. 1989a, b). The laminated structure of these vertically stratified benthic ecosystems is the result of the development of the constituent microflora along physical and chemical gradients. Thus, in a typical intertidal marine microbial mat dense populations of diatoms, cyanobacteria, purple and green sulfur bacteria, sulfate reducing bacteria and colourless sulphur oxidising bacteria co-exist and compete for light, CO_2 and sulfide in the top few millimetres of the beach surface. In these systems a green layer of oxygenic cyanobacteria and unicellular green algae overlies a pink/red layer of anoxygenic purple sulfur bacteria which in turn is underlain by a black layer composed of precipitated ferrous sulfide (FeS). Cyanobacteria are the key component of most microbial mats since they thrive in nutrient poor, unstable and hostile environments. Filamentous species belonging to the genus *Oscillatoria*, able to fix dinitrogen, are commonly found as the initial colonisers in such environments (Stal and Krumbein 1981). They not only enrich intertidal marine sediments with low molecular weight organic compounds as a consequence of photosynthate excretion and cell lysis but also play an important role in stabilising sediments and reducing evaporative water loss (Stal et al. 1985). The released organic carbon is rapidly mineralised by heterotrophic bacteria resulting in the development of anoxic conditions. In the absence of oxygen populations of sulfate reducing bacteria become established. These heterotrophic bacteria respire anaerobically using sulfate as the terminal e^- acceptor with the concomitant production of sulfide. Sulfide generated by dissimilatory sulfate reduction can then be used by phototrophic purple sulfur bacteria such as *Thiocapsa roseopersicina* and *Chromatium vinosum* as e^- donor in anoxygenic photosynthesis or as an energy source by colourless sulfur bacteria. In this type of laminated mat system organic matter required for sulfide production is generated *in situ* principally by cyanobacteria and unicellular green algae and can therefore be self-sustaining for several years (Van Gemerden 1993).

In contrast, in some temperate marine intertidal laminated mat systems e.g., those found on sheltered beaches in Scapa Flow, the Orkney Islands, Northern Scotland derive their organic carbon primarily from the degradation of allochthonous organic matter (buried macro-algal

NATO ASI Series, Vol. G 35
Microbial Mats
Edited by Lucas J. Stal and Pierre Caumette

debris). The consequence of this high input of organic carbon into the sub-surface sediments is that sulfate reduction is stimulated and during the summer months the O_2-H_2S interface is close to the sediment surface enabling mass blooms of purple sulfur bacteria to develop at the beach surface (Herbert 1985). These mass blooms of purple sulfur bacteria are dependent upon a continual supply of S^{2-} which in turn requires a continuous supply of organic matter. Since the quantity of allochthonous material present in these systems is finite it may explain why these mass blooms are annual events. In this chapter we discuss the survival capacity of *Thiocapsa roseopersicina*, the environmental factors which initiate these annual blooms of purple sulfur bacteria and the metabolic versatility of this non-motile phototrophic bacterium.

Survival capacity of *Thiocapsa roseopersicina*

Cultures of *Thiocapsa roseopersicina* OP-1 were grown phototrophically in the medium of Pfennig (1965) under anaerobic conditions until a cell density of $> 10^7$ viable cells.ml^{-1} was achieved. The cultures were then incubated as follows: semi-aerobic (50-60 μM O_2) in the light (75 μE.m^{-2}.s^{-1}); semi-aerobic (50-60 μM O_2) in the dark at 20°C and anaerobically in the dark at 4°C. To maintain semi-aerobic conditions the cultures were continuously sparged with a filter sterilised (0.2 μm pore size) gas mixture of air and nitrogen regulated to maintain a dissolved O_2 concentration in the range 50-60 μM. At designated time intervals samples were aseptically removed and the total and viable number of cells determined. Data presented in Fig. 1a and b show that cultures of *T. roseopersicina* OP-1, when incubated in the absence of S^{2-}, retained

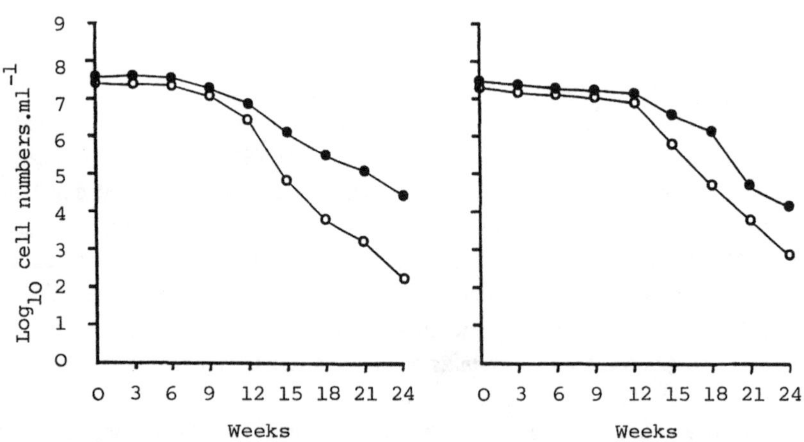

Fig. 1. Survival of *Thiocapsa roseopersicina* OP-1 when incubated in Pfennig's medium under semi-aerobic conditions (50-60 μM O_2) in the light (a) and dark (b). Viable cell numbers (open circles) were determined using the agar shake method (Van Niel 1931) and total cell numbers by phase-contrast microscopy.

viability even after extended exposure to oxygen. Irrespective of whether the cultures were incubated in the light or the dark a high cell viability was maintained over the initial phase (9 weeks) of these experiments. Extended incubation however led to a progressive decline both in total cell numbers and viability especially in cultures incubated in the light. Nonetheless even after 24 weeks incubation viable *T. roseopersicina* OP-1 cells were still present and rapidly initiated growth when inoculated into Pfennig's medium and incubated anaerobically in the light. Cultures maintained in the light under semi-aerobic conditions progressively lost their photopigments and developed a milky-white appearance. In contrast, cultures maintained semi-aerobically in the dark retained their photopigments for the duration of the experiments. Cultures of *T. roseopersicina* OP-1 incubated anaerobically in the dark at 4°C (data not presented) retained a higher viability (4.7×10^3 viable cells.ml^{-1} after 24 weeks) than those incubated semi-aerobically in the dark at 20°C. Similarly, when beach sediment from Swanbister Bay in Scapa Flow inoculated with a dense population of *T. roseopersicina* OP-1 ($> 10^7$ viable cells.g^{-1} sediment) and stored in the dark at 4°C for 30 weeks was then perfused with Pfennig's medium and incubated anaerobically in the light the surface layer developed a pink coloration within 5 days. These data are consistent with field data presented in Fig. 4c which show that significant populations are present in the sub-surface layers of Swanbister Bay beach in Spring. These overwintering populations are then able to initiate a summer bloom when appropriate environmental conditions develop in the surface sediment layers of the beach.

Physico-chemical parameters controlling the development of annual blooms of purple sulfur bacteria

As discussed earlier in this chapter the laminated mat systems found on sheltered beaches in Scapa Flow are unusual in that they are transient annual events, developing in late May and disappearing by early September. In order for these mat systems to develop the appropriate physico-chemical gradients must be generated in the sub-surface sediment layers of the beach and this requires a source of degradable organic matter. Unlike the laminated marine microbial mats that develop on sheltered sandy sediments on islands in the Waddensea, which are dominated by the mat forming cyanobacterium *Microcoleus chthonoplastes*, the principal source of organic matter is derived from moribund macroalgae, deposited on the beaches during winter storms (Herbert 1985). Data presented in Fig. 2 show seasonal changes and depth distribution of organic matter in Swanbister Bay sediments. The most notable feature of these data is that the highest organic content is found in the 0-5 mm depth horizon during Spring (April) and Summer (July) whilst in Autumn (October) the levels are low (< 5 g.100 g dry wt^{-1} sediment). The increased organic matter content recorded in Summer (22.7 g. 100 g dry wt^{-1} sediment) is due both to the presence of moribund macroalgae and the growth of cyanobacteria and unicellular green algae at the beach surface. The presence of organic matter in the surface layers stimulates heterotrophic processes and since respiratory O_2 consumption is greater than

O_2 diffusion the sub-surface layers become progressively anoxic leading to the establishment of populations of sulfate reducing bacteria and the generation of S^{2-}.

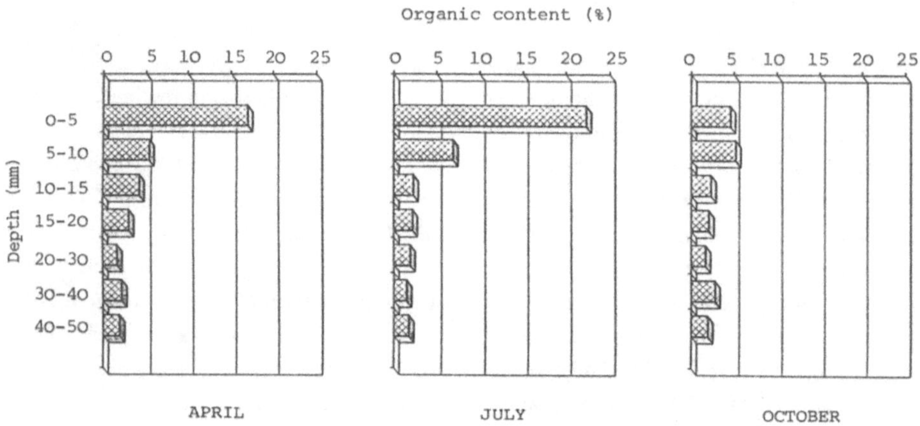

Fig. 2. Seasonal changes in organic matter content of Swanbister Bay beach sediments.

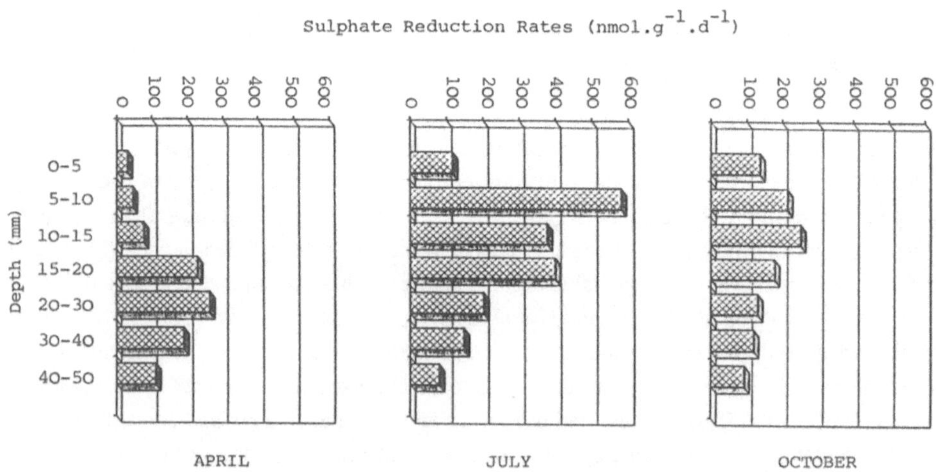

Fig. 3. Seasonal changes in sulfate reduction rates of Swanbister Bay beach sediments. Sulfate reduction rates were determined according to the method of Senior et al. (1982).

Data presented in Fig. 3 show seasonal changes in sulfate reduction rates in the sub-surface layers of Swanbister Beach. Whilst active sulfate reduction occurred throughout the year highest rates were recorded in Summer (July) when ambient temperatures were maximal (16-18°C). Whereas in Spring (April) rates were low due to the prevailing low temperature (6°C) and lack of suitable low molecular weight substrates. Similarly in Autumn (October) sulfate reduction rates were low most probably due to a combination of low temperature (8°C) and non-availability of appropriate carbon substrates since substantial populations of acetate and lactate utilising sulfate reducing bacteria were still present in the sub-surface sediments (Figs. 4a and b). The most significant feature of the data presented in Fig. 3 is that in summer highest sulfate reduction rates were observed in the 5-10 mm depth horizon and are coincident with the population maxima recorded for acetate and lactate utilising sulfate reducing bacteria (Figs. 4a and b). These findings are consistent with those of Jørgensen and Bak (1991) for marine sediments and Visscher and Van Gemerden (1993) for laminated mat systems on the Frisian Island of Texel. With the establishment of anoxia and high sulfate reduction rates and almost continuous daylight during the brief Orcadian summer it is perhaps not surprising that there is a rapid increase in the *Thiocapsa* populations with the concomitant development of transient mass blooms at the beach surface (Fig. 4c).

Metabolic versatility of *Thiocapsa roseopersicina* OP-1

Thiocapsa roseopersicina OP-1, like other members of the genus, is non-motile and in the Orkney mat systems is commonly found as microcolonies attached to sand grains (Van Gemerden et al. 1989b). As a consequence this bacterium is unable to migrate when environmental conditions become unfavourable i.e. when exposed to high O_2 concentrations produced by the overlying layer of cyanobacteria and unicellular green algae during the day. However, *T. roseopersicina* is frequently found as the dominant anoxygenic phototroph in microbial mat systems (Stal et al. 1985; Pierson et al. 1987; Van Gemerden et al. 1989a) implying that this purple sulfur bacterium is well adapted to environments which are subject to rapid and unpredictable changes. Physiological studies have demonstrated that some strains of *T. roseopersicina*, like many members of the Chromatiaceae, are not only able to maintain viability but also grow chemolithotrophically in the presence of molecular O_2 (Bogorov 1974; Kämpf and Pfennig 1980; De Wit and Van Gemerden 1987). In this respect *T. roseopersicina* OP-1 responds in a similar manner when grown in the presence of O_2. Data presented in Table 1 show the cell yield of *T. roseopersicina* OP-1 when grown under different conditions. Cell yields of autotrophically grown cultures were highest when grown in the light under anaerobic conditions (photolithotrophic growth) and decreased to circa $^1/3$rd when grown chemolithotrophically (aerobic/dark). Under semi-aerobic conditions in the light the cell yield was 46% of that achieved under anaerobic conditions. These data indicate that whilst BChl*a*

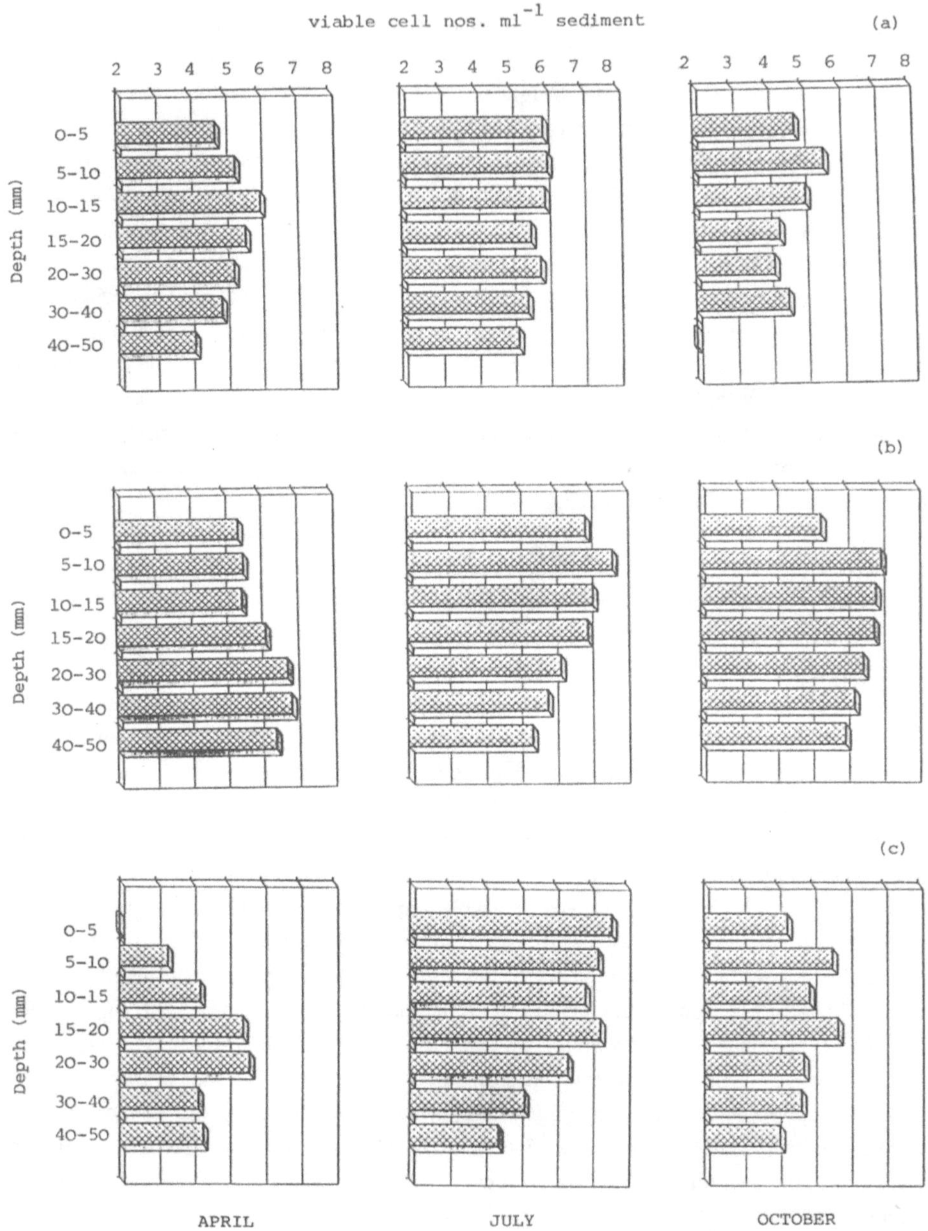

Fig. 4. Seasonal changes in populations of (a) acetate, (b) lactate utilising sulfate reducing bacteria and (c) *Thiocapsa roseopersocina* in Swanbister Bay beach sediments.

Table 1. Growth yield of *Thiocapsa roseopersicina* OP-1 when grown in liquid culture under different conditions (see Kämpf and Pfennig 1980 for details of growth media). * Dissolved O_2 concentration 60 μM.

Growth conditions	Protein concentration ($mg.l^{-1}$)		
	Lithotrophic	Organotrophic	Mixotrophic
Anaerobic/Light	63	147.4	135.4
Semi-aerobic*/Light	29.6	88.2	100.8
Semi-aerobic*/Dark	23.5	81.9	86.3
Aerobic/Dark	20.8	83.8	88.3

synthesis was repressed under these conditions the photopigment content of the inoculum was sufficient to allow phototrophic growth to initially continue in the presence of O_2 thereby accounting for the enhanced cell yield compared with equivalent cultures incubated in the dark. When 0.1% w/v acetate was added to the mineral growth medium growth was considerably stimulated indicating that *T. roseopersicina* OP-1 has an excellent capacity to grow organotrophically. Highest cell yields were recorded under photoorganotrophic conditions. Similarly, when grown mixotrophically (0.05% w/v thiosulfate + 0.05% w/v acetate) growth was substantially stimulated compared with cultures grown photolithotrophically. These data indicate that *T. roseopersicina* OP-1 is metabolically versatile and has the capacity to switch its mode of energy generation according to the prevailing environmental conditions.

One aspect of the metabolism of *T. roseopersicina* that has attracted considerable attention is the ability of this bacterium to grow phototrophically in the presence of O_2 even though BChl*a* and carotenoid synthesis is repressed under aerobic conditions (Cohen-Bazire et al. 1957; Hurlbert 1967). De Wit and Van Gemerden (1990) elegantly demonstrated using chemostat grown cultures that *T. roseopersicina* strain M1, isolated from a microbial mat on the Island of Mellum, was able to grow phototrophically in the presence of 50-60 μM O_2 provided that within each 24 hour time period there was a minimum anoxic-dark period of 3 hours. Those workers showed that upon switching from semi-aerobic conditions BChl*a* synthesis commenced rapidly and even after 3 hours sufficient BChl*a* had been synthesised to enable the cells to grow phototrophically when exposed to semi-aerobic conditions. *Thiocapsa roseopersicina* OP-1 isolated from Orkney microbial mats behaves in a similar manner. Data presented in Fig. 5 show that when *T. roseopersicina* OP-1 was grown in a sulfide limited chemostat (2 mM S^{2-}) at a dilution rate of 0.03 h^{-1} and a dilution rate of 0.03 h^{-1} under alternating aerobic (100 μM O_2) -light and anaerobic-dark conditions the specific BChl*a* concentration decreased progressively during the aerobic-light period and increased rapidly during the anaerobic-dark period. These data indicate unequivocally that BChl*a* synthesis was confined to the dark-anaerobic period and that synthesis was sufficiently rapid to compensate for the decrease during the preceding aerobic-light period. As a consequence *T. roseopersicina* OP-1 was able to grow phototrophically under aerobic conditions in the light even in the absence of photopigment synthesis. This is reflected by the steady increase in protein level under these growth conditions whereas during the dark-anaerobic period there is a progressive decline. Similarly during the dark-anaerobic phase S^{2-} accumulated in the chemostat whereupon switching to aerobic-light conditions the accumulated S^{2-} was rapidly oxidised.

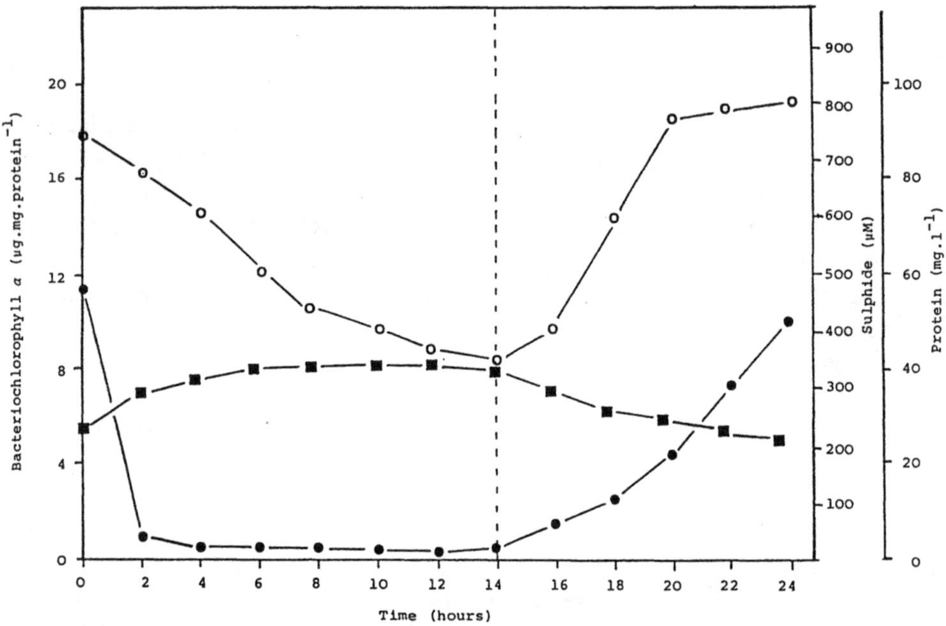

Fig. 5. Effect of alternating light-aerobic (14 h; left in fig.) and dark-anaerobic (10 h; right in fig.) conditions on BChl*a* content, protein level and sulfide of a sulfide (2 mM) limited culture of *T. roseopersicina* OP-1 grown at a dilution rate of 0.03 h^{-1}.

In order to determine the effects of continuous exposure to oxygen and light *T. roseopersicina* OP-1 was first grown anaerobically in a thiosulfate limited chemostat (1.5 mM) at a dilution rate of 0.03 h^{-1} at a light irradiance of 100 $\mu E.m^{-2}s^{-1}$ until steady state conditions were achieved. The culture was then switched to aerobic conditions (100 μM O_2) under the same light regime. Since BChl*a* synthesis was repressed by exposure to O_2 the culture was unable to maintain its growth rate and after six volume changes had washed out completely. Upon repeating the experiment at a lower growth rate ($D = 0.018\ h^{-1}$) viable cell numbers again decreased from 13.6 x 10^7 to 4.1 x 10^7 viable cells.ml^{-1} upon switching to aerobic conditions but then stabilised even though BChl*a* could not be measured in the culture. The culture appeared milky-white due to the accumulation of intracellular sulfur indicating that the cells were growing chemolithotrophically. Data presented in Table 2 show that prior to switching (-24 hours) the chemostat to aerobic conditions the culture had a high BChl*a* content whilst respiratory activity could not be measured. After switching over to aerobic conditions the BChl*a* concentration of the culture decreased at a rate corresponding to the washout rate and the cells began progressively to respire thiosulfate aerobically as demonstrated by the increase in specific respiration rate. These data show that whilst oxygen represses BChl*a* and carotenoid synthesis in *T. roseopersicina* OP-1 and prolonged exposure leads to colourless cells this bacterium is still able to generate sufficient energy to grow, albeit at a slower rate and with a lower biomass yield under these unfavourable environmental conditions.

Table 2. Effect of aerobic conditions (100 μM O_2) on BChl*a* content and respiration rate of *Thiocapsa roseopersicina* OP-1 when grown in the light (100 μEin m^{-2} s^{-1}) in continuous culture under thiosulfate limitation (D = 0.018 h^{-1}, 25°C). At t = 0 culture was switched to aerobic conditions

Time (hours)	BChl*a* concentration ($\mu g.l^{-1}$)	Respiration Rate (nmoles $O_2.min^{-1}$.mg protein)
-48	886	Not Detectable
-24	915	Not Detectable
0	879	Not Detectable
48	504	22.4
96	241	39.3
144	136	45.1
192	55	43.7
240	21	52.9

In summary, data presented in this chapter show that *T. roseopersicina* OP-1 is ideally adapted to grow and survive in the rapidly changing physico-chemical gradients, which are such a characteristic feature of the Orkney laminated mat ecosystems.

References

Bogorov LV (1974) About the properties of *Thiocapsa roseopersicina* strain BBS, isolated from the estuary of the White Sea. Microbiology (Engl Trans of Microbiologya) 43:275-280

Castenholz RW (1984) Composition of hot spring microbial mats: a summary. In: Cohen Y, Castenholz RW, Halvorson HO (eds) Microbial Mats: Stromatolites. Liss, New York, pp 101-119

Cohen Y (1984) The Solar Lake cyanobacterial mats: Strategies of photosynthetic life under sulfide. In: Cohen Y, Castenholz RW, Halvorson HO (eds) Microbial Mats: Stromatolites. Liss, New York, pp 133-148

Cohen-Bazire G, Sistrom WR, Stanier RY (1957) Kinetic studies of pigment synthesis by non-sulfur purple bacteria. Journ of Cell and Comparative Physiology 49: 25-68

De Wit R, Van Gemerden H (1987) Chemolithotrophic growth of the phototrophic sulfur bacterium *Thiocapsa roseopersicina*. FEMS Microbiol Ecol 45:117-126

Herbert RA (1985) Development of mass blooms of photosynthetic bacteria on sheltered beaches in Scapa Flow, Orkney Islands. Proc Roy Soc of Edinburgh 87B:15-25

Hurlbert RE (1967) Effect of oxygen on viability and substrate utilization in *Chromatium*. J Bacteriol 93:1346-1352

Jørgensen BB, Bak F (1991) Pathways and microbiology of thiosulfate transformation and sulfate reduction in a marine sediment (Kattegat, Denmark). Appl Environ Microbiol 57:847-856

Jørgensen BB, Cohen Y (1977) Solar Lake (Sinai). The sulfur cycle of the benthic cyanobacterial mats. Limnol Oceanogr 22:657-666

Kämpf C, Pfennig N (1980) Capacity of Chromatiaceae for chemotrophic growth. Specific respiration rates of *Thiocystis violacea* and *Chromatium vinosum*. Arch Microbiol 127:125-135

Pfennig N (1965) Anreichungskulturen für rote und grüne Schwefelbakterien. Zentr Bakteriol Parasitenk Abt 1 Suppl 1, 197-189

Pierson BK, Oesterle A, Castenholz RW (19874) Pigments, light penetration and photosynthetic activity in the multi-layered microbial mats of Great Sippewisset Salt Marsh, Massachusetts. FEMS Microbiol Ecol 45:365-376

Senior E, Linstrom EB, Banat IM, Nedwell DB (1982) Sulfate reduction and methanogenesis in the sediment of a saltmarsh on the east coast of the United Kingdom. Appl Environ Microbiol 43:987-996

Stal LJ, Krumbein WE (1981) Aerobic nitrogen fixation in pure cultures of a benthic *Oscillatoria* (cyanobacteria). FEMS Microbiol Lett 11:295-298

Stal LJ, Van Gemerden H and Krumbein WE (1985) Structure and development of a benthic marine microbial mat. FEMS Microbiol Ecol 31:111-125

Van Gemerden H (1993) Microbial mats: a joint venture. Mar Geol 113:3-25

Van Gemerden H, Tughan CS, De Wit R, Herbert RA (1989a) Laminated microbial ecosystems on sheltered beaches in Scapa Flow, Orkney Islands. FEMS Microbiol Ecol 62:87-102

Van Gemerden H, De Wit R, Tughan CS, Herbert RA (1989b) Development of mass blooms of *Thiocapsa roseopersicina* on sheltered beaches on the Orkney Islands. FEMS Microbiol Ecol 62:111-118

Van Niel CB (1931) The morphology and physiology of purple and green sulfur bacteria. Arch Mikrobiologie 3:1-112

Visscher PT, Van Gemerden H (1993) Sulfur cycling in laminated marine microbial ecosystems. In: Oremland RS (ed), Biogeochemistry of Global Change: Radiatively Active Trace Gases. Chapman and Hall, New York, pp 672-690

Biological versus inorganic processes in stromatolite morphogenesis: Observations from mineralizing sedimentary systems

Jack D. Farmer and David J. Des Marais

NASA Ames Research Center, MS 239-4
Moffett Field, California (USA) 94035-1000

Introduction

Morphogenesis in stromatolites refers to the pattern of development that creates a megascopic laminated form through repeated cycles of sediment accretion. In general, all sedimentary laminations reflect periodic interruptions in deposition. For example, in stromatolites formed in detrital sedimentary systems, laminations may arise through variations in current velocity and rates of sedimentation (Farmer and Des Marais 1993). In mineralizing sedimentary systems, laminae may form by fluctuations of temperature, pH, or saturation (Guo and Riding 1992), or by seasonal growth cycles of organisms (Chafetz et al. 1991). Laminations can also originate by post-depositional (diagenetic) processes, such as dissolution (Ekdale and Bromley 1988).

The microstructure of laminations provides information about biological, physical and chemical interactions that operated during deposition or early diagenesis. Key biological factors include variations in population density and community structure, competition for resources (light and nutrients), biogeochemical cycles, and the movement (taxis) of microbes in response to environmental change. Important inorganic processes include changes in mineral solubility (due to temperature, pH, or fluid composition), the size and crystal habit of precipitates, sediment porosity and permeability, and diagenesis (cementation, compaction, and mineral phase changes).

Investigations of modern microbial mat systems provide information vital to the interpretation of stromatolites, but there is presently a need for detailed studies of mats in mineralizing systems which will link the origin of microstructure to microscale processes. Of particular interest are studies that distinguish between biological and inorganic contributions to microstructure. Such distinctions are fundamental for interpreting ancient stromatolites, as illustrated by recent controversies about the origin of radial fibrous microstructure in stromatolites and the implications for carbonate saturations in Precambrian oceans (Grotzinger 1990).

NATO ASI Series, Vol. G 35
Microbial Mats
Edited by Lucas J. Stal and Pierre Caumette

Demonstrating biogenicity in stromatolites

Laminations alone provide insufficient proof for a biological origin in stromatolites. Columnar and conical laminations are frequently cited as proof of biological processes, but similar features can also be formed by inorganic precipitation. Many stromatolites have flat-laminations and are recognized only on the basis of the microbial fossil assemblages they contain. However, microfossils tend to be rare in fossil stromatolites and provide an impractical criterion for biogenicity. Because the microstructure and geochemistry of individual laminae in stromatolites reflect microscale interactions between mat organisms and their sedimentary microenvironment, such features may provide the most diagnostic and generally available sources of information for assessing biogenicity.

Microbial mediation of carbonate precipitation

In theory, microorganisms may contribute to carbonate precipitation in a variety of ways, ranging from direct physiological mediation of mineral precipitation (Krumbein 1979), to passive controls where, in the absence of direct physiological mechanisms, organic materials merely provide surfaces of low activation energy for the nucleation of crystals (Pentecost and Bauld 1988). Thus, to understand fully the biological contributions to stromatolite microstructure requires a specific knowledge of the role(s) played by organisms in creating favorable microenvironments for the precipitation of carbonate, and for the construction of mineralized frameworks. Studies of microscale processes in mineralizing mat systems are presently limited by the need for microsensors adapted for use in hard substrates. In this study we address the question of biological and inorganic contributions to microstructure using more classical observational methods of light microscopy and SEM.

Stromatolite morphogenesis in thermal springs

Subaerial thermal springs offer an opportunity to study the formation of laminated sedimentary fabrics where rates of detrital sedimentation are low relative to mineral precipitation. Furthermore, because most prokaryotic mat-forming organisms are intolerant of temperatures higher than $\sim 74^{\circ}$ C (Brock 1978), it is possible to compare inorganically-produced microstructures at high temperatures to those formed in the presence of microbial mats (Farmer and Des Marais 1992). Here we describe laminated calcareous thermal spring sinters (travertines) from two sites in Yellowstone National Park (Figure 1). In the following discussion, we have selected examples to illustrate general differences in microstructure over a range of temperature, pH and degrees of microbial influence. Specific sites for comparison include vent and channel deposits at Asta Spring (Hillside Springs Group, Upper Geyser Basin) where mats are excluded by high temperature, channel deposits associated with non-phototrophic (sulfur-oxidizing) mats, and mid-temperature ponds where phototrophic mats dominate (Angel Terrace, Mammoth Hot Spring).

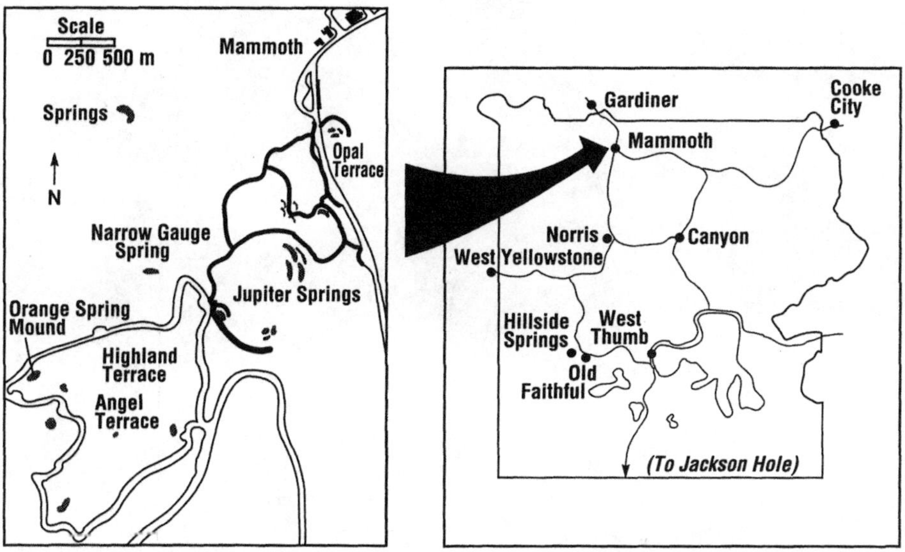

Fig. 1. Map showing study sites in Yellowstone National Park, Wyoming. (After Bargar and Muffler 1975).

Inorganic precipitation at high temperatures

Asta Spring emerges from a high elevation vent on the eastern margin of the Madison Plateau, northwest of Old Faithful (Fig. 1). Terracing is absent and sinters are deposited almost entirely on steep slopes. The maximum vent temperature recorded was 85° C, high enough to exclude microbial mats. MnO_2 (pyrolusite) is the dominant mineral precipitated above 80° C, with aragonite predominating at lower temperatures. Between ~71-67° C sinter surfaces are covered by a thin greenish-yellow mat of *Synechococcus* sp., with a rusty orange *Phormidium* mat dominating below ~67° C. Mats are absent above ~71° C and vent and channel sinter deposits consist of dense botryoids (rounded masses) made up of radiating blades aragonite (Fig. 2A), frequently interlaminated with dendritic MnO_2 (Fig. 2B). Regular banding is visible in some specimens, being defined by alternations of pyrolusite- and aragonite-rich laminae.

Because microbial mats are absent above ~71° C at Asta Spring, we attribute the laminated microstructure of higher temperature botryoids to inorganic precipitation. Periodic changes in temperature, driven by variations in spring discharge, are inferred to be primary

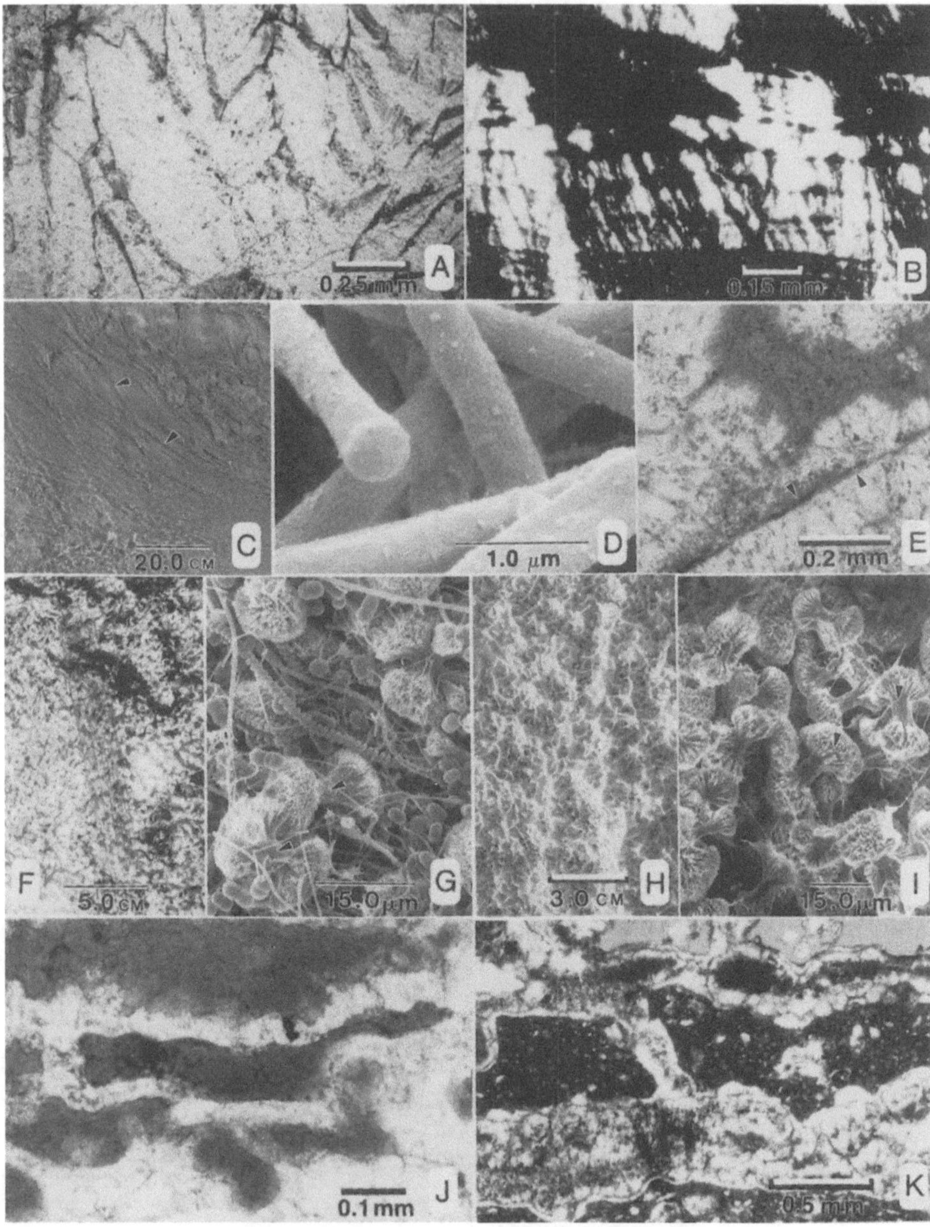

Fig. 2. Microbial mats and sinter microstructures, Asta Spring and Angel Terrace, Yellowstone National Park.

factors in creating laminations. Because MnO_2-rich laminae were deposited at temperatures > 80° C, interlamination with aragonite in sinter samples collected below this temperature may reflect seasonal changes in spring discharge and temperature, or perhaps longer term climatic cycles.

Non-phototrophic mats at high temperature

Angel Terrace Spring is a sulfide-rich system (see Castenholz 1977) with a temperature of 73° C and pH of 6.65 at the vent. Over the range 68-71° C, microbial mats are dominated by *Thermothrix thiopara* (Caldwell et al. 1984; Fig. 2D). This species is reportedly the fastest-growing sulfide oxidizer known (D. Caldwell, personal communication 1993), and forms mats despite high rates of inorganic precipitation (> 20 cm yr^{-1}, based on carborundum layering).

Under oxygen-limited conditions, *Thermothrix* forms elongated, filamentous cells up to 10 cm in length (Caldwell et al. 1984). In areas of rapid flow on Angel terrace, filaments become oriented by currents, forming bacterial "streamers". But channels tend to shift laterally during terrace accretion, and eventually shallow and cool. Along channel margins over a range of ~ 68-70° C, *Thermothrix* filaments are usually overgrown by microfibrous aragonite, which precipitates on cell surfaces, forming mineralized streamers (Fig. 2C, arrows). This produces a sinter fabric composed of tightly-packed rods of microfibrous aragonite, with the long axes of rods being oriented parallel to the original flow direction.

A. Thin section photomicrograph of bladed aragonite from high temperature channel at Asta Spring; **B.** Same as 'A', but showing intergrowths of aragonite and pyrolusite; **C.** Mineralized bacterial "streamers" formed by the precipitation of microfibrous aragonite on filaments of *Thermothrix thiopara*; **D.** Scanning electron micrograph of critical point dried mat-sinter sample from vent area of Angel Terrace showing the sulfur oxidizer, *Thermothrix thiopara*. Note rounded granules of sulfur on cell surfaces; **E.** Photomicrograph of longitudinal thin section through aragonite streamers shown in 'C'; Overgrowths of radial fibrous aragonite surround dark cores (arrows) believed to be fossilized *Thermothrix*; **F.** Ridged networks formed by *Spirulina* mats in mid-temperature ponds at Angel Terrace; **G.** Scanning electron micrograph of critical point dried, mat-sinter sample from site shown in 'F'; View is from a depth of ~ 0.5 mm below mat surface. Mat community (including *Spirulina*, *Synecchococcus*, *Chloroflexus*, and *Oscillatoria*) entraps small "dumbbell"-shaped crystals of aragonite (arrows); **H.** Sinter surface in mid-temperature pond after removal of *Spirulina* mat. Note how the fabric of the sinter mimics the texture of the ridged mat in 'F'; **I.** SEM micrograph of mat-sinter sample in 'G' from a depth of ~ 1.0 mm below mat surface. Note sinter framework of interlocking aragonite 'dumbbells' and general absence of mat organisms; **J-K.** Transverse thin section of sinter in 'H' showing characteristic fenestral microstructure; Note in 'J' that original microfibrous aragonite of fenestral walls has recrystallized to form calcite while fenestrae in 'K' retain original wall structure and were infilled with locally-derived detritus.

At temperatures >68° C on Angel Terrace, and where oxygen coexists with sulfide, *Thermothrix* filaments provide preferred surfaces for the nucleation of aragonite during inorganic precipitation. Streamer fabrics are especially common in such environments. In thin section, *Thermothrix* streamers are composed of "shrub-like" clusters of microfibrous aragonite crystals arrayed radially around dark cores that are inferred to be fossilized *Thermothrix* filaments (Fig. 2E, arrow; see also Pursell and Folk 1990). The filamentous shape of bacterial cells acts as a template for aragonite precipitation and leads to a characteristic streamer fabric recognized in rocks as old as early Carboniferous (~360 Myr).

Phototrophic mats at mid-temperatures

In 1991, downstream from the primary vent at Angel Terrace, several ponds ~10-30 cm in depth had formed behind several "terracette" dams. In this part of the system, over ~45-50° C, accretion rates averaged ~5 cm yr^{-1}, about one-fourth of the rate observed near the vent.

Pond floors at these temperatures were covered by phototrophic mats which showed a variety of ridged and tufted surface textures (Fig. 2F). The origin of such features is debatable, but they may have formed through the photo- and/or chemotactic responses of dominant mat species (see Bauld et al. 1992). A diverse community of photoautotrophs was observed in critical point dried samples (Fig. 2G) which included rod-shaped *Synechococcus* cf. *lividus*, rounded cells of *Synechococcus* cf. *minervae*, and several filamentous forms, including *Spirulina labyrinthiformis*, *Oscillatoria* sp., and the green non-sulfur bacterium, *Chloroflexus* cf. *aurantiacus*, which typically forms an undermat (Ward et al. 1992)

At a depth of ~0.5 mm below the mat surface, organisms are intimately associated with dumbbell-shaped aggregates of microfibrous aragonite (Fig. 2G, arrows). However, by ~1.0 mm depth, a porous sinter framework of interlocking aragonite dumbbells, largely free of microorganisms, is present (Fig. 2I). Thin section studies suggest that the sinter underlying the ridged mat surface was a locus for faster accretion. Sinter ridges grew upward and sideways, branching and arching over, and eventually coalescing to form irregular, enclosed spaces or "fenestrae" (Figs. 2J, K). Fenestrae tend to have cross sections that are somewhat flattened parallel to bedding (imparting a crude lamination) and possess concave-up bottoms and relatively flat roofs. Some fenestrae were apparently infilled with locally-derived detritus, prior to being roofed over by the growth process described above (Fig. 2K). In other cases, fenestral spaces were filled in by thickening of the walls through successive overgrowths of aragonite. Continued accretion of the mat-sinter surface produces a crudely-laminated travertine consisting of vertical stacks of elongate, irregular fenestrae. Similar fabrics are common in Holocene travertines near Mammoth, although the original aragonite has recrystallized to form microgranular calcite.

Observations in mid-temperature ponds on Angel Terrace suggest that inorganic precipitation rates there are low enough for the microenvironment within mats to significantly influence the local rates of carbonate precipitation. The surface morphology of sinter deposits underlying the mats in this setting precisely mimics the surface texture of the overlying mat, exhibiting the same networks of interconnected ridges and knobs (compare Figs. 2F and 2H). This suggests that aragonite precipitation is controlled biologically, occuring primarily within areas of thicker mat. Carbonate precipitation could be enhanced within mats by increases in pH arising through photosynthesis (Krumbein 1979). As noted above, higher precipitation rates at locations of thicker mat, namely under tufts and ridges, produces an accretion surface in the underlying sinter having a similar morphology. Through continued accretion at the mat-sinter surface, a characteristic fenestral fabric is formed. This fabric is considered to be a primary biological feature formed through the interaction of microbial taxis, and mat-mediated carbonate precipitation.

In the systems studied, organic matter is rare in sinters deposited at $>30°$ C, suggesting that decomposition rates in such thermal environments are very high. Prior to cementation, the loss of organic matter may be facilitated by the downward percolation of thermal waters through the typically porous and permeable framework of newly deposited sinters.

Conclusions

In travertine thermal springs at high temperatures, carbonate precipitation is controlled by declining temperature and increasing alkalinity, due to CO_2 outgassing. Under high rates of inorganic precipitation, physiological processes in mats apparently have little influence on carbonate precipitation. However, where mats are present, organic surfaces provide favorable sites for the nucleation of crystals, thus contributing directly to sinter structure as communities are entombed by precipitating minerals.

At temperatures $>74°$ C, mats are absent, being excluded by thermal intolerance. Thus, sinters deposited above this temperature form entirely by inorganic processes. Microstructures commonly observed in inorganically-deposited calcareous sinter include bladed (Fig. 2A, B) and radial fibrous (Fig. 2E) forms of aragonite. Sinters deposited at high temperatures are often dense and finely-laminated. Laminations form by frequent interruptions in deposition and reflect variations in temperature, composition and/or the rate of discharge.

In sulfide springs over a temperature range of ~68-73° C, non-phototrophic, sulfur-oxidizing bacteria (*Thermothrix thiopara*) form filamentous mats (Fig. 2D), despite of high background rates of inorganic aragonite precipitation (>20 cm yr^{-1}). Between ~68-71° C, *Thermothrix* filaments become overgrown by microfibrous aragonite (Fig. 2E), forming characteristic bacterial "streamer" fabrics (Fig. 2C). Although precipitation is driven by inorganic processes, *Thermothrix* filaments influence sinter morphology in providing surfaces of low activation energy, favorable for the nucleation of aragonite.

Over a range of ~40-55° C, the rate of inorganic precipitation is comparatively low (averages < 5.0 cm yr^{-1}), thus allowing for the development of continuous mats and biofilms. In ponded areas at these temperatures, phototrophic mats are well-developed (Fig. 2G) and exhibit a variety of ridged and tufted surface textures (Fig. 2F), probably formed through microbial phototaxis. Mats apparently enhance aragonite precipitation, and the rate of sinter accretion is highest where mats are thickest. This results in the reproduction of mat surface textures (ridges and tufts) in the sinter immediately underlying the mat (Figure 2H). Continued accretion of the mat-sinter surface produces travertine deposits having a characteristic fenestral microstructure (Fig. 2J, K).

Acknowledgments- The authors wish to thank Dr. Richard Castenholz (Univ. of Oregon) and Dr. David Ward (Univ. of Montana) for helpful criticism of the manuscript. Mr. Robert Lindstrom (Res. Office, Yellowstone National Park), Dr. David Blake (NASA Ames), and Dr. Sherry Cady (NASA Ames) provided invaluable assistance in the field. The project was funded by the National Research Council and NASA's Exobiology Program.

References

Bargar KE, Muffler LPJ (1975) Geologic map of the travertine deposits, Mammoth Hot Springs. US Geol Survey Misc Field Studies Map MF-659

Brock TD 1978 Thermophilic microorganisms and life at high temperatures. Springer-Verlag, New York, 465 pp

Caldwell DE, Kieft TL, Brannan DK (1984) Colonization of sulfide-oxygen interfaces on hot spring tufa by *Thermothrix thiopara*. Geomicrobiol J 3: 181-200

Castenholz RW (1977) The effect of sulfide on the blue-green algae of hot springs II. Yellowstone National Park. Microb Ecol 3: 79-105

Chafetz HS, Utech NM, Fitzmaurice SP (1991) Differences in the $S^{18}O$ and $S^{13}C$ signatures of seasonal laminae comprising travertine stromatolites. J Sedim Petrol 61: 1015-1028

Ekdale AA, Bromley RG (1988) Diagenetic microlamination in chalk. J Sedim Petrol 58: 857-861

Farmer JD, Des Marais DJ (1993) Distributional controls and microstructure of laminated microbial mats at Guerrero Negro Baja Mexico . Geol Soc of Amer Abs with Prog 25(8): 294

Farmer JD, Des Marais DJ (1992) Comparative biosedimentology of some terraced travertine deposits. Geol Soc of Amer Abs with Prog 24(7): 53-54

Guo L , Riding R (1992) Aragonitic laminae in hot water travertine crusts, Rapolano Terme Italy. Sedimentology 39: 1067-1079

Grotzinger JP (1990) Geochemical model for Proterozoic stromatolite decline. Amer J of Sci 290-A: 80-103

Krumbein WE (1979) Calcification by bacteria and algae pp 47-68 In: Trudinger PA, Swaine DJ (eds) Biogeochemical Cycling of Mineralforming Elements. Studies in Environ Sci 3. Elsevier , New York, 612 pp

Pentecost A, Bauld J (1988) Nucleation of calcite on the sheaths of cyanobacteria using a simple diffusion cell. Geomicrobiol J 6: 129-135

Pursell VJ, Folk RL (1990) Aragonite calcite dendrites and microfluorite in Mammoth Hot Springs travertine, Yellowstone National Park, Wyoming Geol Soc of Amer Abs with Prog 22: 90

Ward DM, Bauld J, Castenholz RW, Pierson BK (1992) Modern phototrophic microbial mats: anoxygenic, intermittently oxygenic - anoxygenic, thermal, eukaryotic, and terrestrial. In: Schopf JW, Klein C (eds) The Proterozoic Biosphere. Cambridge Univ Press, pp 309-333

Morphological and chemical transformations of *Microcoleus chthonoplastes* during early diagenesis in hypersaline microbial mats

Rutger de Wit[1], Joan O. Grimalt and Mariona Hernandez-Mariné[2]

Dept. of Environmental Chemistry (CID-CSIC), Jordi Girona 18, 08034 Barcelona, Spain

Introduction

In microbial mats, the growth of constituent populations and sediment accretion are accompanied by burial of phototrophic organisms in sediment layers that are anoxic and not reached by light. Nevertheless, cyanobacterial remains are often preserved in the sedimentary column. In the virtual absence of bioturbation, the original stratigraphy remains conserved in the sedimentary record. Hence, sediments with microbial mats often comprise an actively growing mat on top of a laminated system corresponding to buried mats of previous years, as described for Solar Lake (Egypt) by Jørgensen et al (1988). We have studied similar laminated systems in solar saltern ponds that have been exploited for more than one century.

The conservation of the cyanobacterial remains facilitates the geological interpretation of stromatolites. Thus, microfossils that bear close resemblance to actual cyanobacteria and other mat-forming bacteria have been found in geological strata which are up to 3,5 billion years old (e.g., Awramik 1984). Furthermore, a lot of geochemical studies have been performed to decipher the geochemical fate of various cyanobacterial molecular compounds in the sedimentary record (e.g., Boon and De Leeuw 1987; Grimalt et al. 1992).

In this study, we combine microscopic methods including transmission electron microscopy (T.E.M.) with chemical analyses to study the morphological and chemical transformations of the cyanobacterium *Microcoleus chthonoplastes* during early diagenesis in hypersaline microbial mats. Moreover, the slicing of the sediment for lipid analyses has been performed on an unprecedented small scale, which has provided detailed information on stratification patterns relating to the biological stratification of the living mat (3 mm) and the early diagenetic transformations in the sediment (c.f., Grimalt et al. 1992).

---

[1] Present address: Laboratoire d'Océanographie Biologique and C.N.R.S.-U.R.A. 197, Université de Bordeaux I; 2 rue du Prof. Jolyet, Arcachon, France.
[2] Faculty of Pharmacy. University of Barcelona. Diagonal s/n, 08028 Barcelona, Spain.

NATO ASI Series, Vol. G 35
Microbial Mats
Edited by Lucas J. Stal and Pierre Caumette

Description of the microbial mat samples and methods used

Cores of microbial mats dominated by *M. chthonoplastes* have been sampled from solar saltern ponds in "Salines de la Trinitat" (Ebre Delta, S. Catalonia, Spain) and in "Salinas del Cabo de Gata" (Almeria, S.E. Spain). These mats are very similar in both salterns where they occur in ponds with water column salinities of 100 to 140 g l^{-1}. Subsamples from the cores were carefully sliced in recognisable sections of 0.2 to 2 mm .

The stratification of the populations in these mats was studied by optical phase-contrast and *in vivo* epifluorescence microscopy. The actually living microbial mat comprised approximately the top three millimeters of the sediment in which several layers could be distinguished. The top 0.2 mm was yellowish and mainly comprised photobleached *Microcoleus* cells. A consistent blue-green coloured cyanobacterial layer was found from 0.2 mm up to 1.5-2 mm depth, in which *M. chthonoplastes* was predominant, but high numbers of small bacteria were also found. At 2 mm depth, a red coloured layer of variable thickness (0.2-1 mm) was found which mainly consisted of *Thiocapsa* and *Chromatium* spp. Green filamentous phototrophic bacteria formed dense populations from two to three mm depth, but their occurrence in lower numbers was also observed in the depth horizon from 1-2 mm depth. Microscopically, these multicellular filamentous bacteria were characterised by a faint green colour under phase contrast, the absence of visible *in vivo* fluorescence upon excitation with blue and green light, and the presence of chlorosomes (T.E.M.). It still remains unknown whether these bacteria affiliate with *Chloroflexus* or represent multicellular members of the Chlorobiaceae (Stolz 1983; Stolz, This Volume). From three millimeters downwards the sediment showed darkish green laminations corresponding to buried microbial mats from previous years which contained moribund *Microcoleus* and green filamentous phototrophic bacteria together with high numbers of small unidentified bacteria.

Samples for T.E.M. were encapsulated and prepared according to Hernandez-Mariné (1992). Samples for lipid and hydrocarbon determinations were extracted in dichloromethane/ methanol (2:1); subsequently saponified to recover ester-bound fatty acids, fractionated by differential extraction in hexane and by column chromatography, and analysed by gas chromatography coupled to mass-spectrometry (GC-MS), according to Grimalt et al. (1992).

Morphological transformations of *Microcoleus chthonoplastes*

The morphological changes were studied on the optical and ultra-structure level by comparing photosynthetically active *M. chthonoplastes* from the top mat (0.5-1.5 mm) with buried layers at 4-5 mm depth, presumably being 2-3 years old, and at 10-12 mm depth. The optical and phase-contrast photomicrographs from these sections are shown in Figure 1.

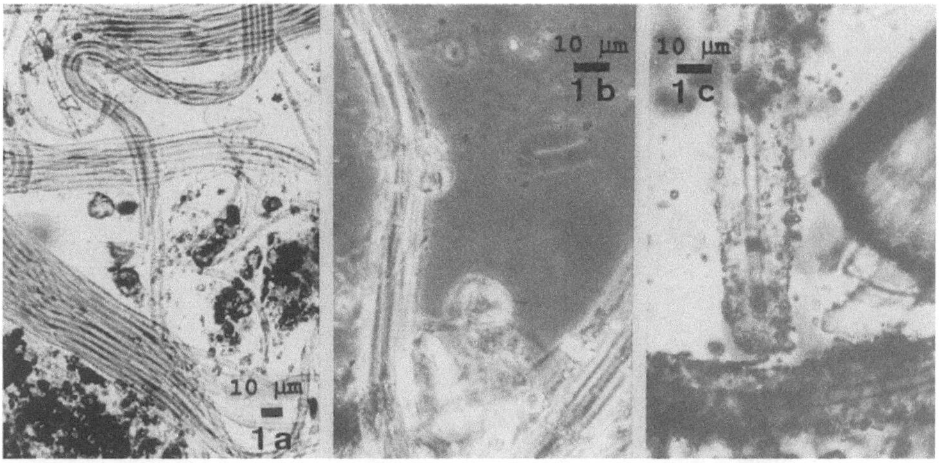

Fig. 1. Photomicrographs of mat-samples of *Microcoleus*-mats: a) the top layer (0.5-1.5 mm depth); b) phase contrast micrograph of 4-5 mm depth; and c) phase-contrast micrograph of 10-12 mm depth.

In the top layer, the *M. chthonoplastes* bundles contained a high number (8-25) of trichomes within the common sheath (Fig. 1a). In contrast in the buried layers, at 4-5 mm depth, the sheaths only contained 1 to 7 trichomes (Fig. 1b); whilst, at 12 mm depth, most sheaths contained one trichome (Fig. 1c). This shows that, although a small part of the trichome population is able to survive for a prolonged period in the anoxic dark sediment, the cyanobacterial sheath material is better conserved in the sedimentary record than the living trichomes.

T.E.M. cross sections of *M. chthonoplastes* bundles from the top mat show the occurrence of healthy trichomes in a common sheath (Fig. 2a). The radial arrangement of the thylakoids is a characteristic feature of this species (Stolz 1983). Fig. 2b represents a cross section at higher magnification showing the typical gram-negative cell wall structure which consists of a cytoplasmic membrane, peptidoglycan layer and outer membrane. In addition, Fig. 2b shows a carboxysome surrounded by the cytoplasm and the electron-dense polyphosphate granules. A longitudinal section represented in Fig. 2c, shows bleb formation at the cell wall surface and the apparent incorporation of such types of blebs into the sheath material, interpreted as the current formation of new sheath material (Butler and Allsopp 1972).

A cross-section of a buried *M. chthonoplastes* bundle from 4-5 mm depth (Fig. 3a) shows a common sheath with few trichomes. In the remaining trichomes, the thylakoids are highly vacuolised and partly desintegrated (Fig. 3b); therefore, they are spatially more separated than in the photosynthetically active layer. However, the structures are dissimilar

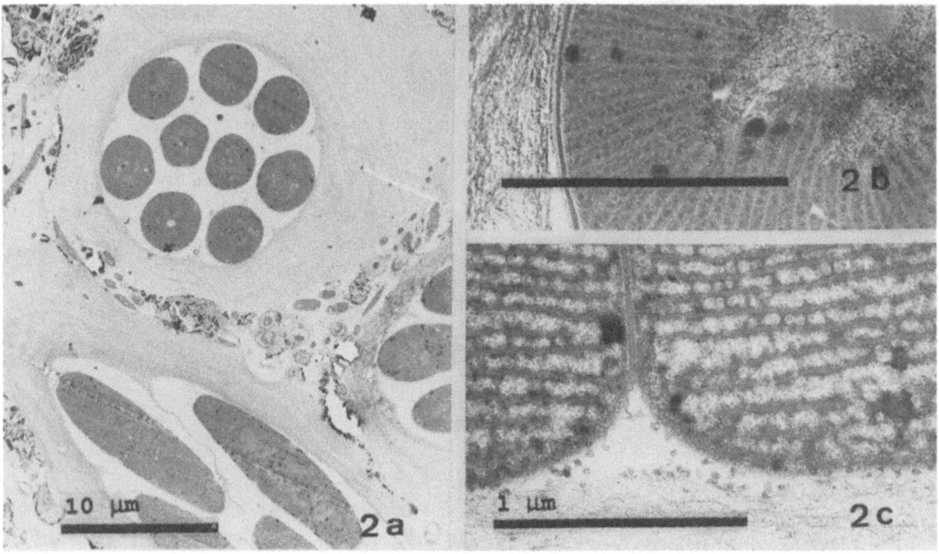

Fig. 2. T.E.M. photomicrographs of mat-samples of *Microcoleus*-mats from the top layer (0.5-1.5 mm depth): a) Cross and oblical sections of *M. chthonoplastes* bundles; b) Enlarged cross-section of *M. chthonoplastes* trichome showing the radial arrangement of the thylakoids; c) Transversal section of *M. chthonoplastes* trichome showing formation of blebs at the outer cell-membrane and their migration towards the sheaths.

from the ones found in photo-bleached trichomes. In the buried surviving cells, the polyphosphate granules have disappeared, and many electron transparent inclusion bodies were found. In comparison with the top mat, the buried *M. chthonoplastes* bundles have sheaths that are characterised by an increased electron density, which is related to a decrease of the water content due to compaction (c.f. Figs. 2a,b and 3a,b). Furthermore, the bleb formation at the cell wall surface and the apparent incorporation of such types of blebs into the sheath material was also observed in these buried *M. chthonoplastes* bundles (Fig. 3b), suggesting that these photosynthetically non-active trichomes continued to form sheath material.

In summary, the early diagenetic morphological changes of *M. chthonoplastes* in hypersaline microbial mats encompass the following phenomena: 1) desintegration of a part of the trichome population resulting in a decreasing amount of trichomes within the common sheaths; 2) in surviving bundles: the maintenance and perhaps novel synthesis of sheath material; 3) in surviving cells: morphological changes (vacuolisation) and partial desintegration of thylakoids.

In Solar Lake microbial mats, up to 8 years survival of buried trichomes was detected by the potential photosynthetic activity measured upon incubation in the light (Jørgensen et al. 1988). The reduced photosynthetic potential observed in the buried layers can be explained as

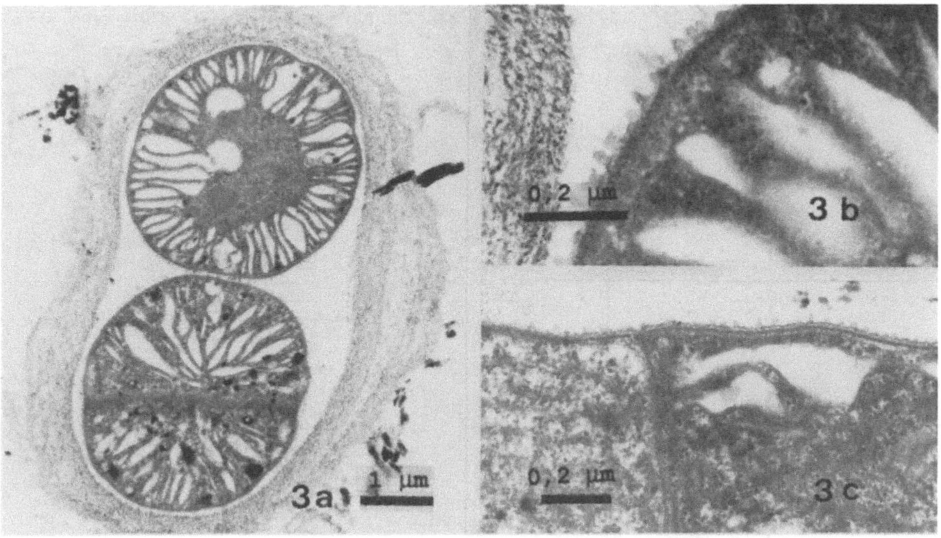

Fig.3. T.E.M. photomicrographs of mat-samples of *Microcoleus*-mats from a buried layer (4-5 mm depth): a) Cross sections of *M. chthonoplastes* bundles showing highly vacuolised and partly desintegrated trichomes; b) Enlarged cross-section of a *M. chthonoplastes* trichome showing the vacuolisation of the thylakoids and blebs formation at the outer cell-membrane; c) Transversal section of *M. chthonoplastes*.

the combined effect of the decreasing number of surviving trichomes and a decrease of the specific potential activity due to partial desintegration of thylakoid structures within the cells. On the other hand, several geochemical studies have postulated that sheath material is more refractory than cell material in the sedimentary record (e.g., Boon and De Leeuw 1987), and immunological techniques have shown that antigenic determinants of *M. chthonoplastes* sheaths are recognised in sub fossil samples from Solar Lake sediments (Muyzer et al. 1987). Microscopical observations have now documented the morphological features showing that sheath material is better preserved in the sedimentary record than trichomes.

Chemical transformations in *Microcoleus chthonoplastes* mats

Lipid compounds have been determined in sediment slices throughout the sedimentary column in these microbial mats (Grimalt et al. 1992); depth profiles are shown for a selection of compounds in Figure 4. The typical cyanobacterial fatty acids n-hexadecanoic acid, n-octadec-9(Z)-enoic acid (Fig. 4) and n-hexadec-9(Z)-enoic acid were the predominant fatty acids in the top mat from 0-2 mm with a maximum in the 0.2-0.7 mm depth layer. Below 2 mm depth, the concentration of these compounds fell sharply: in buried mats, their concentrations were always less than 4 % of the value found in the 0.2-0.7 mm depth layer.

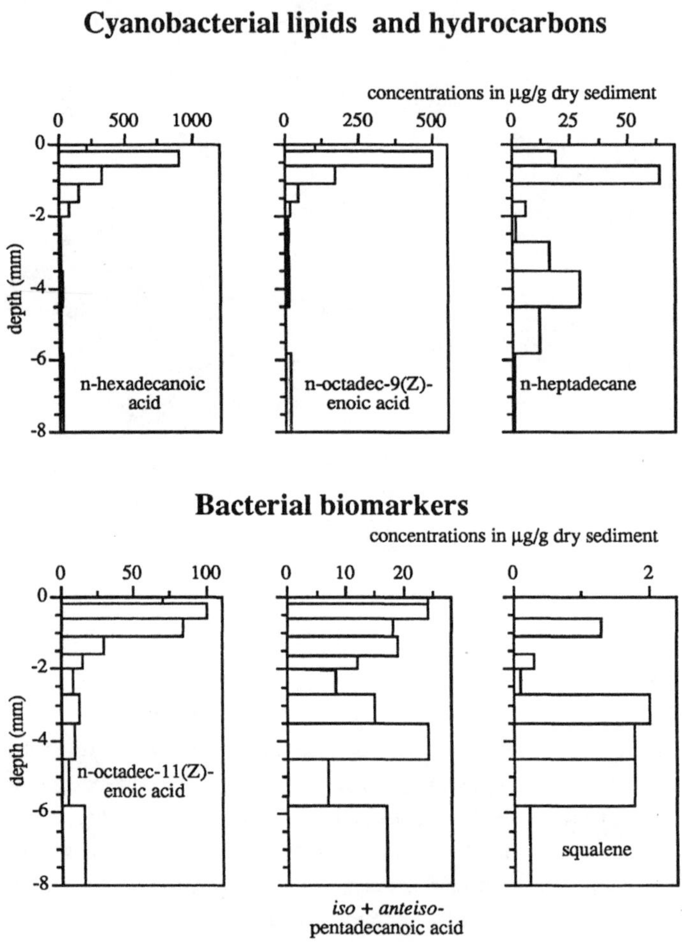

Fig. 4. Depth distributions of some cyanobacterial lipids, hydrocarbons and bacterial lipid biomarkers in *Microcoleus*-mats.

The majority of the fatty acids in the assay has been recovered from the phospholipid membranes upon saponification. In the top mat, *M. chthonoplastes* typically contains 60 to 90 radial thylakoids as observed on cross section (see Fig. 2a-b). From this observation we estimated that the surface of the total thylakoid membrane does exceed the cytoplasmic membrane by a factor in between 10 and 20. Thus, the largest mass of phospholipids in the cyanobacteria from the top mat is allocated in the thylakoid membranes. Therefore, the strong decrease of the cyanobacterial fatty acids below 2 mm depth, is due to loss of trichomes by desintegration and by partial degradation of the thylakoids within the surviving cells.

Hydrocarbons do occur in cyanobacteria and in microscopic algae, and n-heptadecane is the predominant hydrocarbon found in *M. chthonoplastes* (Winters et al. 1969). The depth profile of this latter compound is shown in Fig. 4. The peak concentration (64 μg/mg) does occur in the depth layer from 0.7 to 1.2 mm and values still amounting to 30-45 % of the maximum are found in the buried mats between 3 and 5 mm depth. Thus, n-heptadecane is less rapidly degraded than the typical cyanobacterial fatty acids. The low concentrations found between 1.5 and 3 mm depth relate to low densities of *M. chthonoplastes*, since these layers were dominated by purple sulphur bacteria and multicellular green filamentous bacteria.

Figure 4 also gives the depth profiles of some of the typical non-cyanobacterial prokaryotic biomarkers. n-Octadec-11(Z)-enoic acid, which is a typical fatty acid found in members of the branch of the gram-negative purple eubacteria (Oliver and Colwell 1973; Lechevalier and Lechevalier 1988), exhibited a profile which is very similar to the one found for the cyanobacteria, but the concentrations in the buried mats were still 5 to 16 % of the maximum values. In contrast, the *iso*- and *anteiso*-pentadecanoic acids showed much less variation along the sedimentary column studied. These branched fatty acids occur in cyanobacteria only in very small amounts, but are common and of widespread occurrence among the rest of the eubacteria; especially in gram positive bacteria and in some gram-negative bacteria including sulphate reducers (Kaneda 1991). These acids have been found in enrichment cultures including sulphate-reducers and a *Thiomicrospira* sp. obtained from the mat (Grimalt et al. 1992). Squalene has been found in many genera of both gram-negative and gram-positive eubacteria and in methanogenic archeabacteria (e.g., Lechevalier and Lechevalier 1988).

Conclusion

The cyanobacterial lipids are predominant in the top mat and do decrease strongly below 2 mm, whilst other prokaryotic lipids are present throughout the sedimentary column. As a result, the solvent-extractable lipid compositions in buried mats are predominated by the non-cyanobacterial lipids, and the cyanobacteria only leave a minor signal in this sedimentary record (Grimalt et al. 1992). The presence of non-cyanobacterial lipids in the buried mats must be attributed mainly to the heterotrophic populations growing on the cyanobacterial remains. Nevertheless, a small part of the cyanobacterial trichomes population survives for longer periods; a phenomenon which is associated with the strong morphological and chemical transformations observed. On the other hand, cyanobacterial sheaths are maintained and even the apparent incorporation of new material into the sheath is observed in buried layers. As a result, the cyanobacterial sheaths are more refractory and its component biopolymers are more likely to be preserved in the sedimentary record.

Acknowledgements

Thanks are due to INFOSA and UNION SALINERA DE ESPAÑA for permission to sample the salterns in the Ebre Delta and in Cabo de Gata, respectively. R.d.W. has been supported by a fellowship of the European Environmental Research Organisation (E.E.R.O.). Partial financial support from EEC (SCIENCE Project No. SC1-CT91-0736) is acknowledged.

References

Awramik SM (1984) Ancient stromatolites and microbial mats. In: Cohen Y, Castenholz RW, Halvorson HO (eds) Microbial Mats: Stromatolites. Liss, New York, pp 1-22

Boon JJ, De Leeuw JW (1987) Organic geochemical aspects of cyanobacterial mats. In: Fay P, Van Balen Ch (eds) The Cyanobacteria. Elsevier, Amsterdam , pp 471-492

Butler RD, Allsopp A (1972) Ultrastructural investigations on the Stigonemataceae (Cyanophyta). Arch Mikrobiol 82: 283-299

Grimalt JO, De Wit R, Teixidor P, Albaigés J (1992) Lipid biogeochemistry of *Phormidium* and *Microcoleus* mats. Org Geochem 19: 509-530

Hernandez-Mariné M (1992) A simple way to encapsulate small samples for processing for TEM. J of Microscopy 168: 203-206

Jørgensen BB, Cohen Y, Revsbech NP (1988) Photosynthetic potential and light-dependent oxygen consumption in a benthic cyanobacterial mat. Appl Environ Microbiol 54: 176-182

Kaneda T (1991) Iso- and anteiso-fatty acids in bacteria: biosynthesis, function, and taxonomic significance. Microbiol Rev 55: 288-302

Lechevalier H, Lechevalier MP (1988) Chemotaxonomic use of lipids - an overview. In: Ratledge C, Wilkinson SG (eds) Microbial Lipids Volume 1. Academic Press, London, pp 869-902

Muyzer G, De Koster, S, Van Zijl Y, Boon JJ, Westbroek P (1987) Immunological studies on microbial mats from Solar Lake (Sinai) - A contribution to the organic geochemistry of sediments. Org Geochem 10: 697-704

Oliver JO, Colwell R (1973) Extractable lipids of gram-negative marine bacteria: Fatty acid composition. Int J Syst Bacteriol 23: 442-458

Stolz JF (1983) Fine structure of the stratified microbial community at Laguna Figueroa, Baja California, Mexico. I. Methods of *in situ* study of the laminated sediments. Precambrian Research 20: 479-492

Winters K, Parker PL, Van Baalen Ch (1969) Hydrocarbons of blue-green algae: geochemical significance. Science 163: 467-468

On the significance of solar ultraviolet radiation for the ecology of microbial mats

Ferran Garcia-Pichel and Richard W. Castenholz[1]

Max-Planck-Institut für marine Mikrobiologie,
Fahrenheitstrasse. 1, D-28935 Bremen, F.R.G.

The shortest wavelengths of the solar spectrum reaching the ground, known as the solar ultraviolet (UV, 280-400 nm), amount to only a small percentage of the total incident irradiance (Table 1). However, they may produce disproportionately large biological effects because they activate many photochemical reactions involving molecules of central biological importance. The consequence of these photoreactions is usually detrimental to the cellular metabolism. Mutation and lethal DNA and RNA damage, lipid peroxidation, photoinhibition of photosynthesis and respiration, and specific enzyme inactivation can all be a result of exposure to UV (Jagger 1985). These physiological effects are not always a laboratory curiostity but have ecological implications under natural conditions: the solar UV has been recognized as an important environmental stress factor in a variety of microbially dominated ecosystems. UV has been shown to cause sustained inhibition of both primary productivity and bacterial production in planktonic communities and to drive marked changes in species composition in periphyton communities (see Worrest 1982), but not much is known about the significance of ultraviolet radiation in benthic ecosystems. We have gathered some direct experimental evidence, as well as strong indirect inference from ecophysiological studies in diverse mat communities, suggesting that UV may be of importance, at least for the top phototrophic layers.

It is our purpose here to elaborate on some characteristics of the benthic or mat environment that are likely to play a significant role in their UV-photobiology. We also present some evidence implicating UV as a stress factor in the ecology of microbial mats and attempt to point out specific aspects where research efforts should be directed in order to assess the extent to which UV can be considered a major ecological factor for microbial mats.

UV and the mat environment

The impact of UV radiation on a microbial mat will naturally be a function of the levels and spectral quality of the incident fluxes. These vary in space and time, intensities increasing with

[1] Department of Biology, University of Oregon, Eugene, Oregon 97403, USA

NATO ASI Series, Vol. G 35
Microbial Mats
Edited by Lucas J. Stal and Pierre Caumette

altitude and decreasing with latitude, and peaking during noon and in the summer season. Some ilustrative values are shown in Table 1. Because microbial mats are usually found in shallow waters or semi-exposed, the effect of water absorption is likely to be less important than in planktonic environments, but even shallow water depths may create significant spectral shifts towards longer wavelengths. For reference, the mean 1% depths in coastal Antarctic waters (Vosjan and Pauptit 1992) are given: 44 m for visible, 36 m for UV-A and 22.5 m for UV-B.

Table 1. Typical values of noon (vector downwelling) irradiance (W m^{-2}). All data obtained with the same model radiometer and sensors. (1 W m^{-2} equals approximately 2.7, 3, and 4.5 μmol m^{-2} s^{-1} of UV-B, UV-A and Visible+IR, respectively.)

	Visible+IR (400-1000 nm)	UV-A (315-400nm)	UV-B (280-315nm)
Yellowstone Ntal. Park (USA); July	1245	47.2	4.2
Hunter's Hot Springs, OR (USA); June	858	41.3	3.5
Guerrero Negro, BCS (México); May	1000	38	3.0

The penetration and distribution of UV within the mat will also be of importance in determining the true doses received by the organisms. The description of radiative fields within benthic communities and substrates (see chapter by M. Kühl et al., this volume) has contributed a framework within which benthic photophysiological phenomena can be accurately interpreted. Unfortunately those measurements have been restricted to the visible and the infrared (IR). At this point, even the most basic information on UV penetration into sedimentary ecosystems is still lacking. In mats, specular and diffuse reflectance at the surface may contribute significant losses to the effective UV doses. Measurements of UV diffuse reflectance indicate that microbial mat assemblages can be very diverse in their UV optical properties, as diverse as they are in the visible (Table 2). Even though the data suggest that the penetration of UV is on average less than that of the visible, commonly held views that UV does not penetrate benthic environments may be in need of revision. Several other factors specific to the microbial mat habitat may in principle affect how UV will interact with the biological processes. Foremost among those that will tend to decrease its importance with respect to that in neritic ecosystems is the attenuation of UV within the mat by absorbers that are neither targets of UV photodamage nor photosensitizers of UV-mediated photodynamic action. This should be important in microbial mats with high contribution by inert organic and inorganic matter (i.e.: sand, silt) to UV attenuation. However, other factors will tend to increase the effectiveness of UV action. Foremost among these is the density of the microbial mat assemblages. Compact photosynthetic layers usually bring about steep gradients of oxygen with maxima close to the surface, where concentrations of O_2 several times saturation are not uncommon. The effects of near UV (UV-A ; 315-400 nm) are mostly sensitized, i.e., caused indirectly through excitation of absorbing but non-target molecules, and strongly dependent on mediation by reactive oxygen species formed in the process. The simultaneous midday maxima of oxygen concentration within, and UV incident on microbial mats, in which high contents of potential photosensitizers (tetrapyrroles, quinones, alkaloid secondary metabolites, etc) occur, constitute the right mixture for strong UV actions.

Table 2. Surface Diffuse Reflectance (in percent; 60° incidence angle, measured over the vertical ±7°) of various types of microbial mats and crusts from Yellowstone National Park (USA) against a MgO reflectance standard. Low values of diffuse reflectance are an indication of strong absorption, high values indicate high scatter and low absorption.

Mat Type	Visible+IR (400-1000 nm)	UV-A (315-400 nm)	UV-B (280-315 nm)
Phormidium mat (Octopus Spring)	34.0	10.1	6.7
Synechococcus mat (Octopus Spring)	17.3	7.2	4.8
Calothrix sp. mat (West Thumb Springs)	9.1	5.2	3.6
Stigonema crust	6.4	1.6	1.1

UV as a physiological stress: oxygenic photosynthesis

On the basis of current physiological knowledge, photosynthesis, the means of primary productivity in most microbial mats, is probably one of the prime targets for UV photodamage. Studies with cultures of both cyanobacteria and diatoms, some of benthic origin, leave little doubt that sustained photodamage can be caused by UV-dosages (or dosage rates) well below natural levels, even though a substantial capability for photoacclimation has also been encountered in some isolates (see Garcia-Pichel et al. 1993). The question remains then not so much if UV will have an effect on *in situ* photosynthetic rates in microbial mats but how severe this effect will be and how deep within the euphotic zone will the effects be significant. We have carried out some simple experiments illustrating the potential effects of UV on primary productivity, even though a thorough quantitative assessment of the long-term importance of UV-mediated depression on primary productivity is not yet available. In a laboratory acclimated (room temperature, 100 μmol photons m^{-2} s^{-1} PAR), *Microcoleus chthonoplastes* dominated sandy mat, 1 hour of exposure to 8 μmol photons m^{-2} s^{-1} UV-B radiation from suntanning tubes (peak emission at 310 nm) added to the background visible illumination, photoinhibed the depth-integrated photosynthetic rates by 66 % (Fig. 1) with no substantial recovery for 3 more hours. The

UV-B fluence rates used were probably close to maximal yearly values (see Table 1), and the organisms in the mat had not experienced UV-B for about 1 week, so this figure must be regarded as an estimate of the maximal effect of UV-B. More significant, perhaps, is the depth-distribution of the UV impact. As expected, inhibition was strongest at the mat surface, but substantial inhibitory effects spanned the major part of the euphotic zone down to 0.7 mm. The profile of inhibition did not reflect a theoretical, quasi-exponential decay in dosages; hence factors other than merely the dosage received *in situ* must have been at play. Both an increase with depth in the inherent sensitivity of the organisms, or, perhaps, diffusion to deeper layers of long-lived reactive oxygen species (i.e., hydrogen peroxide) formed close to the surface may have been important.

In natural summer populations of *Synechoccocus lividus* from thermal mats in Oregon, using [^{14}C]- Na HCO_3 uptake rates as a measure of photosynthesis, increases in the photosynthetic rates at least two-fold were found upon exclusion of UV (A+B) from the incident solar radiation by means of long-pass optical filters. The effects of exclusion of UV-B alone were insignificant or much smaller (Miller and Castenholz, unpubl.). These data clearly implicate UV-A as the most important spectral region for UV- photoinhibition and demonstrate that UV can be a significant stress factor even in field populations acclimated to natural UV doses.

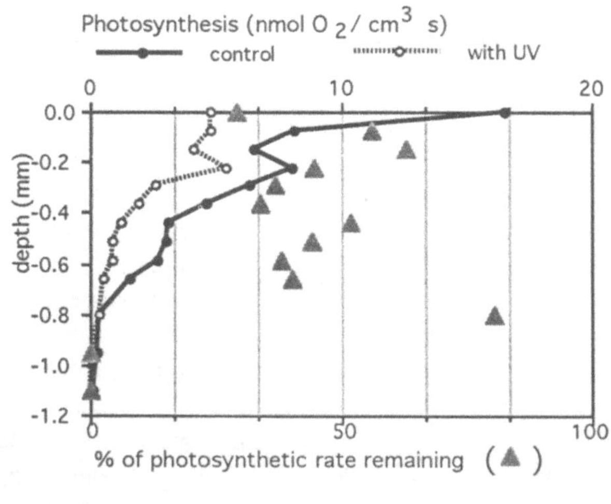

Fig. 1. Depth profiles of photosynthesis (O_2 evolution) and photoinhibition caused by inclusion of ca. 8 μmol m^{-2} s^{-1} of UV-B radiation in the incident light field (100 μmol m^{-2} s^{-1} PAR) during 1 h. Sandy *Microcoleus chthonoplastes* mat from Nordeney, FRG.

Primary productivity may be affected indirectly through the effect of UV on photosynthetic pigment synthesis as well. We have observed, for example, that in thermal mats formed by *Calothrix thermalis* , the exclusion of UV (A+B) for ca. 1 week, resulted in elevated contents of phycoerythrin with respect to the rest of the photosynthetic pigments. This is in agreement with physiological data from cultures, in which it was found that UV-A inhibits a-aminolevulinate synthetase (the central enzyme in the synthesis of tetrapyrroles) and has marked effects on the synthesis of phycoerythin in particular (Hirosawa and Miyachi 1983; Ohki et al. 1982).

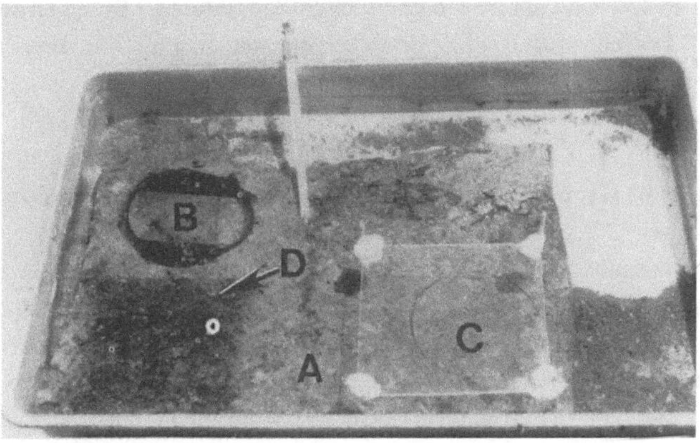

Fig. 2. Some effects of 1 h long modifications of the incident irradiance on the migration of a *Oscillatoria/Spirulina*-dominated hypersaline mat around midday. In the light areas cyanobacteria are at depth, in the dark areas migration to the surface has occurred. A: control area, full irradiance. B: exposed only to UV-A portion of spectrum (36 W m^{-2}). C: UV-B excluded, 2.5 W m^{-2} of UV-A and full visible and IR. D: exposed to 66 W m^{-2} of visible and IR, 0.3 W m^{-2} of UV-A and no UV-B.

UV as a cue for organismal behavior: diel migrations

Diel migration patterns are a common phenomenon among soft or sandy mats dominated by oscillatorian cyanobacteria. Usually the migration involves downward population movements of the cyanobacteria during the daytime, with a return to the surface at dusk. Cyanobacterial migrations have been studied in detail in a hypersaline mat from Guerrero Negro (Mexico) dominated by *Oscillatoria* sp. and *Spirulina subsalsa*. Migration was chiefly a response to incident light intensity, spanned depth ranges of 0.5 - 0.7 mm , and resulted in a very effective avoidance of exposure to photoinhibiting irradiances (Garcia-Pichel et al. 1993, submitted). Using optical filters (Fig. 2; not all treatments shown) it could be determined that UV-A was the portion of the solar spectrum to which the cyanobacteria responded. Filters were placed over the mat around noon, when the cyano—bacteria were at depth, and the responses were recorded after 1 h of exposure. Calculations show that as little as 1.3 W m^{-2} of UV-A alone was enough to keep the cyano-bacteria at depth. Such levels are reached early in the daytime and represent approximately 3 to 4 % of the maximal levels reached at noon. UV-A was over an order of magnitude more effective than visible light in preventing the ascent of the cyanobacteria to the surface. Solar UV-B has been shown to impair motility in cyanobacteria (Donkor and Häder 1991). However, additional experiments showed that UV-A was also very effective in promoting downward migrations in our mats, which suggests that the UV-effects on migration are due to the true usage of UV as an environmental cue by the organisms rather than to an impairment of motility, since active motility is needed for the response to occur. These responses to UV may act in concert with or be modulated by other cues, such as the concentration of key chemical species to determine the overall migration patterns of the organisms.

Synergistic effects of UV with other stress factors: sunscreen protected mats

The ecological consequences of UV-exposure in mats must be understood as a continuous balance between the rates of photodamage and those of biological repair. Whenever the rates of biological repair are hindered, photodamage acquires increased importance for the ecology of the organisms. This is often the case in microbial mats in which environmental conditions (desiccation, freezing, lack of nutrients) allow for only episodic biological activity, maintaining the ecosystem in a dormant (but sunlit!) state during the rest of the time. Under these conditions, active repair mechanisms such as *de novo* synthesis of damaged constituents, refurbishing of photooxidized components, and enzymatic DNA repair, are not possible and, thus, the ability to obtain photoprotection by a passive mechanism becomes a major asset for the survival of the organisms.

We have studied the passive photoprotection by UV-sunscreen compounds in cyanobacteria in some detail (Garcia-Pichel and Castenholz 1991, 1993; Garcia-Pichel et al. 1992, 1993). Several findings from these studies are relevant for the microbial mat environment. Cyanobacteria may produce two types of sunscreens: extracellular, lipid-soluble sheath pigments, such as scytonemin, and intracellular, water -soluble mycosporine-like susbtances. Among mat-forming cyanobacteria one finds species with both types of compounds (i.e., *Lyngbya aestuarii* from intertidal mats), with only one type (i.e., *Nostoc microscopicum* from Antarctic ponds; with scytonemin only), and with no sunscreens (i.e., *Synechococcus lividus* from thermal mats). An ecological correlation exists between dominance by sunscreen-bearing cyanobacteria and habitats with episodic growth events, so that species with abundant sunscreens become dominant as, for instance, the desiccation events become longer going up a tidal gradient. These compounds, the synthesis of which is triggered by UV, can shield the cells from a significant portion of the incoming UV, but leave a long wavelength window in the visible suitable for photosynthetic activity. Because of increased pathlengths, they become very effective in colony- or mat-forming species. This is, however, at the expense of heavy investments in sunscreen, up to 5-10 % of the dry biomass in some cases, which may not pay up under conditions of balanced or rapid growth. It is interesting to note that, since the deposition of sunscreen takes place at the mat surfaces, it is the whole community and not only the cyanobacteria that become shielded from photodamage. The widespread presence of these compounds, and the dominance that sunscreen bearing cyanobacteria attain speaks for a major synergistic role of UV in controling community composition, when additional stress factors exist in the environment.

Conclusions and outlook

The available evidence suggests that UV may be an important factor for the ecology of microbial mats, since significant harmful effects on primary productivity have been measured, which not only occur at the mat surface but also at depth. UV appears to be important as an environmental cue for microbial behavior in mats. UV also seems to play a role in determining cyanobacterial community composition in habitats in which other stress factors are encountered. Biochemical processes other than photosynthesis, such as respiration or dinitrogen fixation are likely to be affected by UV as well, if either sufficient doses penetrate to the active zones or if reactive UV-photoproducts can diffuse to those zones in sufficient quantities. Additional effects, either photochemical or photobiological, on biogeochemical processes known to be afected by UV radiation or its photoproducts may include iron and sulfur cycling. Systematic studies of UV effects on mats and measurements of the UV radiative fields within microbial communities are needed to better understand these questions.

References

Donkor V, Häder DP (1991) Effects of solar ultraviolet radiation on motility, photomovement and pigmentation in filamentous gliding cyanobacteria. FEMS Microbiol Ecol 88: 159-168

Garcia-Pichel F, Castenholz RW (1991) Characterization and biological implications of scytonemin, a cyanobacterial sheath pigment. J Phycol 27: 395-409

Garcia-Pichel F, Castenholz RW (1993) Occurrence of UV-absorbing, mycosporine-like compounds among cyanobacterial isolates and an estimate of their screening capacity. Appl Environ Microbiol 59: 163-169

Garcia-Pichel F, Sherry ND, Castenholz RW (1992) Evidence for a UV sunscreen role of the extracellular pigment scytonemin in the terrestrial cyanobacterium *Chlorogloeopsis* spp. Photochem Photobiol 56: 17-23

Garcia-Pichel F, Wingard CE, Castenholz RW (1993) Evidence regarding the UV-sunscreen role of a mycosporine-like compound in the cyanobacterium *Gloeocapsa* sp. Appl Environ Microbiol 59: 170-176

Hirosawa T, Miyachi S (1983) Effect of long-wavelength ultraviolet (UV-A) radiation on the growth of *Anacystis nidulans*. Plant Sci Lett 28: 291-298

Jagger J (1985) Solar-UV actions on living cells. Praeger.

Ohki K, Watanabe M, Fujita Y (1982) Action of near UV and blue-light on the photocontrol of phycobiliprotein formation; a complementary chromatic adaptation. Plant Cell Physiol 23: 651-656

Vosjan JH, Pauptit E (1992) Penetration of photosynthetically available light (PAR), EV-a and UV-b in Admiralty Bay, King George Island , Antarctica. Circumpolar J 1-2: 50-58

Worrest RC (1982) Review of the literature concerning the impact of UV-B radiation on marine organisms. In: Calkins J (ed) The role of solar radiation in marine ecosystems. Plenum Press

Environmental factors controlling the development of microbial mats in inland saline lakes; the granulometric composition of the sediment

Maria-Carmen Guerrero, Ana-Belen Tadeo and Rutger de Wit[1]

Laboratory of Limnology, Dept. of Ecology, Universidad Autónoma de Madrid, 28049 Madrid, Spain.

Microbial mats in Spanish continental salt lakes

The presence of inland saline and alkaline lakes is one of the most original limnological features of the Iberian Peninsula. These waterbodies are associated with endorheic phenomena that occur in arid and semi-arid climates. On a global basis, inland saline lakes are common (Williams 1986), but in Western Europe, these lakes are confined to the Iberian Peninsula. The Spanish continental salt and soda lakes lakes occur in groups known as endorheic centers (see Fig. 1). These lakes encompass a broad variety of different physico-chemical conditions, seasonal fluctuations and sediment features. Lake Chiprana (Aragon) is the only deep permanent hypersaline lake in Spain (Guerrero et al. 1991). In contrast, all other inland saline and alkaline lakes in Spain are temporal water basins which contain water during 3 to 10 months of the year. During the inundation periods, the water level and concomitantly the salinity is subjected to extreme temporal fluctuations. The ionic composition of these lakes has been studied previously (Montes and Martino 1987; Comin and Alonso 1988). In summary, the Spanish inland lakes encompass a wide spectrum of ionic types: in the Andalusian lakes, sodium predominates with either chloride or sulphate as the dominant anion; the lakes in La Mancha mainly contain magnesium sulphate; and the lakes in Aragon encompass a whole range from sodium chloride to magnesium sulphate.

Since 1989, the Spanish inland saline and akaline lakes have been screened for the occurrence of microbial mats. The inventory is depicted in Fig. 1 and the results have been described previously (Guerrero and De Wit 1992). In this study, four different microbial mat communities have been described in more detail.

---

[1] Laboratoire d'Océanographie Biologique, Université de Bordeaux I; 2, Rue du Prof. Jolyet 33120 Arcachon, France.

NATO ASI Series, Vol. G 35
Microbial Mats
Edited by Lucas J. Stal and Pierre Caumette

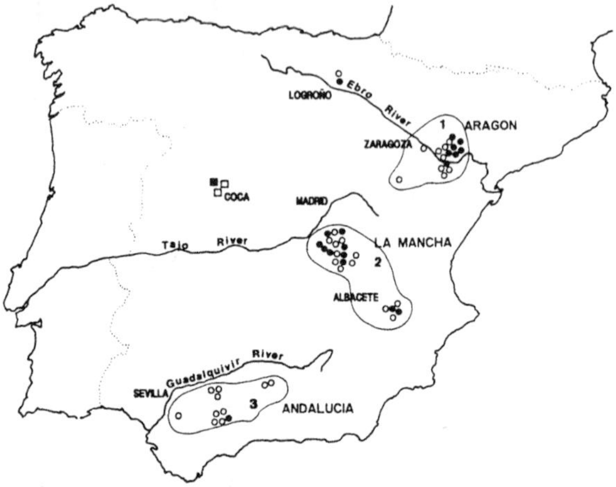

Fig. 1. Map of the Iberian Peninsula showing the distribution of microbial mats in saline and alkaline inland lakes (closed circles = saline lake with conspicuous microbial mats; open circles = saline lakes in which no microbial mats have been found so far; closed squares = subsaline soda lake (pH > 9.5) with microbial mats; open squares = subsaline soda lakes without conspicuous microbial mats). After Guerrero and De Wit, 1992.

The inventory depicted in Fig. 1 shows that microbial mats have been found in about 35 % of the inland saline and alkaline lakes in Spain. The high abundance of these lakes in small geographic areas and their great variability of environmental features allow for the study of micro-climatological, biological, geochemical and sedimentory factors favouring or repressing the development and growth of microbial mats. Actually, we are working on a project which involves the monitoring of mat development and the study of abiotic parameters in saline lakes from the La Mancha and Aragon endorheic centers. Statistical analysis has been used to cluster the lakes according to their environmental features and to find correlations with the microbial mat communities. In this paper we present some preliminary data that indicate how the absence or presence of microbial mats and community structure is correlated with the granulometric composition of the sediment.

Microbial mats and sediment granulometry

Nine inland saline lakes from the La Mancha endorheic center (Fig. 1), were studied intensively during the hydrological cycle of 1991-1992. In four lakes, microbial mats developed, while in the other five lakes, microbial mats were absent throughout the cycle or only developed for very short periods on small localized patches. The granulometric composition of the sediment was determined using a cascade of sieves according to Casagrande (1948) and the sediment classification terminology proposed by Way (1973) has been used. The organic matter content of the sediment was determined by calcination at 450 °C (Hall et al. 1963).

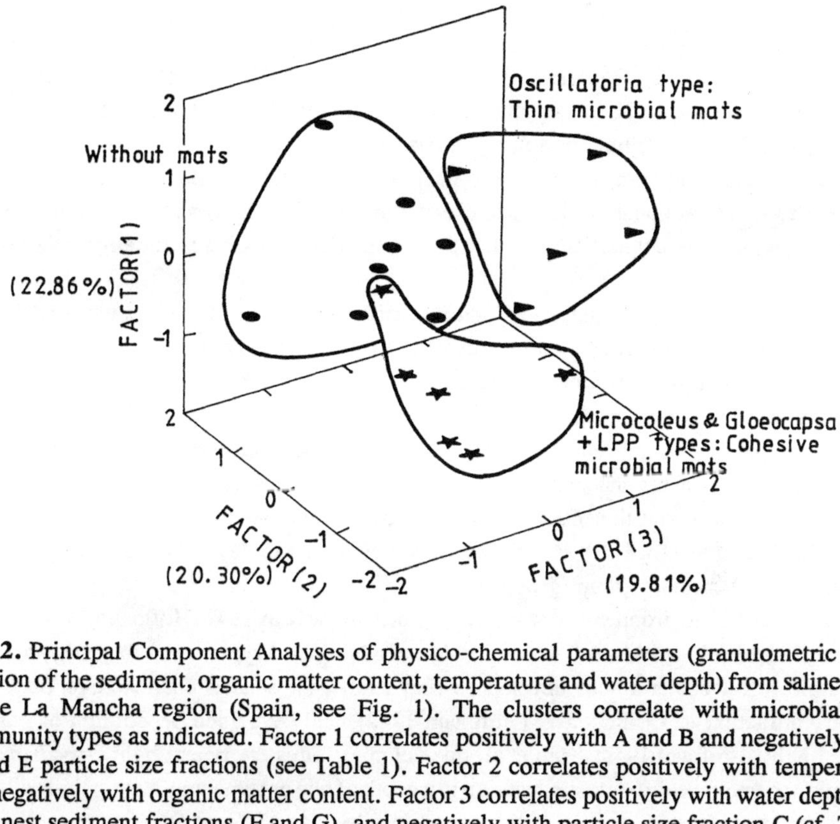

Fig. 2. Principal Component Analyses of physico-chemical parameters (granulometric composition of the sediment, organic matter content, temperature and water depth) from saline lakes in the La Mancha region (Spain, see Fig. 1). The clusters correlate with microbial mat community types as indicated. Factor 1 correlates positively with A and B and negatively with D and E particle size fractions (see Table 1). Factor 2 correlates positively with temperature and negatively with organic matter content. Factor 3 correlates positively with water depth and the finest sediment fractions (F and G), and negatively with particle size fraction C (cf. Table 1).

Principal Component Analysis (P.C.A.) (Harman 1976) was performed on a data matrix with 19 data sets encompassing the granulometric composition, water depth, temperature and organic matter content. A three-dimensional representation of the P.C.A.-analyses for the Factors 1, 2 and 3 is shown in Fig. 2. Three clusters (groups) have been separated that correlate with the microbial mat community structure:

Group I: Sites with complex and compact mats with interwoven sediment particles. *Microcoleus chthonoplastes* or *Gloecapsa* sp. together with species belonging to the LPP-B group were the predominant mat-building species.

Group II: Sites in lakes without mats.

Group III: Sites with a thin mat occurring as a biological matrix on top of the sediment. Characteristically, these mats developed on exposed surfaces with sufficient moisture, and became detached and floated to the water surface when submerged during the flooding periods. The main mat-building species were *Oscillatoria* sp.

In the P.C.A. (Fig. 2), the first three factors contributed with almost equal weight (approximately 20 % each) to the total variance observed. Factor 1 was mainly related to sediment particle size distribution, notably the size classes E, D, B and A (see Fig. 2, Table 1). Factor 2 was related to temperature and organic matter content; while, Factor 3 was related to water depth and sediment particle size distribution, notably the G and F fractions. Hence, it is clear that the sediment particle size distribution is a determinative factor among the factors studied.

Table 1 gives the mean values (%) and standard deviations of the sediment particle size classes for the different groups. Analysis of variance (ANOVA) has been applied to study the significance of the differences among the groups (Sokal and Rohlf 1987). Highly significant differences were found for the fine sand (0.125 to 0.25 mm) and the silt + clay (< 0.063 mm) size classes ($p < 0.001$). Group I with the *M. chthonoplastes* or *Gloeocapsa* sp. mats, have been found on sediments that were richest in fine sand (24.4 %) and poor in silt + clay (7.6 %). In contrast, the thin *Oscillatoria* mats were found on top of sediments that were poorest in fine sand (13.8 %) and richest in silt + clay (15 %). The lakes without mats represent an intermediate group (group II).

For the coastal environment, it was already noted as early as 1942 (Hofmann 1942) that the so-called "Farbstreifensandwatt" which consists of a laminated system with *M. chthonoplastes* and purple sulfur bacteria does develop on fine sand. Fine sand is characterised by a high capillary attraction of freatic water (Hofmann 1942), and by a good penetration of red and infrared light (Hoffman 1949; Kühl et al. this volume). Thus, sufficient light is available for oxygenic and anoxygenic photosynthesis up to a depth of one or some mm. within the sediment. In contrast, light is very rapidly extinguished in silty-clay. In contrast, however, the thin *Oscillatoria* type mats typically developed on top of the more silty sediments; hence light availability for photosynthesis is thus only determined by the biological matrix. These mats developed often in lakes that had a high input of organic material from waste water disposal, which was reflected in a high sedimentary organic matter content (data not shown). So far, we do not have a mechanistic explanation why microbial mats are not found in lakes that have an intermediate sediment composition.

Table 1. Granulometric composition of the sediment for three different groups (cf. Fig. 2) of inland saline lakes (La Mancha, Spain; see Fig. 1)

Size class	Diameter (mm)	Sediment Classification	Mean Values (%) and standard deviation (in brackets)						
			Group I *Microcoleus*and *Gloeocapsa* mats		Group II without mats		Group III thin *Oscillatoria* type mats		
A	> 2	Gravel	6.7	(3.1)	5.4	(4.1)	9.8	(2.0)	n.s
B	1 to 2	Coarse sand	12.5	(5.9)	16.7	(6.2)	18.2	(2.4)	n.s.
C	0.5 to 1	Coarse sand	16.5	(3.8)	24.4	(10.7)	14.1	(10.4)	n.s.
D	0.25 to 0.5	Medium sand	21.3	(3.5)	16.8	(7.5)	16.7	(10.8)	n.s.
E	0.125 to 0.25	Fine sand	24.4	(4.5)	16.8	(4.1)	13.8	(4.0)	***
F	0.063 to 0.125	Very fine sand	10.3	(3.4)	8.7	(2.0)	12.5	(1.7)	n.s.
G	< 0.063	Silt + Clay	7.6	(2.9)	10.9	(2.8)	15.0	(1.1)	***

n.s. = not significant

*** $p < 0.001$

References

Casagrande A (1948) Classification and Identification of Soils, Transportation 113
Comin F, Alonso M (1988) Spanish salt lakes. Their chemistry and biota. Hydrobiologia 158: 237-245
Guerrero M-C, Balsa J, Pascual B, Martinez B, Montes C (1991) Caracterización limnológica de la laguna salada de Chiprana (Zaragoza, España) y sus comunidades de bacterias fototróficas. Limnetica 7: 83-96
Guerrero M-C, De Wit R (1992). Microbial mats in the inland lakes of Spain. Limnetica 8: 197-204
Hall CE, Safranko J, Stenger VA (1963) Rapid combustion method for the determination of organic matter in aqueous solutions. Anal Chem 35: 315
Harman HH (1976) Modern Factor Analysis, 3rd Ed. University of Chicago Press, Chicago
Hoffmann C (1942) Beiträge zur Vegetation des Farbstreifen-Sandwatt. Kieler Meeresforsch 4: 85-105
Hoffmann C (1949). Über die Durchlässigkeit dünner sandschichten für Licht. Planta 37: 48-56
Montes C, Martino P (1987) Las lagunas Españolas. In: Seminario sobre bases científicas para la protección de los humedales en España. Real Academia de Ciencias Exactas, Físicas y Naturales de Madrid, Madrid. pp 95-145
Sokal R, Rohlf FJ (1987) Introduction to biostatistics. 2nd Ed. W.H. Freeman and Company, San Francisco.
Way DS (1973). Terrain Analysis. A guide to site selection using aerial Photographic Interpretation. Community development Series. Hutchinson and Ross, Stroudsburg
Williams WD (1986) Limnology, the study of inland waters: a comment on perception of studies of salt lakes, past and present. In: De Decker P, Williams WD (eds) Limnology in Australia. Dr. W. Junk Pub, Boston, pp 471-496

Structure of the sediment at depositional saline environments

Ester Clavero, Joan O. Grimalt, Vidal Merino[1] and Mariona Hernández-Mariné[1]

Department of Environmental Chemistry (CID-CSIC), Jordi Girona 18, 08034 Barcelona, Spain

Introduction

Hypersaline evaporitic systems are one of the places where microbial mats usually develop. In such environments the evaporation of seawater leads to a mixture of salt deposits, in which NaCl predominates and $MgCl_2$, $MgSO_4$, $CaSO_4$, KCl, NaBr, $CaCO_3$ are present as minor components. These salts precipitate at different stages of seawater concentration due to their different solubilities. Calcite is the first salt that precipitates, followed by gypsum and finally halite. In solar salterns the process of salt precipitation is spatially distributed in different ponds with increasing salinities. Generally, the salinity in the different ponds is kept constant and the environmental conditions are well defined (Barbé et al. 1990).

Complex, stratified communities develop in the sediments of these ponds. The composition and distribution of these communities is closely related with the main ecologic parameters (Des Marais et al. 1989; Javor and Castenholz 1981; Jørgensen et al. 1983; Guerrero et al. 1987).

This study describes the diatom and cyanobacterial species composition and distribution in the sediment at the different salinities and sediment depths.

Sampling site description and methods

The sampling sites were located at "La Trinitat" salterns in the delta of Ebro River (Fig. 1). Each precipitation domain comprised different ponds. The saltern receives water from Alfacs Bay. The ponds in which $CaCO_3$, sulfates and NaCl precipitated are called respectively 'depositors', heaters and crystallizers.

Microalgae, cyanobacteria, and community distribution in depositor 1 and 2, and in heater 7 were studied. These are shallow water pools with maximum depths of approximately 60 cm.

The sediment was sampled by taking cores. The cores were sliced in sections of 1 mm, which corresponded with the thickness of the different layers.

---

[1] Faculty of Pharmacy, University of Barcelona, Diagonal s/n, 08028 Barcelona, Spain

NATO ASI Series, Vol. G 35
Microbial Mats
Edited by Lucas J. Stal and Pierre Caumette

Samples for transmission electron microscopy were processed according to Hernández-Mariné (1992) and observed using a Philips 200 transmission electron microscope.

Diatoms were identified in samples cleaned with 30% hydrogen peroxide (Renberg 1990) and thoroughly rinsed with distilled water. Material for light microscopy was air-dried on a coverslip which was subsequently mounted with naphrax on microscope slides. They were observed using Nomarsky optics on a Nikon Optiphot-2 light microscope. For scanning electron microscopy (SEM), clean material was air-dried on 0.2 μm membrane filters, and mounted onto aluminum stubs. Stubs containing dry material were sputter-coated with ca. 60 nm of gold and viewed on a Hitachi S-2300 scanning electron microscope.

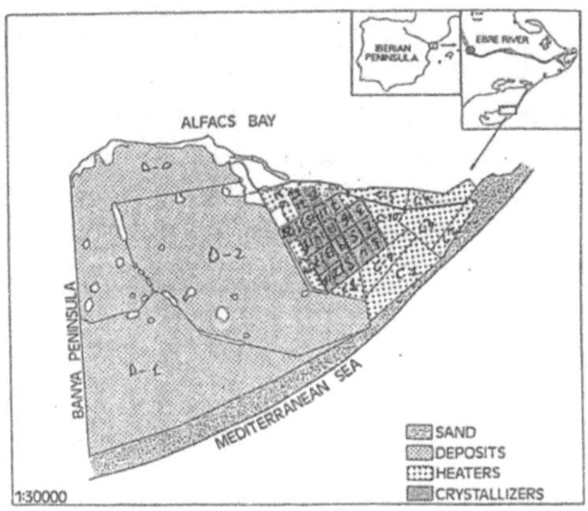

Fig. 1. Location of "La Trinitat" salt ponds.

Results

In depositor 1 (salinity 7.58°Bé,σ=1.7) *Cladophora* sp. masses were found covered by epiphitic diatoms, mainly *Cocconeis bardawilensis*. The interface water-sediment was inhabited by diatoms and dinoflagellates (Table 1), which were resuspended and sedimented periodically due to convective movements of water. The diatom community was dominated by species of the genus *Amphora (A. coffeaeformis, A. acutiuscula, A. angusta*), and sometimes appeared local small blooms of *Pleurosigma elongatum*. Filamentous cyanobacteria (*Oscillatoria limnetica, Phormidium valderianum*) and colourless sulfur bacteria (*Beggiatoa* sp.) were found in the same layer (Fig. 2). At greater depth cyanobacteria and diatoms were replaced by phototrophic and chemotrophic bacteria, which was a result of changes in ambient light and oxygen levels. At this depth we found only the remains of diatoms i.e. broken empty valves which were colonized by high numbers of bacteria (Fig. 5).

Table 1. Dominant species or genera observed in "La Trinitat" salt ponds

Depositor 1	Depositor 2	Heater 7
Cyanobacteria: *Microcoleus chthonoplastes* Thuret, *Oscillatoria limnetica* Lemm., *O.limosa* Agardh, *Phormidium valderianum* (Delp.) Gom., *Spirulina subsalsa* Oerst	**Cyanobacteria:** *Microcoleus chthonoplastes* Thuret, *Spirulina subsalsa* Oerst	**Cyanobacteria:** *Microcoleus chthonoplastes* Thuret
Diatoms: *Achnantes biasolettiana* (Kütz.) Grun., *Amphora angusta* (Greg.) Cl., *A.acutiuscula* Kütz., *A.coffeaeformis* Ag., *A.exilissima* Giffen, *Cocconeis bardawilensis* Ehrlich, *C.placentula* Ehr., *Gyrosigma spencerii* (Quekett) Grif. *et* Henfr., *Mastogloia braunii* Grun., *Navicula cincta* (Ehr.) Ralfs, *N.ramosissima* (Ag.) Cl., *Nitzschia closterium* (Ehr.) W.Smith, *N.constricta* (Kütz.) Ralfs, *N.ovalis* Arnott, *N.punctata* (W.Smith) Grun., *N.sigma* (Kütz.) W.Smith, *N.supralitorea* Lange-Bertalot, *Pleurosigma elongatum* W.Smith, *Scoliopleura tumida* (Breb.) Rabenhorst, *Stauroneis wislouchii* Poretzky *et* Anisimova, *Surirella fastuosa* (Ehr.) Kütz., *S.striatula* Turpin, *Synedra tabulata* (Ag.) Kütz.	**Diatoms:** *Amphora angusta* (Greg.) Cl., *Amphora* ssp., *Nitzschia closterium* (Ehr.) W.Smith, *Nitzschia sigma* (Kutz.) W.Smith, *Nitzschia* ssp., *Scoliopleura tumida* (Breb.) Rabenhorst, *Surirella striatula* Turpin	**Photosynthetic bacteria:** *Chlorobium* sp., *Chromatium* sp., *Thiocapsa* sp.
Green algae: *Chlorella* sp.	**Green algae:** *Chlorella* sp.	
Dinoflagellates: *Prorocentrum micans* Ehr., *Gymnodinium* sp.		

In depositor 2, where salinity was higher (11.64°Bé, $\sigma=1.2$), diatom population decreased both in number of species and in abundance (Table 1). The mat was tougher and constituted a layer that prevented from water filtration, favouring the operation of the salt pond (Orti et al. 1984). The population of cyanobacteria population increased. Some species were replaced and others species became dominant (*Microcoleus chthonoplastes*, *Spirulina subsalsa*) (Fig. 8). At greater depth we found broken frustules (Fig 7,9) and phototrophic bacteria but other organisms were not observed.

In heater 7, the gypsum domain, diatoms were not found. There was a well-developed mat and *Microcoleus chthonoplastes* was the dominant organism of the top layer (Fig. 10). At greater depth anoxygenic phototrophic bacteria (*Chlorobium* sp., *Chromatium* sp., *Thiocapsa* sp.) were present as well as remains of *M. chthonoplastes* (Fig 11).

Conclusions

The studied community consisted of microorganisms related to saline environments (Noël 1984; Tomás 1988; Stolz 1990). Motile biraphid diatoms dominated on the superficial sediment. Movement has been described as a strategy to avoid burial, high irradiances or predation (Admiraal 1984). In deep levels of the sediment diatoms appeared to be damaged, broken or partially dissolved. This could be due to environmental conditions such as high salinities (Round et al. 1990).

Salinity has been regarded as a key factor influencing the distribution of diatom species (Admiraal 1984). In the present study a quantitative and qualitative estimation of the composition of diatom populations has been presented in terms of an increasing salinity. Although salinity seemed to be the predominant factor determining diatom species composition, other factors such as temperature may also be important.

The development of a tough microbial mat was also related with salinity. *Microcoleus chthonoplastes* was the dominant cyanobacterium at higher salinity. Mat formation may be promoted at high salinity because of the exclusion of grazing.

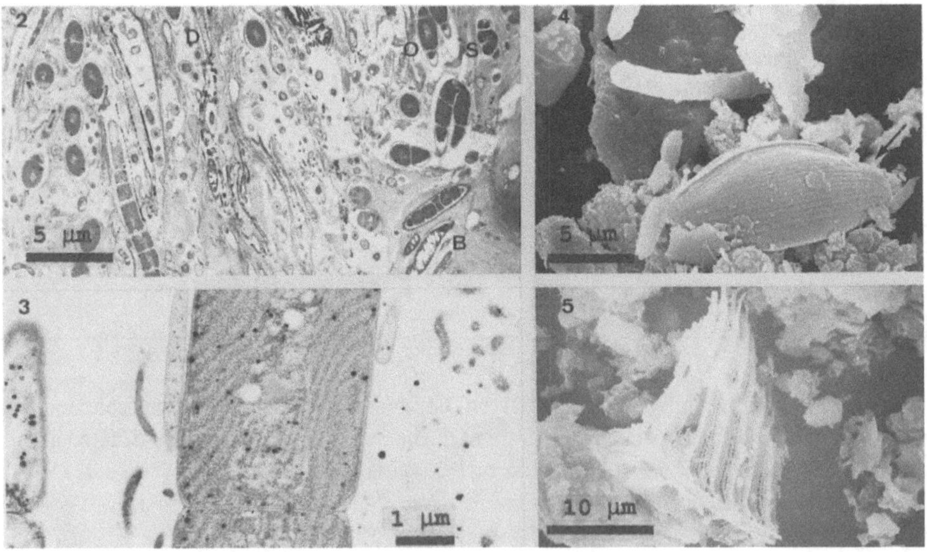

Figs 2-5. Depositor 1; Fig 2, 4. View of the sediment surface. Fig 2. TEM. **s:** *Spirulina* sp., **arrowhead:** diatom frustule, **o:** *Oscillatoria limnetica*, **b:** *Beggiatoa* sp., Fig 4. SEM. **arrowhead:** *O. limnetica*, **arrow:** *Amphora acutiuscula*; Fig 3, 5. Sediment at 5-7mm depth. Fig 3. TEM. *Microcoleus chthonoplastes*; Fig 5. SEM. *Surirella striatula* in an advanced stage of degradation.

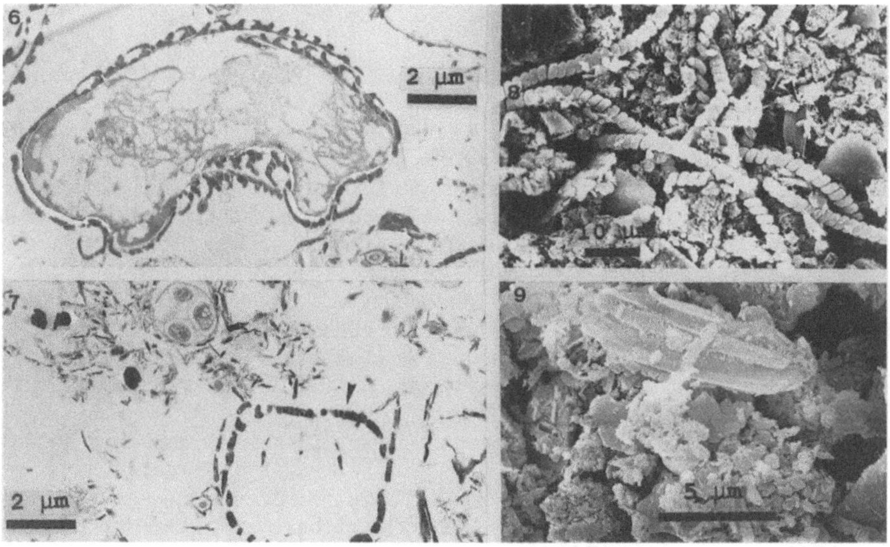

Figs 6-9. Deposit 2; Fig 6, 8. View of the sediment surface. Fig 6. TEM. *Amphora* sp., Fig 8. SEM. *Spirulina subsalsa* filaments, and arrowhead: diatom; Fig 7,9. Sediment at a depth of 7-8 mm. Fig 7. TEM. **arrowhead:** Cross section of a *Nitzschia* sp. empty frustule; Fig 9. *Amphora acutiuscula* in an advanced stage of degradation.

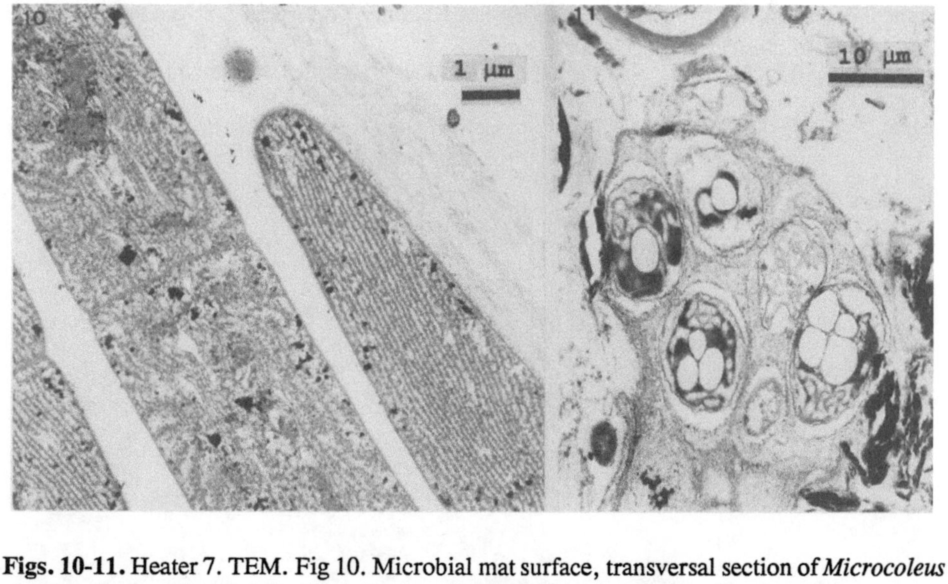

Figs. 10-11. Heater 7. TEM. Fig 10. Microbial mat surface, transversal section of *Microcoleus chthonoplastes* ;Fig 11. Microbial mat at a depth of 12 mm, purple sulfur bacteria.

References

Admiraal W (1984) The ecology of estuarine sediment-inhabiting diatoms. Progress Phycol Res 3:269-322

Barbé A, Grimalt JO, Pueyo JJ, Albaiges J (1990) Characterization of model evaporitic environments through the study of lipid components. Org. Geochem. 16:815-828

Des Marais DJ, Cohen Y, Nguyen H, Cheatham M, Cheatham T, Munoz E (1989) Carbon isotopic trends in the hypersaline ponds and microbial mats at Guerrero Negro, Baja California Sur, Mexico: Implications for Precambrian stromatolites. In: Cohen Y, Rosenberg E (eds) Microbial mats: Physiological Ecology of Benthic Microbial Communities. American Society for Microbiology, Washington DC, pp 191-203

Guerrero R, Pedros-Alio, Esteve I, Mas J (1987) Communities of phototrophic bacteria in lakes of the Spanish Mediterranean region. Acta Acad. Aboensis 47:125-151

Hernández-Mariné M (1992) A simple way to encapsulate small samples for processing for TEM. J Microscopy 168:203-206

Javor BJ, Castenholz RW (1981) Laminated microbial mats, Laguna Guerrero Negro, Mexico. Geomicrobiol. J. 2:237-273

Jørgensen BB, Revsbech NP, Cohen Y (1983) Photosynthesis and structure of benthic microbial mats: microelectrode and SEM studies of four cyanobacterial communities. Limnol. Oceanogr. 28:1075-1093

Noël D (1984) Les diatomées des saumures et des sédiments de surface du Salin de Bras del Port (Santa Pola province d'Alicante, Espagne) Rev. Inv. Geol. 38/39:79-107

Ortí F, Pueyo JJ, Truc G (1984) Las salinas marítimas de Santa Pola (Alicante, España). Breve inroducción al estudio de un medio natural controlado de sedimentación evaporítica somera. Rev. Inv. Geol. 38/39:9-29

Renberg I (1990) A procedure for preparating large sets of diatom slides from sediment cores. J Paleolimnol 4:87-90

Round FE, Crawford RM, Mann DG (1990) The Diatoms. Cambridge University Press (eds), Cambridge, pp 747

Stolz JF (1990) Distribution of phototrophic microbes in the flat laminated microbial mat at Laguna Figueroa, Baja California, Mexico. BioSystems 23:345-357

Tomás X (1988) Diatomeas de las aguas epicontinentales saladas del litoral mediterráneo de la Penísula Ibérica. Tesis Doctoral Univ. Barcelona, pp 714

Microbiological mediation of sediment structure and behaviour

David M. Paterson

Gatty Marine Laboratory, Dept of Biological and Medical Sciences, University of St Andrews St Andrews Fife, KY16 8LB, Scotland

Introduction

Almost all natural sediment deposits support the growth of microbial assemblages. On occasions these assemblages become sufficiently developed to be recognisable as microbial mats or biofilms but even before the development of such extensive structure, microbial colonisation can influence the microstructure of the bed and the critical erosion threshold for the resuspension of particles. The entrainment of particles is a critical event in transport of sediments and as yet there is still insufficient knowledge to accurately predict the response of natural sediments to hydrodynamic stress. Early attempts to predict sediment behaviour were for the most part based on the relationship between hydrodynamic stress and particle size (see Miller et al. 1977). This type of relationships, such as the Shield's curve, have proved to be reasonably reliable for abiotic well-sorted sandy sediments. However, the erosional behaviour of sediment mixtures, small cohesive sediments and, to a large extent, natural sediments *in situ* defies adequate description. Many workers now recognise that predictions based purely on physical parameters are insufficient to explain even the erosional behaviour of non-cohesive sediments *in situ* (Paterson and Daborn 1991; Daborn et al. 1993). It has long been suggested that the biological "status" of the sediment may influence transport processes (Carter 1933), but there are still relatively few studies that deal directly with this problem. There is increasing evidence in the literature that as Montague (1986) stated:

> *"The physical and chemical properties of sediments that contain a significant biotic community are radically different from those of abiotic sediments".*

It is, however, problematic to assess with accuracy the influence of the varied microbial communities that inhabit sediment on the response of individual particles or the sediment bed to erosive stress. The biological stabilisation of a bed (biogenic stabilisation) can be defined as "*an increase in the threshold stress required to initiate the erosion of a given sediment caused as a direct or indirect result of biotic activity*". This contribution will briefly review the concept of biogenic stabilisation and the way in which it has been studied.

NATO ASI Series, Vol. G 35
Microbial Mats
Edited by Lucas J. Stal and Pierre Caumette

The measurement of sediment erodibility

Examination of the natural biotic mediation of sediment response to erosive stress has been hindered by the difficulty in making measurements of sediment stability. Many different approaches have been tried, both direct measures of sediment erodibility in flow channels (eg flumes systems) and of parameters thought to be correlated with the erodibility of the bed (surrogate methods).

Flume studies

Flumes vary greatly in design (Nowell and Jumars 1987) from small circular flumes of less than 1 m to linear channels of over 30 m in length. The broad definition of "flume" employed here includes stirring devices such as those used by Holland et al. (1974) and the recently developed "erosion bell" (Williamson and Ockenden 1993). Each flume has individual characteristics determined by its dimensions and design. The hydraulic conditions must be carefully characterised before realistic comparisons can be made between systems. There are three basic approaches to the use of flumes for the determination of biological effects on sediment erosion: laboratory flumes using prepared or resettled sediments; laboratory flumes using cores of natural sediments taken from the field; and flumes designed for deployment *in situ*.

Laboratory studies on prepared sediments

Among the advantages of fully-controlled laboratory systems are that test sediments can be characterised, microbial growth manipulated, and control systems more easily established (e.g. Dade et al. 1990). Most work has been done with non-cohesive sandy sediments possibly because more is known about their erosion and transport. Glass beads are convenient substitutes for fine sediments and the influence of grain size can be examined with sorted beads (Rhoads et al. 1978). Cohesive sediments are more difficult to examine because of the problems of compaction, cohesion and the determination of incipient scour. However, some workers have used clay mineral sediments such as kaolinite as analogs of natural sediments (Holland et al. 1974; Parchure 1984). The major disadvantage of these systems is the difficulty in relating the measurements obtained to natural conditions. While acting as an important stage in the understanding of the processes and mechanisms of biogenic stabilisation extrapolation to field conditions must be tested *in situ*.

Laboratory studies on natural sediments

Sediment cores taken from the field are often used to give a measure of the erodibility of natural sediments (Grant and Gust 1987; Daborn et al. 1993). The method of obtaining and treating samples and the delay before testing is critical. Microbial assemblages can respond within minutes or even seconds to environmental change and disturbance may lead to behavioural responses from infauna such as the reconstruction or clearing of burrows. Such changes cannot be avoided but may be minimised by careful treatment of the cores after sampling (eg. cooling). Conditions within the flume should adhere closely to environmental conditions since temperature, salinity and pH all affect the erosion threshold of a particular sediment (Montague 1986). Appropriate control systems can be made for non-cohesive sediments by cleaning the natural sediment grains of organic constituents (Manzenrieder 1983; Dade et al. 1990). Control systems for cohesive sediments are more complex because the surface attraction between cohesive particles has a considerable influence. Particle surface charge is mediated by the physico-chemical micro-environment around the particle which responds rapidly to any chemical or physical change.

In situ determination of erosive stress

The use of *in situ* flumes is becoming more widespread (Black 1989) and different systems have been deployed specifically to measure biological influences on sediment stability (Manzenrieder 1983; Nowell et al. 1987; Fonseca 1989; Paterson 1989; Amos et al. 1992). Designs vary from large and complex systems to small portable devices (Paterson 1989). The latter allows many measurements within a single tidal cycle (Paterson *et al.* 1990), although scaling problems exist when comparing results (Williamson and Ockenden 1993). Selection of appropriate equipment very much depends on the objective of the study but the most important advantage of *in situ* devices is the measurement of sediments under almost natural conditions. This is of paramount importance when biological influences on sediment erosion are under investigation.

Surrogate methods

Fall cone devices (Hansbo 1957) and shear vanes (Kravitz 1970) are the most common surrogate methods used to give a measure of the strength or "stability" of sediments. These devices do not measure the erodibility of the sediment and the relationship of any single parameter to actual sediment resuspension has yet to be fully substantiated, especially in natural sediments. The reliability of these techniques in terms of variation and repeatability has also been questioned, especially when applied to weak, saturated cohesive sediments (Kravitz 1970).

Influence of mats and biofilms on sediment fabric

The expected packing structures of both sandy (non-cohesive) and muddy (cohesive) sediment can be found in standard sedimentology literature (e.g., Osipov and Solokov 1978). Few of these texts discuss the relative influence of microbial colonisation on early sediment fabric and often the pre-treatment of sediment for SEM or TEM analysis removes or distorts organic material (Chang and Rittman 1986). The analysis of microbial mat or biofilm structure tends to be more sophisticated (Oppenheim and Paterson 1990; Wachendorfer and Krumbein 1991; Stolz, this volume) but these techniques are not generally applied in sedimentological studies and, indeed, are not appropriate in all cases. It is certain, however, that variations in the void ratio of non-cohesive sediments (Fig. 1) or the open, card house structure expected of unconsolidated cohesive sediments (Osipov and Sokolov 1978) can be influenced by microbial activity (Fig. 2).

Both microbes and sediment infauna are capable of producing copious amounts of polymeric material and the pore-void ratio of sediments deposited under identical hydrodynamic conditions but under different biological influences can be quite different. "Void space" may be largely or entirely filled by microbial cells and polymeric materials of various origins (Figure 1c). This infilling of void space has a fundamental effect on the flux of material through the sediment since diffusion rates through a mucilaginous matrix may be reduced by up to 50% (Jørgensen, this volume).

Mechanisms of biogenic stabilisation

Several descriptions of the mechanisms of biogenic stabilisation have been reported (Fürböhter 1983; Montague 1986; Heinzelmann and Wallisch 1991). These can be broadly classified into flow effects, network effects, boundary layer effects, matrix effects and cohesive effects (Table 1).

Flow effects

While not strictly a mechanism of stabilisation *per se*, flow effects have a significant influence on the likelihood of sediment erosion. Organisms or structures that reduce water flow above the sediment bed also reduce the stress at the sediment surface. The more obvious examples are not microbial (e.g., sea grasses) but less noticeable forms also exist including the green alga *Vaucheria* which forms a carpet-like mat close to the sediment surface (Fig. 2d and e).

Network effects

Filamentous microalgae and bacteria can form extensive networks among sediment particles decreasing sediment erosion (Neumann et al. 1970; Fürböhter 1983). Sediment inhabited by the green alga *Vaucheria* is up to three times more difficult to erode than uncolonised sediment at a similar intertidal level (up to 13 Nm^{-2}: George, unpublished data). The green filamentous alga *Enteromorpha* can also form effective stabilising mats that withstand currents up to four times greater than control sediments (Neumann et al. 1970). Cyanobacterial filaments often form a dense network of strands among the surficial sediment (Fig. 1f).

Table 1. General mechanisms of biogenic stabilisation. It should be noted that mechanisms of stabilisation are not mutually exclusive and specific populations may influence the response of sediments to erosive stress in a number of ways.

Mechanism	Example organism	Selected references
Resistance to flow above the bed	Mangroves, sea grasses algae Tube forming infauna	Fonseca 1989 Scoffin 1970 Montague 1986
Physical network	Filamentous forms eg cyanobacteria, microalgae	Neumann et al. 1970 Paterson and Daborn 1991 Vos et al. 1988
Boundary layer roughness		
a) smoothing of surface topography ,	Infauna	Nowell et al. 1981
b) reduction of roughness elements by infilling	Microbiota: bacteria, cyanobacteria, algae	Manzenrieder 1983 Heinzelmann and Wallisch 1991
Particle cohesion	Organisms associated with particles producing extracellular polymeric substances eg bacteria, fungi algae.	Vos et al., 1988 Paterson 1990 Dade *et al.* 1990
Organic matrix	Organisms producing extracellular polymeric substances which are not immediately hydrolysed including infauna, bacteria, algae (particularly diatoms)	Manzenrieder 1983 Paterson 1990 Underwood and Paterson 1993

Boundary layer effects

The shear stress at the sediment boundary is influenced by the roughness of the bed surface. Therefore, a reduction in roughness reduces shear stress for a given flow. This can be achieved in two ways: the surface can be smoothed by organisms passing over it and "resculpting" the surface topography; or by covering the natural surface with a network of filaments (Neumann et al. 1970), a smooth mucilaginous layer (Vos et al. 1988; Neumann et al. 1970) or a matrix (Paterson et al. 1986).

Matrix effects

Binding by extra cellular polymeric substances (EPS) seems to be most effective when sufficient material is present to form a coherent matrix among the surface sediment particles (Vos et al. 1988; Paterson 1989). The presence of this matrix can be demonstrated by histochemical staining and by SEM. However, SEM micrographs must be interpreted with care since most standard sample preparation techniques for SEM (e.g., freeze drying or critical point drying) require dehydration of the sample which greatly changes the structure of a hydrated mucilaginous matrix. The threads or strings of EPS depicted in many SEM micrographs are only the remnants of a more extensive mucilaginous matrix which can be visualised by low-temperature SEM (Paterson et al. 1986; Jeffree and Read 1990).

Cohesive effects

The "cohesive" effect is likely to be the most common biological effect on sediment transport. It is caused by the coating of particles with any organic material that has a tendency to adhere. The origin of such organic coverings is highly variable (see Decho 1990). Increased cohesion has often been cited as a cause of increased sediment stability, but in reality few attempts have been made to examine the mechanism of such microbial stabilisation closely although more work has been done with aggregates in suspension (Eisma et al. 1990).

Combined effects

The stabilising effect of a given population may be brought about by a combination of the above mechanisms. For example, filamentous microalgae and cyanobacteria are capable of forming networks, but some also exude mucilage (Decho 1990). The binding effect of diatoms has been particularly studied and two mechanisms have been demonstrated that reflect the ecology of the species involved. Those cells which become attached to sand grains (epipsammic) increase the cohesion of grains by their organic secretions and also by attaching to more than one grain at a time (Vos et al. 1988). The motile (epipelic) forms, which dominate cohesive sediments, move by extruding mucilage that helps to bind the sediments into an organic matrix (Paterson 1989; Paterson et al. 1990).

Fig. 1. Low-temperature scanning electron micrographs of natural non-cohesive sediments. **A.** Apparently "clean" non-cohesive sandy sediment. Bar marker = 200 μm. **B.** On close inspection inter-grain boundaries may still support organic attachments. Sand grain supported by organic bond (O). Note surface diatoms. Bar marker =50 μm. **C.** Fracture face of sandy sediment core. Sediment fabric varies with depth in relation to microbial development. Bar marker = 500 μm. **D.** Particles below interface region. Void spaces are still apparent although reduced by organisms and polymeric substances. Bar marker = 200 μm. **E.** Surface region. Inter-particle void spaces completely infilled by organic matrix. Bar marker = 200 μm. **F.** The surface of a cyanobacteria/diatom assemblage. The biogenic nature of the interface and the network formation by cyanobacterial filaments is apparent. Bar marker = 10 μm.

Measurements of biogenic stabilisation

Since no standard approach to the measurement of biogenic stabilisation has been widely adopted, either within the laboratory or under field conditions, much of the available data are not directly comparable. Results are reported as threshold values for erosion in terms of bed shear stress (Manzenrieder 1983), critical erosion velocity (Grant and Gust 1987), resuspension of sediments (Parchure 1984), rpm of mixing systems (Holland et al. 1974) and other variables.

Table 2. The influence of biota on the response of sediment to induced erosive stress. Information from the literature expressed in terms of the relative increase in bed stability over control measurements, expressed as a percentage of control values. Experimental data cannot be directly compared since units are not standardised.

Study	Date	Substratum	Biota	Relative binding (as % increase over control)
Non-cohesive				
Neumann et al.	1970	Non-cohesive	Algae/cyanobacteria	500
Holland et al.	1974	Non-cohesive	Diatoms	(100)
Manzenrieder	1983	Non-cohesive	Bacteria/algae	300-770
Grant and Gust	1987	Non-cohesive	Bacteria/cyanobacteria	25
Vos et al.	1988	Non-cohesive	Diatoms	100
Dade et al.	1990	Non-cohesive	Bacteria	200
Paterson	1990	Non-cohesive	Diatoms	400
Madsen et al.	1993	Non-cohesive	Diatoms/bacteria	300
Cohesive				
Holland et al.	1974	Cohesive	Diatoms	(>350)
Rhoads et al.	1978	Cohesive	Microbes	45
Parchure	1984	Cohesive	Microbes	300
Paulic	1986	Cohesive	Microbes (autotrophes)	200
Black	1991	Cohesive	Diatoms/bacteria	500
Underwood and Paterson	1993	Cohesive	Diatoms/bacteria	200-300

( ) best estimate based on available data

Comparison can be made on the basis of ratio measurements such as the "Biological Stabilisation Coefficient" which is an expression of the measured critical erosion threshold over that predicted on the basis of grain size (Manzenrieder 1983) or against a "control" system, but these data should be viewed with caution because the scales are often ordinal.

Studies of biogenic stabilisation

Early studies of biogenic stabilisation provided data supporting the hypothesis that biota enhance sediment stability (Table 2). More recently the emphasis has switched towards a more mechanistic approach and the desire to provide a predictive capability: the search for a biological parameter to describe sediment stability.

One of the most comprehensive *in situ* flume studies (Neumann et al. 1970) in the Bight of Abaco, Bahamas began to apply a mechanistic approach. Three basic mat types were identified according to the dominant structural organism (*Cladophoropsis*, *Lyngbya* and *Schizothrix*). Each mat exhibited different responses to erosion stress which were explained as functions of mat structure and the mechanisms of biogenic stabilisation. Control systems were established by chemical treatment of the natural sediments and in comparison to these controls, the threshold for erosion of some mats was increased 5-fold. Holland et al. (1974) cited this as one of the first quantitative investigations to support previous speculation (e.g., Carter 1933) on the importance of biota in sediment transport. Holland and co-workers adopted an autecological approach using selected diatom species, and found that the efficiency of the binding was qualitatively related to the ability of the diatom species to produce mucilage under the culture conditions. Grant and Gust (1987) continued the work in the field by demonstrating a correlation between the photopigment content of mats of purple sulphur bacteria and the critical erosion velocity of natural cores measured in a laboratory flume.

They stressed that photopigments only indicate the biomass of potentially binding organisms and are not a direct measure of sediment stabilisation. They also suggested a more direct assay of EPS would be a better indicator of the binding mechanism and, in support, showed a correlation between colloidal carbohydrate and chl *a* within samples. More recently Dade and co-workers (Dade et al. 1990) successfully isolated EPS from a marine bacterium (*Pseudomonas* sp.) and altered the threshold for the erosion of a test sediment by simply adding the bacterial EPS to sterile sediment. However, the extent of the binding was not as great for the natural bacterial populations.

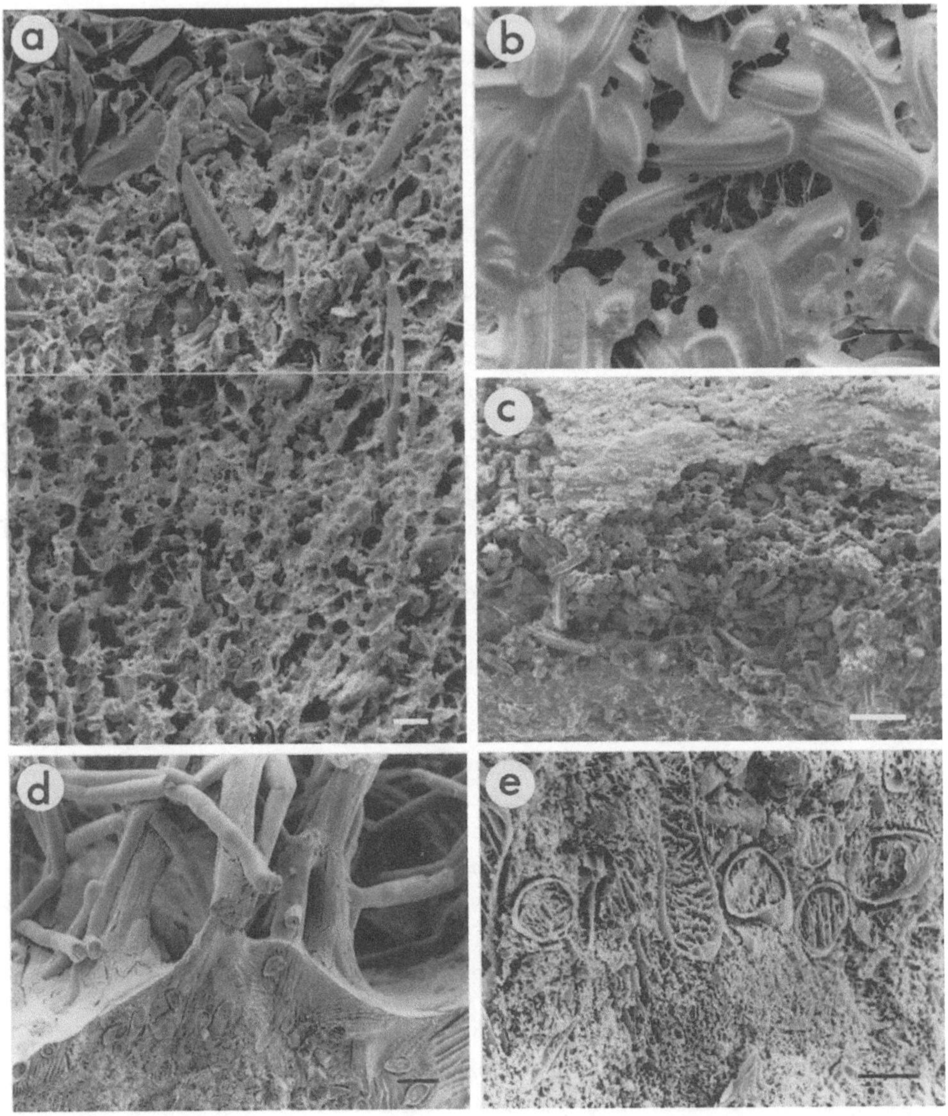

Fig. 2. Low-temperature scanning electron micrographs of natural cohesive sediments. **A.** Fracture face of a cohesive sediment core. Open porous fabric under a surface biofilm. Bar marker = 20 μm. **B.** Surface of cohesive sediment biofilm with. diatom cells (*Nitzschia epithemiodes*) associated in a mucilaginous matrix. Bar marker = 20 μm. **C.** Fracture of surface biofilm dominated by *N. epithemiodes* from cohesive salt marsh sediment. Bar marker = 50 μm. Organisms may influence stability by several mechanisms. **D.** The green alga, *Vaucheria sp.*, forms a forest of filaments above the sediment surface. Bar marker = 100 μm. **E.** In addition to the surface "felt", filaments ramify through the sediment considerably strengthening the matrix. Bar marker = 50 μm.

Conclusions

Sediment-inhabiting microbial assemblages have been shown to increase the stability of intertidal and subtidal sediments, both cohesive and non-cohesive. Effects have been measured in the laboratory and in the field but the importance of seasonally varying processes of biogenic stabilisation on the annual sediment budget have rarely been considered (Frostick and McCave 1979) or fully determined for any ecosystem. The study of biogenic stabilisation would benefit from a more consistent approach to methods of measuring sediment stability. The importance of field studies to determine biological influences is emphasised and the use of multi-disciplinary approaches seems most likely to produce meaningful results. While the search for a biological or biochemical parameter for biogenic stabilisation is understandable it is fraught with difficulty. The mixed nature of microbial assemblages and the variability in the physical nature, turnover and rheology of EPS in sediments may preclude such a simple relationship from being reliable. A more fruitful approach might be based on a "suite" of variables including a component for community structure.

Acknowledgements

I would like to thank NATO for financial support for the advanced workshop and the EC for support towards some of the work reported which was part of the EC MAST programme, project 0027-C.

References

Amos CL, Grant J, Daborn GR and Black K (1992) Sea Carousel- A benthic, annular flume. Est Coastal and Shelf Sci 34: 557-577

Black K (1989): The *in situ* measurement of sediment erodibility: a review. - Unpublished Report submitted to ETSU, Department of Energy, Harwell UK

Black K (1991) The erosion characteristics of cohesive estuarine sediments: some *in situ* experiments and observations. PhD Thesis University of Wales pp 313

Carter N (1933) A comparative study of the algal flora of two salt marshes. Part III. J Ecol 21: 385-403

Chang HT and Rittman BE (1986) Biofilm loss during sample preparation for scanning electron microscopy. Water Res 20(11): 1541-1456

Daborn GR, Amos C, Brylinski HC, Drapeau G, Faas RW, Grant J, Long B, Paterson DM, Perillo GME, Piccolo, MC (1993) An ecological "cascade" effect: Migratory birds affect stability of intertidal sediments. Limnol and Oceanog 38: 225-231

Dade WB, Davis JD, Nichols PD, Nowell ARM, Thistle D, Trexler MB, White DC (1990) Effects of bacterial exopolymer adhesion on the entrainment of sand. Geomicrobiol J 8: 1-16

Decho AW (1990): Microbial exopolymer secretions in ocean environments: Their role(s) in food webs and marine processes. Oceanogr Mar Biol Ann Rev 28: 73-153

Eisma D, Bernard P, Cadee GC, Ittekkot V, Kalf J, Laane R, Martin JM, Mook WK, Put A van, Schuhmacher T (1990) Suspended-matter particle size in some West-European estuaries. Part 1: Particle size distribution. Neth J Sea Res 28: 193-214

Fonseca MS (1989) Sediment stabilisation by *Halophita decipiens* in comparison to other sea grasses. Est Coastal and Shelf Science 29: 501-507

Frostick LE and McCave IN (1979) Seasonal shifts of sediment within an estuary mediated by algal growth. Est Coastal and Mar Sci 9: 569-576

Führböter A. (1983): Über mikrobiologische Einflusse auf den Erosionsbeginn bei Sandwatten. Wasser und Boden 3: 106-116

Grant J , Gust G (1987) Prediction of coastal sediment stability from photopigment content of mats of purple sulphur bacteria. Nature 330: 244-246

Hansbo S (1957) A new approach to the determination of the shear strength of clay by the fall-cone test. Proc R Swed Geotech Inst 14: 1-49

Heinzelmann C, Wallisch S (1991) Benthic settlement and bed erosion. A review. J Hydraulic Res 29: 355-371

Holland AF, Zingmark RG, Dean JM (1974) Quantitative evidence concerning the stabilisation of sediments by marine benthic diatoms. Mar Biol 27: 191-196

Jeffree CE, Read ND (1990) Ambient and Low-Temperature Scanning Electron Microscopy. In: Hall JL, Hawes C (eds) Electron Microscopy of Plant Cells. Academic Press, pp 313 - 413.

Kravitz JM (1970: Repeatability of three instruments used to examine the undrained shear strength of extremely weak saturated cohesive sediments. J Sed Petrol 40: 1026-1037

Manzenrieder H (1983) Retardation of initial erosion under biological effects in sandy tidal flats. -Leichtweiss Inst, Tech University Braunschweig 469-479

Madsen, KN, Nilsen P and Sundback K (1993) The influence of benthic microalgae on the stablility of a subtidal sediment. J Exp Mar Biol Ecol. 170: 159-177

Miller MC, McCave IN, Komar PD (1977) Threshold of sediment motion under unidirectional currents. Sedimentology 24: 507-527

Montague CL (1986) Influence of biota on erodibility of sediments. In: Mehta AJ (ed.) Lecture notes on coastal and estuarine studies. 14: 251-269

Neumann AC, Gebelein CD and Scoffin TP (1970) The composition, structure and erodibility of subtidal mats, Abaco, Bahamas. J Sed Petrol 40: 274-297

Nowell AR, Jumars PA (1987) Flumes: Theoretical and experimental considerations for simulation of benthic environments. Oceanogr Mar Biol Ann Rev 25: 91-112

Nowell AR, Jumars PA, Eckman JE (1981) Effects of biological activity on the entrainment of marine sediments. Marine Geol 42: 133-153

Oppenheim DR, Paterson DM (1990) The fine structure of an algal mat from a freshwater maritime Antarctic lake. Can J Bot 68: 174-183

Osipov VI and Solokov VN (1978) Microstructure of recent clay sediments examined by scanning electron microscopy. In: Whalley WB (ed) Scanning electron microscopy in the study of sediments. Geol abs Ltd

Parchure TM (1984) Erosional behaviour of deposited cohesive sediments. Ph.D. Thesis, University of Florida

Paterson DM (1989) Short-term changes in the erodibility of intertidal cohesive sediments related to the migratory behaviour of epipelic diatoms. Limnol Oceanogr 34: 223-234.

Paterson DM (1990) The influence of epipelic diatoms on the erodibility of an artificial sediment. In: Simola H (ed) Proceedings of the 10th Int Symp on Living and Fossil Diatoms. Joensuu 1988 Koenigstein 345-355

Paterson DM, Daborn GR (1991) Sediment stabilisation by biological action: Significance for coastal engineering. In: Peregrine DH , Loveless JH (eds) Developments in Coastal Engineering . University of Bristol Press, pp 111-119

Paterson DM, Crawford RM, Little C (1986) The structure of benthic diatom assemblages: A preliminary account of the use and evaluation of low-temperature scanning electron microscopy. J Exp Mar Biol Ecol 96: 279-289

Paterson DM, Crawford, RM, Little, C (1990) Sub-aerial exposure and changes in the stability of intertidal estuarine sediments. Est Coastal and Shelf Sci 34: 223-234

Paulic MJ, Montague CL, Mehta AJ (1986) The influence of light on sediment erodibility. In: Shen HW (ed) Third Int Symp River Sedimentation . Univ Mississip pp 1758-1764

Rhoads DC, Yingst JY, Ullman WJ (1978) Seafloor Stability In Central Long Island Sound: Part I. Temporal Changes In Erodibility of Fine-Grained Sediment. In: Wiley ML (ed) Estuarine Interactions. Academic Press, pp 221-244

Scoffin TP (1970) The trapping and binding of subtidal carbonate sediments by marine vegetation in Bimini Lagoon, Bahamas. J Sediment Petrol 40: 249-273

Underwood GJC, Paterson DM (1993) Recovery of intertidal benthic diatoms after biocide treatment and associated sediment dynamics. J mar biol Ass UK 73: 25-45

Vos PC, De Boer PL, Misdrop R (1988) Sediment stabilization by benthic diatoms in intertidal sandy shoals. In: De Boer PL (ed) Tide-influenced Sedimentary Environments and Facies. Reidel, pp 511-526

Wachendorfer V, Krumbein WE (1991) The fluorescent sediment thin section technique: Spatial distribution of microorganisms in North Sea microbial mat systems. Kieler Meeresforschungen Sonderheft 8, Kiel: 381-388

Williamson HJ, Ockenden MC (1993) *In situ* erosion of cohesive sediment. Report to Energy Technology Support Unit. ETSU TID 4112, pp 61

Field and cultivated *Microcoleus chthonoplastes*: The search for clues to its prevalence in marine microbial mats

Lee Prufert-Bebout and Ferran Garcia-Pichel

Max-Planck-Institut für marine Mikrobiologie
Fahrenheitstraße 1, D-28359 Bremen, F.R.G.

Introduction

The cyanobacterial species "*Microcoleus chthonoplastes*" is defined by the presence of many typically oscillatorian filaments with tapered, conically shaped terminal cells within a common sheath both in the botanical (Geitler 1932) and, more recently in the bacteriological literature (Castenholz 1988). The significant and often dominant ecological role of *Microcoleus* species in the formation and stabilization of intertidal and hypersaline microbial mat systems around the world is well recognized. However, it is not known if geographically disjunct populations referred to as "*Microcoleus chthonoplastes*" on the basis of gross morphological traits alone constitute a close genetic unit, or are in fact examples of evolutionary convergence driven by the adaptive advantages of a "microcoleus-like" morphology in benthic marine environments. What adaptive advantages this morphology may confer has been difficult to ascertain since these features, most notably the defining features of bundle formation, and presence of conically-shaped terminal cells, are not observed in currently available cultures.

We therefore undertook to isolate new strains from a variety of environments where *Microcoleus chthonoplastes* is a dominant morphotype, with special emphasis on generating cultures which would maintain the morphological features observed in the field populations. Consequently a systematic, comparative study of *Microcoleus chthonoplastes* from new isolates as well as from their parent field samples was begun. Preliminary observations and results from these analyses, reported here, suggest that *Microcoleus chthonoplastes* is indeed a phenotypically coherent group of oscillatorians. Further comparative investigations of the physiology and genetic relatedness of cultures from a wide range of locations will hopefully shed light on some of the many questions concerning the distribution and abundance of *Microcoleus* species in microbial mat systems worldwide.

NATO ASI Series, Vol. G 35
Microbial Mats
Edited by Lucas J. Stal and Pierre Caumette

Isolation and cultivation

Enrichment cultures were purposefully avoided in our isolation attempts. Intact bundles were removed from mat samples, either directly, or following incubation of a small core of microbial mat in a marine medium (BG11:Rippka et al. 1979 or F/2:Guillard and Ryther 1962) agarose (0.25%) plate. Pre-incubation allowed bundles to glide out from mat cores, thus physically separating themselves from much of the associated sediment-microorganism matrix. Individual bundles were then removed and thoroughly cleaned by dragging them through 1.0% seawater agar plates using a glass hook constructed from a heat pulled pasteur pipet (Figs. 1a and b). All cleaning steps were conducted under a dissecting microscope to insure by visual monitoring that diatoms, other cyanobacterial species and visible bacterial aggregates were not transferred to new medium. Clean bundles were then inoculated onto 0.25% agarose F/2 plates and incubated in a directional light gradient. Gliding single filaments and bundles were collected from the plate region distal to the inoculation point and transferred to fresh medium. All populations of *Microcoleus chthonoplastes* isolated in this manner thus far, showed retention of bundling in culture, even when started from individual trichomes. We failed to obtain axenic isolates of these organisms with this method. Potentially axenic transfers carried out in this manner were repeatedly observed to die off within 2 to 3 doublings after transfer.

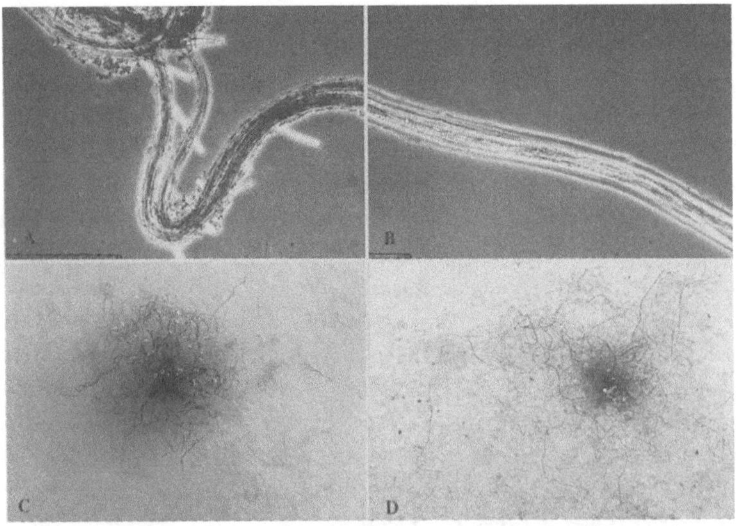

Fig. 1a. and b Photomicrographs of *Microcoleus chthonoplastes* bundles from an intertidal mat in North Carolina, USA **a:** taken directly from mat **b:** following cleaning procedure **1c:** MEL-1 strain grown in soft (0.1%) agarose medium in petri-plate. Soft agarose media allow *Microcoleus* bundles to remain neutrally buoyant which facilitates extensive bundle formation and retention and **d:** grown in liquid medium on sterile sand. Growth on a "hard" (0.25%) agarose plate is shown in Fig 2.

We cultivated *M. chthonoplastes* isolates in a variety of different growth media and/or conditions. Our F/2 medium (Guillard and Ryther 1962) is made from GF/F filtered offshore North Sea water. The trace metal addition is omitted, however an Fe-EDTA solution (final concentrations $FeCl_3$ 4.0 μM; Disodium EDTA 20.0 μM) is added. All isolation and cultivation trials, both in diffuse and unidirectional light fields, were carried out under light intensities ca. 10-20 μmol photons (400-700 nm)$\cdot m^{-2}$ s^{-1} provided by standard warm white indoor fluorescent fixtures.

Maintenance of bundling morphology in culture from a wide range of environments is possible. Neither a directional light field, nor dessication, nor solid substrate is required for formation of bundles in culture. A semisolid medium (0.1% agarose) adequately supported the formation, growth and movement of bundles (Fig 1). Cultivation in liquid medium in glass vessels or on nucleopore filters supported bundle formation initially, but eventually filaments clumped together to form a spherical mass, with apparent cessation of growth. Liquid medium overlaying sterile sand also showed good bundle maintenance.

Characteristics of strains: observations and analyses

We report here on the characteristics observed thus far of both field and cultivated populations from 7 different sites [Mellum Is., FRG (Strain MEL-1); Nordeney Is., FRG (NOR-1); Ebre Delta, Spain (EBD-1); Rachel Carson Estuarine Reserve, NC, USA (NCR-1); Pond 5 Guerrero Negro Salterns, BCS, Mexico (GN5-1); Ojo de Liebre Lagoon, BCS, Mexico (GNL-1); Solar Lake, Sinai Peninsula, Egypt (SOL-1)]. Cell widths ranged from 3.1 to 5.6 μM. All strains had conically shaped terminal cells. Different bundle morphologies were observed in different strains. Isolates MEL-1 and GNL-1 both formed uniformly thick rope type bundles. Isolates NCR-1 and EBR-1 showed typically shorter and thinner bundles with tapering ends. The isolate from Solar Lake, SOL-1, showed on average fewer trichomes per bundle but with a nearly constant tube-sheath width throughout the bundle (Fig 2). Unlike in most Oscillatorians, gliding motility in these strains always occurred without concomitant rotation of trichomes. Noticeable production of geosmin occurred in the MEL-1 strain only. Hormogonia were seen moving in all directions out and away from the ends of bundles in the strain EBR-1. Sheaths were generally well-defined tubular structures, with the exception of the SOL-1 strain sheath which exhibited a more confluent consistency.

Visible/UV *in vivo* absorption spectra were recorded using a filter technique (after Amesz et al. 1961) miniaturized to allow spectral analysis of a few intact live trichomes. Measurements were made using a Shimadzu model UV-160A UV/VIS recording spectrophotometer. In addition to phycocyanin, all strains and field samples contained the more rarely occurring phycoerythrocyanin (a phycobiliprotein containing both phycocyanobilin and cryptoviridin chromophores). Phycoerythrocyanin was thought to be restricted to some subgroups of heterocystous cyanobacteria (Bryant 1982). All strains tested so far contain also a mycosporin-like compound absorbing strongly at 332 nm, similar to compound V (*sensu*

Garcia-Pichel and Castenholz 1993). This compound may be involved in protection from UV radiation by acting as a sunscreen.

Figure 2d. shows the typical response of *M. chthonoplastes* bundles to a unidirectional light gradient. Bundles from strain MEL-1 were inoculated randomly (at top of photo) with respect to the light gradient . After several days a distinct alignment of bundles away from and parallel to the incident light was observed. The ability to orient in this manner was found in all isolates. Alignment appears to result from both new growth and movement occurring in a highly directed manner away from incident light, rather than from random phototactic movement as commonly found in many other cyanobacterial species. Motility of the NCR-1 strain was examined in more detail (see Ramsing and Prufert-Bebout this volume) and showed that a variety of movement strategies may be utilized by this organism. A radical change in direction appears to neccesitate the formation of new sheath tubes. The threshold light intensity for bundle alignment and growth away from the light source is rather low, between 15 and 20 μmol photons m^{-2} s^{-1}. Below this threshold random growth occurs. Alignment towards the light source has not yet been documented.

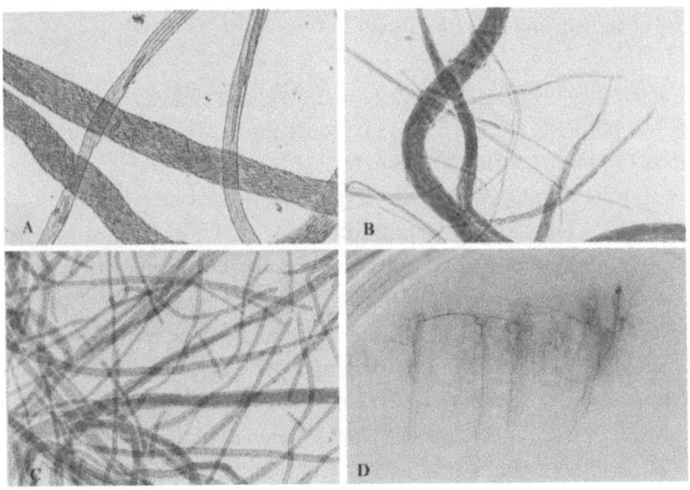

Fig. 2. Photomicrographs of isolated *Microcoleus chthonoplastes* strains from **a:** Mellum Island F.R.G., (MEL-1) **b:** Rachel Carson Estuarine Reserve, North Carolina, U.S.A., (NCR-1) **c:** Solar Lake Sinai, Sinai Peninsula, Egypt, (SOL-1) **d:** Photoalignment of *Microcoleus chthonoplastes* (MEL-1) bundles in response to unidirectional light.

Catalase activity was measured as the rate of O_2 production in response to the addition of two drops of 0.03% H_2O_2 solution to agar immobilized *M. chthonoplastes* bundles using an oxygen microelectrode with the sensing tip inserted within the bundles. Bundles were removed from mats by pulling with tweezers. Some were measured directly while others were subjected to the cleaning procedure already described before measurement. Data were normalized to bundle width. Bundles and associated matrix community showed activity ranging from 0.020 to 0.800 μmol H_2O_2 removed $\mu m^{-3} s^{-1}$, whereas cleaned bundles showed activies between 0.00 to 0.005 μmol H_2O_2 removed $\mu m^{-3} s^{-1}$. Considerable catalalase activity could be removed by simply cleaning the bundles thoroughly indicating that most of the catalase activity *in situ,* if significant, was provided by the bacteria around the sheaths. A possible dependence of *M. chthonoplastes* on its associated sheath bacterial flora for H_2O_2 detoxification may help explain failure to obtain axenic isolates. The absence of catalase is known in other cyanobacteria but remains a rare character, which may prove to be of taxonomic importance.

Discussion

The procedure of selecting and cleaning bundles (in effect removing microalgal and eucaryotic competition and/or grazing pressure), as compared to enrichment techniques, yields a higher success rate of isolation for unialgal populations of *Microcoleus chthonoplastes* with natural-type morphologies. The slow growth rates and inability to spread through the culture medium without some form of physical support that we have observed in our cultures suggest that traditional enrichment cultures tend to select for fast-growing, non-adherent oscillatorians, some of which strongly resemble non-bundling forms of *Microcoleus.* Having demonstrated growth of bundle forming *M. chthonoplastes* in relatively simple, non-defined media and conditions, we hypothesize that *Microcoleus spp.* dominance in complex microbial mats may in part be due to their superior tolerance and metabolic flexibility in those complex environments rather than to a requirement for particular aspects of those regimes (i.e. fluctuating pH, DIC, and H_2S levels). Further testing of this possibility awaits production of axenic bundling isolates, and tests for growth on defined media. Adaptation to the light environment in microbial mats may well also be a major contributing factor to the success of this group, and is currently under investigation.

Of the analyses done thus far, populations from widely disparate locations are remarkably similar in certain unusual properties including: low level or lack of catalase activity, presence of relatively rare phycoerythrocyanin, presence of mycosporin compounds with absorption maxima at the same wavelength, photoalignment, and lack of rotation during movement. Other traits however (cell dimensions, bundle morphology, sheath consistency, presence of hormogonia, presence of geosmin) show noticeable variation between strains. Work is underway to determine the molecular phylogeny of these isolates to provide a more accurate estimate of their degree of relatedness utilizing a newly developed denaturing gradient gel electrophoresis (DGGE) technique for analysis of polymerase chain reaction-amplified genes coding for 16S rRNA (Muyzer et al. 1993).

Thus, there seems to be some validity in regarding "Microcoleus chthonoplastes" as a phylogenetic and functional unit within microbial mat ecosystems. The task at hand is to explore the physiological features that distinguish this group from other cyanobacteria in the search for clues to its preponderance in microbial mats. The biological meaning of photoalignment, of their radial thylakoid arrangement (D'Amelio et al. 1989; De Wit et al. this volume), motility responses (Ramsing and Prufert-Bebout, this volume) and their metabolic responses to hydrogen sulfide are all aspects worthy of further study.

Acknowledgements

We gratefully acknowledge R. Castenholz, M. Villbrandt, B.B. Jørgensen, J. Mir, B. Bebout, U. Karsten and others for providing us with field materials from various locations. We are thankful to Y. Cohen for introducing us to the observations of G. Zavarzin concerning the lack of catalase activity in *Microcoleus spp.*.

References

Amesz J, Duysens LN, Brandt M (1961) Methods for measuring and correcting the absorption spectrum of scattering suspensions. J Theoret Biol 1: 59-74

Bryant DA (1982) Phycoerythrin and phycocyanin: properties and occurrence in cyanobacteria. J Gen Microbiol 128: 835-844

Castenholz RW, Waterbury JB (1988) Oxygenic photosynthetic bacteria. In: Bergey's Manual of Systematic Bacteriology. Vol 3, Williams and Wilkins, Baltimore Maryland, pp 1710-1799

D'Amelio ED, Cohen Y, Des Marais D (1989) Comparative Functional Ultrastructure of two hypersaline submerged cyanobacterial mats: Guerrero Negro, Baja California Sur, Mexico, and Solar Lake, Sinai, Egypt. In: Cohen Y, Rosenberg E (eds) Microbial mats: Physiological Ecology of Benthic Microbial Communities. Am Soc Microbiol, Washington, pp 97-113

Garcia-Pichel F, Castenholz RW (1993) Occurrence of UV-absorbing, mycosporine-like compounds among cyanobacterial isolates and an estimate of their screening capacity. Appl Environ Microbiol 59: 163-169

Geitler L (1932) Cyanophyceae. In: Kolkwitz (ed), Kryptogamenflora von Deutschland, Osterreich und der Schweiz. Akadamische Verlagsgesellschaft, Leipzig, 14: 1-1196

Guillard RRL, Ryther JH (1962) Studies of marine planktonic diatoms. I. *Cyclotella nana* Hustedt and *Detonella confervacea* (Cleve) Gran. Can J Microbiol 8: 229-239

Muyzer G, De Waal EC, Uitterlinden AG (1993) Profiling of complex microbial populations by denaturing gradient gel electrophoresis analysis of polymerase chain reaction-amplified genes coding for 16S rRNA. Appl Environ Microbiol 59: 695-700

Rippka R, Deruelles J, Waterbury JB, Herdman M, Stanier RY (1979) Generic Assignments, strain histories and properties of pure cultures of cyanobacteria. J Gen Microbiol 111: 1-61

The sensitivity for salinity increase in the drought resistant cyanobacterium *Crinalium epipsammum* SAB 22.89

Ben de Winder and Luuc R. Mur

*Laboratory for Microbiology, University of Amsterdam,
Nieuwe Achtergracht 127, 1018 WS Amsterdam, The Netherlands*

Introduction

In the coastal dunes of the Netherlands phototrophic organisms are colonising the sand surface (De Winder et al. 1989a). In previous studies the response of the three representative organisms, the cyanobacteria *Crinalium epipsammum* and *Tychonema sp.* and the green alga *Klebsormidium flaccidum*, to re- and dehydration (De Winder et al. 1989b, 1990b) was investigated. Equilibration with water-saturated air did only enable photosynthesis when this was granted by the water-retention characteristics of the physical environment of the samples. Rehydration enabled the recovery of photosynthesis of desiccated samples on substrata with a good water retention only. Photosynthetic activity is thus ascertained at environmental conditions of non-liquid water when this is enabled by a "compatible matrix". The cyanobacterial strains showed a recovery of photosynthesis instantaneously. In contrast, rewetted cells of the green alga showed a recovery of photosynthesis only after a time-lag. During controlled dehydration *K. flaccidum* and to a lesser extent *C. epipsammum* proved to be able to maintain full carbon-fixing activity until the water content of its surroundings approached zero. *Tychonema* sp. appeared to reduce its activity in response to becoming dehydrated. The experiments described substantiated the view on the ecology of these drought tolerant organisms. The cyanobacteria as the initial colonizers inhabiting the barren sand are able to react quickly to changing water availability, by an on and off switching of their metabolism. As a result of the growth of the cyanobacteria the water retention of the top soil is improved to conditions were the green alga *K. flaccidum* is adapted.

The organisms which are living in the upper millimetres of the soil have to withstand conditions of changing water availabiliy and a consequent change in ion concentrations because of evaporation and rehydration. For adaptation to salinity increase two basic mechanisms have been suggested: (i) a low permeability of the plasmamembrane for Na^+ in combination with an energy demanding process for the extrusion of entering sodium ions. This process decreases the rate of photosynthesis and increases the rate of respiration (Blumwald et al. 1983; Fry et al. 1986) (ii) A temporary increase in cellular K^+ concentration, followed by the synthesis of low molecular organic compounds to cope with the unbalanced osmotic pressure (Reed and Stewart 1988). The Na^+ extrusion mechanism is supposed to be involved both in the short and in the long term response. The synthesis mechanism of osmoregulatory compounds is likely to

NATO ASI Series, Vol. G 35
Microbial Mats
Edited by Lucas J. Stal and Pierre Caumette

occur in long term exposure to high salinity. It is speculated that especially in the case of terrestrial cyanobacteria, drastic and sudden changes of ionic and osmotic forces will be tackled by a short-term correction mechanism, in order to secure adaptation of the cytoplasm to the outside water-potential.

In this paper we describe the physiological tolerance of the terrestrial cyanobacterium *Crinalium epipsammum* SAB 22.89 (De Winder et al. 1990a). The results presented offer understanding of the strategies for survival of this terrestrial phototrophic organism in the harsh ecological niche of coastal dunes.

Materials and methods

Batch cultures of *C. epipsammum* were grown in a light-dark cycle of 8:16 h, at a photon flux density (PFD) of 40 μmol m^{-2} s^{-1}, continuously sparking with air (approximately 40 l h^{-1}), at pH 8.0-8.5 and at 20 °C (growth rate = 0.029 h^{-1}). Continuous cultures (D=0.011) of *C. epipsammum* were grown at an average PFD of 20 μmol m^{-2} s^{-1}, pH 8.1, otherwise as batch cultures. Growth rates at different salinities were determined by using full strength BG-11 medium (-90 kPa) and 10- (-9 kPa), 5- (-18 kPa) and 2- fold diluted BG-11 (-45 kPa), or BG-11 with 20 (-180 kPa), 100 (-540kPa) and 200 mmol.kg^{-1} NaCl (-990 kPa) respectively. The iron and phosphate additions were in all cases adjusted to half of the original concentration of full strength BG-11 medium. The water potentials (between brackets) of the media were calculated according to the formulas given by Lang (1967). For the determination of low molecular sugars, *C. epipsammum* was grown in batch cultures in 10-fold diluted BG-11 medium supplemented with NaCl to a final concentration of 200 and 400 mmol kg^{-1} respectively. The growth rate was calculated from the increase in biomass versus time.

CO_2 fixation rates at concomitant dehydration in the presence of buffered BG-11 medium supplemented with either 0, 50, 100, 200, 400 and 800 mmol kg^{-1} NaCl (pH 8.2) were carried out as described in De Winder et al. (1990b).

O_2 production and consumption: Samples were taken from the continuous cultures, concentrated by centrifugation (600 g) to OD_{750} = 1 (10 ± 1.0 mg chl l^{-1}) and resuspended in the 10-fold diluted TAPS-buffered BG-11 medium as outlined above. The rate of O_2 evolution and O_2 consumption as a function of the concentration of NaCl was determined with of a Clark-type O_2 electrode (Yellow Springs) at 20 °C. The readings were corrected for the osmotic effect on the diffusion-impedance of the oxygen electrode-membrane. Oxygen production was measured at saturating light intensities. Illumination was provided by a glass fibre connected to a halogen lamp (KS 1500, Schott).

Low molecular carbohydrates. After 48 h of growth the cells from batch cultures were centrifuged (10 min, 12000 g) and washed twice with the growth medium. The cell pellet was extracted overnight in the dark with boiling ethanol (80% v/v) and subsequently washed free of chlorophyll with cold ethanol (80% v/v) by centrifugation (4 min, 3000 g). The supernatants were pooled and the ethanol was evaporated under slightly lowered pressure at 50 °C, care was taken to prevent the solution from boiling. Total carbohydrate in whole cells was assayed by the anthrone method (Herbert et al. 1971) in the extracted and lyophilized cell pellet, and in the evaporated ethanol extracts, using glucose as a reference

Results and discussion

In nature the terrestrial cyanobacterium *C. epipsammum* is subjected to strongly changing water availability, and accompanying changes in the osmotic potential outside the cell. Therefore, it was investigated whether this organism possessed a high resistance towards temporary or permanent exposure of high salt concentrations. *C. epipsammum* shows an absolute upper limit of growth in BG-11 medium supplemented with 100 mmol kg^{-1} NaCl (0.4 osmol kg^{-1} or -990 kPa; Table 1).

Table 1. Growth rate of *Crinalium epipsammum* in media of different ion strength.

growth medium	water potential (kPa)	growth rate
BG-11 10-fold diluted	-9	0.0266 ± 0.0018
BG-11 5-fold diluted	-18	0.0371 ± 0.0015
BG-11 2-fold diluted	-45	0.0350 ± 0.0025
BG-11 full strength	-90	0.0295 ± 0.0014
BG-11 + 20 mmol kg^{-1} NaCl	-180	0.0198 ± 0.0026
BG-11 + 100 mmol kg^{-1} NaCl	-540	0.011 ± 0.002
BG-11 + 200 mmol kg^{-1} NaCl	-990	0

This is well within the limits of the least salt permissible group, which is defined to tolerate up to 1.42 osmol kg^{-1} NaCl (Reed et al. 1986). The sensitivity of *C. epipsammum* for salt is reflected by optimal growth in 5-fold diluted BG-11 medium. *C. epipsammum* synthesised low molecular carbohydrates (Table 2), but only after prolonged incubation time. Without a sufficient time span for adaptation, which to our impression is the case for cyanobacteria suffering complete and fast de- or rehydration, a mechanism involving formation of compatible solutes may be less favourable.

Table 2. Carbohydrate content (mg glucose equivalents per mg chl) in cells of *C. epipsammum* maintained at different salinities during 48 h.

medium	carbohydrates	
	low molecular weight	high molecular weight
1/5 BG11	6 ± 0.2	12 ± 2.2
+200 mmol kg^{-1}	11 ± 1.4	11 ± 0.7
+400 mmol kg^{-1}	22 ± 2.2	15 ± 0.5

In this work the instant and short term reactions as increased photosynthetic (Fig 1) and respiratory (Fig 2) flow, enabling extended capacity for ATP production or proton gradient maintenance across the cytoplasmic membrane are concluded to be of greater importance.

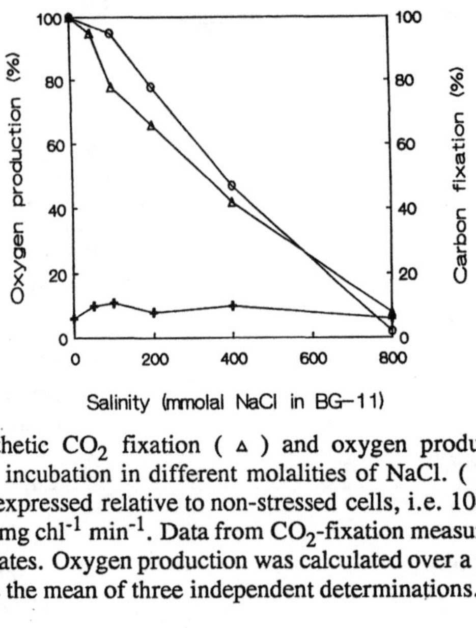

Fig. 1. Photosynthetic CO_2 fixation (▵) and oxygen production (○) of cells of *C. epipsammum* after incubation in different molalities of NaCl. (+), DCMU toxicated CO_2-fixation. Data are expressed relative to non-stressed cells, i.e. 1040 nmol CO_2 mg chl^{-1} min^{-1} and 1200 nmol O_2 mg chl^{-1} min^{-1}. Data from CO_2-fixation measurements are the mean of four independent triplicates. Oxygen production was calculated over a period of 10 min steady state production and are the mean of three independent determinations. In all cases the SD was less than 5%.

The experiments indicated that accelerated electron flow, especially as reflected by an increased dark respiration may rank to about 200 mmol kg^{-1} NaCl. Salt shocks of up to 800 mmol kg^{-1} gave rise to a temporary increase in respiration above the control rate, which bounces off to a low rate demonstrating deficient electron supply to the respiratory chain.

These data pointed to a gradual response pattern by which *C. epipsammum* handles salt stress during dehydration. An increase of respiration may be due to enhanced activity of the plasma membrane cytochrome oxidase given the functional link between this enzyme and the Na^+/H^+ antiporter (Molitor et al. 1986). By consequence, the enhancement of respiration in the experiments may reflect partial compensation for the reduced level of linear photosynthesis in the stressed cells. Recently direct evidence for increased PS I cyclic and respiratory activity in order to provide additional ATP for sodium extrusion has been documented (Jeanjean et al. 1993). The data of our experiments indicate that salt stress above 100 mmol kg^{-1} NaCl cannot be counterbalanced by increased bioenergetic activity and that the energy demanding process of ion extrusion is slowing down growth. Finetuning of metabolism in salt stress as well as in drought stress has recently been the subject of various studies. Synthesis of osmolytes has been documented (Reed et al. 1986). Increased activity of dark respiration and actual biosynthesis of cyt-aa_3 in the cytoplasmic membrane has been extensively illustrated (Blumwald et al. 1983; Fry et al. 1986; Molitor et al. 1986). Data on changes of PS-II and PS-I (involving cyclic photophosphorylation) activity are relatively scarce, although salt stress has recently been associated with increased PS-I content in various cyanobacteria (Schubert and Hagemann 1990) and drought stress with increased PS-II to PS-I spill-over (Jensen & Feige 1987). Other adaptations such as morphological modification (Campbell 1979; Peat and Potts 1987) or synthesis of drought-stress proteins (Scherer and Potts 1989) may contribute as well.

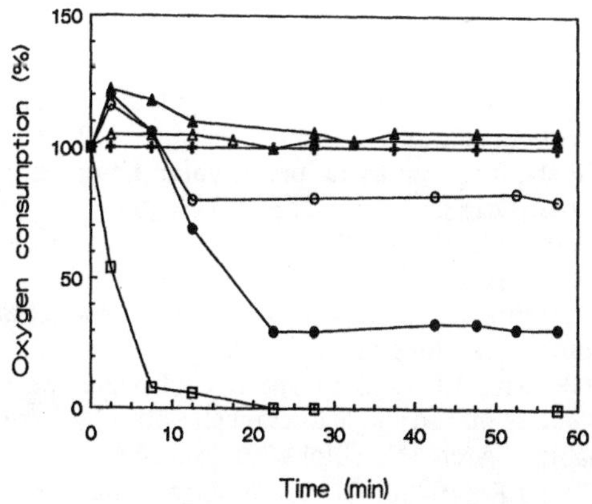

Fig. 2. Respiration rates of *C. epipsammum* immediately after upshock with different concentrations of NaCl. Cells were grown in 5-fold diluted BG-11 medium. Final concentrations of NaCl: (▵) 100 mmol kg^{-1}; (▴) 200 mmol kg^{-1}; (○) 400 mmol kg^{-1}; (●) 800 mmol kg^{-1}; (□) 1200 mmol kg^{-1}. (+), activity in growth medium.

In the case of *C. epipsammum* (and other organisms subjected to de- and rehydration cycles) several of the above mentioned alternatives may come into play. It should be questioned whether synthesis of low-molecular weight compounds, which is a widespread phenomenon during continuous salt stress, is applicable for terrestrial cyanobacteria. Evidently, *C. epipsammum* needs to keep low amounts of intracellular water attracting compounds as these otherwise will invoke cell lysis when the first drops of rain will rewet the desiccated cyanobacterial cell. Although *C. epipsammum* is drought tolerant and able to react quickly to renewing water-availability, the organism is rather sensitive towards increasing salinity. The studied organism rather seems to have advantage in relying on a short term mechanism and making maximum energy available for Na^+ extrusion through increased respiration and photophosphorylation until desiccation and salt stress pressure exceed the counterbalancing capacity. The time span between actual water potential and total desiccation in nature is presumably short enough to prevent massive osmolyte generation. *C. epipsammum* in its natural harsh environment turns into a resting state while becoming desiccated which ensures optimal and fast recovery after rehydration.

Acknowledgements

Thanks are due to the the skilful help of Andien van den Heuvel and to Hans Matthijs for helpful discussions

References

Blumwald E, Mehlhorn RJ, Packer L (1983) Ionic osmoregulation during salt adaptation of the cyanobacterium *Synechococcus* 6311. Plant Physiol 73:377-380.

Campbell SE (1979) Soil stabilization by a prokaryotic desert crust: implications for precambrian land biota. Origins of Life 9: 335-348

De Winder B, Pluis J, De Reus L, Mur LR (1989a) Characterization of a cyanobacterial, algal dune crust in the coastal dunes of The Netherlands. In: Cohen Y, Rosenberg E (eds) Microbial Mats, Physiological ecology of benthic microbial communities, American Society of Microbiology, Washington pp 11-18.

De Winder B, Matthijs HCP, Mur LR (1989b) The role of water retaining substrata on the photosynthetic response of three drought tolerant phototrophic micro-organisms isolated from a terrestrial habitat. Arch Microbiol 152: 458-462.

De Winder B, Stal LJ, Mur, LR (1990a) *Crinalium epipsammum sp. nov.*: a filamentous cyanobacterium with trichomes composed of elliptical cells and containing poly-ß-(1-4)-glucan (cellulose). J Gen Microbiol 136: 1645-1653.

De Winder B, Matthijs HCP, Mur LR (1990b) Dehydration and ion stress effects on the carbon fixation in drought tolerant phototrophic micro-organisms. FEMS Microbiol Ecol 74: 33-38

Fry IV, Huflejt M, Erber WWA, Peschek GA, Packer L (1986) The role of respiration during adaptation of the freshwater cyanobacterium *Synechococcus* 6311 to salinity. Arch Biochem Biophys 244: 686-691.

Herbert D, Phipps PJ, Strange RE (1971) Chemical analysis of microbial cells. In: Norris JR, Ribbons DW (eds) Methods in microbiology, Academic Press, London, pp 209-344

Jensen M, Feige GB (1987) The effect of desiccation and light on the 77K chlorophyll fluorescence properties of the lichen *Peltigera aphtosa*. Prog Prob in Lichenol in the Eighties. Bibl Lichenol 25:325-330.

Jeanjean R, Matthijs HCP, Onana B, Haveaux M, Joset F (1993) Exposure of the cyanobacterium *Synechocystis* PCC6803 to salt stress induces concerted changes in repiration and photosynthesis. Plant Cell Physiol 34:1073-1081

Lang ARG (1967) Osmotic coeficients and water potentials of sodium chloride solutions from 0-40 °C. Austr J Chem 20:2017-2023.

Molitor V, Erber W, Peschek GA (1986) Increased levels of cytochrome oxidase and sodium-proton antiporter in the plasmamembrane of *Anacystis nidulans* after growth in sodium-enriched media. FEBS Letters 204:251-255.

Peat A, Potts M (1987) The ultrastructure of immobilized desiccated cells of the cyanobacterium *Nostoc commune* UTEX584. FEMS Microbiol Lett 43:223-227

Reed RH, Borwitzka LJ, Mackay MA, Chudek WA,Foster M, Warr SCR, Moore DJ & Stewart WDP (1986) Organic solute accumulation in osmotically stressed cyanobacteria. FEMS Microbiol Rev 39:51-56.

Reed RH, Stewart WDP (1988) The response of cyanobacteria to salt stress. In Rogers Gallon. Biochemistry of algae and cyanobacteria, Proc Phytochem Soc Europe, 28 Clarendon press Oxford. p 217-231.

Scherer S, Potts M (1989) Novel water stress protein from a desiccant-tolerant cyanobacterium -purification and partial characterization. J Biol Chem 264:12546-12554.

Schubert H, Hagemann M (1990) Salt effects on 77K fluorescence and phototsynthesis in the cyanobacterium Synechococcus sp. PCC 6803. FEMS Microbiol Lett 71:169-172

Osmotic adaptation of microbial communities in hypersaline microbial mats

Aharon Oren[1], Uri Fischel, Zeev Aizenshtat[1,2], Eitan B. Krein[2] and Robert H. Reed[3]

Institute of Life Sciences, The Hebrew University of Jerusalem, 91904 Jerusalem, Israel

Microorganisms that live in hypersaline environments need to maintain a high osmotic pressure in their cytoplasm to balance the osmotic pressure caused by the high salt concentrations. Most halophilic and halotolerant eubacteria and eukaryotic protists exclude salt, and synthesize or accumulate high concentrations of organic solutes that do not interfere with intracellular enzymatic activities ("osmotic solutes" or "compatible solutes").

A great variety of organic osmotic solutes have been detected in different halophilic microorganisms. The list of such compounds encomprises polyols (glycerol, mannitol and others), carbohydrates (e.g., sucrose, trehalose, glucosylglycerol), amino acids (proline, glutamate and others) and amino acid derivatives, and tetrahydropyrimidines such as ectoine (1,4,5,6-tetrahydro-2-methyl-4-pyrimidine carboxylic acid) and 5-hydroxyectoine (Severin et al. 1992; Trüper and Galinski 1989).

Some years ago it was claimed that glycine betaine is the universal compatible solute of halophilic eubacteria (Imhoff and Rodriguez-Valera 1984). However, glycine betaine is synthesized by a limited number of bacteria only, but when grown in complex media containing high concentrations of yeast extract they take up betaine or its precursor choline from the medium. Ectoine is now recognized to be widespread as the major compatible solute of heterotrophic eubacteria when these are grown in defined media (Severin et al. 1992).

The information available on organic osmotic solutes in halophilic microorganisms is based on laboratory studies of pure cultures, and very little is known about the relative abundance of these solutes in nature. Thus, it is unknown to what extent natural communities of non-photosynthetic eubacteria accumulate glycine betaine produced by other members of the microbial community. The importance of glycine betaine in the microbial processes in hypersaline environments may be inferred from studies on the formation of methane in these ecosystems. In most anaerobic hypersaline environments methane is formed, and trimethylamine is its most likely precursor. The most probable source of trimethylamine is the anaerobic breakdown of glycine betaine (Oremland and King 1989; Oren 1990). Though little

[1] The Moshe Shilo Center for Marine Biogeochemistry.
[2] Institute of Chemistry, The Hebrew University of Jerusalem, 91904 Jerusalem, Israel
[3] Department of Chemical and Life Sciences, University of Northumbria at Newcastle, Newcastle upon Tyne NE1 8ST, UK.

NATO ASI Series, Vol. G 35
Microbial Mats
Edited by Lucas J. Stal and Pierre Caumette

information exists on the bacteria responsible for the anaerobic degradation of glycine betaine at high salt concentrations, the occurrence of anaerobic degradation of betaine in hypersaline ecosystems can be deduced from the rate of methanogenesis.

Well-developed laminated microbial mats characteristically develop on the bottom of shallow hypersaline water bodies (salt flats, saltern ponds). Such mats may form a suitable experimental system to investigate the distribution of the different osmotic solutes in natural systems in which dense communities of phototrophic and heterotrophic prokaryotes live together. It may be expected that in such complex microbial communities, in which compounds produced by one organism may become available to others, glycine betaine produced by photosynthetic bacteria and cyanobacteria may be available to the eubacteria present, relieving them from the need to produce ectoine or other osmotic solutes. If this indeed happens, it would to a certain extent rehabilitate the former claim that glycine betaine is the dominant osmotic solute in halophilic and halotolerant prokaryotes (Imhoff and Rodriguez-Valera 1984). Hardly anything is known on the distribution of compatible solutes in natural communities of halophilic microorganisms such as often develop on the bottom of hypersaline aquatic systems. To our knowledge the only published attempt to quantify glycine betaine in a hypersaline environment relates to a *Spirulina*-dominated algal mat at a salinity of 180‰, in which total betaine concentrations of up to 100 μmol g^{-1} dry weight were reported (King 1988).

Therefore we initiated a study on the contribution of glycine betaine, ectoine, and other osmotic solutes to the total osmotic solute content of microbial communities in hypersaline environments of different salinities. Benthic microbial mats from two aquatic hypersaline systems were studied:

1. The experimental salt ponds on the grounds of the Interuniversity Institute of Eilat, Israel, ponds with salinities between approximately 80-120‰, with well-developed layered mats, containing layers of diatoms, filamentous cyanobacteria (dominated by *Microcoleus chthonoplastes*), different types of anoxygenic photosynthetic bacteria, and other microorganisms.

2. Hamei Mazor, a hypersaline sulfur spring on the shore of the Dead Sea near Ein Gedi, with a salinity of about 170‰, and an abundance of filamentous cyanobacteria (*Oscillatoria*-type) and purple bacteria (*Thiocapsa*) (Oren 1989).

Osmotic solutes were extracted from microbial mat samples or from pure cultures of bacteria isolated from them. To identify and quantify osmotic solutes three different techniques were used:

1. Natural abundance ^{13}C-nuclear magnetic resonance spectroscopy (NMR), using the pulsed Fourier transform mode in a Bruker NMR spectrometer, operating at 400 MHz.

2. ^{1}H-NMR, using a Bruker HX90 continuous wave NMR spectrometer operating at 90MHz. For both modes of NMR analysis the samples were dissolved in D_2O.

3. High performance liquid chromatography (HPLC), using a Lichrospher 100 NH_2 column, eluted with 75% acetonitrile, and a UV detector operating at 190 nm. This method

enables the detection of osmotic solutes such as glycine betaine, amino acids and derivatives, and ectoine, but carbohydrates and polyols are not detected as their absorption in the ultraviolet range is too low.

An extract of a microbial mat developing in a salt pond at the Interuniversity Institute of Eilat with a salinity of 100‰ was analyzed by means of ^{13}C-NMR. Resonances at 60.8, 60.9, 61.8, 70.0, 71.9, 72.4, 73.3, 79.2 and 98.3 ppm (Fig. 1) prove the presence of glucosylglycerol as the sole or major osmotic solute (Borowitzka et al. 1980).

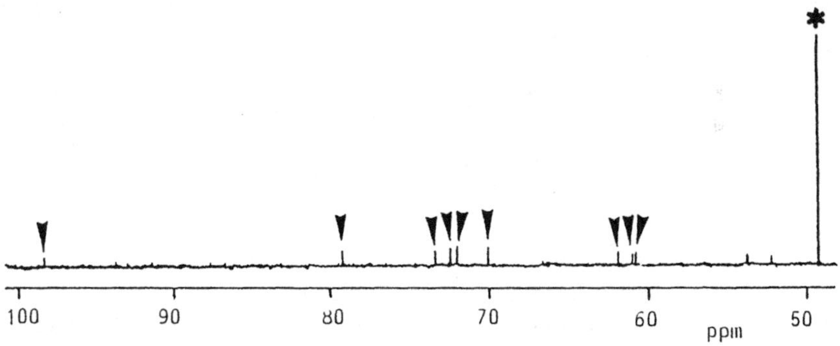

Fig. 1. Natural abundance ^{13}C-NMR spectrum of osmotic solutes extracted from a microbial mat collected from a salt pond at the Interuniversity Institute of Eilat (salinity 100‰). Peaks due to glycosylglycerol are marked by arrows. The resonance at 49 ppm (*) is due to the presence of ^{13}C-methanol added as standard.

Glucosylglycerol is widespread as an osmotic solute in cyanobacteria that live at intermediate salinites. Species that grow at lower salinities are characterized by the presence of sucrose or trehalose as osmotic solutes, while the most halophilic types produce glycine betaine (Mackay et al. 1984). The cyanobacterium that dominated in the mat examined was *Microcoleus chthonoplastes*, a species reported to contain glucosylglycerol as compatible solute (Kevbrin et al. 1991). No sign of the presence of significant concentrations of glycine betaine was found in this microbial mat growing at a salinity of 100‰.

For the analysis of the nature of the osmotic solutes present in the *Oscillatoria*-dominated benthic microbial mat in the hypersaline sulfur spring of Hamei Mazor on the shore of the Dead Sea near Ein Gedi, with a salinity of about 170‰, we used ^{1}H-NMR. Mat material was collected near the spring, where the cyanobacteria exist almost in pure culture (Oren 1989). The main peaks in the spectrum (Fig. 2) were identical with that of glycine betaine, with resonances at 3.27 ppm due to the three methyl groups, and an additional resonance at 3.87 ppm due to the CH_2 group. This result is in agreement with the expected dominance of glycine betaine in a cyanobacterial community existing at extremely high salt concentrations (Mackay et al. 1984).

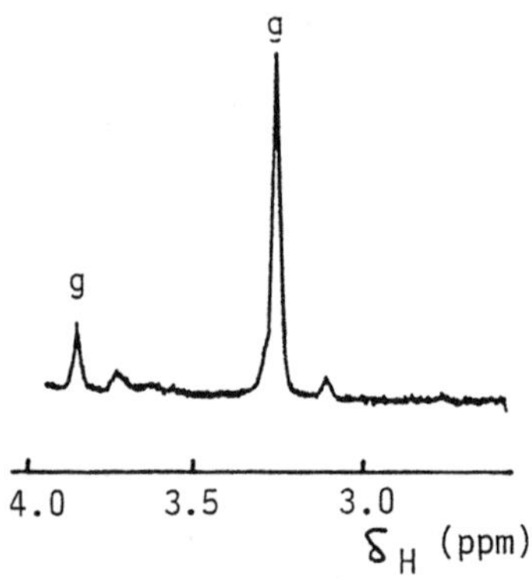

Fig. 2. ^{1}H-NMR spectrum of an extract of a microbial mat, consisting mainly of *Oscillatoria*-type cyanobacteria, collected from Hamei Mazor, a hypersaline sulfur spring on the shore of the Dead Sea. Resonances are due to the presence of glycinebetaine (g).

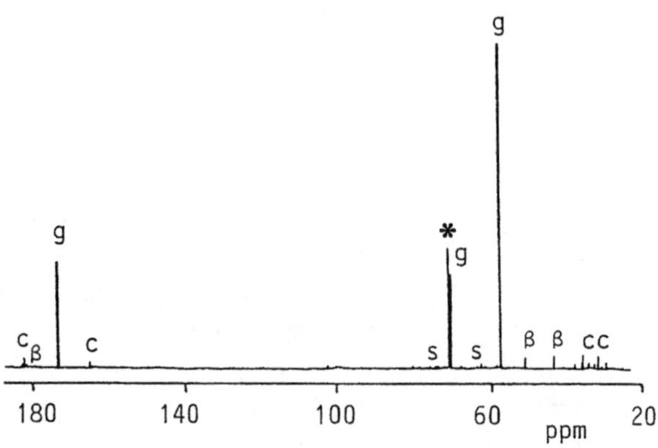

Fig. 3. Natural abundance ^{13}C-NMR spectrum of osmotic solutes extracted from *Ectothiorhodospira marismortui*, grown in the presence of 10% NaCl. Resonances can be attributed to the presence of glycine betaine (g), N^{α}-carbamoyl-L-glutamine 1-amide (c), *β*-glutamate (*β*), and sucrose (s). The resonance at 71 ppm (*) is of ^{13}C-dioxane used as standard.

The area around the sulfur spring of Hamei Mazor has been the source of inoculum for the isolation of two interesting novel red anoxygenic photosynthetic bacteria:

Ectothiorhodospira marismortui (Oren et al. 1991), and *Rhodospirillum sodomense* (Madigan MT, University of Southern Illinois at Carbondale, unpublished results). Techniques of HPLC and natural abundance ^{13}C-NMR yielded information on the compatible solute content of these species. *Ectothiorhodospira marismortui* was found to contain glycine betaine as the principal osmotic solute, a minor amount of sucrose, and a relatively high concentration of the novel compound N^{α}-carbamoyl-L-glutamine 1-amide (Galinski and Oren 1991; Oren et al. 1991) (Fig. 3). Additional resonances were detected, which may be due to the presence of β-glutamate.

The newly isolated *Rhodospirillum sodomense* was found to contain two osmotic solutes: glycine betaine and ectoine (Fig. 4). A similar combination of osmotic solutes was reported earlier in the related halophilic species *Rhodospirillum salinarum*, isolated from a saltern in Portugal (Severin et al. 1992).

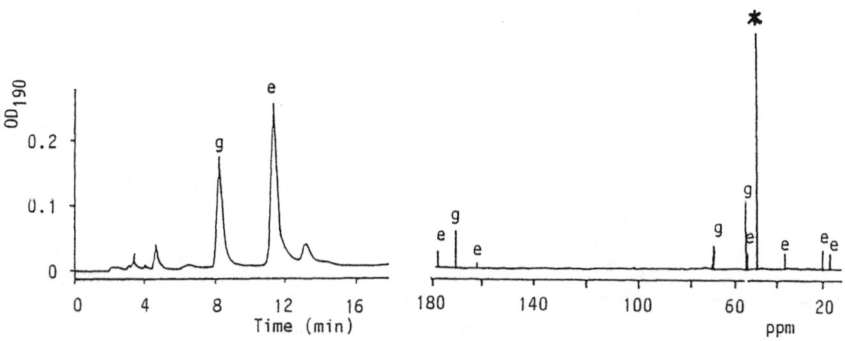

Fig. 4. HPLC elution pattern (left panel) and natural abundance ^{13}C-NMR spectrum (right panel) of osmotic solutes extracted from *Rhodospirillum sodomense*. UV absorption peaks and NMR resonances can be attributed to the presence of glycine betaine(g) and ectoine (e). Based on the use of standards in the HPLC experiments, a molar ratio of glycine and ectoine of 6.7:1 was calculated. The NMR resonance at 49 ppm (*)is due to ^{13}C-methanol used as standard.

The results presented above show that techniques such as NMR and HPLC can be used to characterize compatible solutes in natural microbial assemblages in benthic mats developing in hypersaline environments, and that hypersaline microbial mats may yield novel types of microorganisms, with new and interesting osmotic solutes.

Acknowledgements

We thank M.T. Madigan (University of Southern Illinois at Carbondale) for a gift of a culture of *Rhodospirillum sodomense*. This study was supported by the G.I.F., the German-Israeli Foundation for Scientific Research and Development, and by the Moshe Shilo Center for Marine Biogeochemistry, BMFT-Minerva Gesellschaft für Forschung, München, Germany.

References

Borowitzka LJ, Demmerle S, Mackay MA, Norton RS (1980) Carbon-13 nuclear magnetic resonance study of osmoregulation in a blue-green alga. Science 210: 650-651

Galinski EA, Oren A (1991) Isolation and structure determination of a novel compatible solute from the moderately halophilic purple sulfur bacterium *Ectothiorhodospira marismortui*. Eur J Biochem 198: 593-598

Imhoff JF, Rodriguez-Valera F (1984) Betaine is the main compatible solute of halophilic eubacteria. J Bacteriol 160: 478-479

Kevbrin VV, Dubinin AV, Osipov GA (1991) Osmoregulation in the marine cyanobacterium *Microcoleus chthonoplastes*. Microbiology 60: 407-410

King GM (1988) Methanogenesis from methylated amines in a hypersaline algal mat. Appl Environ Microbiol 54: 130-136

Mackay MA, Norton RS, Borowitzka LJ (1984) Organic osmoregulatory solutes in cyanobacteria. J Gen Microbiol 130: 2177-2191

Oremland RS, King GM (1989) Methanogenesis in hypersaline environments. In: Cohen Y, Rosenberg E (eds) Microbial mats: physiological ecology of benthic microbial communities. American Society for Microbiology, Washington, pp 180-190

Oren A (1989) Photosynthetic and heterotrophic benthic bacterial communities of a hypersaline sulfur spring on the shore of the Dead Sea (Hamei Mazor). In: Cohen Y, Rosenberg E (eds) Microbial mats: physiological ecology of benthic microbial communities. American Society for Microbiology, Washington, pp 64-76

Oren A (1990) Formation and breakdown of glycine betaine and trimethylamine in hypersaline environments. Antonie van Leeuwenhoek 58: 291-298

Oren A, Simon G, Galinski EA (1991) Intracellular salt and solute concentrations in *Ectothiorhodospira marismortui*: glycine betaine and N^{α}-carbamoyl glutamine amide as osmotic solutes. Arch Microbiol 156: 350-355

Severin J, Wohlfarth A, Galinski EA (1992) The predominant role of recently discovered tetrahydropyrimidines for the osmoadaptation of halophilic eubacteria. J Gen Microbiol 138: 1629-1638

Trüper HG, Galinski EA (1989) Compatible solutes in halophilic phototrophic procaryotes. In: Cohen Y, Rosenberg E (eds) Microbial mats: physiological ecology of benthic microbial communities. American Society for Microbiology, Washington, pp 342-348

Panel discussion: Colonisation and early development of microbial mat communities

David T. Welsh (rapporteur) and Rodney Herbert (chairman)

Department of Biological Sciences, University of Dundee,
Millers Wynd, Dundee. DD1 4HN, Scotland.

Studies of microbial mat ecosystems have concentrated primarily on community structure, microbial interactions and nutrient cycling within established mat systems. Consequently relatively little is known of the events involved in the initial colonisation and early development of these communities. In this session some of the factors which may be important in regulating these processes were discussed.

The primary constraint influencing the development of microbial mat communities is the physical environment, with factors such as pH, temperature, light and nutrient availability considered to be particularly important. The nature of the sediment substrate is also considered to be of major significance due to its influence on light transmission and spectral quality, nutrient availability and diffusional boundary layers and pathlengths, which are important for the establishment of physicochemical gradients within these systems. Heterogeneitity of the sediment environment plays a significant role in determining both the number and types of niche available for colonisation and thus exerts a major influence upon species diversity and the complexity of the resultant microbial community.

The types of organisms which initially colonise and then become established on a newly exposed sediment surface will be determined by a number of factors such as the species diversity of the colonising microflora derived from the surrounding environment. This led to a lively discussion of the quotation of Van Niel "that everything is everywhere". This concept was challenged particularly for organisms found only in specialised environments, such as hydrothermal springs which are geographically isolated and whose microflora may be sensitive to environmental conditions away from their specialised habitat. For example *Synechococcus lividus*, which dominates the high temperature region of hot spring benthic microbial mats in the U.S., is absent from similar Icelandic hot springs where this niche remains unoccupied. This observation would suggest that physical distance is a barrier to the dispersion of this cyanobacterium, since *S. lividus* has been shown to be capable of growth in water samples collected from representative Icelandic hot springs.

Several discussants considered that the ability to attach to the sediment substrate was significant for both initial colonisation and establishment of microorganisms within benthic mat systems, since unattached organisms may be washed out of these systems particularly in

NATO ASI Series, Vol. G 35
Microbial Mats
Edited by Lucas J. Stal and Pierre Caumette

hydrologically active marine or spring environments. Thus the possession of a polysaccharide capsule or hydrophobic surface groups which promote interaction with sediment particles, may be an important prerequisite for successful colonisation. Equally, the converse is true and attached microorganisms must also be able to detach from sediment particles in order to disperse and colonise other surfaces. This may be achieved by alternation of the cell cycle between mature adherent reproductive cells and a freely motile dispersal stage e.g. motile swarmer cells (*Caulobacter* sp. and *Rhodomicrobium vannielii*), baeocytes (pleurocapsalean cyanobacteria) and hormogonia (benthic cyanobacteria). These specialised cell types may be important early colonisers of recently exposed surfaces. Additionally, the attachment of such organisms may represent a significant step in the development of benthic microbial mats, since their growth stabilises the sediment due to the formation of cross links between individual sediment particles.

The organisms which initially colonise a new environment will be strongly influenced by the nutrient status of the environment. Generally, nutrients in newly exposed environments are limiting and thus photosynthetic microorganisms and in particular N-fixing cyanobacteria are considered by most authorities to be the key primary colonisers. This view was confirmed by a number of speakers who reported that diatoms together with heterocystous and non-heterocystous cyanobacteria were commonly observed primary colonisers of natural and artificial substrates such as glass slides or nylon screens. This primary community is thought to be important in determining the subsequent development of the mat community since its activity modifies the environment and therefore influences which other organisms can establish within the system.

In the final session of the discussion methods for studying the initial events in the colonisation and early development of microbial mat communities were discussed. It was proposed that these processes could be followed using a gel stabilised model system, inoculated with a limited range of mat forming microorganisms. However, the general consensus of opinion was that such a model system is an oversimplification and that natural systems are too variable and complex to be modelled successfully in this manner. Therefore, simpler recolonisation or colonisation experiments utilising natural or artificial substrates placed in natural environments, where changes in the colonising microbial populations could be followed microscopically or using the new techniques of molecular biology were considered potentially more fruitful. Additionally, laboratory based physiological investigations to determine more fundamental questions, such as, which factors make an organism an effective coloniser or why one species of phototroph dominates in a particular mat ecosystem were thought worthy of consideration.

II. New methods in microbial mat research

Analysis of microbial mats by use of electrochemical microsensors: Recent advances

Niels Peter Revsbech

Department of Microbial Ecology, Institute of Biological Sciences, University of Aarhus, Bd. 540, DK-8000 Aarhus C, Denmark

Introduction

During the recent few years there has been a considerable expansion of our capabilities for microsensor analysis of chemical microenvironment and rates of microbial metabolism within microbial mats. Since 1987 we have thus seen published accounts of new electrochemical microsensors for N_2O, NO_3^-, and NH_4^+; the O_2 microsensor has been dramatically improved, and computerized equipment for remote operation of microsensors at great water depths has been developed. It now also seems that the electrochemical microsensors for analysis of chemical species in microbial mats may get competition from microsensors with optical signal transduction. The technological advances have enabled us to perform detailed studies of the nitrogen cycle within intact sediments and microbial mats. It has also been possible to perform very detailed studies including total oxygen budgets of photosynthetic microbial mats, and we have thereby verified the microsensor method for determining photosynthetic rates. This chapter will describe both the recent technological advances and the new insights we have gained by the new technologies.

Current status of microsensor technology

Oxygen

Polarographic Clark-type oxygen microsensors have been used through the last decade (Revsbech and Jørgensen 1986). The performance of such microsensors (Fig. 1) was recently improved by introduction of a guard cathode and use of more insulating glasses (Revsbech 1989). The oxygen microsensors now have stable signals which often exhibit less than 2% drift per day, and they can thus be used for routine applications where long-term stability is essential. The introduction of the guard cathode has also made it much easier to construct the microsensors, as it is now not necessary with a close fit between the outer casing with the oxygen-permeable membrane and the oxygen-reducing cathode. One problem with the use of a guard cathode is, however, the high current in the guard cathode circuit and the corresponding high rate of AgCl formation on the inner Ag/AgCl anode. This high rate of Cl^- consumption

NATO ASI Series, Vol. G 35
Microbial Mats
Edited by Lucas J. Stal and Pierre Caumette

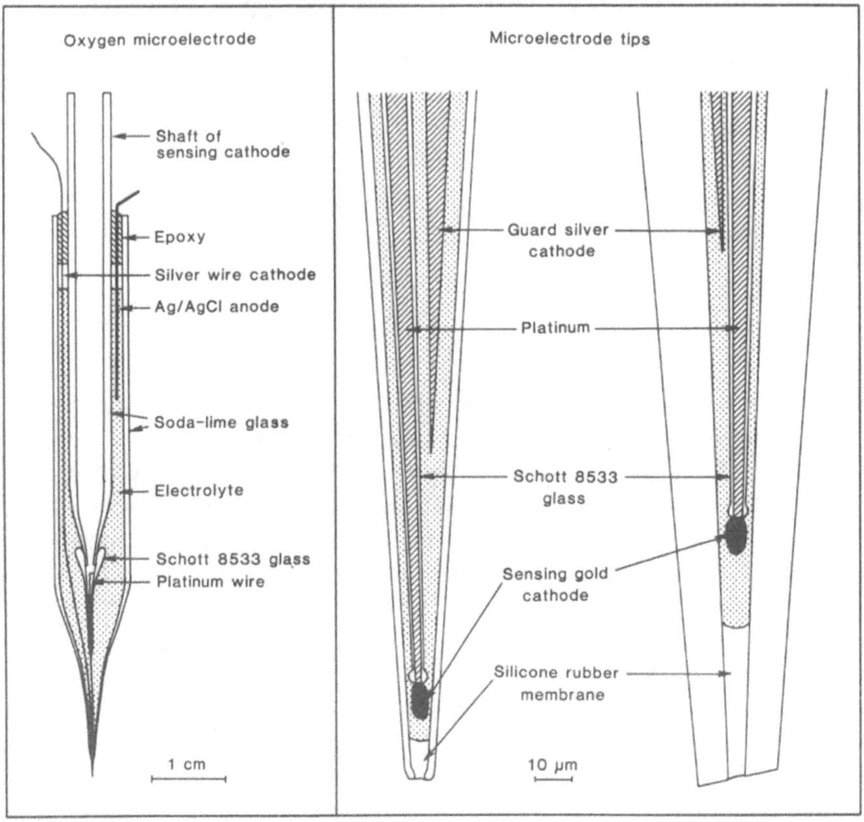

Fig.1. Oxygen microsensor with guard cathode. The tips illustrate a thin and rapidly responding sensor and a sturdy but slower sensor. From Revsbech (1989).

eventually leads to a significant drop in Cl^- concentration and thus a change in the potential of the anode. The change in reference potential results in a change in the oxygen sensitivity of the microsensor, and the otherwise large benefit of a stable oxygen signal is thus time-limited. The problem of Cl^- change can, however, be reduced by a few precautions. First of all, the entry of oxygen gas to the bulk electrolyte should be minimized by having a complete seal of a virtually oxygen-impermeable epoxy resin between cathode shaft and outer casing. For highest stability, this precaution may, however, not be enough, as some of the current originating from the large surface area of guard cathode as originally described (Revsbech 1989) is independent of oxygen. An approach with a small exposed metal area immediately behind the sensing cathode as used in the guard cathode of the combined O_2/N_2O microsensor (Revsbech et al. 1988) may then be used instead. The problems of Cl^- consumption described here are not pure theory: An oxygen microsensor with a fully exposed guard cathode and open air access from behind will be destroyed within 24 h.

The high stability of the guard cathode equipped oxygen microsensors may be combined with an extreme speed of response. As the guard cathode will prevent diffusion of oxygen from behind to the exposed metal surface of the sensing cathode, it is possible to use exposed metal surfaces of considerable extension. Sensors with 90% response times of about 0.1 s have thus been made by etching (hydrofluoric acid, extreme caution!) the glass away from the distant 0.1 mm of thin cathodes. As the majority of the total diameter of the cathodes is glass, this procedure makes it possible to make much thinner cathodes than otherwise possible, and thus also to make sensors with thinner tips and with short ($<10\ \mu m$) diffusion paths from the exterior to the reducing metal surface. The fast responding sensors made this way have been used to determine microprofiles of photosynthetic rates with high spatial resolution as described later in this chapter. It should be stressed, however, that such extremely thin cathodes are only necessary if it is required to have response times of less than 0.2s and stirring artifacts (Revsbech 1989) of less than 2%. Sensors with extremely thin tips but with relatively slow response can easily be made by having conical outer casings where the cathode may be situated up to 150 μm behind the tip (then about 10 s for 90% response).

Sulfide

Sulfide electrodes are made of silver metal with a coating of Ag_2S. The potential of the electrode is dependent on the solubility of the Ag_2S and thus on the concentration of S^{2-}. Most of the dissolved sulfide is at pH values near neutral found as HS^- and H_2S, and at pH values below 13 it is necessary to perform simultaneous determinations of p[S^{2-}] and pH to calculate the actual concentration of total dissolved sulfide. The sulfide microsensor was first used by Revsbech et al. (1983) who described rather non-ideal calibration curves with a response of 37 mV for a change from 10^{-3} to 10^{-4} M sulfide (pH 7) and absence of a reproducible reading below 10^{-5} M. It seems, however, that these early findings of non-ideal behavior were largely due to depolarization of the electrode by oxygen. When oxygen is excluded from the calibration solution, linear calibration curves down to below 10^{-6} can often be obtained (Fig. 2). It has lately been demonstrated that it is possible to determine microprofiles of sulfate reduction from measured microprofiles of sulfide (Kühl and Jørgensen 1992; B.B. Jørgensen this volume). The microprofiles of sulfate reduction determined so far have been obtained in trickling filter biofilms, but it should be possible to make similar determinations in natural microbial mats if these are low in sulfide-precipitating iron minerals.

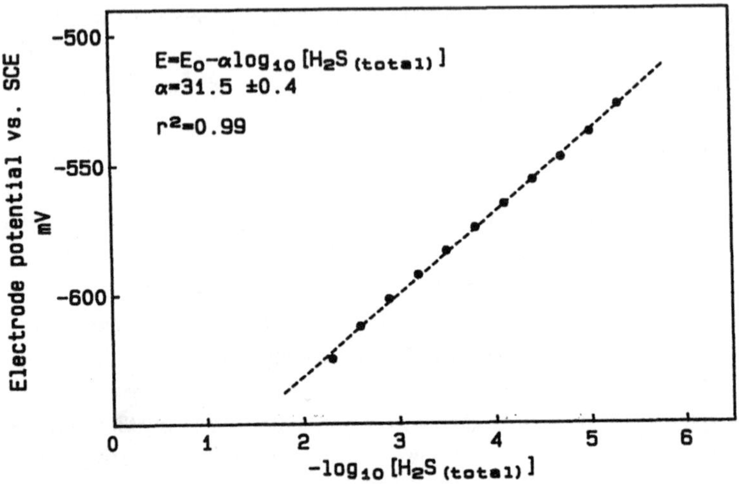

Fig. 2. Calibration curve of sulfide microsensor with a tip diameter of 40 μm. The calibration was performed in a phosphate buffer at pH 7.2. From Kühl and Jørgensen (1992).

pH

It is possible to construct microsensors for various ions based on ion-selective membranes which selectively allow passage of the specific ion through the material. As only one charged species is transported, the process results in a difference in charge (potential) across the membrane. For pH the ion-conducting material may consist of H^+-ion selective glass membranes (Revsbech and Jørgensen 1986) and various compositions of liquid ion exchangers (Ammann et al. 1981). Because of the short lifetime and fragility of liquid ion exchanger (LIX) microsensors, glass-based microsensors are preferable for most environmental applications. If there is a demand for extreme spatial resolution, the possibility for using a LIX-type pH microsensor should, however, be investigated, as these can be made with tips of less than 1 μm.

Nitrous oxide

The combined O_2/N_2O microsensor (Revsbech et al. 1988) is probably the most complicated microsensor yet constructed, and its complexity has limited its application. The sensor contains an O_2-sensing gold cathode and a N_2O-sensing silver cathode, and in addition there is also a guard cathode made of gold. One of the problems during construction has been that the two gold cathodes within the sensor were dissolved during internal plating with silver of the N_2O cathode. We have now realized that this dissolution can be prevented by doing all electroplatings in the absence of oxygen (Dalsgaard and Revsbech 1992). By doing so it is possible to apply smaller amounts of gold and thus also to make more rapidly responding sensors with a smaller distance between silicone membrane and N_2O cathode.

Nitrate and ammonium

The microsensors for nitrate and ammonium are LIX-type microsensors originally described by de Beer and van den Heuvel (1988) and de Beer and Sweerts (1989). Their ease of use was improved by surrounding the capillary containing the LIX with a glass casing so that an efficient electrical shielding could be made by an electrolyte between the two glass layers (Jensen et al. 1993). The electrolyte shielding alleviates the need for a large Faraday cage around the setup and the microsensors can thus be used in a much more convenient and flexible way in the laboratory. A conventional Faraday cage is also less efficient than the electrolyte shielding which makes it possible to record potentials with an accuracy of $\pm 10\ \mu V$, even when the work is done in an ordinary laboratory environment with moving persons, etc. There are two large problems with LIX-based microsensors, however. The worst problem is interference from other ions. Even when the selectivity coefficient is very advantageous, interference may be a serious problem for practical work. The nitrate microsensor has, as an example, a selectivity coefficient for NO_3^- over HCO_3^- or Cl^- of about 10^3, but the interference from these ions still mean that the lowest detectable NO_3^- concentration in freshwater environments in most cases is around 10 μM whereas it is 1 μM in distilled water, and any NO_3^- analysis in a marine environment with high Cl^- is impossible. There is furthermore always a major inaccuracy near the detection limit, so a detection limit around 10 μM means considerable inaccuracy on NO_3^- determinations below 30 μM. Another problem is caused by direct contact between the LIX and the environment, which often causes signal noise and instability. These problems are especially serious when analysis is attempted in media with hydrophobic surfaces (such as cyanobacterial mats) which interact with the hydrophobic LIX. The author of this chapter has tried to circumvent the interface problem by placing a hydrophilic and ion-permeable membrane in the tip of the microsensor so that there is no direct contact between the LIX and the analyzed medium, but the preliminary results have not been encouraging.

Carbon dioxide

Microsensors for CO_2 have been used for physiological applications for many years (Voipio and Kaila 1993 and references therein). Recently there has been considerable interest for use of similar microsensors in investigations of sediments, biofilms, and microbial mats. The sensitivity of the described microsensors is, however, just on the borderline for practical use in natural waters approaching atmospheric equilibrium with respect to CO_2 (about $2*10^{-5}$ M). In our laboratory we have constructed microsensors based on glass-type pH microsensors surrounded by a 10 mM bicarbonate buffer and placed behind a CO_2-permeable silicone membrane. These sensors could be calibrated in the range from 2-20*10^{-5} M (corresponding to saturation with gas mixtures containing from 0.03 - 0.3% CO_2), which is the environmentally most relevant range, but the response is very slow at such low CO_2 concentrations and subject to deterioration within a few days of microsensor assembly. Furthermore CO_2 concentrations per se are not as interesting as total dissolved inorganic carbon, and it is therefore essential to have combined measurements of CO_2 and pH along the same vertical profile. It is, however, possible that the development of microsensors with optical signal transduction (optrodes) will enable us to perform much more accurate CO_2 measurements on a routine basis (see below).

Hydrogen

Reliable electrochemical microsensors for hydrogen have been described, and such microsensors have been used to study the symbiosis between legumes and Rhizobia (Witty 1991). The sensitivity of electrochemical microsensors is, however, not sufficient to quantify the hydrogen concentrations present in natural non-symbiotic environments which even under anaerobic conditions rarely exceed 50 nM. Recently, a sensor equipped with a gas permeable membrane and connected to a highly sensitive gas chromatograph has been described (Rothfuss et al. this volume). Although not in a real microscale it has by this technique for the first time been possible to measure relatively detailed profiles of hydrogen and methane in sediments.

Optrodes

Sensors with optical signal transduction form a heavily funded area of analytical chemistry, and there are now macroscopic optrodes available which may be miniaturized and applied in investigations of microbial mats and similar substrates. Although optrodes are not covered by the title of this chapter, it should be mentioned that highly promising optrodes for HCO_3^- and NH_4^+ have been developed (O.S. Wolfbeiss 1991; personal communication).

Equipment for microsensor analysis

A review of microsensor construction and equipment for microsensor analysis was published by Revsbech and Jørgensen (1986). Since then portable computers with Analog-Digital conversion and remote control of the micromanipulator motor have facilitated operation and data acquisition. It should be stressed, however, that computerized micromanipulator control and data collection are an advantage only if large amounts of data are collected. Construction of such automated systems should thus be discouraged until all other aspects of the analysis works on a routine basis.

In our laboratory we have had the best possible experiences with the highly versatile Keithley 617 electrometer for handling signals from all types of electrochemical microsensors. When used for amperometric microsensors it is necessary to use the Keithley 617 with an external polarization unit (Revsbech and Ward 1984), or in the case of a guard-cathode equipped oxygen microsensor two polarization units. It is not necessary to use a coaxial cable for the connection to the guard cathode, and the current from this circuit should not be passed through the meter (i.e., the positive side of the polarization unit must be connected directly to the built-in reference through "ground" on the meter).

Deep-sea landers for operation of microsensors down to about 6000 m water depth have been developed by several groups (Reimers 1987; Gundersen and Jørgensen 1990). These landers are able to operate the microsensors with computerized depth control and data acquisition. The depth resolution may be as fine as 25 μm, and extremely detailed data can thus be obtained. The chapter by Bo Barker Jørgensen in this volume will present data obtained in deep-sea microbial mats by use of such a deep-sea lander.

Recent major advances in our understanding of microbial mats obtained by use of electrochemical microsensors

There are four major areas where microsensor analyses have resulted in major advances in our understanding of stratified microbial communities during the last 6 years, i.e., since the last microbial mat meeting. One of these areas is the hydrodynamics and diffusion characteristics around the benthic interface, and this topic is covered in this volume by Bo Barker Jørgensen who also presents the first microsensor determinations of sulfate reduction. Significant advances have also been made in the areas of benthic photosynthesis and nitrogen transformations around the oxic-anoxic interface. These topics are reviewed below.

Microsensor analysis of microbenthic photosynthesis

The routinely used methods for measuring microbial photosynthesis are far more controversial than usually realized. It is now 40 years ago that Steeman Nielsen (1952) first described the ^{14}C method for measuring planktonic photosynthesis, and he was the first also to realize its limitations. Still the values obtained by the routine procedures are widely accepted as "primary production," but primarily because of the interest in "microbial loop" research there is now growing awareness of the shortcomings of the method (e.g., Williams 1993). To have a better definition of primary production it is now common to define "new primary production" as a parameter which tells how much of the photosynthetically fixed carbon is available for export away from the photosynthetic community. When microsensors are used for determination of photosynthetic rates in a benthic community of microorganisms, two parameters can be obtained: (i) The net flux of oxygen out of the photosynthetic community which can be defined as net photosynthesis and (ii) rates of gross photosynthesis as determined by the light-dark shift technique (Revsbech and Jørgensen 1986). The net photosynthesis is a well defined parameter, as it directly tells how much oxygen is in surplus and thus also (assuming an O/C ratio of 1.2) how much the net fixation of carbon is. Unfortunately, the net photosynthesis can in practice only be calculated for the whole photosynthetic layer, and the often poor definition of diffusion characteristics may also make the flux calculations by which the net photosynthesis is determined rather inaccurate. The gross photosynthesis is a more tricky but also very valuable parameter, as it can be determined at <0.1 mm spatial resolution and together with the net photosynthesis gives a much better description of the photosynthetic community than net photosynthesis alone. However, until recently there was one very basic assumption in the theory of the method which was not satisfactorily checked: That the respiration rate was constant during the period (up to 4 s) through which the rate of oxygen decrease in the dark was monitored. To verify this assumption, Glud et al. (1992), covered a microbial mat with a 0.5 mm thick layer of agar and afterwards incubated the mat in the light until the oxygen profile was stable. The oxygen concentration was then continuously recorded during separate light-dark cycles (5 min light, 3.8 s dark) at 0.1 mm depth intervals (4 examples of the recordings shown to the right in Fig. 3). Oxygen profiles derived from these continuous data are shown to the left in Fig. 3. The time evolution of oxygen content within the microbial mat and its coating of agar can be calculated from these data, and the results are plotted in Fig. 4. It can be seen that the rate of decrease in oxygen content is perfectly linear, and the assumption of a constant rate of respiration was thus correct.

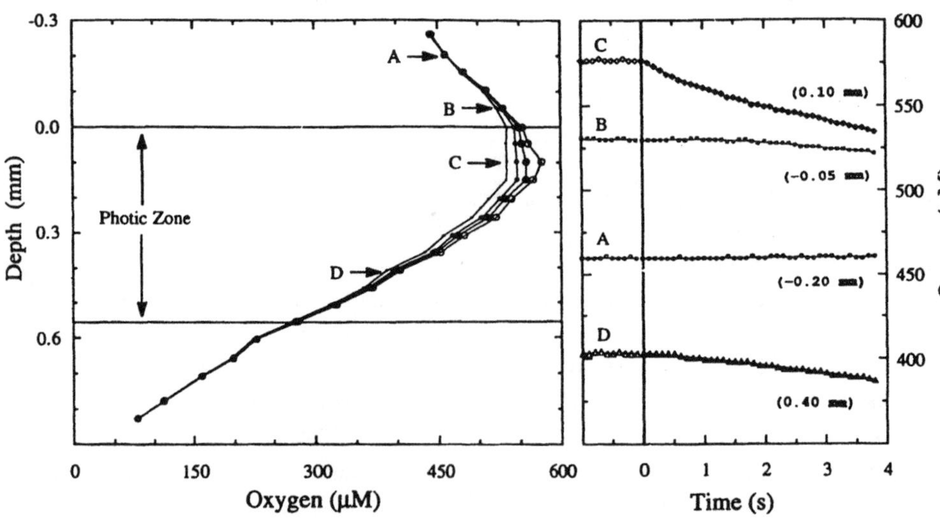

Fig. 3. Left: Oxygen profiles in a microbial mat during the first 3.8 s of a dark period. Right: Four examples of the continuous curves from which the profiles were obtained. From Glud et al. (1992).

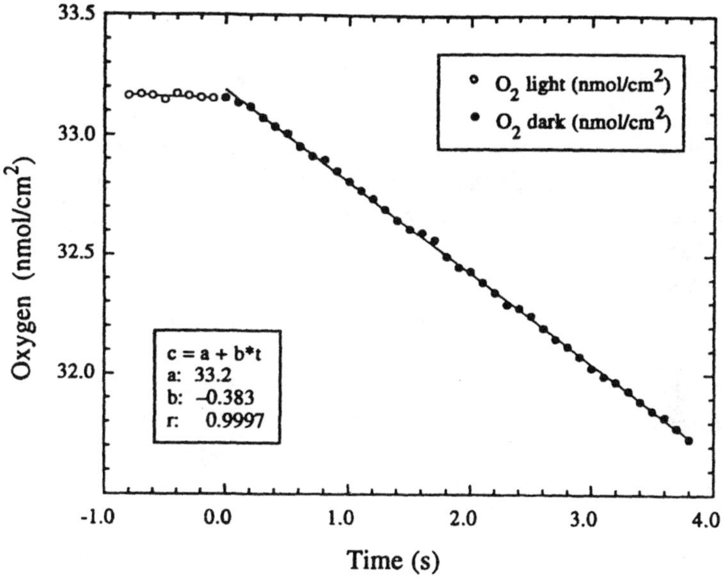

Fig. 4. Oxygen content of a microbial mat as a function of time after darkening. The oxygen contents were calculated from the data shown in Fig. 3. From Glud et al. (1992).

It has previously been shown (Jensen et al. 1989) that respiration within microbial mats, especially those dominated by diatoms, is affected by changes in O_2/CO_2 ratios in the same way as photorespiration should cause changes, and we thus expect that oxygen, which is almost immediately consumed through the oxygenase capabilities of RUBP-carboxylase, is included in our values for gross photosynthesis. The gross photosynthesis as determined by oxygen microsensors may thus not be an environmentally very relevant parameter, but together with estimates of net photosynthesis it still gives us the best possible understanding of photosynthetic processes in microbial mats, and the quality of the data compare very well with those which can be obtained from the planktonic environment.

Nitrification and denitrification

Denitrification has been thoroughly studied in the photosynthetic microbial mats (or biofilms) which grow on the light-exposed surface layer of trickling filters. These mats are especially well suited for such a study as they are easily accessible, relatively homogeneous, and are characterized by high rates of denitrification due to a continuous supply of NO_3^- and organic matter with the wastewater. The combined microsensor for O_2 and N_2O can be used for such a study, as depth profiles of denitrification can be modeled from the obtained N_2O profiles when N_2O reduction is inhibited by addition of acetylene (Dalsgaard and Revsbech 1992). The combined measurement of O_2 and N_2O is a great advantage as it makes it possible to position the denitrifying activity correctly with respect to the oxic-anoxic interface. The relatively homogeneous mat below the oxic-anoxic interface makes the calculation of a denitrification profile rather accurate, whereas a calculation of an activity profile from the oxygen data is virtually impossible because of small-scale heterogeneity and badly defined diffusion (or rather transport) characteristics of such heterogeneous substrates. The oxic surface layer of the biofilms was 0.1-0.3 mm thick during darkness and up to 1.5 mm thick during illumination (Dalsgaard and Revsbech 1992). No denitrification occurred in layers of the biofilms where the oxygen concentration exceeded 10-20 μM, and the spatial resolution of the activity determinations was not sufficiently good to verify any aerobic denitrification. Although we were not able to demonstrate aerobic denitrification in these mixed communities of microorganisms, careful investigations of some pure cultures showed a measurable activity when the oxygen concentration was low (Robertson et al. 1989; Robertson et al. in prep). Heterotrophic nitrification by artificial biofilms of *Thiosphaera pantotropha* could be shown by N_2O formation during incubation with acetylene in medium without added NO_3^- or NO_2^- but with NH_4^+ (T. Dalsgaard et al. in prep). Heterotrophic nitrification in the naturally occurring biofilms was, however, not detected. The rate of denitrification in biofilms and mats is governed by the supply of electron donors and acceptors. Examples of denitrifying activity in a biofilm incubated with 5 mM glucose and various NO_3^- concentration are shown in Fig. 5.

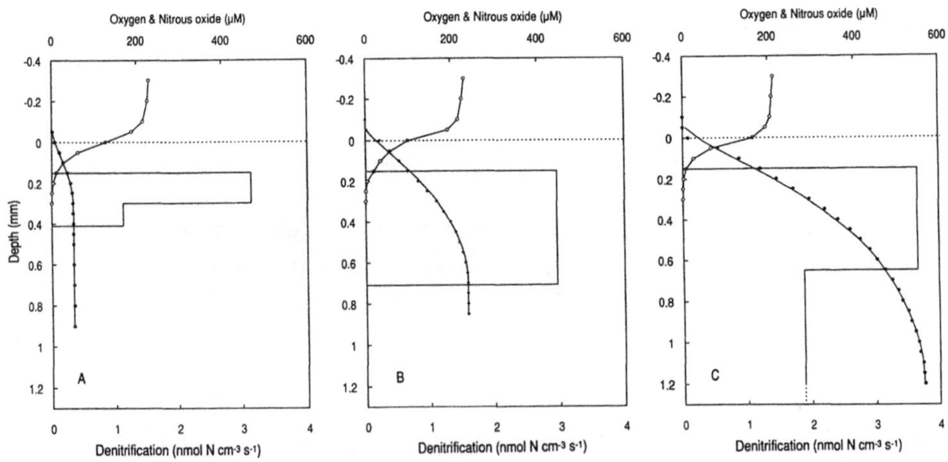

Fig. 5. Effect of NO_3^- concentrations on denitrification in a biofilm incubated with 0.8 mM NH_4^+, 5 mM glucose, and A: 125 μM NO_3^-; B: 500 μM NO_3^-; C: 1250 μM NO_3^-. Symbols represent measured steady-state concentrations of O_2 (○) and N_2O (●). Boxes indicate specific denitrification activity. Lines through symbols for N_2O indicate the N_2O profile that was calculated from the specific denitrification activity shown. Dotted lines indicate the biofilm-water interface. From Dalsgaard and Revsbech (1992).

The denitrification zone was less than 0.3 mm thick at 125 μM NO_3^- whereas denitrifying activity was present in the deepest analyzed layer at 1250 μM NO_3^-.

Although very detailed microprofiles of denitrification can be obtained by use of the combined O_2/N_2O microsensor, the approach has one very serious drawback: It is necessary to add acetylene to inhibit reduction of N_2O, but this acetylene also inhibits nitrification, which in many environments is the most important source for NO_3^-. The coupled process of nitrification-denitrification (see also chapter by Lars Peter Nielsen and Niels Peter Sloth in this volume) is thus not accounted for by analysis of N_2O profiles, but the coupled process may be investigated by analysis with NO_3^- microsensors. An example of such analysis is shown in Fig. 6, which shows data from a sediment incubated with no NO_3^- and NH_4^+ (Fig. 6A) or with 300 μM NH_4^+ (Fig. 6B) in the overlying water. The figure legend should be consulted for a thorough explanation of the figure. Although there was no added NO_3^-, nitrification in the oxic surface sediment resulted in a small NO_3^- concentration in the overlying water. Most of the NO_3^- consumption in the anoxic layers was, however, due to the direct flux of NO_3^- from the

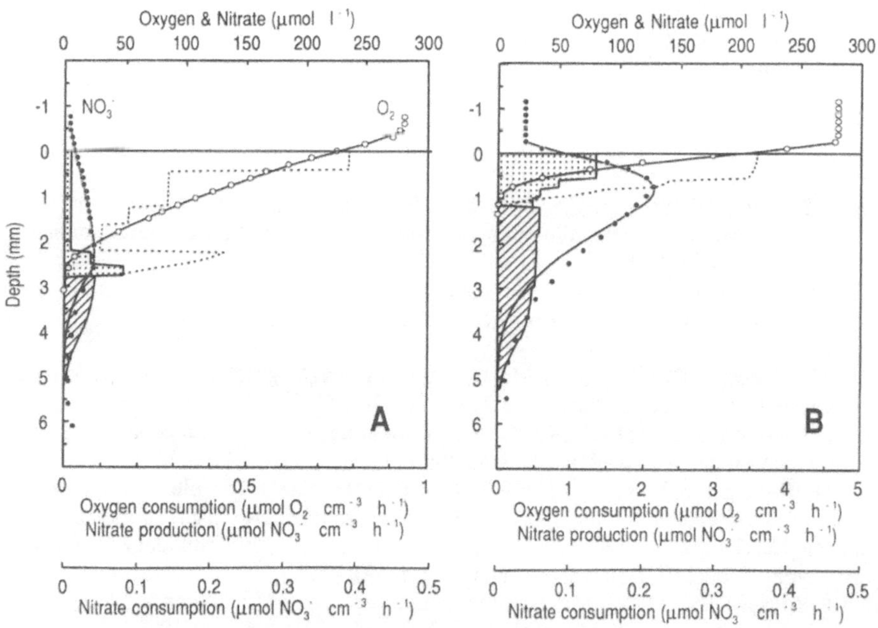

Fig. 6. Profiles of O_2 (○) and NO_3^- (●) in a sediment incubated without added NH_4^+ (A) or with 300 μM NH_4^+ added to the overlying water (B). Modelled profiles of O_2 and NO_3^- are shown by solid lines. Depth profiles of O_2 consumption, NO_3^- production, and NO_3^- consumption are illustrated by the white (enclosed by broken line), the shaded, and the hatched areas, respectively. Note that the scales for O_2 consumption and NO_3^- production are different in A and B. From Jensen et al. (1993).

nitrification zone and down to the anoxic layers. There are, however, also problems with the interpretation of data as those shown in Fig. 6. First of all, we cannot tell whether the NO_3^- consumption is due to denitrification, assimilation, or dissimilatory reduction to ammonium. The worst problem is, however, the poor accuracy of the NO_3^- determinations at low concentrations due to interference from HCO_3^-. It can be seen that the measured NO_3^- values (filled circles) never reach zero in deeper layers, and to model the depth profile of NO_3^- consumption it was necessary to force the NO_3^- profile to zero at the depth where no further decrease in apparent NO_3^- concentration was measured. Such an approach does not give a superb resolution of activity, but it still gives us an outstanding possibility for observing near-quantitative distributions of NO_3^- transformations which could not be obtained by any other means.

Acknowledgements

I thank T. Dalsgaard, R. Glud, K. Jensen, B.B. Jørgensen, M. Kühl, and N.B. Ramsing for permission to use their published figures. R. De Wit contributed with helpful suggestions for changes in the original manuscript. The developmental work in our laboratory was supported by the Danish Center for Microbial Ecology.

References

Ammann D, Lanter F, Steiner, RA, Schulthess P, Shijo Y, Simon W (1981) Neutral carrier based hydrogen ion selective microelectrode for extra- and intracellular studies. Anal Chem 53: 2267-2269

Dalsgaard T, Revsbech NP (1992) Regulating factors of denitrification in trickling filter biofilms as measured with the oxygen/nitrous oxide microsensor. FEMS Microbiol Ecol 3:

de Beer D, Sweerts J-PRA (1989) Measurement of nitrate gradients with an ionselective microelectrode. Anal Chim Acta 219: 351-356

de Beer, D van den Heuvel (1988) Response of ammonium-selective microelectrodes based on the neutral carrier nonactin. Talanta 35: 728-730

Glud RN, Ramsing NB, Revsbech NP (1992) Photosynthesis and photosynthesis-coupled respiration in natural biofilms quantified with oxygen microsensors. J Phycol 28: 51-60

Gundersen JK, Jørgensen BB (1990) Microstructure of diffusive boundary layers and the oxygen uptake of the sea floor. Nature 345: 604-607

Jensen J, Revsbech NP (1989) Photosynthesis and respiration of a diatom biofilm cultured in a new gradient growth chamber. FEMS Microbiol Ecol 62: 29-38

Jensen K, Revsbech NP, Nielsen LP (1993) Microscale distribution of nitrification activity in sediment determined with a shielded microsensor for nitrate. Appl Environ Microbiol 59: 3287-3296

Kühl M, Jørgensen BB (1992) Microsensor measurements of sulfate reduction and sulfide oxidation in compact microbial communities of aerobic biofilms. Appl Environ Microbiol 58: 1164-1174

Reimers CE (1987) An *in situ* microprofiling instrument for measuring interfacial pore water gradients: methods and oxygen profiles from the North Pacific Ocean. Deep Sea Res 34: 2019-2035

Revsbech NP (1989) An oxygen microelectrode with a guard cathode. Limnol Oceanogr 34: 472-476.

Revsbech NP, Jørgensen BB (1986) Microelectrodes: Their use in microbial ecology. Adv Microb Ecol 9: 293-352

Revsbech NP, Ward DM (1984) Microprofiles of dissolved substances and photosynthesis in microbial mats measured with microelectrodes. In: Cohen Y, Castenholz RW, Halvorson HO (eds) Microbial mats: Stromatolites. Alan R. Liss, New York, pp 171-188.

Revsbech NP, Jørgensen BB, Blackburn, TH, Cohen Y (1983) Microelectrode studies of the photosynthesis and O_2, H_2S and pH profiles of a microbial mat. Limnol Oceanogr 28: 1062-1074

Revsbech, NP, Nielsen LP, Christensen PB, Sørensen J (1988) Combined oxygen and nitrous oxide microsensor for denitrification studies. Appl Environ Microbiol 54: 2245-2249

Robertson LA, Cornelisse R, De Vos P, Hadioetomo R, Kuenen JG (1989) Aerobic denitrification in various heterotrophic nitrifiers. Antonie van Leeuwenhoek 56: 289-299.

Steemann Nielsen E (1952) Use of radioactive carbon (C^{14}) for measuring organic production in the sea. J Cons Cons Int Explor Mer 18: 117-140

Voipio J, Kaila K (1993) Interstitial PCO_2 and pH in rat hippocampal slices measured by means of a novel fast CO_2/H^+-sensitive microelectrode based on a PVC-gelled membrane. Pflügers Arch 423: 193-201

Williams, PJleB (1993) Chemical and tracer methods of measuring plankton production. ICES Mar Sci Symp 197: 20-36

Witty JF (1991) Microelectrode measurements of hydrogen concentrations and gradients in legume nodules. J Exp Bot 42: 766-771

Wolfbeis OS (ed) (1991) Fiber optic chemical sensors and biosensors. CRC Press, Boca Raton, Ann Arbor, Boston, London

Optical properties of microbial mats: Light measurements with fiber-optic microprobes

Michael Kühl, Carsten Lassen[1], and Bo Barker Jørgensen

Max Planck Institute for Marine Microbiology, Fahrenheit Str. 1, D-28359 Bremen, Germany

Introduction

Photosynthetic microbial mats occur as dense stratified communities in top layers of sediments, where microorganisms and sediment particles are embedded in an extracellular polymer matrix (e.g., *Farbstreifensandwatt*, Stal et al. 1985) or growing as a thin photosynthetic biofilm on solid substrata, e.g., stone or plant surfaces (Kühl 1993). In extreme environments, like hypersaline salt marshes and hot springs, regular mats several mm' to cm' thick and composed of almost pure biomass and exopolymers develop (Cohen and Rosenberg 1989). Photosynthetic microorganisms are the predominant component of these microbial mats, which often exhibit a vertical stratification of different colored layers due to the presence of photosynthetic microalgae and bacteria containing different photopigments with depth (Nicholson et al. 1987; Pierson et al. 1990). High metabolic rates due to the high density of microorganisms in mats combined with molecular diffusion acting as the major transport mechanism result in steep chemical gradients with depth as has been demonstrated by microelectrode measurements at <50-100 μm spatial resolution (Revsbech and Jørgensen 1986; Revsbech this volume). Production and consumption of the major electron acceptors can be calculated from measured microprofiles and different functional layers can thus be identified in microbial mats from microelectrode measurements. The typical sequence found is an upper oxygenic photosynthetic layer with concurrent oxygen respiration and a lower anoxic layer with denitrification and sulfate reduction as the predominant respiratory processes and with anoxygenic photosynthesis, provided sufficient light is penetrating from above (Jørgensen et al. 1983; Jørgensen and Des Marais 1986b; Revsbech et al. 1989; Kühl 1993). At the oxic-anoxic interface various chemoautotrophic bacteria (e.g., *Beggiatoa* or *Thiobacillus spp.*) oxidize reduced chemical species (e.g., H_2S) diffusing upwards from deeper mat layers, with oxygen or nitrate from above (Kühl and Jørgensen 1992a).

While a technique for measuring anoxygenic photosynthesis at high spatial resolution still has to be developed, oxygenic photosynthesis can be measured in microbial mats at <100 μm spatial resolution with the oxygen microelectrode light-dark shift technique

[1] Institute of Biological Sciences, Department of Microbial Ecology, University of Aarhus, Ny Munkegade, DK-8000 Aarhus C, Denmark

NATO ASI Series, Vol. G 35
Microbial Mats
Edited by Lucas J. Stal and Pierre Caumette

(Revsbech and Jørgensen 1983). Most studies of benthic photosynthesis have lacked information about the *in situ* spectral composition and intensity of light available for the microphytobenthos (Admiraal 1984). Thus, oxygenic photosynthesis has mostly been related to the incident quantum flux from above per unit surface area of the mat by measuring the downwelling 400-700 nm quantum irradiance with a large cosine collector. The relatively few measurements of light penetration in sediments reported in the literature, were done mostly by covering a large cosine collector with increasing amounts of sediment or by inserting miniprobes with non-ideal light collecting properties into the mat or sediment (e. g., Fenchel and Straarup 1971, Haardt and Nielsen 1980, Colijn 1982, Pierson et al. 1990).

Table 1. Basic optical parameters for light measurements in microbial mats.

Parameter	Definition	Microscale measuring technique
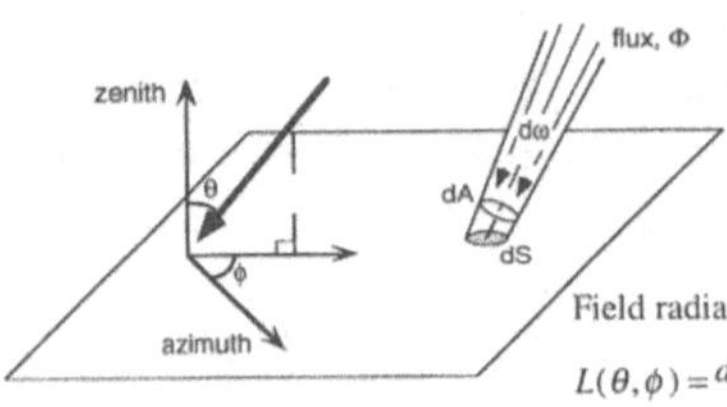 Field radiance: $L(\theta,\phi) = d^2\Phi/dA\,d\omega$	The radiant flux, Φ, from a certain direction (θ, φ) in a spherical coordinate system per unit solid angle, $d\omega$, per unit area perpendicular to the direction of light propagation, dA	Measured by a simple, flat-cut untapered or tapered optical fiber. The radiance fiberprobe has a directional response defined by the acceptance angle of the optical fiber. Tip diameter 10-125 µm. (Jørgensen and DesMarais, 1986; Kühl and Jørgensen 1992)
 Scalar irradiance: $E_0 = \int_{4\pi} L(\theta,\phi)\,d\omega$	The integral radiant flux incident from all directions about a point in the sediment. E_0 consists of a downwelling, E_{0d}, and an upwelling, E_{0u}, component corresponding to the integral flux incident from the upper or lower hemisphere, respectively	Measured by a coated and tapered optical fiber with a diffusing sphere fixed on the tip. The fiberprobe has an isotropic response for light and a tip diameter of 50-100 µm. (Lassen et al. 1992; Kühl and Jørgensen 1992)
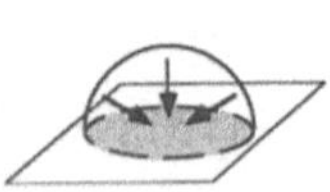 Downwelling irradiance $E_d = \int_{2\pi} L(\theta,\phi)\cos\theta\,d\omega$	The integral radiant flux incident from the upper (downwelling irradiance, E_d) or lower hemisphere (upwelling irradiance, E_u) per unit area of a horizontal surface element	Measured by a coated optical fiber with a diffusing disk fixed at the flat cut end of the fiber. The ir-radiance fiberprobe weights the incident radiance, $L(\theta, \phi)$, with the cosine of the incident zenith angle, θ. Tip diameter 40-125 µm. (Lassen et al. in prep.)

In order to study the optical properties and photosynthesis-light interactions in sediments and microbial mats at a spatial resolution comparable to oxygen microelectrode measurements, i.e. <100 μm, we have developed fiber-optic microprobes with defined light collecting properties for spectral measurements of the fundamental light parameters: radiance, irradiance, and scalar irradiance (see definitions in Table 1) (Jørgensen and Des Marais 1986a; Lassen et al. 1992a, Kühl and Jørgensen 1992b, Lassen et al. in prep.). The microprobes were used in several types of marine sediments and microbial mats in purely optical studies and in studies, where concurrent microscale measurements of light and photosynthesis were used to investigate the photosynthetic performance of microbenthic communities (Jørgensen and Des Marais 1988; Jørgensen et al. 1987; Lassen et al. 1992b; Kühl 1993; Kühl and Jørgensen 1993; Ploug et al. 1993). In the following, we summarize and discuss some results of these recent studies with focus on the optical properties of microbial mats. A detailed discussion of the photosynthetic performance of a marine microbial mat is presented elsewhere in this volume (Lassen et al. this volume).

Fiber-optic microprobes

Optical fibers are well-suited for building optical microprobes for several reasons: i. They have defined light collecting properties. ii. The raw fiber has a small diameter (typically 50-200 μm), and it is relatively easy to taper the fiber down to a 10-15 μm tip diameter. iii. Optical fibers are flexible and relatively sturdy when used in sediments and other dense materials.

The basic design of our optical microprobes is simple (Fig. 1B). The protective coating of a 1-1.5 m long single strand optical fiber cable is stripped of at both ends. One end of the fiber is then cut with a diamond knife to obtain a perfectly flat output end of the fiber, which is inserted into the light detection system. The other end is inserted through a hypodermic needle and fixed with a drop of epoxy resin. This light collecting end of the fiber can be modified according to the optical parameter to be measured.

For use in microbial mats and sediments it is essential to connect the fiberprobes to a sensitive detector system. Presently we are using an intensified diode array system for 3-5 nm spectral resolution light measurements of visible and infrared light (up to 900 nm) (Fig. 1A, described in Kühl and Jørgensen 1992b) and a simple battery operated photomultiplier system equipped with a heat reflecting mirror for integral measurements of 400-700 nm light (PAR).

Radiance microprobes

The most fundamental light parameter for studies of the light field is the field radiance, which is the radiant flux per unit area per unit solid angle from a certain direction (Table 1). A radiance microprobe should thus have a well-defined directional response with a narrow acceptance angle for light. The directional sensitivity is specified by the numerical aperture, NA = $n_0 \sin \Theta_a$, where n_0 is the refractive index of the surrounding medium and Θ_a, is the acceptance half angle of the optical fiber (Senior 1985). Radiance fiberprobes thus have a higher acceptance half angle in air (n_0 = 1) than in water (n_0 = 1.33).

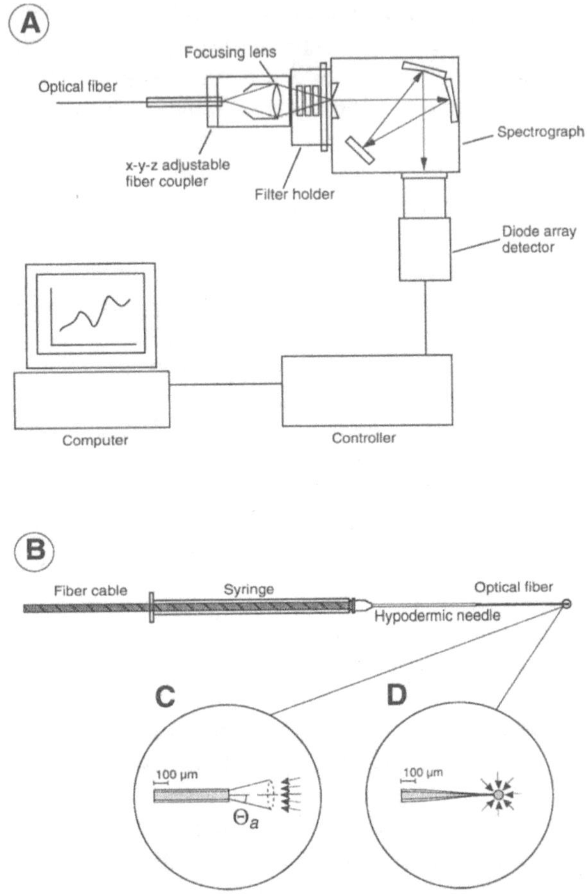

Fig.1. Diode array detector system for fiber-optic microprobes. A. Detector system. B. Fiber-optic microprobe. C. Tip of microprobe designed to measure field radiance. D. Spherical tip of microprobe designed to measure scalar irradiance (From Kühl and Jørgensen 1992b)

A simple radiance probe can be constructed by using a flat cut untapered optical fiber for collecting light (Fig. 1C). We used multimode graded-index optical fibers with an outer diameter of 125 μm and a NA of 0.2 (Newport Corp, F-MSD) giving an acceptance angle of 11.5° in air and 8.6° in water. Smaller radiance probes were made by tapering the optical fiber down to 10-50 μm in a small acetylene flame and then cutting the fiber with a diamond knife (Fig. 2A). Tapered fibers have a broader acceptance angle due to light entering through the tapered sides of the fiber tip. This problem can be solved by coating the fiber tip with an opaque enamel or a Cr-metal film (Vogelmann et al. 1991). We are able to make coated radiance probes with tip diameters down to <15 μm and acceptance angles <15-20° in air.

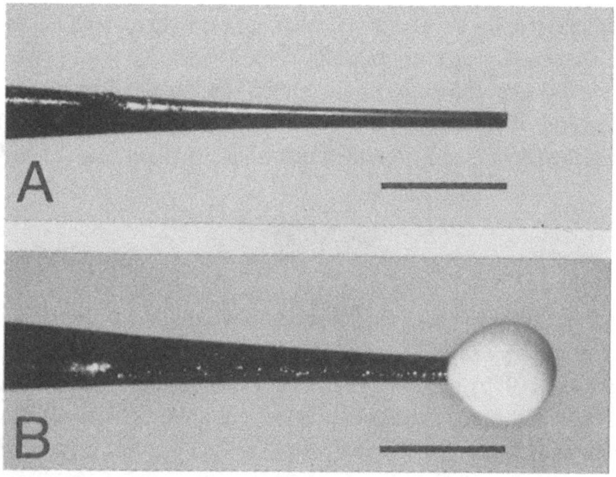

Fig. 2. Fiberoptic microprobes for field radiance (A) and scalar irradiance (B). Length of bar is 100 μm.

Scalar irradiance microprobes

Scalar irradiance is an integral measure of the total radiant flux from all directions about a point in the sediment, i. e. the spherically integrated radiance from all directions (Table 1). The scalar irradiance is the most relevant measure of total quantum flux available for the microphytobenthos which receive light from all directions around the cell in sediments and microbial mats (Jørgensen and Des Marais 1988; Kühl and Jørgensen 1992b). A scalar irradiance microprobe should thus weight incident light from all directions equally, i. e. have an isotropic angular response. This is accomplished by casting a small diffusing sphere of TiO_2-doped methacrylate onto the tip of a tapered and coated optical fiber (Fig. 1D, Fig. 2B). The construction details are described by Lassen et al. (1992a). The microprobes exhibit an isotropic ($\pm$10%) response in air and in water. Scalar irradiance microprobes can be made with sphere diameters from 50 to 100 μm and can thus be used for measurements of total quantum flux in microbial mats at $<$0.1 mm spatial resolution, i. e. the same spatial resolution as obtained by the oxygen microelectrode technique for measuring oxygenic photosynthesis.

Irradiance microprobes

Irradiance is an integral measure of the incident radiant flux per unit horizontal surface area from above (downwelling irradiance) or below (upwelling irradiance) the surface (Table 1). Light travelling at oblique angles is weighted with the cosine of the incidence angle, and irradiance thus weights scattered light less than vertically incident light. An irradiance microprobe with the described cosine response for light can be made by fixing a small droplet of TiO_2-doped methacrylate on the end of a coated optical fiber, which is then totally covered with black opaque enamel. By careful grinding of the coated fiber end under the microscope, the white methacrylate is exposed at the tip forming a small disk surrounded by black enamel, i. e. a miniature version of the large irradiance cosine collector which is commonly used in aquatic ecology. Further details are presented elsewhere (Lassen et al. in prep.).

Radiance measurements

Simple radiance microprobes connected to a single photodiode for use in microbial mats and sediments were introduced by Jørgensen and Des Marais (1986a). Studies with this system are reviewed by Jørgensen (1989) and will not be included in the present overview. Here we present data obtained with the more sensitive (but less portable) light detection system described by Kühl and Jørgensen (1992b).

Spectral light penetration and zonation of microbial mats

Due to the directional sensitivity of radiance microprobes they are useful in studies of the dynamics and regulation of the vertical zonation of phototrophs in mats. By measuring spectral downwelling radiance by penetrating a mat sample from below or by measuring spectral upwelling radiance coming with the fiber probe from above the mat, the vertical zonation of major photopigments can be described (Jørgensen and Des Marais 1988; Kühl and Jørgensen 1992b). An example is given in Fig. 3, where downwelling spectral radiance was measured at 50 μm resolution in a laminated hypersaline microbial mat from Eilat, Israel. The radiance transmittance spectra exhibit throughs at absorption wavelengths of the major photopigments in the mat (Fig. 3A). In the upper 0.0-1.5 mm, Chl *a* (440nm and 675 nm) and carotenoids (450-550 nm) were the dominant photopigments due to the predominance of diatoms in this layer. At 1.5-2.0 mm below the mat surface, a dense cyanobacterial layer (*Microcoleus chthonoplastes* and *Phormidium spp.*) was found exhibiting strong spectral signals from Chl *a* (675 nm) and phycocyanin (620-625 nm). Below 2.5 mm, spectral signals from Bchl *a* (807 nm and 865 nm) and Bchl *c* (746 nm) were found due to the presence of green filamentous bacteria (*Chloroflexus sp.*) and purple sulfur bacteria in the bottom layer of the mat. A more detailed picture of this vertical zonation in the mat was obtained from analyzing radiance depth profiles for the absorption wavelengths mentioned above (Fig. 3B). By fitting functions (6'th or 7'th order polynomia) to the log-transformed radiance profiles it was possible to calculate depth profiles of the vertical attenuation coefficient of radiance, K_L. The attenuation coefficient is defined as the rate of change of log-transformed radiance values with depth, $K_L = -d(\ln L)/dz$ (Kirk 1983). Depth profiles of attenuation coefficients could thus be calculated by taking the first derivative with respect to depth of polynomia fitted to log transformed radiance depth profiles (Fig. 3C). In the upper 0-1 mm of the mat, the apparent high attenuation of radiance at all wavelengths was due to the scattering of photons out of the downwelling path of the

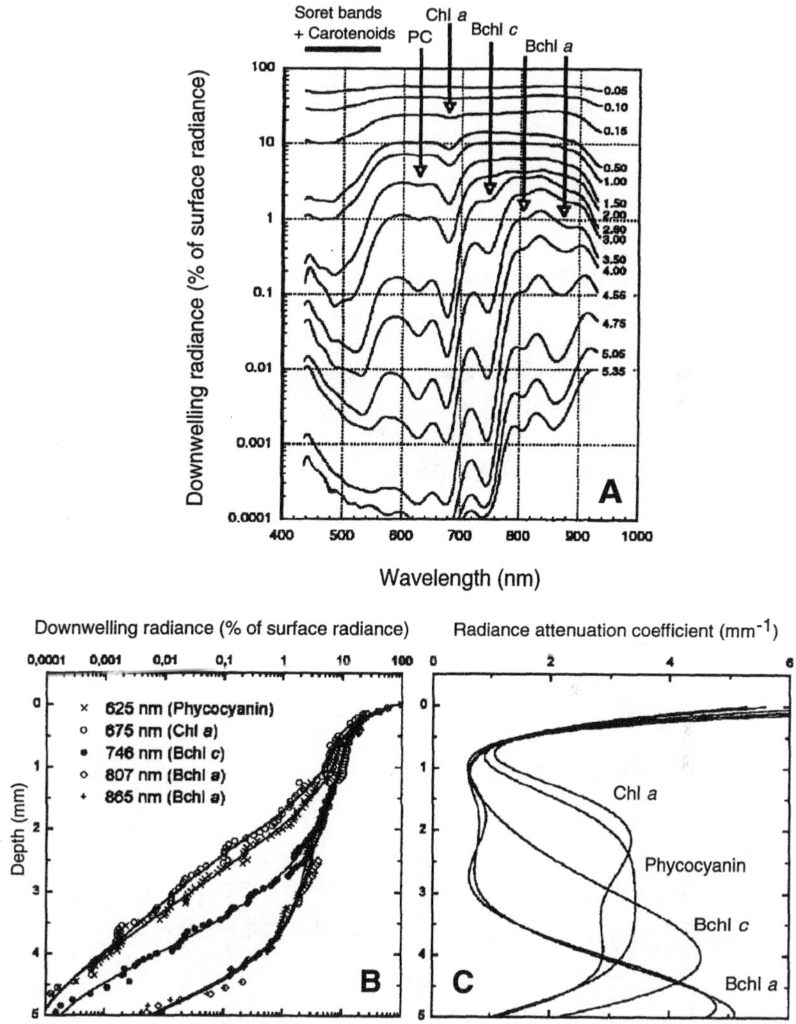

Fig.3. Measurements of downwelling radiance in a cyanobacterial mat from Eilat, Israel.
A. Radiance spectra from selected depths in the mat (indicated by numbers on graph).
B. Depth profiles of radiance at absorption wavelengths of major photopigments in the mat. Solid lines represent 6'th order polynomia fitted to the data.
C. Depth profiles of attenuation coefficients calculated from the first derivative of ln-transformed data.
(Kühl et al. unpublished data)

incident collimated light as well as light absorption by photopigments. This effect of scattering out of the light path is less important in deeper mat layers with a more diffuse light field (see also later in this chapter). Although the results are thus unclear in the upper 0-0.5 mm of the mat, the attenuation coefficient depth profiles demonstrate the vertical zonation of the major photopigments in the microbial mat as a continuous mixture of photopigments throughout the mat, with highest Chl *a* and phycocyanin concentrations in the upper mat layers (0-3 mm) and gradually increasing amounts of Bchl *c* and Bchl *a* in the deeper mat layers (2.5-5 mm) (Fig. 3C).

Reflectance spectra and migration in microbial mats

The vertical zonation of a microbial mat is not static as many of the phototrophic microorganisms migrate in response to changes in light and in chemical gradients during a diel cycle (Jørgensen 1982). It is possible to map this migration by measuring either down- or upwelling radiance spectra with fiber-optic microprobes. Figure 4 shows the ratio between the upwelling radiance from a white $BaSO_4$ reflection standard and the reflected radiance from the

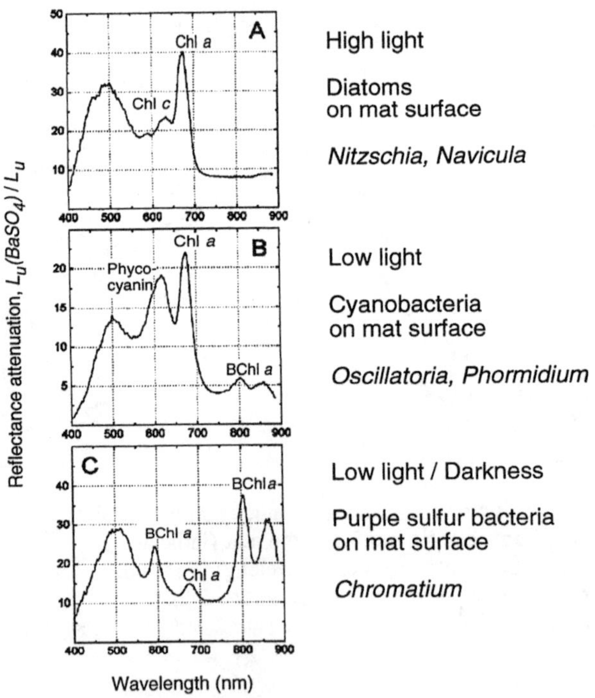

Fig. 4. Changes in attenuation of upwelling radiance due to the migration of photosynthetic microorganisms in a marine microbial mat from Limfjorden, Denmark. Data are presented as the ratio of upwelling radiance from a reflectance standard and the upwelling radiance from the mat. (Kühl et al. unpublished data)

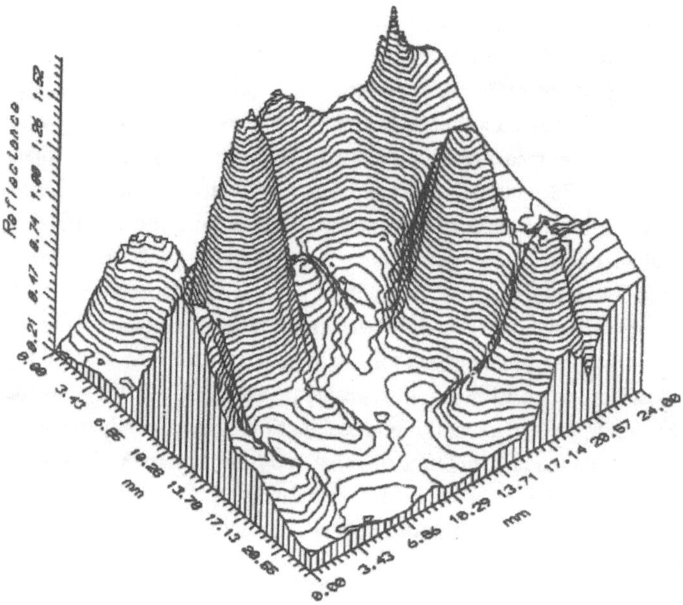

Fig. 5. Radiance reflectance of 620 nm light measured by a radiance microprobe within a 25 x 25 mm surface area of a cyanobacterial mat (Pond 5, Guerrero Negro, Baja California, Mexico). Data are expressed in % of reflected radiance from a $BaSO_4$ reflectance standard (B. B. Jørgensen and L. L. Richardson unpublished data).

surface of a marine microbial mat from Limfjorden, Denmark. The dominant migrating phototrophs were diatoms (mostly *Nitzschia spp.* and *Navicula spp.*), cyanobacteria (*Oscillatoria spp.*), and purple sulfur bacteria (*Chromatium sp.*). Under high light (> 100-200 μEinst. m^{-2} s^{-1}) the surface of the mat was covered with diatoms (Fig. 4A) exhibiting clear light absorption peaks of Chl *a* (675 nm), Chl *c* (630-35 nm), and broad absorption between 400-560 nm due to chlorophylls and carotenoids. In low light (<50 μEinst m^{-2} s^{-1}) cyanobacteria migrated within a few hours up to the mat surface (Fig. 4B) exhibiting absorption peaks of Chl *a* (675 nm), phycocyanin (620 nm) and other phycobilins (560- 620nm). Small peaks of BChl *a* absorption (790-810 nm, and 830-880 nm) also showed up indicating that the purple sulfur bacteria had also come closer to the mat surface. After extended periods in low light or in darkness (> 6 hours) the purple sulfur bacteria migrated up to the mat surface (Fig. 4C), now exhibiting clear absorption peaks of BChl *a* at 590 nm, 805 nm and 865 nm.

By doing many point measurements of reflectance, the horizontal distribution of photopigments, i. e. the *reflectance topography* of microbial mats, can be studied at high resolution. An example is shown in Fig. 5, where the surface radiance reflectance at 620 nm (phycocyanin absorption maximum) was measured within a 25 x 25 mm area of a hypersaline cyanobacterial mat at 1 mm spatial resolution (Pond 5, Guerrero Negro, Baja California, Mexico). The lowest reflectance values correspond to the highest concentrations of phycocyanin. The 620 nm reflectance topography of the mat is thus a mirror image of the surface distribution of cyanobacteria. In combination with spectral surface scans of larger mat areas by a CCD camera, measurements of the reflectance topography of mats with radiance microprobes could provide detailed information on the horizontal distribution of photosynthetic microorganisms and their migratory patterns (Richardson and Jørgensen, unpublished data).

In conclusion, radiance microprobes are powerful tools for investigations of the finescale vertical and horizontal distributions of photosynthetic microorganisms in microbial mats. Fiber-optic microprobes are, however, also able to resolve changes in the vertical zonation due to migration of the phototrophs in microbial mats. The method has the advantage of being non-destructive and allowing for repeated measurements in the same mat sample under different light conditions. Especially in combination with recently developed minicoring techniques for simultaneous collection of microscopy and pigment samples at high spatial resolution (Garcia-Pichel et al. 1993), fiber-optic microprobes can now be used for detailed studies of migratory behavior of microorganisms in microbial mats.

The light field in microbial mats

Angular radiance distributions

Knowledge of the light intensity in microbial mats is essential to understand the regulation of microbenthic photosynthesis. Quantification of the light intensity in a microbial mat is, however, not possible by measuring only radiance depth profiles in a single direction. This is evident from measurements of radiance depth profiles taken at different zenith angles relative to the incident collimated light (Kühl and Jørgensen 1993). Figure 6 shows such an angular radiance distribution measured at 650 nm in a marine microbial mat exhibiting a yellow-brown surface layer of sand with a few diatoms, and a dark green subsurface cyanobacterial layer on top of a pink layer of purple sulfur bacteria.

At the sediment surface the light field was highly directional and composed of the incident collimated light (0°-20° zenith angle) and diffuse backscattered light (90°-180° zenith angle) from the mat. The little forward scattered light detected at the mat surface was due to internal reflection at the water-air interface back to the sediment surface of some backscattered light from the sediment surface. Already 0.5-1.0 mm into the sand layer the angular light field changed dramatically. The collimated component of the light field decreased strongly due to intense scattering of light out of the downwelling light path, resulting in much forward scattered light. Thus, only a few tenths of a mm below the mat surface the spatial light distribution had changed from a collimated to a diffuse light field with scattered light being the major light source for photosynthetic microorganisms in the mat. In deeper mat layers the angular light distribution was affected by the lamination of the mat. At >1 mm depth the backscattered light was attenuated more strongly due to the underlying dense cyanobacterial layer and the light distribution became more forward biased.

Thus, the laminated structure of microbial mats exhibiting varying scattering to absorption ratios with depth can interact with the general trend of the light field to become more diffuse with depth. These effects of layers with different absorption and scattering properties on the spatial light distribution are not present in non-stratified coastal sediments, where the angular light field approaches an asymptotic radiance distribution which depends on the inherent absorption and scattering properties of the sediment (Kühl and Jørgensen, 1993). The magnitude of these changes in the light field with depth also depend strongly on the ratio of scattering versus absorption at different wavelengths so that the light field is relatively more forward biased for wavelengths subject to high absorption than for wavelengths subject to low absorption. Further details on the analysis of the angular light field can be found in Kühl and Jørgensen (1993).

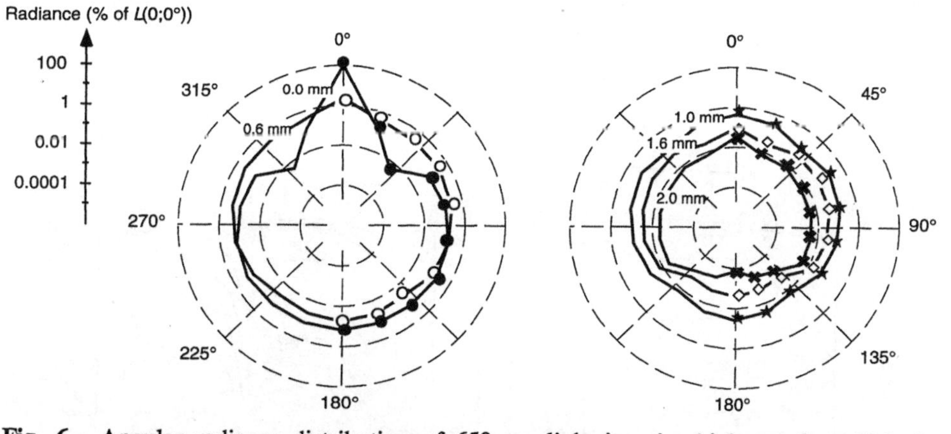

Fig. 6. Angular radiance distribution of 650 nm light in microbial mat from Kaløvig, Denmark. Numbers on curves indicate depth. Symbols indicate measured radiance values expressed on a log-scale in % of the downwelling radiance from 0° zenith angle at the mat surface (Kühl and Jørgensen unpublished data).

Irradiance and scalar irradiance

Irradiance and scalar irradiance are both common measures of light intensity. Irradiance is defined as the incident light from above (downwelling irradiance, E_d) or below (upwelling irradiance, E_u) a horizontal surface (Table 1) and weights light incident at oblique angles less than collimated light. Measurements of the angular radiance distribution (Fig. 6) show, however, that microorganisms in microbial mats live in a highly diffuse light field with primarily scattered light coming from all directions around the cell. Irradiance is therefore not the optimal measure of total light intensity in sediments, microbial mats or other highly light scattering media. A more useful measure of light intensity is the scalar irradiance, E_0, which is an integral measure of light incident from all directions around a certain point in the mat (Table 1).

The significance of scattered light in microbial mats is illustrated in Fig. 7 and Fig. 8. In Fig. 7, spectral measurements of downwelling radiance and scalar irradiance are compared in the same marine microbial mat. The changes in spectral composition of light with depth corresponded well with the observed zonation of phototrophic microorganisms: a sandy surface layer with diatoms (Chl *a* absorption at 440 nm and 675 nm), a dense cyanobacterial layer (Chl *a* and phycocyanin absorption at 620 nm), and a bottom layer of purple sulfur bacteria (BChl *a* absorption at 805 nm and 860-880 nm). Downwelling spectral radiance was attenuated much more strongly with depth than scalar irradiance. At 0.8 mm below the mat surface, downwelling radiance was attenuated to <2-3% of the incident light throughout the spectrum (Fig. 7A), whereas scalar irradiance was only attenuated to 20-60% in the visible part of the spectrum and still exhibited 100% of the incident IR light at 0.8 mm depth (Fig. 7B). Thus, the difference between the attenuation of downwelling radiance and scalar irradiance was an order of magnitude or more.

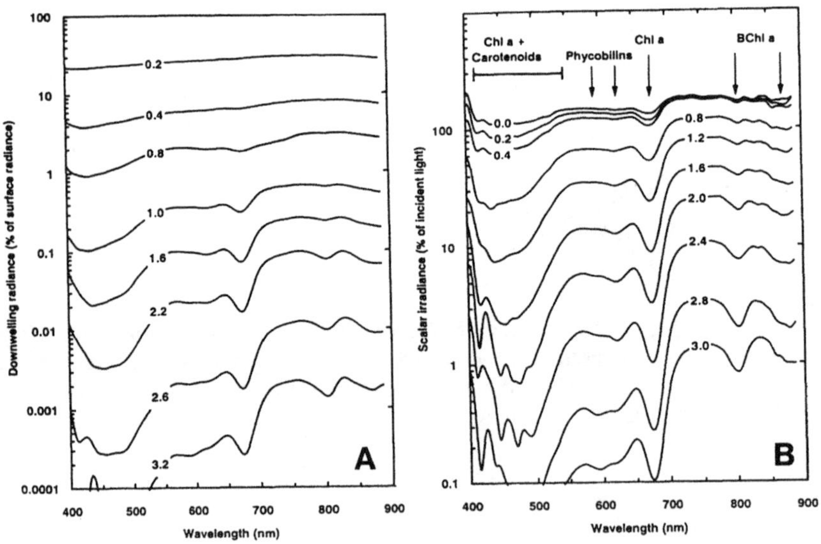

Fig. 7. Spectral light measurements in a marine microbial mat from Kaløvig, Denmark. A. Downwelling radiance. Data are normalized to downwelling radiance at the mat surface. B. Scalar irradiance. Data are normalized to incident downwelling irradiance at the sediment surface. Numbers on curves indicate depth in mm below the mat surface. Arrows indicate absorption wavelengths of major photopigments. (From Kühl and Jørgensen 1992b).

The spectral signals of photopigments were more pronounced in the scalar irradiance spectra than in the radiance spectra due to the spectral contribution of scattered light from sediment layers at some distance from the actual measuring position. The scattered light had been filtered through mat layers and was thus relatively enriched with respect to wavelengths subject to low absorption, thereby enhancing spectral differences in the light field.

A surprising result was the significant alteration of the light intensity and the spectral composition of light near the mat surface as compared to the incident light (Fig. 7B). Intense scattering resulted in a surface or subsurface maximum of scalar irradiance, i. e. of the total quantum flux to each microalgal cell, of up to 200% of the incident light intensity in the IR region of the spectrum. The spectral composition of the surface light was altered due to the spectral signal of backscattered light from deeper mat layers thereby enhancing light intensity at wavelengths with low absorption relative to absorption wavelengths of the photopigments in the mat.

The surface maximum of scalar irradiance was even more pronounced in pure quartz sand with a high scattering to absorption ratio (Fig. 8). Integrated visible (400-700 nm) light reached 200% of incident scalar irradiance at the sediment surface. The light intensity was

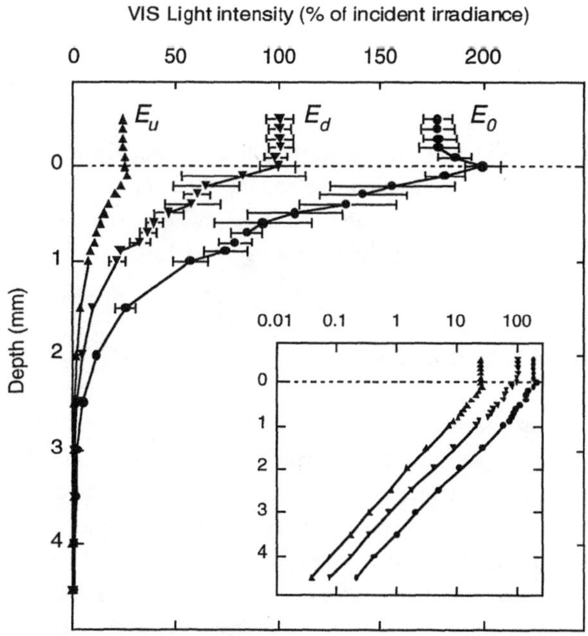

Fig. 8. Depth profiles of upwelling irradiance, E_u, downwelling irradiance, E_d, and scalar irradiance, E_0, in pure quartz sand with 125-250 μm particle size. Data are expressed in % of the downwelling irradiance at the sediment surface. Inset shows log-transformed data. (Data from Kühl et al. submitted).

attenuated exponentially with depth but remained above 100% of incident scalar irradiance throughout the upper 0.0-0.5 mm of the sediment. When comparing scalar irradiance with the irradiance, which is also shown in Fig. 8, it is evident that downwelling irradiance underestimated the total light intensity by 100% throughout the sediment. Although the measurements in pure sand represent an extreme situation with high scattering intensity and little absorption, similar scalar irradiance maxima have been calculated also from measured radiance distributions or have been measured directly in microbial mats and biofilms (e.g., Jørgensen and Des Marais 1988; Lassen et al. 1992b; Kühl 1992; Kühl and Jørgensen 1993) as well as in plant and animal tissue (e.g., Star et al. 1987; Vogelmann et al. 1991). This phenomenon of a near surface maximum of scalar irradiance thus seems to be an inherent property of compact light-scattering media in which multiple scattering is important for the radiative transfer.

Although a build-up of light intensity to a value higher than that incident onto the surface intuitively seems to be a violation of the laws of thermodynamics, the phenomenon can be explained as an effect of scattering and internal reflections at optical boundaries (e.g., Anderson et al. 1989; Seyfried 1989). Light attenuation in microbial mats is due to both absorption and multiple scattering, where scattering enhances the probability of absorption. If absorption is low, however, strong scattering retains a high flux density at a given depth in the mat (i. e., light is travelling a longer distance per vertical distance traversed). This effect of multiple scattering is especially enhanced near optical boundaries, e.g., the mat-water interface or at interfaces between different layers in a mat, where differences in refractive index could result in internal reflection at the boundaries. This results in apparent light trapping phenomena such as the local maximum of light intensity relative to the incident light from above. A more detailed discussion of the optical mechanisms behind the observed scalar irradiance maximum is beyond the scope of this overview and can be found elsewhere (e. g., Anderson et al. 1989; Seyfried 1989; Vogelmann et al. 1991; Kühl and Jørgensen 1993).

Modelling the light field in microbial mats

With the present microsensor techniques it is now possible to obtain detailed data for the optical properties of microbial mats and sediments, which can be used to model the light field in microbenthic communities. Such models must take into account the optical complexity of microbial mats with respect to multiple scattering, refractive index mismatch and the microheterogeneity in mat structure and composition. Models taking these factors into account have recently been developed for studies of light propagation in plant tissues which have an optical complexity similar to microbial mats (e.g., Fukshansky et al. 1993). These models are also to a large extent based on detailed light measurements with fiber-optic probes. A similar approach thus seems very promising for modelling the light field in microbial mats and in other compact phototrophic communities

Light and photosynthesis

The presence of a near-surface zone in microbial mats and sediments, where the light field exhibits a different spectral composition and a significantly higher light intensity has of course important experimental implications for photosynthesis studies in these systems. In studies of photosynthesis versus light intensity (P vs. I curves) or wavelength (action spectra) it is essential to use scalar irradiance as the measure of light intensity in microbial mats and sediments. By the use of scalar irradiance microprobes in combination with oxygen microelectrodes it is now possible to study the photosynthetic performance at high spatial resolution in microbial mats. A typical dataset showing depth profiles of oxygen concentration, photosynthesis, and photon scalar irradiance (400-700 nm) in a coastal microbial mat is presented in Fig. 9. The first studies in microbial mats using this type of data for measuring photosynthetic efficiency and action spectra related to the in situ measured scalar irradiance were recently published (Lassen et al. 1992b, Ploug et al. 1993). More details are presented by Lassen et al. in this volume. The technique has been applied also in other microbial habitats

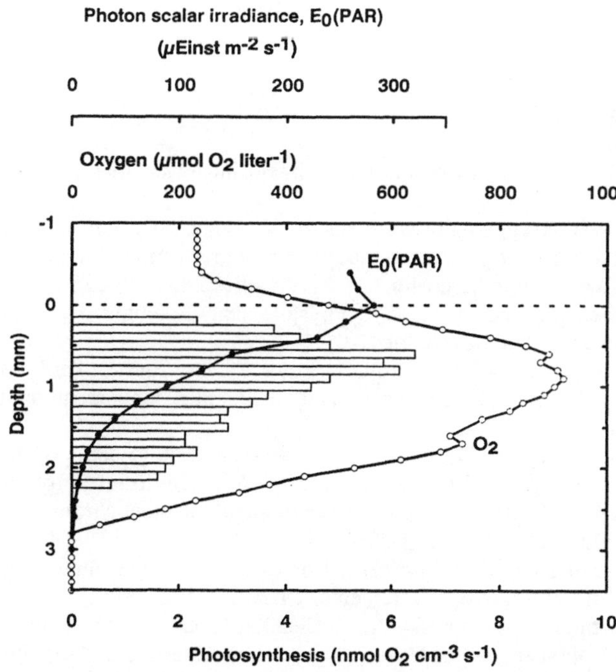

Fig. 9. Depth profiles of oxygen concentration, photosynthesis, and photon scalar irradiance in a marine microbial mat from Kaløvig, Denmark. (Data from Kühl and Jørgensen 1992b)

such as epiphytic communities of red algae and cyanobacteria on Codium macroalgae (Lassen et al. unpublished data) and in preliminary studies of symbionts in the coral tissue and of the endolithic algae in the sceleton of stony corals (Kühl et al. unpublished data).

Conclusion

The present fiber-optic microsensors used in connection with a sensitive detector system now enable studies of the light field in microbial mats at the same high spatial resolution as that with which the chemical microenvironment can be characterised by microelectrodes. From combined radiance, irradiance, and scalar irradiance measurements, the optical properties of mats and sediments can be described. Thus, sediment optics can now be analyzed at a comparable detail as in aquatic optics. The modelling of the light field in microbial mats should take into account, however, the higher optical complexity due to multiple scattering, fluorescence and microheterogeneity through which the optical properties of mats and sediments differ from those of water. Furthermore, the possibility to relate photosynthesis measured by oxygen microelectrodes to the *in situ* light intensity and spectral composition in microbial mats and sediments by concurrent measurements of scalar irradiance now makes it possible to investigate microbenthic photosynthesis at a level previously reached only in plankton research.

References

Admiraal W (1984). The ecology of estuarine sediment-inhabiting diatoms. Progr Phycol Res 3: 269-322

Anderson RR, Beck H, Bruggemann U, Farinelli W, Jaques SL, Parrish JA (1989). Pulsed photothermal radiometry in turbid media: Internal reflection of backscattered radiation strongly influences optical dosimetry. Appl Optics 28: 2256-2262

Cohen Y, Rosenberg E (eds.) (1989). Physiological Ecology of Benthic Microbial Communities. Am Soc Microbiol.

Colijn F (1982). Light absorption in the waters of the Ems-Dollard estuary and its consequences for the growth of phytoplankton and microphytobenthos. Neth J Sea Res 15: 196-216

Garcia-Pichel F, Mechling M, Castenholz RW (1993). Diel migrations of micro-organisms within a benthic, hypersaline mat community. Appl Environ Microbiol, in press

Haardt H, Nielsen GÆ (1980) Attenuation measurements of monochromatic light in marine sediments. Oceanol Acta 3: 333-338

Fenchel TM, Straarup BJ (1971) Vertical distribution of photosynthetic pigments and the penetration of light in marine sediments. Oikos 22: 172-182

Fukshansky L, Martinez v Remisowsky A, McClendon J, Ritterbusch A, Richter T, Mohr H (1993) Absorption spectra of leaves corrected for scattering and distributional error: a radiative transfer and absorption statistics treatment. Photochem Photobiol 57 (3)

Jørgensen BB (1982) Ecology of the bacteria of the sulphur cycle with special reference to anoxic-oxic interface environments. Philos Trans R Soc London Ser B 298: 543-561

Jørgensen BB, Revsbech NP, Cohen Y (1983) Photosynthesis and structure of benthic microbial mats: Microelectrode and SEM studies of four cyanobacterial communities. Limnol Oceanogr 28: 1075-1093

Jørgensen BB, Des Marais DJ (1986a) A simple fiber-optic microprobe for high resolution light measurements: Application in marine sediment. Limnol Oceanogr 31: 1376-1383

Jørgensen BB, Des Marais D J (1986b) Competition for sulfide among colorless and purple sulfur bacteria in a cyanobacterial mat. FEMS Microbiol Ecol 38: 179-186

Jørgensen BB, Cohen Y, and Des Marais DJ (1987) Photosynthetic action spectra and adaptation to spectral light distribution in a benthic cyanobacterial mat. Appl Environ Microbiol 53: 879-886

Jørgensen BB and Des Marais DJ (1988) Optical properties of benthic photosynthetic communities: Fiber-optic studies of cyanobacterial mats. Limnol Oceanogr 33: 99-113

Jørgensen BB (1989) Light penetration, absorption and action spectra in cyanobacterial mats. In: Cohen Y, Rosenberg E (eds) Physiological Ecology of Benthic Microbial Communities. Am Soc Microbiol, pp 123-137

Kirk JTO (1983) Light and photosynthesis in aquatic ecosystems. Cambridge.

Kühl M (1993) Photosynthesis, O_2 respiration and sulfur cycling in a cyanobacterial biofilm. In: Guerrero R, Pedrós-Alió C (eds) Trends In Microbial Ecology. Proceedings of the 6'th International Symposium on Microbial Ecology, Barcelona Sept. 6-11 1992. Spanish Society for Microbiology, pp 163-167

Kühl M, Jørgensen BB (1992a) Microsensor measurements of sulfate reduction and sulfide oxidation in compact microbial communities of aerobic biofilms. Appl Environ Microbiol 58: 1164-1174

Kühl M, Jørgensen BB (1992b) Spectral light measurements in microbenthic phototrophic communities with a fiber-optic microprobe coupled to a sensitive diode array detector. Limnol Oceanogr 37: 1813-1823

Kühl M, Jørgensen BB (1993) The light field of microbenthic communities: Radiance distribution and microscale optics of sandy coastal sediments. Limnol Oceanogr, in press

Lassen C, Ploug H, Jørgensen BB (1992a) A fibre-optic scalar irradiance microsensor: Application for spectral light measurements in sediments. FEMS Microbiol Ecol 86: 247-254

Lassen C, Ploug H, Jørgensen, BB (1992b) Microalgal photosynthesis and spectral irrradiance in coastal marine sediments of Limfjorden, Denmark. Limnol Oceanogr 37: 760-772

Nicholson AM, Stolz JF, Pierson BK (1987) Structure of a microbial mat at Great Sippewisset Marsh, Cape Cod, Massachusetts. FEMS Microbiol Ecol 45: 343-364

Pierson BK, Sands VM, Frederick JL (1990) Spectral irradiance and distribution of pigments in a highly layered marine microbial mat. Appl Environ Microbiol 56: 2327-2340

Ploug H, Lassen C, Jørgensen BB (1993) Action spectra of microalgal photosynthesis and depth distribution of spectral scalar irradiance in a coastal marine sediment of Limfjorden, Denmark. FEMS Microbiol Ecol 102: 261-270

Revsbech NP, Jørgensen BB (1983) Photosynthesis of benthic microflora measured by the oxygen microprofile method: capabilities and limitations of the method. Limnol Oceanogr 28: 749-756

Revsbech NP, Jørgensen BB (1986) Microelectrodes: their use in microbial ecology. Adv Microb Ecol 9: 293-352

Revsbech NP, Christensen PB, Nielsen LP (1989) Microelectrode analysis of photosynthetic and respiratory processes in microbial mats. In: Cohen Y, Rosenberg E (eds) Physiological Ecology of Benthic Microbial Communities. Am Soc Microbiol, pp 153-162

Senior JM (1985) Optical fiber communications: Principles and practice. Prentice-Hall

Seyfried M (1989) Optical radiation interactions with living tissue In: Diffey BL (ed) Radiation Measurement in Photobiology. Academic Press, London , pp 191-223

Stal LJ, van Gemerden H, Krumbein WE (1985) Structure and development of a benthic marine microbial mat FEMS Microbiol Ecol 31: 111-125

Star WM, Marijnissen JPA, van Gemert MJC (1987) Light dosimetry: Status and prospects J Photochem Photobiol Ser B 1: 149-167

Vogelmann TC, Martin G, Chen C, Buttry D (1991) Fibre optic microprobes and measurement of the light microenvironment within plant tissues Adv Bot Res 18: 255-295

Gas diffusion probe for measurement of CH_4 gradients

Franz Rothfuss, Peter Frenzel and Ralf Conrad

Max-Planck-Institut für terrestrische Mikrobiologie, Karl-von-Frisch-Straße, D-35043 Marburg, Germany

Introduction

The usual procedure to measure CH_4 concentration profiles in sediments or submerged soils is by extraction of the porewater and analysis of the extracted CH_4 by gas chromatography (e.g., Martens and Berner 1977). However, this method has a very low spatial resolution of usually >1 cm, is destructive for the environment, and is also laborious so that per day only a few vertical profiles can be analyzed. An alternative to extraction is the application of peepers (Hesslein 1976) which, however, have to be left for equilibration in the sediment for at least one week and furthermore, have a limited spatial resolution and require special precautions if used in anoxic sediments (Adams and Fendinger 1986; Brandl et al. 1990).

Therefore we developed a series of probes, based on gas diffusion, which allowed the measurement of *in situ* concentrations of dissolved CH_4 with a high spatial and temporal resolution. In the following, we describe the principle of the gas diffusion probes and compare the advantages and disadvantages of several construction types.

Design and operation of the gas diffusion probes

The designs of 4 different types of gas diffusion probes are shown in Fig. 1. From the principle, the probes were based on the diffusion of dissolved CH_4 through a silicone membrane into a gas space at the end of the probe. This gas space was flushed with N_2 and analyzed gaschromatographically for CH_4. For measurement of dissolved CH_4, the probe was inserted into the sediment using a micromanipulator. A dissecting microscope was used to determine when the membrane-covered opening of the probe was just at the surface (0 mm depth). The probe was allowed to adjust to the environment for at least 1 min. Then, the gas space of the probe was flushed with 9 ml N_2. The probe was left for a defined time (usually 2 min) during which the CH_4 dissolved in porewater of the sediment could diffuse over the membrane into the N_2-filled gas space of the tip of the probe. Then, the probe was again flushed with 1 ml N_2, which was recovered and analyzed by gas chromatography.

The probe was calibrated in aqueous solutions of known CH_4 concentrations soaking an artificial sediment of glass beads (100 μm size) to overcome the sensibility of the probe for water movements (see below) and to account for tortuosity in the real sediment. The calibration

NATO ASI Series, Vol. G 35
Microbial Mats
Edited by Lucas J. Stal and Pierre Caumette

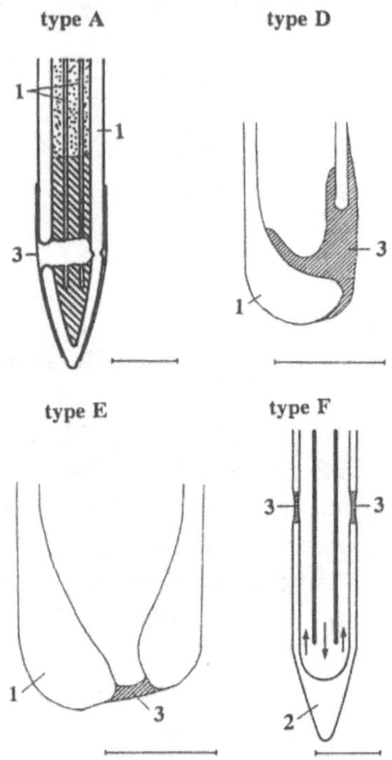

Fig. 1. Schemes of the tips of 4 different types of gas diffusion probes. The scale at type A indicates a length of 5 mm, those at type D, E, and F of 1 mm. 1 = glass; 2 = stainless steel; 3 = silicone.

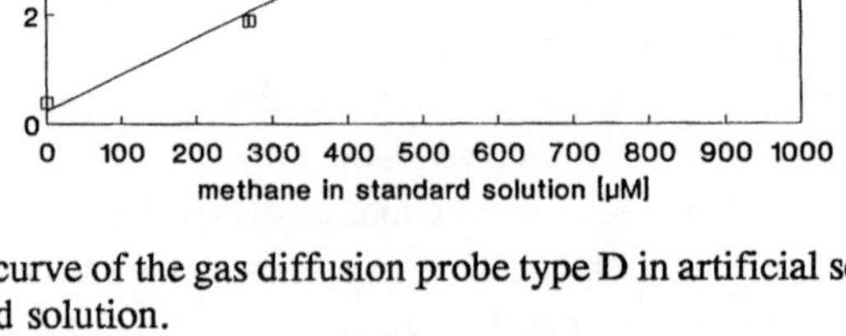

Fig. 2. Calibration curve of the gas diffusion probe type D in artificial sediment of glass beads soaked with standard solution.

curves resulted in signals of CH_4 (ppmv CH_4) in the gas space of the probe which increased linearly with the concentrations of CH_4 (μM CH_4) dissolved in the standard solutions (Fig.2). The calibration curves were reproducible for standardized measurement conditions exhibiting a precision of better than $\pm 10\%$.

The probe of type D was used to measure vertical profiles of dissolved CH_4 in a microcosm of anoxic paddy soil (Fig.3). The vertical CH_4 profile measured with the probe compared well with that measured by extraction. The profiles clearly showed that CH_4 was produced in deeper layers and diffused upwards to be consumed in the oxic top soil layers of usually 2 mm thickness.

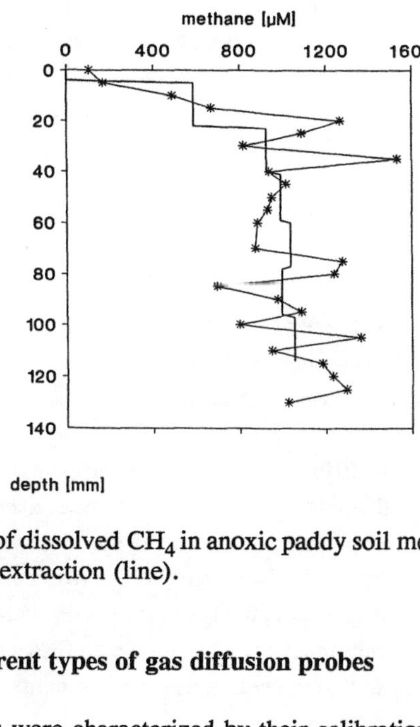

Fig. 3. Vertical profiles of dissolved CH_4 in anoxic paddy soil measured with the gas diffusion probe type D (*) and by extraction (line).

Characteristics of different types of gas diffusion probes

The gas diffusion probes were characterized by their calibration curves giving the detection limit and the sensitivity. Both characteristics were a function of the type of construction of the probe, mainly of the surface area of the membrane separating the gas space at the end of the probe from the environment. The surface area of the window limited the spatial resolution of the probe. The characteristics of the different types of the probes are summarized in Table 1. Probes of type A having a relatively large membrane window (ca. 8.3 mm^2) showed a spatial resolution of 3000 μm with a detection limit of about 2 μM CH_4 and a relatively high sensitivity of 0.29 ppmv/μM CH_4. Increasingly smaller probes of the type D and E reached spatial resolutions of the order of 300 μm but, on the other hand, exhibited larger detection limits (>14 μM CH_4) and lower sensitivities (<0.0067 ppmv/μM CH_4).

Table 1. Main characteristics of 4 different types of gas diffusion probes

type of probe	sensitivity [ppmv/μm]	detection limit [μm]	spatial resolution [μm]	membrane area [mm^2]
A	0.029	2	1600-3000	8.3
D	0.0067	14	600-1300	0.34
E	0.0018	160	350	0.0064
F	0.0034	10	500-1200	0.39

The data in Table 1 are all for an equilibration time of 2 min. Higher temporal resolutions could be achieved, however, at the cost of loss of sensitivity. On the other hand, shorter exposure times would result in a slightly increased spatial resolution, as it is given by the diameter of the window plus the diffusion range of CH_4 in front of it. A rough estimate of the diffusion range (s) in the water surrounding the tip of the probe is given by Adam et al. (1988):

$$s = \sqrt{(2tD_s)}$$

where t = exposure time and D_s = diffusion coefficient determined for the environment. With an exposure time of 2 min and a value for $D_s = 5 \times 10^{-6}\ cm^2s^{-1}$ (determined by us for CH_4 in submerged paddy soil) the diffusion range is about 0.35 mm. Adding this range twice to the diameter of the membrane gives the spatial resolution of the probe assuming no overlap between two sampling points.

For measurement, the gas diffusion probes consume part of the CH_4 dissolved in the vicinity of the membrane-covered window. This effect became evident when standard solutions were stirred. The signal of the probe increased with the speed of stirring. Similar effects are known from some types of polarographic O_2 sensors which also consume part of the O_2 to be analyzed (Gust et al. 1987). Stirring effects in O_2 microsensors were overcome, however, by making the diffusion over the membrane the rate-limiting step (Revsbech 1989). Because of the sensitivity to stirring or even to small convective water movements, calibration was routinely done in artificial sediments (see above).

The amount of CH_4 consumed should increase with the area of the membrane and the exposure time. At constant exposure time it thus should be larger for the probe of type A > D = F > E. This effect became evident when the CH_4 concentration at a particular spot in a submerged paddy soil was measured repeatedly without moving the gas diffusion probe. The signal of the type A probe with the largest membrane area decreased to 66% at the second measurement, whereas the signals of the probes with smaller membrane areas stayed fairly constant with coefficients of variation of 2.4%, 13%, and 3.8% for type D, E, and F, respectively.

Type E probes had the smallest membrane-covered windows and thus, the highest spatial resolution. The small membrane area could only be constructed by locating the window at the distal end of the probe. The necessity of this construction had the following disadvantage. By inserting the probe, the sediment matter may be slightly compacted in front of the tip of the probe. As the window of type E probes was also at the tip, it was thus exposed to the possibly compacted environment. This obviously created a too low signal due to diffusional constraints. The signal became again higher, when the probe was slightly moved back thus relieving the compaction of the sediment and probably creating a water-filled pore space in front of the probe (Fig.4).

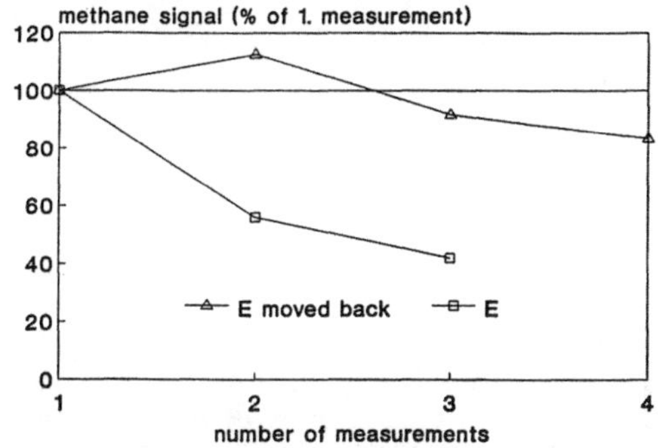

Fig. 4. Effect of repeated measurements with the gas diffusion probe type E which was inserted into about 10 mm depth of an anoxic paddy soil microcosm () or was inserted and then moved back by 2mm ().

Stainless steel needle probes of type F with membrane-covered side-port windows were constructed to overcome some of the difficulties involved in fabrication and application of glass-made probes. The needles were manufactured from stainless steel tubes (1 mm o.d.; 0.78 mm i.d.) with a solid tip and two lateral windows of 0.5 mm diameter, positioned about 6 mm from the tip and covered with silicone membranes. By this arrangement sediment compaction is avoided and the risk to damage the membrane by contact with sand grains is highly reduced. By using stainless steel tubes as thin as 0.4 mm for flushing the gas space, very durable and pressure-resistent probes could be manufactured. The needle probes will be especially suited for measurements in coarse-grained sediments and in all other environments that are resistent or erosive.

Conclusions

Use of gas diffusion probes is feasible for measurement of vertical microprofiles of CH_4 with a resolution of about 100 μm in sediments, submerged paddy soil or other methanogenic environments where CH_4 concentrations are expected to be in a range of 0.01 to 1.5 mM. Similar probes may be generally useful for measurement of gases dissolved in sediments, e.g. H_2, N_2O, CO, $(CH_3)_2S$ etc. They will be limited by the concentration range of the trace gases and the availability of sensitive gas chromatographic analyzers. Otherwise, larger membrane-covered windows must be constructed to increase the recovery of the trace gas. Recently, for example, we described a gas diffusion probe for the mesurement of dissolved H_2 (Krämer and Conrad 1993). Since H_2 is typically present in nanomolar concentrations, the consequently larger windows of the probe allowed only a spatial resolution of 5-10 mm.

References

Adam G, Läuger P, Stark G (1988) Physikalische Chemie und Biophysik, 2nd ed, Springer, Berlin

Adams DD, Fendinger NJ (1986) Early diagenesis of organic matter in the recent sediments of Lake Erie and Hamilaton Harbor. I. Carbon gas geochemistry. In: Sly PG (ed) Sediments and Water Interactions. Springer, New York, pp 305-318

Brandl H, Hanselmann KW, Bachofen R (1990) *In situ* stimulation of bacterial sulfate reduction in sulfate-limited freshwater lake sediments. FEMS Microbiol Ecol 74: 21-32

Martens CS, Berner RA (1977) Interstitial water chemistry of anoxic Long Island sound sediments. 1. Dissolved gases. Limnol Oceanogr 22: 10-25

Hesslein RH (1976) An in situ sampler for close interval pore water studies. Limnol Oceanogr 21: 912-914

Krämer H, Conrad R (1993) Measurement of dissolved H_2 concentrations in methanogenic environments with a gas diffusion probe. FEMS Microbiol Ecol 12: 149-158

Gust G, Booij K, Helder W, Sundby B (1987) On the velocity sensitivity (stirring effect) of polarographic oxygen microelectrodes. Neth J Sea Res 21: 255-163

Revsbech NP (1989) An oxygen microsensor with a guard cathode. Limnol Oceanogr 34: 474-478

Light and electron microscopy in microbial mat research: An overview

John F. Stolz

Department of Biological Sciences, Duquesne University, Pittsburgh, PA. 15282-1502 USA

A variety of traditional techniques for light microscopy (e.g., differential interference contrast, phase contrast, fluorescence) and electron microscopy (transmission and scanning) have been used to obtain valuable information on the structure and species composition of microbial mat communities. Recent advances in techniques and instrumentation for microscopy are providing new tools for studying these communities *in situ*. Innovations in immunohistochemistry and molecular biology allow in situ labeling of specific microorganisms with fluorophores or gold particles which have been conjugated to antibodies or oligonucleotides. Digital image analysis software and confocal laser microscopy have made serial sections and 3-D reconstruction possible, providing structural data on biofilms, microbial mats, and microfossils in petrographic sections. This review will highlight a few examples of these new technologies with emphasis on those which have potential use in microbial mat studies.

Transmission electron microscopy

Microbial mat systems

Electron microscopy gives a microbes eye view of microbial ecosystems. Capable of resolving ultrastructural detail at the subnanometer scale, it provides the means to identify the smallest of microorganisms as well as bacteriophage and viruses. Early studies of microbial mat communities used scanning electron microscopy (SEM, Cohen et al. 1977; Cohen 1984). SEM was used partly because of the ease of sample preparation and partly because of the mineral matrix (e.g., calcium carbonate, ditrital sand grains). These studies began to reveal the complexity of microbial mat communities and their stratified nature (e.g., different populations at different depths), as well as provide critical clues about the processes involved in mineralization (Cohen 1984). The major draw back of SEM is that it provides little ultrastructural detail other than that which is apparent externally. The application of transmission electron microscopy (TEM) to mat studies was hindered by the difficulty of obtaining thin sections of mat material heavily intercalated with sediment. Fortunately, the initial attempts were successful (Stolz 1983, 1984) and these studies have facilitated the identification of the species of many microbial mat communities by their ultrastructure. To date microbial mats from Laguna Figueroa (Margulis et al. 1980; Stolz 1983, 1984, 1990),

NATO ASI Series, Vol. G 35
Microbial Mats
Edited by Lucas J. Stal and Pierre Caumette

Sippewisett Marsh (Nicholson et al. 1987), Solar Lake (Cohen 1984; D'Amelio et al. 1989), Guerrero Negro (D'Amelio et al. 1987, 1989), and Abu Dhabi (Stolz 1991) have been studied in this manner. Those species which have distinctive ultrastructural characteristics are most ammendable to identification. Photosynthetic prokaryotes, for example, can be distinguished by their light harvesting structures (Stolz 1984, 1991). Cyanobacteria typically have thylakoids dotted with phycobilisomes, prochlorophytes have thylakoids lacking phycobilisomes, purple phototrophs have intracytoplasmic chromatophore membranes (ICMs), and green phototrophs have chlorosomes (Stolz 1991). A comparison of *Microcoleus* dominated laminated mats from Laguna Figueroa, Abu Dhabi, Guerrero Negro, and Solar Lake reveil that they share similar key species (Fig. 1). Although *M. chthonoplastes* is the dominant cyanobacteria, ultrastructurally similar species of filamentous green and purple phototrophic bacteria are also abundant (Stolz 1983, 1984, 1990; D'Amelio et al. 1987, 1989). These organisms represent new species as they cannot be ascribed to existing species. Not only are filamentous green bacteria a major constituent of these mats, but they exhibit a great variety of light harvesting structure arrangement as seen in the purple phototrophs. Different species have been identified with vacuoles lined with chlorosomes, single membrane invaginations lined with chlorosomes, and stacked lamella with chlorosomes, in addition to the more commonly recognized arrangement of chlorosomes lining the inner cell membrane (e.g., *Chloroflexus aurantiacus*) and prostheca (e.g. *Prosthecochloris estuarii*) (Sprague and Fuller 1991). Whether these bacteria are members of the Chloroflexaceae or Chlorobiaceae await immunohistochemical analysis (see below) and culture.

Phototrophic bacteria are not the only denizens of microbial mats which can be identified by TEM. Examination of microbial scums and associated laminated mats of Laguna Figueroa has resulted in the discovery of a predatory prokaryote feeding on microcolonies of *Thiocapsa pfennigii* (Stolz 1991), spirochetes (Margulis et al. 1990), and bacteriophage (Margulis et al. 1980). Studies of Solar Lake and Guerrero Negro mats have identified sulfur bacteria (e.g., *Beggiotoa alba*, *B. leptomitiformis.*, *Thioploca nigra*) and spirochetes (D'Amelio et al. 1989). Distinctive intracellular inclusions also aid in the indentification of species (Shively et al. 1988). The increased resolution provided by scanning transmission electron microscopes (STEM) has allowed the detailed elemental mapping of these inclusions. Magnetotactic bacteria with their distinctive magnetosomes, have been shown to precipitate not only magnetite (Fe_3O_4) but also the iron sulfide greigite (Fe_3S_4) (for a review see Stolz 1993).

Immunohistochemistry

Two major breakthroughs in recent years have lead to the simplication of immunohistochemical (IHC) studies at the ultrastructural level. The first was the development of commercially available gold conjugated secondary antibodies and the second, acrylic based embedding media such as LR White. Gold conjugated secondary antibodies are now available in different sizes (e.g., 5 nm, 10 nm), for different primary antisera (e.g., mouse, rat, rabbit) and from several

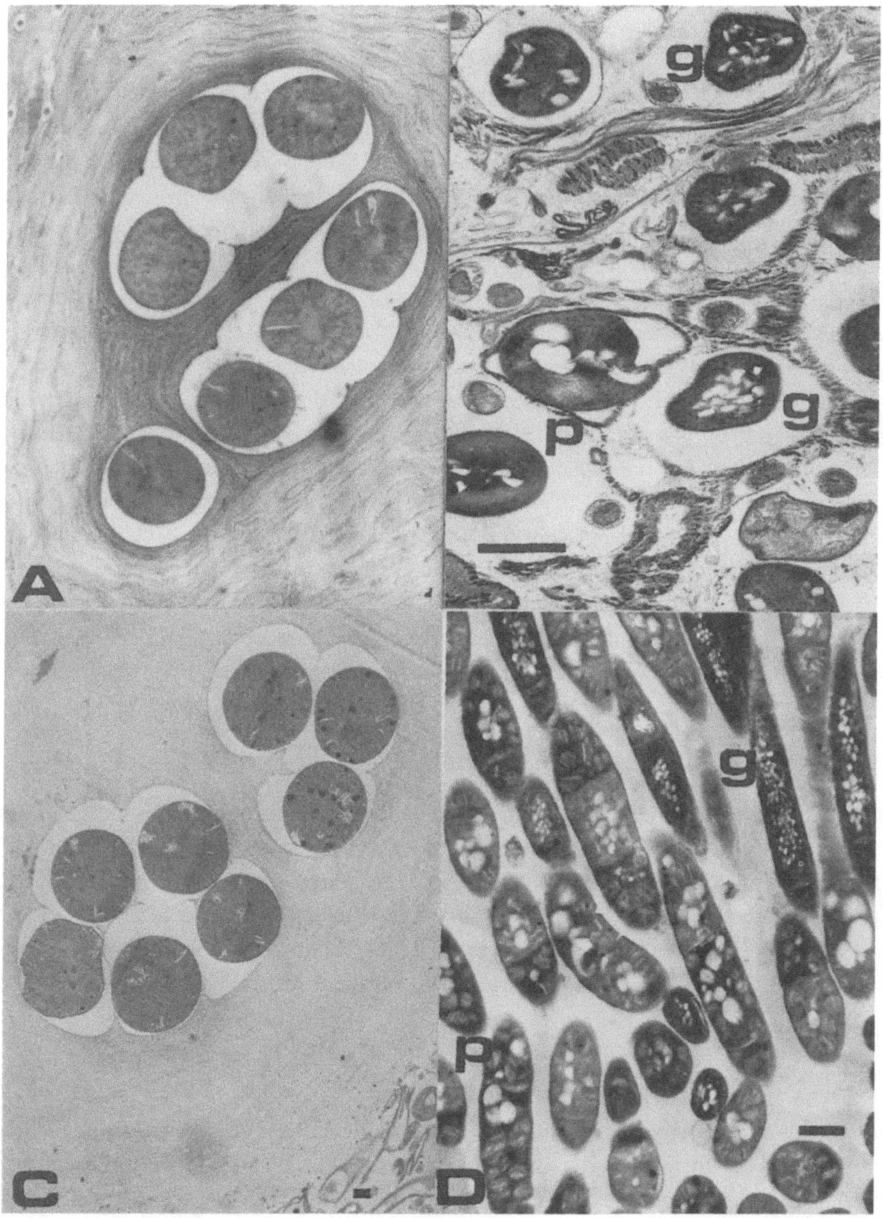

Fig. 1. Transmission electron micrographs of *Microcoleus* dominated laminated mat communities from Laguna Figueroa, Bajà California, Mexico, and Abu Dhabi, the Persian Gulf. A) Cross section through a bundle of *M. chthonoplastes* showing radial thylakoids, Abu Dhabi. B) filamentous green (g) and purple (p) phototrophs, Abu Dhabi, C) Cross section through a bundle of *M. chthonoplastes*, Laguna Figueroa, D) green and purple phototrophs, Laguna Figueroa. All bars 1μm.

supply houses (e.g., Janssen, Sigma, Fisher). LR White is a water miscible acrylic which can be polymerized with heat (60°C), or cold with an accelerator. Although it tends to be a little more brittle than epoxy resins, no special precautions have to be taken when preparing the material for thin sections. It is advised that postfixation with osmium tetroxide should be eliminated if the accelerator is used. Examples of IHC are the study nitrogen fixation by the marine cyanobacterium *Trichodesmium* sp. using antibodies to glutamine synthetase (Carpenter et al. 1992) or nitrogenase (Paerl et al. 1989) and by the mat forming cyanobacterium *Oscillatoria limosa* (Stal and Bergman 1990).

IHC can also aid in the identification of species where ultrastructure alone may be misleading. One such case is the green bacteria. The two families of green bacteria, the Chlorobiaceae and the Chloroflexaceae, share the common taxonomic characteristic of light harvesting structure, the chlorosome, however, their physiology, biochemistry, and photochemistry are quite different (Stolz et al. 1991; Sprague and Fuller 1991). In the aforementioned studies of microbial mat composition, the indentification of filamentous green bacteria as species of mesophilic *Chloroflexus* (Stolz 1983, 1984; D'Amelio et al. 1989) may have been premature. Using the protein diversity found in the chlorosome, *Chlorobium* specific probes have been developed (Cahill and Stolz, manuscript in preparation). Antibodies have also been raised against three chlorosome proteins from *Chloroflexus aurantiacus* (Feick and Fuller 1984). Thus by using these antibodies and gold conjugated secondary antibodies, it will be possible to distinguish between *Chlorobium* and *Chloroflexus* species in thin sections.

Light microscopy

Brightfield, differential interference contrast, phase contrast, and fluorescence microscopy have been the mainstay of microbial mat studies for decades (see edited volumes by Walter 1976; Cohen et al. 1984; Cohen and Rosenberg 1989). Epifluorescence microscopy has been used routinely to detect, enumerate, and size planktonic species (Sieracki et al. 1985, Viles and Sieracki 1992). Computer enhanced and confocal laser microscopy have been used in studies of biofilms (Caldwell et al. 1992) and have great potential for applications in microbial mat studies.

Fluorescence Microscopy

Fluorescence microscopy has traditionally been used for the detection and enumeration of autofluoresent organisms (e.g., picoplankton, Sieraki et al. 1985). The great number of fluorophores and their commercial availability has facilitated the development of a variety of molecular probes which can be used to label cells that lack autofluorescence. The most recent edition of Molecular Probes lists sixteen fluorophores of pratical importance (Johnson 1992). For routine counting of bacterial cells, for example, acridine orange or 4'6-diamino-2-phenylindole (DAPI) can be used to fluorescently stain the cells (Parsons et al. 1984). Immunofluorescence (IF) using antibodies and fluorophore conjugated oligonucleotides have been used as group and species specific probes . It is also important to realize the potential for new product development in that many of these fluorophores can be covalently crosslinked with a variety of biomolecules (Haugland 1992).

Polyclonal and monoclonal antibodies have been used to identify both species of bacteria as well as polypeptides by immunofluorescence microscopy (IFM, Macario and Conway de Macario 1992; Lillebaek 1992). IF probes have several advantages over oligonucleotide probes. Polyclonal antibodies can be raised simply and in a variety of animal hosts (e.g., mice, rats, rabbits). The same antisera can be used for IHC studies with light microscopy or electron microscopy (see above). The option of labeling sectioned material is particularly valuble for *in situ* studies of microbial mat communities where permeability is restricted by cell densities and sheath material. Secondary antibodies conjugated to different fluorochromes (e.g., fluorescein, rhodamine) are commercially available and if the immunogen is a surface protein, permeabilization of the cells is unneccessary. As mentioned above, polyclonal antisera raised to chlorosome proteins have been used to identify species of Chlorobiaceae. Fig. 2 shows one example, cells of *Chlorobium tepidum* labeled with fluorescein. The cells were initially fixed in 4% formalin and permeabilized with lysozyme. The primary antiserum was raised against whole chlorosomes from *Cb. tepidum* and *Cb. limicola* and the secondary antiserum was fluorescein conjugate goat anti-rabbit (Cahill and Stolz, manuscript in preparation).

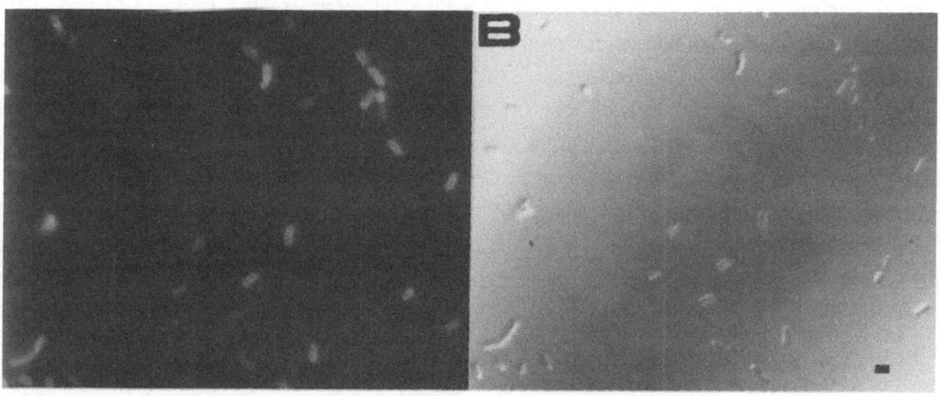

Fig. 2. Cells of *Chlorobium tepidum* labeled with fluorescein using polyclonal antisera raised to whole chlorosomes of *Cb. tepidum* and *Cb. limicola.* A) IFM B) differential interference contrast. Bar 5 μm.

In less than a decade the use of 16S ribosomal RNA sequences for the design of species specific oligonucleotide probes has increased dramatically (Olson et al. 1986; Giovannoni et al. 1988; DeLong et al. 1989; Amann et al. 1990; Tsien et al. 1990). A recent sampling of the literature turns up probes for sulfate reducers (Devereux et al. 1992; Ramsing et al. 1993), magnetotactic bacteria (Spring et al. 1992; DeLong et al. 1993), and rumenal bacteria (McSweeny et al. 1993). The probes are typically 18 to 23 nucleotides in length and are synthesized with an aminoethylphosphate linker coupled to the 5' end (Amann et al. 1990). This linker allows the probe to be labeled with a fluorochrome (eg fluorescein, rhodamine; Amann et al. 1990) or an enzyme (eg horseradish peroxidase; Amann et al. 1992). Although the polymerase chain reaction and universal primers have been used with great success in determining the 16S rRNA sequences of microbial mat species (Weller et al. 1992; Ward et al. 1990, 1992) the efficacy of oligonucleotide probes for *in situ* labeling has yet to be determined.

Digitally enhanced microscopy

Digitally enhanced microscopy has become more prevalent as a result of developments in high resolution video cameras, software and hardward for image analysis, and confocal laser microscopes. The essential components to an image processing system are a sensitive video camera (eg CCD - charged couple device; SIT - silicone intensification target), frame grabber board (with software) in a compatible computer (eg IBM or compatible PC), and a microscope (Caldwell et al. 1992). In the case of confocal laser microscopy, a laser microscope is used. The images generated are particularly clear because confocal pinholes at the laser and detector eliminate stray light from above, below and beside an object (Caldwell et al. 1992). The precise optical sections that are made and stored can then be reconstructed into a 3-D image by computer. Computer enhanced microscopy (CEM) uses brightfield, phase contrast, or fluorescent images from a conventional light microscope (Caldwell et al. 1992). The video analog image from the light microscope is converted into digital format. In digital format the image is divided into pixels, each with a given grey scale value (e.g., 0 - 255) ranging from black to white. The image can then be analysed by a variety of functions dependent on the software package. Two common features in image analysis software (e.g., Image 1, Universal Imaging Corporation) are the image enhancement and morphometrics functions. Image enhancement includes subtraction of background, alteration of grey scale, and filtration operations (e.g., low pass filter, contrast enhancement). Deconvolution, the process in which the 3-D point spread function (or haze factor) is used to generate an inverse function, can be used to remove out of focus objects from the image (Image 1 Insights 1992). Pseudocolor, which converts the grey scale values into a range of colors can be used to aid visualization of low contrast images. The morphometrics function, once calibrated for distance, measures parameters such as area, percent area, diameter, length, automatically providing the minimum, maximum, mean, and standard deviation for each parameter. Enumeration of species can be done either with fluorescence (where the cells are autofluorescent or have been stained with a fluorophore) and thresholding of intensity (eg only those things fluorescent are counted), or by setting morphometric filters (eg only those objects of particular size or shape are counted). Image-analyzed epifluorescence microscopy has been used to detect, enumerate and size picoplankton (Sieracki et al. 1985, 1989; Viles and Sieracki 1992).

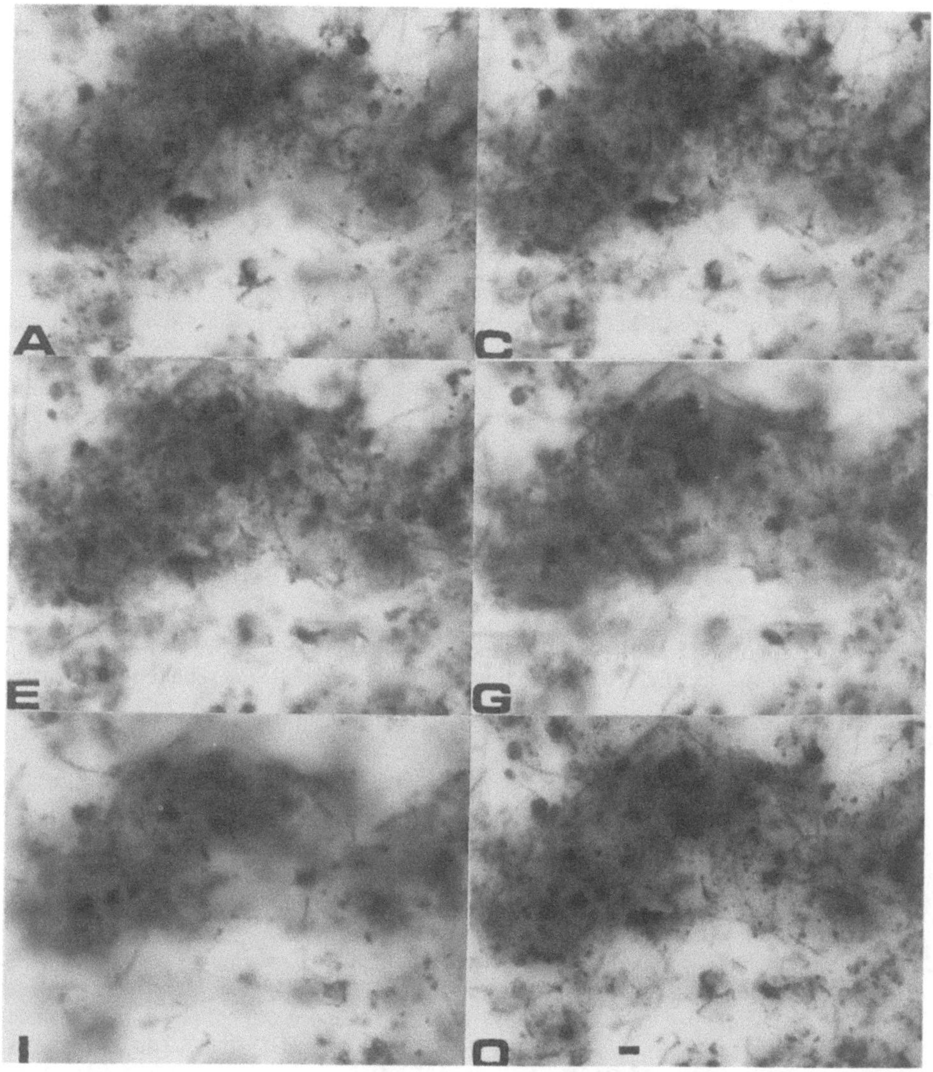

Fig. 3. Petrographic section of Gunflint chert from the Schreiber Beach local. The image marked 0 in the lower right is the 3-D reconstruction generated from nine images 1 μm apart. Of the nine images only A,C,E,G and I representing the tif files Afossil, Cfossil, Efossil, Gfossil and Ifossil are shown. Bar 5μm.

Computer enhanced microscopy can help in the study of microbial communities by imaging 3-D objects, such as a filamentous bacterium or a microcolony, which are not in focus in a single plane. A simple application is the imaging of microfossils suspended in a chert matrix (Fig. 3). Traditional methods still in use today (see Schopf 1993), call for multiple photographic exposures of the microfossil at different planes of focus. The exposures are then assembled in a montage in order to show the whole microfossil in focus. By using the 3-D reconstruction option on the Image 1 image analysis system (Universal Imaging), it is possible to sum images taken at numerous focal planes and view them at the 0° angle. This has the effect of showing everything within the total range of focal planes. The image in figure 3-O is a petrographic section of Gunflint material from the Schreiber Beach local collected by S. Golubic (Boston University). It was generated from nine images of focal lengths 1 mm apart taken with a Javalin CCD camera mounted onto a Nikon Microphot SA with brightfield illumination and stored as the individual tif files Afossil through Ifossil (A,C,E,G, and I are shown). Then using the 3-D reconstruction function of the Image 1 image analysis system (Universal Imaging) the images were reconstructed at the 0° angle.

References

Amann RI, Krumholz L, Stahl DA (1990) Fluorescent oligonucleotide probing of whole cells for determinative, phylogenetic, and environmental studies in microbiology. J Bacteriol 172: 762-770

Amann RI, Zarda B, Stahl DA, Schleifer K-H (1992) Identification of individual prokaryotic cells by using enzyme-labeled rRNA-targeted oligonucleotide probes. Appl Environ Microbiol 58: 3007-3011

Carpenter EJ, Bergman B, Dawson R, Siddiqui PJA, Soderback E, Capone DG (1992) Glutamine synthetase and nitrogen cycling in colonies of the marine diazotrophic cyanobacterium *Trichodesmium* spp. Appl Environ Microbiol 58: 3122-3129

Caldwell DE, Korber DR, Lawrence JR (1992) Confocal laser microscopy and digital image analysis in microbial ecology. Adv Microbiol Ecol 12: 1-67

Cohen Y, Krumbein WE, Shilo M (1977) Solar Lake (Sinai) 2. Distribution of photosynthetic microorganisms and primary production. Limnol Oceanogr 22: 609-610

Cohen Y (1984) The Solar Lake cyanobacterial mats: strategies of photosynthetic life under sulfide. In: Cohen Y, Castenholz RW, Halvorson HO (eds) Microbial mats: Stromatolites. Alan R Liss Inc, NY, pp 133-148

Cohen Y, Castenholz RW, Halvorson HO (eds) (1984) Microbial mats: Stromatolites. Alan R Liss Inc, New York

Cohen Y, Rosenberg E (eds) (1989) Microbial mats: physiological ecology of benthic microbial communities. ASM Publishing, Washington DC

D'Amelio ED, Cohen Y, Des Marais DJ (1987) Association of a new type of gliding filamentous phototrophic bacterium inside bundles of *Microcoleus chthonoplastes* in hypersaline cyanobacterial mats. Arch Microbiol 147: 528-534

D'Amelio ED, Cohen Y, Des Marais DJ (1989) Comparative functional ultrastructure of two hypersaline submerged cyanobacterial mats: Guerrero Negro, Baja California Sur, Mexico, and Solar Lake, Sinai, Egypt. In: Cohen, Y, Rosenberg, E (eds) Microbial mats: physiological ecology of benthic microbial communities, ASM, Washington DC, pp 97-113

DeLong EF, Wickham GS, Pace NR (1989) Phylogenetic strains: ribosomal RNA based probes for the identification of single cells. Science 243: 1360-1363

DeLong EF, Frankel RB, Bazylinski DA (1993) Multiple evolutionary origins of magnetotaxis in bacteria. Science 259: 803-806

Devereux R, Kane MD, Winfrey J and Stahl DA (1992) Genus-specific and group-specific hybridization probes for determinative and environmental studies of sulfate-reducing bacteria. Syst Appl Microbiol 15: 601-609.

Feick RG, Fuller RC (1984) Topography of the photosynthetic apparatus of *Chloroflexus aurantiacus*. Biochem 23: 3693-3700

Giovannoni SJ, DeLong DE, Olsen GJ, Pace NR (1988) Phylogenetic group-specific oligonucleotide probes for the identification of single microbial cells. J Bacteriol 170: 720-726

Haugland RP (1992) Molecular probes, handbook of fluorescent probes and research chemicals. Molecular Probes Inc, Eugene

Johnson I (1992) Optical properties of fluorescent probes. In: Haugland RP (ed) Molecular probes, handbook of fluorescent probes and research chemicals. Molecular Probes Inc, Eugene, pp 1-4

Lillebaek R (1992) The distribution of sulfate-reducing bacteria (SRB) in estuarine/marine sediments as revealed by specific antibodies. Abstracts of the Sixth International Symposium on Microbial Ecology 150

Macario AJL, Conway de Macario E (1992) Antibody probes and immunotechnology for direct identification of microbes in complex ecosystems. Abstracts of the Sixth International Symposium on Microbial Ecology 18

Margulis L, Barghoorn ES, Ashendorf D, Banerjee S, Chase D, Francis S, Giovannoni S, Stolz J (1980) The microbial community in the layered sediments at Laguna Figueroa, Baja California, Mexico: does it have precambrian analogues. Precamb Res 11: 93-123

Margulis L, Hinkle G, Stolz J, Craft F, Esteve I, Guerrero R (1990) *Mobilifilum chasei*: morphology and ecology of a spirochete from an intertidal stratified microbial comunity. Arch Microbiol 153: 422-427

McSweeny CS, Mackie RI, Odenyo AA, Stahl DA (1993) Development of oligonucleotide probe targeting 16S rRNA and its application for detection and quantitation of the ruminal bacterium *Synergistes jonesii* in a mixed-population chemostat. Appl Environ Microbiol 59: 1607-1612

Nicholson JAM, Stolz JF, Pierson BK (1987) Structure of a microbial mat at Great Sippewissett Marsh, Cape Cod, Massachusetts. FEM Microbiol Ecol 45: 343-364

Olson GJ, Lane DJ, Giovannoni SJ, Pace NR, Stahl DA (1986) Microbial ecology and evolution: a ribosomal RNA approach. Ann Rev Microbiol 40: 337-365

Paerl HW, Priscu JC, Brawner DL (1989) Immunochemical localization of nitrogenase in marine *Trichodesmium* aggregates: relationship to N_2 fixation potential. Appl Environ Microbiol 55: 2965-2975

Parsons TR, Maita Y, Lalli CM (1984) A manual of chemical and biological methods for seawater analysis. Pergamon Press, Oxford 173 p

Ramsing BR, Kühl M, and Jørgensen BB (1993) Distribution of sulfate reducing bacteria, O_2, and H_2S in photosynthetic biofilms determined by oligonucleotide probes and microelectrodes. Appl Environ Microbiol 59: 3840-3849.

Stal L, Bergman B (1990) Immunological characterization of nitrogenase in the filamentous non-heterocystous cyanobacterium *Oscillatoria limosa*. Planta (Berlin) 182: 287-291

Schopf JW (1993) Microfossils of the early archean apex chert: new evidence of the antiquity of life. Science 260: 640-646

Shively JM, Bryant DA, Fuller RC, Konopka AE, Stevens SEJr, Strohl WR (1988) Functional inclusion bodies in prokaryotic cells. Int Rev Cytol 113: 35-79

Sieracki ME, Johnson PW, Sieburth, JM (1985) Detection, enumeration, and sizing of planktonic bacteria by image-analyzed epifluorescence microscopy. Appl Environ Microbiol 49: 799-810

Sieracki ME, Reichenbach SE, Webb KL (1989) Evaluation of automated threshold selection methods for accurately sizing microscopic fluorescent cells by image analysis. Appl Environ Microbiol 55: 2762- 2772

Sprague SG, Fuller RC (1991) The green phototrophic bacteria and Heliobacteria. In: Stolz JF (ed) Structure of phototrophic prokaryotes. CRC Press, Boca Raton, pp 79-103

Spring S, Amann R, Ludwid W, Schleifer K-H, Petersen N (1992) Phylogenetic diversity and identification of nonculturable magnetotactic bacteria. System Appl Microbiol 15: 116-122

Stolz JF (1983) Fine structure of the stratified microbial community at Laguna Figueroa, Baja California, Mexico. I. Methods of *in situ* study of the laminated sediments. Precamb Res 20: 479-492

Stolz JF (1984) Fine structure of the stratified microbial community at Laguna Figueroa, Baja California, Mexico. II. Transmission electron microscopy as a diagnostic tool in studying microbial communities *in situ*. In: Cohen Y, Castenholz RW, Halvorson HO (eds) Microbial Mats: Stromatolites. Alan R Liss Inc, NY pp 23-38

Stolz JF (1990) Distribution of phototrophic microbes in the flat laminated microbial mat at Laguna Figueroa, Baja California, Mexico. BioSyst 345-357

Stolz JF (ed) (1991) Structure of phototrophic prokaryotes. CRC Press, Boca Raton.

Stolz JF (1993) Magnetosomes. J Gen Microbiol 139: 1663-1670

Tsien HC, Bratina BJ, Tsuji K, Hanson RS (1990) Use of oligonucleotide signature probes for identification of physiological groups of methylotrophic bacteria. Appl Environ Microbiol 56: 2858-2865

Universal Imaging Corporation (1992) Deconvolution, Image 1 Insights 4: 1-2

Viles CL, Sieracki ME (1992) Measurement of marine picoplankton cell size by using a cooled charge-coupled device camera with image-analyzed fluorscence microscopy. Appl Environ Microbiol 58: 584-592

Walter MR (ed) (1976) Stromatolites. Elsevier, Amsterdam

Ward DM, Weller R, Bateson MM (1990) 16S rRNA sequencing reveal uncultured inhabitants of a well-studied thermal community. FEMS Microbiol Rev 75: 105-116

Ward DM, Bateson MM, Well R, Ruff-Roberts A (1992) Ribosomal RNA analysis of microorganisms as they occur in nature. Adv Microb Ecol 12: 219-286

Weller R, Bateson MM, Heimbuch BK, Kopczynski ED, Ward DM (1992) Uncultivated cyanobacteria, *Chloroflexus*-like inhabitants, and spirochete-like inhabitants of a hot spring microbial mat. Appl Environ Microbiol 58: 3964-3969

Motility of *Microcoleus chthonoplastes* subjected to different light intensities quantified by digital image analysis

Niels B. Ramsing and Lee Prufert-Bebout

Max-Planck Institute for Marine Microbiology,
Fahrenheitstraße 1, D-28359 Bremen, Germany

Introduction

Light dependent movement of gliding cyanobacteria has been studied for more than a hundred years and several types of response to light have been observed (Häder 1987). The following concepts have been recommended by an *ad hoc* committee on behavioral terminology, (Diehn et al. 1979): *Phototaxis :* Orientation of movement with respect to a directional light field; *photokinesis:* by which the steady-state speed of movement is related to the total light intensity and a *photophobic response:* where the movement is altered by a spatial or temporal change in light intensity. Most photosynthetic organisms display one or more of these behaviours in their search of suitable light conditions (Castenholz 1982).

The recent use of fiber optic light microsensors with fine scale resolution has revealed that light fields within microbial mats are remarkably uniform. Directional incident light is strongly reduced in the upper few hundred μm's of microbial mats due to a combination of attenuation and scattering effects, leaving only low intensity homogeneous light in deeper photosynthetically active layers (Kühl and Jørgensen 1993; Kühl et al. 1993; Lassen et al. 1992a, 1992b). Over diurnal and seasonal timescales light available at any given depth will undoubtedly change necessitating movement of phototrophs to more favourable conditions. However, the homogeneity of the light field may make it difficult for photosynthetic organisms living within mats to use the movement mechanisms described above, i.e. little or no directional light is available to trigger a phototactic response. Even a photophobic response is likely to be of limited value, as the total light intensity changes gradually rather than abruptly inside the mat. Nevertheless, besides the above mentioned photokinesis, a range of other responses can be envisioned to enable the organisms to position themselves in the mat (Table 1). Some of these rely on decreasing mobility under optimal light conditions, i.e. by decreasing the speed of movement (photokinesis) or by moving less frequently. Another possibility is to reverse the direction of movement more often under agreeable light conditions, or to bend in a random direction that could ultimately lead to curling of the filaments. It is likely that several of these hypothetical behaviours are important for different organisms.

Microcoleus chthonoplastes is a cosmopolitan species commonly found in abundance in microbial mats. Little is known of their motility regulation in natural type morphs (Prufert-Bebout and Garcia-Pichel, this volume). We here present data for a shade adapted (approx.

NATO ASI Series, Vol. G 35
Microbial Mats
Edited by Lucas J. Stal and Pierre Caumette

10 μE m^{-2} s^{-1}) *Microcoleus* strain isolated from microbial mats in North Carolina, USA. Short movies of the movements of individual trichomes at different light intensities were acquired and subjected to digital image analysis. Through analysis of differences between consecutive picture frames it was possible to determine which of the above mentioned strategies were utilised by the cyanobacteria.

Materials and methods

The gliding cyanobacterium investigated here was isolated from microbial mats in North Carolina, USA. It was a uni-algal, non-axenic culture, which has retained its capacity to form bundles in a common sheath (Prufert-Bebout and Garcia-Pichel, this volume). Cyanobacterial filaments were placed in a 0.3% agar on a microscope slide, covered with a coverslip and sealed at the edges with petroleum jelly to prevent dehydration during measurement. An Axioplan microscope, Zeiss, Oberkochen was used both to observe the filaments and provide a uniform light field perpendicular to the agar slide using the built-in illumination after removing the condenser. Through insertion of different neutral density grey filters and regulation of the halogen lamp voltage, we could adjust the light intensity to any value between 0.4 μE and 4000 μE, and yet obtain good pictures.
The absolute light intensity at the position of the microscope slide was measured with a Biospherical Instruments Inc. Quantum Scalar Irradiance meter (QSL-100).
High resolution phase contrast pictures were obtained with a Peltier cooled slow scan CCD camera, from Photometrics . Seven images (each 1035 x 1317 pixels x 12 bit) were acquired with 40 s intervals at eight different light intensities from 0.4 μE to 150 μE. The slide was allowed to acclimatise for about 30 min at each light intensity. A 5x Neofluar objective was used which gave a resolution of about 1.36 μm pr pixel and enabled simultaneous examination of about 500 - 600 trichomes. The number of trichomes remained constant throughout the experiment, i.e. at all light intensities, and covered about 10 % of the observed field. The images were scaled to a 8-bit grayscale and further analysed with the program NIH image ver. 1.52 as described in the following sections.

Results and discussion

Movement was quantified using digital image analysis of differences between consecutive images in each movie series. The process is illustrated in fig. 1 (which only displays a small cut-out comprising < 3% of a whole image). The difference image between two succeeding frames fig. 1a and fig. 1b is shown in fig. 1c. Tracks of moving trichomes can be seen as white paths in the direction in which the trichomes are moving, leaving black paths behind. Locating the black pixels with an intensity below a threshold level gives the trailing ends forward

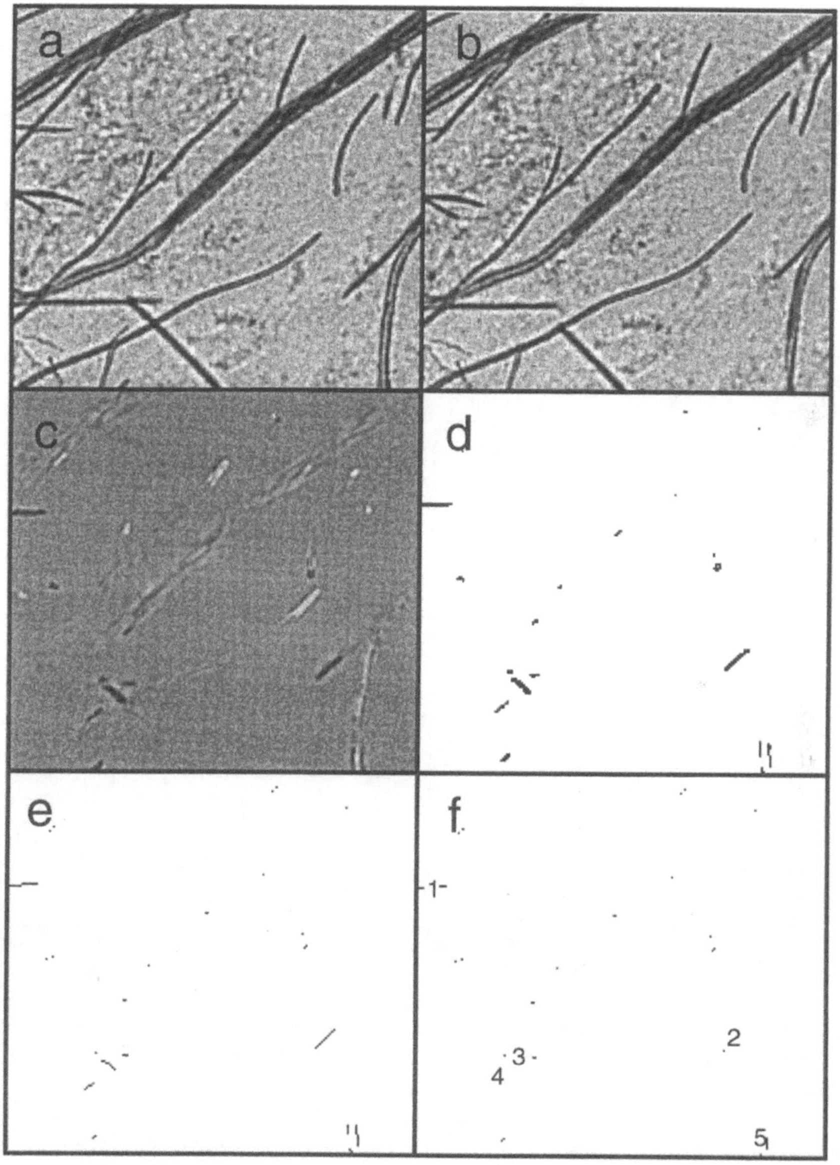

Fig. 1. Image analysis of consecutive images in a movie. (a) and (b) subsections of images taken 40 s apart. (c) difference between (a) and (b). (d) binary image of (c) obtained by thresholding all black areas. (e) skeletonization of (d) to obtain one pixel wide tracks. (f) automatic labelling and counting of tracks longer than 5 pixels (equivalent to 0.17 μm/s)

movement ("tail" movement, Fig. 1d), whereas locating all white pixels gives the front ends forward movement referred to as "head" movement (not shown). The total amount of movement can now be quantified at a given light intensity by counting how many black or white pixels are found after thresholding "tail-" or "head-" movement respectively. By analysing all six independent difference-pictures in each picture series at a given light intensity, we could obtained 95% confidence limits for the parameters being investigated.

Fig. 2 shows the total movement at various light intensities. Fig. 2a shows the "head" and the "tail" movement separately. As they are not significantly different they have been pooled in Fig. 2b, which includes all detected movement. A general trend towards reduced total movement at light intensities around 4 $\mu E\ m^{-2}\ s^{-1}$ can be observed.

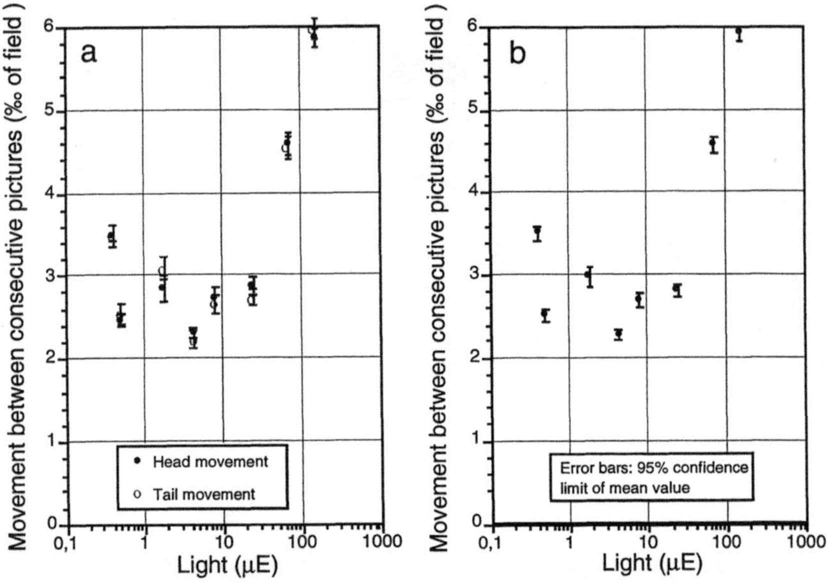

Fig. 2. Total movement of trichomes. (a) all detectable movement (i.e. both "head" and "tail" movement. (b) Comparison of "head" movement (white pixels) and "tail" movement (black pixels).

A light induced decrease in total movement can be due to either lower speed of movement or less frequent movement of individual trichomes. To investigate this we etched the black paths shown in Fig. 1d by a process known as skeletonization leaving one-pixel wide tracks (Fig. 1e), whose length and curvature could be measured (Fig. 1f). A total of 48 difference pictures were analysed which together contained more than 10,000 tracks.

The number of tracks longer than 5 pixels (i.e. corresponding to a speed in excess of 0.17 $\mu m/s$) showed a clear minimum at light intensities between 4 - 20 $\mu E\ m^{-2}\ s^{-1}$, (Fig. 3a).

Thus fewer trichomes seemed to be moving under these light conditions. The track length (i.e. speed of movement) was insignificantly altered by a change in illumination intensity. Neither average speed (track-length) nor maximum observed speed displayed any large dependence on light intensity (Fig. 3b). Photokinesis therefore seem to play, at most, a minor role in positioning *Microcoleus* in a suitable lightfield.

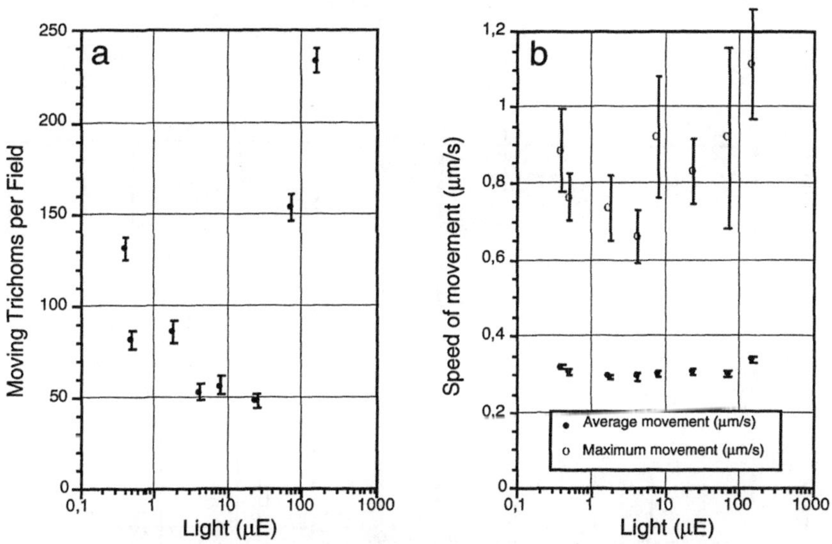

Fig. 3. Frequency and speed of movement. (a) Number of moving trichomes per field of view, as a function of light intensity. The total number of trichomes in the field of view remained constant throughout the experiment. (b) Average and maximum movement speed at different light intensities

To evaluate the number of trichomes reversing their direction of movement two consecutive difference images are evaluated. The first difference image is thresholded for movement of the front-end of the filaments (white pixels = "heads"), the second image is thresholded for movement of the trailing-end of the filaments (black pixels = "tails"). Pixels that are highlighted in both thresholded images indicate a trichome that has first moved in one direction, then reversed its direction of movement. The pixels highlighted in both thresholded images are found by applying an .AND. operation to the two binary images. The resulting image is skeletonized and tracks longer than 2 pixels are counted. Fig. 4a show the number of trichomes that are reversing their direction of movement at different light intensities. It shows a fairly constant number of reversals per field up to a light intensity of 10 μE m^{-2} s^{-1} above which the number of reversals increase rapidly. However, as more trichomes are moving at higher light intensities (Fig. 3a), the ecologically relevant parameter is the ratio of the number of reversals per number of moving trichomes i.e. the frequency of reversal for each moving trichome. The frequency of reversal is depicted in fig. 4b. A clear maximum at light intensities between 8 - 30 $\mu E\ m^{-2}\ s^{-1}$, indicating that the *Microcoleus* reverses its direction of movement more frequently in this intensity range. This response should not be confused with that of photophobic responses (step-up or step-down), which can occur under spatial or temporal light gradients. In these experiments each light intensity corresponds to one separate treatment.

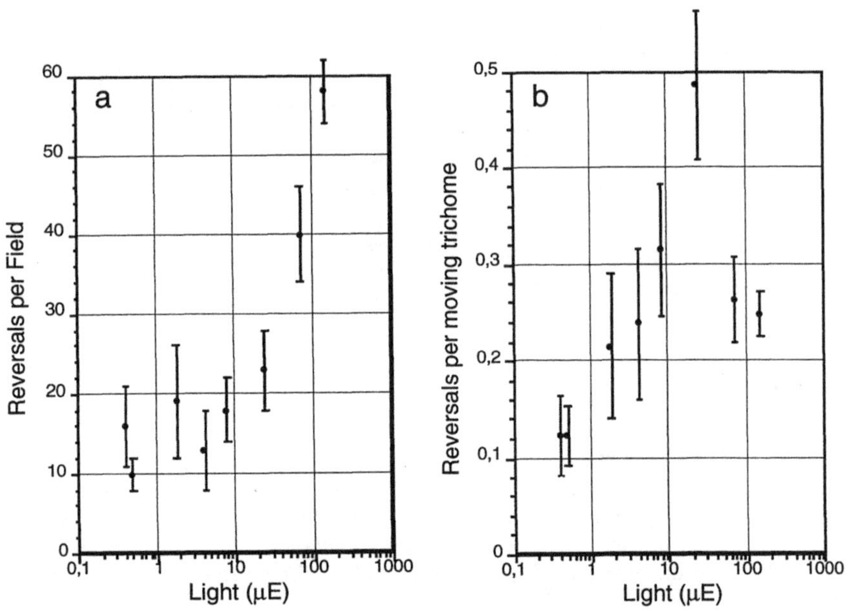

Fig. 4. Reversals. Number of filaments reversing their direction of movement within a 80 s time period per field. (a) total number of reversals and (b) number of reversals divided by the number of moving filaments at that light intensity. i.e. frequency of reversal.

Another motility strategy that could ultimately position *Microcoleus* under allegedly suitable light conditions, involves bending more frequently under agreeable light conditions which would ultimately lead to a curling of the filaments if the light intensity is in a proper range. The process is likely to be a relatively long-term adaptation for bundles as the movement of trichomes in bundles is restricted by the surrounding sheath and adjacent trichomes except at bundle ends where new sheath material can be formed. Single trichomes with or without distinct sheath have more flexibility with regards to possible curling motion. We have attempted to evaluate the degree of curling by approximating each skeletonized track (>5 pixel long) with the best fitting ellipsoid. The average length of the minor axis of the best fitting ellipsoid is shown in Fig. 5a. It is fairly constant, but shows a slight increase at high light intensities. However, this is likely due to a slight increase in the average track length (Fig. 3b), which would increase the average length of both ellipsoid axes. A more appropriate number is the average ratio of the minor axis divided by the major axis of the best fitting ellipsoid. This ratio is directly related to the curvature of the track. The ratio depicted in Fig 5b. is likewise fairly constant. Nevertheless, a highly significant maximum curvature can be observed at intermediate light intensities, 2 - 20 $\mu E\ m^{-2}\ s^{-1}$. It thus seems likely that bending is a long term strategy for filaments to maintain a favourable position in a microbial mat, devoid of a directional light field. Curling could ultimately lead to development of a dense intertangled mat of filaments at agreeable light conditions with the highest possible biomass per area.

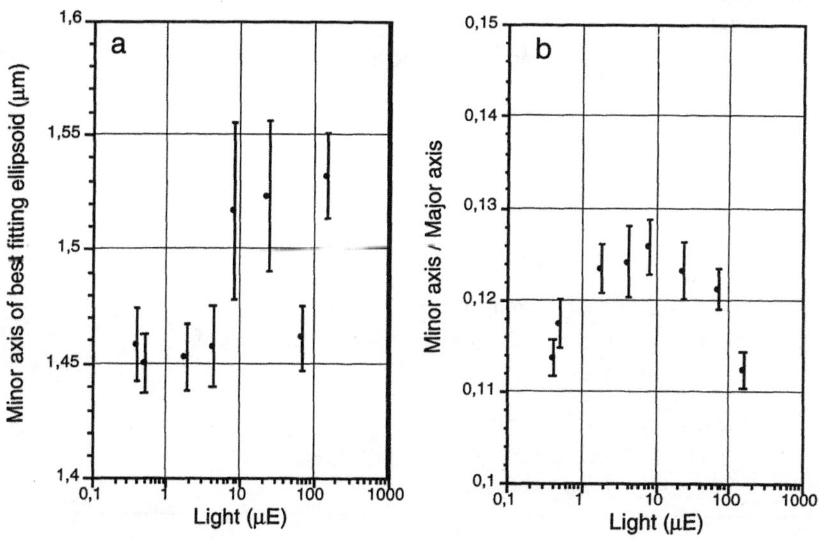

Fig. 5. Average curvature of movement tracks. (a) average length of minor axis of best fitting ellipse for each track. (b) ratio of minor axis divided by major axis of best fitting ellipse. The ratio is related to the curvature of the track.

We also attempted to evaluate whether individual trichomes clustered together when exposed to high light intensities in order to benefit from mutual shading, and likewise spread out to avoid shading at low light intensities as has been shown in other species (Castenholz 1968). We were unable to detect such behaviour during our observations. However, clustering could be too slow a response to be detectable in our relatively short time courses (about 30 min at each light intensity).

Conclusion

Using computer aided image analysis it was possible to evaluate several proposed strategies that could enable trichomes of *Microcoleus chthonoplastes* to position themselves in a uniform lightfield. The result is summarised in Table 1. It was found that *Microcoleus* utilised more frequent rests, more frequent reversals and increased bending under intermediate light conditions (4 - 10 μE m^{-2} s^{-1}) to optimise its position in the mat with respect to light. Only minor changes in their speed of movement was observed at various light intensities.

Table 1. Light dependent movement of *Microcoleus* sp.

Potential strategies to find optimal light conditions	Usage
1) Total movement Minimise movement if conditions are favourable	+++
1a) Frequency of movement Move less frequently if light conditions are favourable	+++
1b) Speed of movement Move slower if conditions are favourable	-
2) Reversing direction of movement Reverse direction of movement more frequently if conditions are favourable.	++
3) Change direction (Curling up) Change direction more frequently (create bends) if conditions are favourable.	+

It should be noted, however, that this type of analysis is biased towards the movement of single filaments. The movement of trichomes in bundles is more likely to be obscured by countermovement of other filaments in the same bundle. In future investigations we will try to evaluate the behavior of different strains adapted to different light intensities based on the algorithms and programs developed in this study. We have demonstrated here the value and potential of computer aided image analysis for such investigations.

Acknowledgements
We would like to thank Ferran Garcia-Pichel for collaboration in isolation and cultivation of the *Microcoleus* strain and for helpful discussions of the phototactic behavior of gliding cyanobacteria.

References

Castenholz RW (1968). The behavior of *Oscillatoria terebriformis* in hot springs. J Phycol 5, 123-139

Castenholz RW (1982). Motility and taxes. In Carr NG, Whitton BA (eds) The biology of Cyanobacteria, Blackwell Scientific Publications Oxford pp 413-439

Diehn B, Feinleib ME, Haupt W, Hildebrand E, Lenci F, Nultsch W (1979) Terminology of behavioral responses of motile microorganisms. Photochem photobiol 26: 559-560.

Häder D-P (1987) Photomovement. In Fay P, van Baalen C (eds), The Cyanobacteria Elsevier, Amsterdam

Kühl M, Jørgensen BB (1993) Spectral light measurements in microbenthic phototrophic communities with a fiber-optic microprobe coupled to a sensitive diode array detector. Limnol Oceanogr 37: 1813-1823

Kühl M, Lassen C, Jørgensen BB (1993) Light penetration and light intensity in sandy sediments measured with irradiance and scalar irradiance fiber-optic microprobes. Mar Ecol Progr Series, in press

Lassen C, Ploug H, Jørgensen BB (1992a). A fibre optic scalar irradiance microsensor: Application for spectral light measurements in sediments. FEMS Microbiol Ecol 86: 247-254.

Lassen K, Ploug H, Jørgensen BB (1992b). Microalgal photosynthesis and spectral irradiance in coastal marine sediments of Limfjorden, Denmark. Limnol Oceanogr 37: 760-772

Application of molecular genetics to the study of microbial communities

David A. Stahl and William C. Capman

Department of Veterinary Pathobiology, University of Illinois, Urbana, Illinois 61801 USA

Introduction

The character of studies of microbial mats is being changed by the use of molecular genetic techniques. Since these techniques are not dependent upon the vagaries of enrichment and pure culture isolation, they promise a complete accounting of community structure and direct access to the study of microorganisms at the levels of populations and single cells. Thus, the integration of molecular techniques with more standard techniques (e.g. microscopy, microelectrodes, stable isotopes, radiotracers, and analytical chemistry) should prove a powerful synthesis. Our goal is not to provide a review of all the recent literature (and we apologize for any omissions in this regard) but rather to provide perspective on available and developing techniques.

There have been several excellent recent reviews on molecular approaches to microbial ecology and the reader is referred to these for more specific information (Sayler and Layton 1990; Olsen 1991; Stahl and Amann 1991). In this contribution we will frame discussion around three basic questions in microbial ecology and discuss the utility (and/or shortcomings) of available molecular techniques to address them. Although we attempt to provide a general overview of nucleic-acid based techniques, specific examples shown in the accompanying figures are primarily taken from our own studies. The central questions we will address are: 1) what microbial populations are present, 2) which (or what fraction) are active and in what processes, and 3) what other processes are those populations capable of mediating?

Community structure: microbial identification and tracking in the environment

Population identification is the first step in establishing the relationships between the whole (community) and its parts (populations). The question of population identity can not be reasonably answered without a robust system of classification. The identification of a microorganism is usually based upon a collection of identifiers/characteristics that circumscribe the unit of species. Having circumscribed an organism (species) by both classical and molecular criteria, it should be possible to use a subset of those identifiers to identify the organism in the environment. For example, specific signatures of identity might include nucleic acid sequences,

NATO ASI Series, Vol. G 35
Microbial Mats
Edited by Lucas J. Stal and Pierre Caumette

and fatty acid or antigenic profiles. An important consideration is undescribed microbial diversity that is the backdrop against which specific identification must be made. The reader is referred to an excellent review by Bull and Hardman on this subject (1991).

The character of the comparative data set is changing with the increasing use of nucleic acid sequence for microbial identification. The rapid accumulation of nucleic acid sequence information, has also imposed a much needed evolutionary/phylogenetic framework on microbial systematics. However, this often contradicts existing determinative schemes and knowledge of genealogy alone has not clarified the concept of microbial species. It is likely that only within an environmental and ecological context will microbial species be reasonably defined.

The question of populations not represented in the culture collection

Two major obstacles to studying microbial populations in the environment are that many (or most) microorganisms remain unexamined because they have yet to be cultured. Even for those microorganisms in culture, identification based on selected phenotypic properties is often imprecise. This underscores the close connection between research in microbial systematics and environmental microbiology. A fundamental advance, therefore, has been the development of new and more specific biomarkers in microbial systematics. However, the key to open system application is that identification can be made within the context of uncharacterized environmental diversity. Of available determinative techniques, only those that are phylogenetically based provide sufficient information to do so. Of available phylogenetic methods, those using the rRNAs as markers of identity and phylogenetic relationship are best established. For this reason, we provide a relatively comprehensive discussion of the use of rRNAs for environmental studies. Additional aspects of rRNA-based studies of a microbial mat community are provided by Ward et al. (this volume).

Nucleic acid sequence information

The use of nucleic acids in environmental studies includes the use of the polymerase chain reaction (PCR) (to amplify specific DNA or RNA sequences), comparative sequencing, and nucleic acid hybridization (to identify and quantify genes/organisms and infer function). In addition, nucleic acid sequences are increasingly easier to obtain and exploit through centralized databases, not only through GenBank, but also through more specialized efforts such as the Ribosomal RNA Database Project at the University of Illinois (Larsen et al. 1993). The rRNA database serves as a central repository for the vast and growing collection of ribosomal RNA sequences, as a resource for sequence analysis and provides assistance in nucleic acid probe development. These techniques are becoming common enough in environmental microbiology that they have been well-covered in recent reviews (Sayler and Layton 1990; Stahl and Amann 1991; Olsen 1991; Steffan and Atlas 1991; Ward et al. 1992).

Isolation of nucleic acid from environmental samples

For amplification, cloning and/or hybridization of nucleic acids from environmental samples, the nucleic acids must first be isolated and, to some extent, purified (Sayler and Layton 1990; Steffan and Atlas 1991). The point of emphasis is that failure to recover nucleic acids from some microbial populations represents a major potential bias in environmental studies. Although a variety of methodological variations are in use, we emphasize the need for continued development. Even, those techniques already in use have yet to be fully evaluated.

Polymerase chain reaction: General cautions

The use of the polymerase chain reaction (PCR) is now widely used to recover gene sequences from environmental samples (Steffan and Atlas 1991). As noted below, this technique has greatly expanded the collection of ribosomal RNA sequences recovered from a variety of environments, without the requirement for culture. However, several aspects of the use of the polymerase chain reaction for identification and tracking of microbes (and/or their genes) need to be more critically addressed. For example, our own experience has demonstrated that even when using nucleic acids extracted from pure cultures, PCR reactions are sometimes inconsistent (Stahl, unpublished). Amplification of rRNA sequences has been shown in some cases to exclude important environmental populations (Amann et al. 1992a, Spring et al. 1991). In addition, PCR amplification from samples containing several variations of the gene of interest (e.g. rRNA genes) may lead to the production of sequence hybrids that are derived from two separate populations (Liesak et al. 1991). Problems in PCR repeatability and selectivity also call into question the use of PCR for quantitative purposes. Although PCR has not yet been used in an environmental context to quantify mRNA, this has been attempted with pure cultures (Steffan and Atlas 1991).

Ribosomal RNA sequencing

Of the commonly used nucleic acid-based techniques, comparative sequencing of the rRNAs now is the single most powerful method for characterizing natural microbial populations (Pace et al. 1986, Ward et al. 1992). This is because the technique is technically straightforward and sequences can be analytically related to an extensive collection of reference sequences. For simplicity of discussion, an rRNA sequence (16S-like, 23S-like, or 5S rRNA sequence) derived directly from the environment, without intermediate isolation and characterization of the parent organism, will be referred to as a "molecular isolate". Any molecular isolate now can be compared to an extensive collection of sequences representing the major divisions of life on this planet. Identification of close relatives is dependent upon the size and composition of the reference collection. In this regard, the importance of sequence databases can not be overstated.

Because of the greater information content of the larger rRNA (16S and 23S) most recent environmental studies have used the larger molecules as a basis to describe community population diversity.

Recovery of rRNA sequences from environmental samples

A variety of methods have been used to recover rRNA sequence information from environmental samples. These include direct sequencing of 5S rRNAs recovered from environmental samples (Stahl et al. 1985), cDNA cloning and sequencing (Ward et al. 1992 and this volume, Weller et al. 1991), and PCR amplification, cloning and sequencing using either rRNA or DNA as template. The use of PCR amplification and sequencing has dominated recent environmental studies. This is because the technique is exceptionally well-suited to the analysis of conserved biopolymers such as the ribosomal RNAs. Regions of conserved sequence common to all rRNAs, or to rRNAs of specific phylogenetic groups of organisms, can be used as priming sites for the selective amplification of either their 16S-like or 23S-like rRNA genes (Schmidt et al. 1991, Britshgi and Giovannoni 1991, DeLong 1992). This technique, in combination with the use of oligonucleotide probes, provides a very powerful approach to the systematic inspection of microbial community structure.

Ribosomal RNA-directed oligonucleotide hybridization probes: Quantification of populations and single cells

The ribosomal RNAs have, for many of the reasons already discussed, proved exceptionally well-suited as targets for determinative hybridization probes (and PCR primers). Although they are highly conserved, they exhibit great variation in regional sequence conservation. Some nucleotide positions and locales have remained virtually unchanged since the divergence of all existing life (universal sequences), whereas other regions vary so quickly that they can be used to differentiate among species or subspecies of bacteria. The reader is referred to a recent chapter by Stahl and Amann (1991) for a detailed description of the design and application of oligonucleotides probes for determinative and environmental studies.

An important aspect of using rRNA-directed oligonucleotide probes is that such probes can be used in a "phylogenetically nested" manner. That is to say, different probes can be designed to encompass large assemblages of organisms (e.g. the primary domains, archaea, eukarya and bacteria), smaller cohesive phylogenetic groups (for example spirochetes or methanogens), genera, species and even individual strains. Recent microbial groups addressed in this manner include Gram-negative sulfate-reducing bacteria (Devereux et al. 1992), methylotrophic bacteria (Tsien et al. 1990, Alvarez-Cohen et al. 1992), members of the genus *Xenorhabdus* (Putz et al. 1990), subdivisions of the proteobacteria (Manz et al. 1993), and

Frankia strains (Akkermans et al. 1991). However, this is an area of very active research and the applications of radioactive and fluorescent probes to characterize natural populations are too numerous to cover in this contribution.

Experimental format

Two basic formats have been used to quantify target populations: hybridization of DNA oligonucleotide probes to the total rRNA isolated from an environment or alternatively, hybridization of fluorescent probes to whole cells. The standard approach to the former measure is to immobilize total RNA on nylon membranes prior to hybridization. The probes are generally labeled with ^{32}P or an appropriate reporter group for non-radioactive (e.g. enzymatic or antibody-based) detection. The reader is referred to Stahl and Amann (1991) for detailed discussion of general methods and alternative labeling and detection techniques.

It is important to recognize that these two approaches provide different kinds of information. Hybridization to total nucleic acid provides a measure of rRNA abundance, expressed as either the amount of target rRNA or fraction of the total rRNA recovered from the sample. Total rRNA is quantified using universal hybridization probes that are complementary to highly conserved regions of sequence (common to all three domains) in the rRNAs. This value can not be directly converted to cell numbers (since ribosome content per cell is variable within and between populations). However, since cells increase their ribosome content with increasing growth rate, rRNA abundance might be better interpreted as a measure of the relative activity of a specific population. In contrast, the use of fluorescent probes in combination with microscopy or flow cytometry can provide a count of individual cells (Amann et al. 1990). Enzyme-conjugated probes (Amann et al. 1992b), indirect enzymatic (Zarda et al. 1991) and indirect fluorescence detection formats (Lim et al. 1993) have also been used to visualize single cells using rRNA targeted probes. Difficulties associated with single cell identification in environmental samples include low signal intensity, due to low ribosome content of slowly growing cells, and interference from autofluorescence of the cells or associated materials. It is likely that methods to reduce noise and/or increase specific signal must be developed before this technique can be generally applied to a variety of environments.

Spatial relationships

Microbial mats and biofilms are conspicuous examples of microbial communities that have a spatial dimension. Other examples include aggregates and flocs as occur in aquatic systems and promoted in wastewater treatment facilities. We have recently used group-specific radioactive probes to demonstrate the stratification of sulfate-reducing bacteria in a saline microbial mat. The use of digital imaging in combination with fluorescent probes is discussed in the contribution by Ramsing and Prufert-Bebout to this volume. In addition, ongoing work in our laboratory is using fluorescent probes to examine the distribution of individual cells in biofilms, flocs, and microbial mats (Amann et al. 1992a; Poulsen et al. 1993). An example of the use of fluorescent probes to characterize the spatial relationships of archaea (methanogen) and bacteria in microbial floc obtained from an anaerobic bioreactor system is shown in Figure 1.

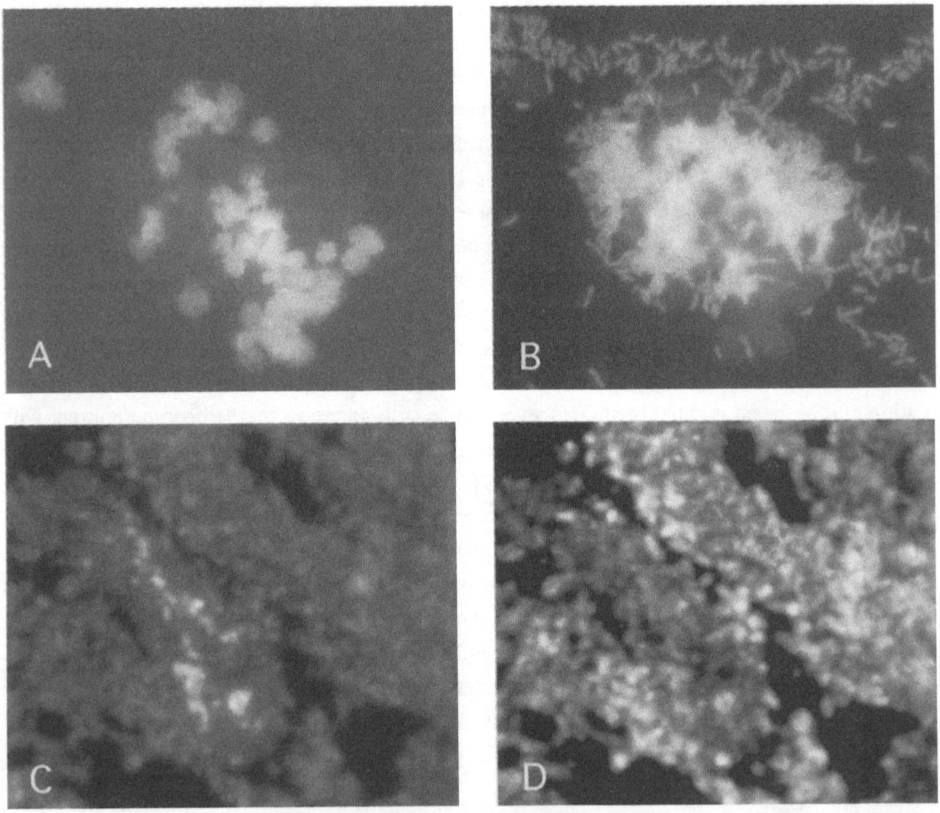

Fig. 1. Dual probing examples. (A) - (B) Mixture of *Methanosarcina acetivorans* and *Desulfovibrio desulfuricans* from pure cultures, demonstrating the ability to distinguish cell types using *in-situ* hybridization techniques. (A) Rhodamine labeled archaeal probe (915 probe, Stahl and Amman 1991), and (B) fluorescein labeled bacterial probe (Eub338 probe, Stahl and Amman 1991). (C) - (D) Floc material from a methanogenic fixed-bed biofilm reactor. The floc material was fixed in paraformaldehyde, embedded in paraffin, and sectioned 2 μm thick. Paraffin was removed with xylene prior to *in-situ* hybridization. (C) Fluorescein labeled archaeal probe 915 and (D) rhodamine labeled bacterial probe Eub338.

Integrated rRNA studies

As noted above, the same stretches of rRNA sequence that offer varying probe specificity also can be used to design PCR primer sets to retrieve 16S rRNA sequences of representatives of specific phylogenetic groups (above). In this regard, the work of Amann et al. (1992a) and Kane et al. (1993) are instructive. These studies combined the techniques of group-specific PCR amplification, 16S rRNA comparative sequence analysis, and oligonucleotide probe hybridization to phylogenetically characterize, microscopically visualize, and direct the cultural isolation of sulfate-reducing bacteria (SRB) in/from multispecies biofilm communities. Specifically, a region of 16S rRNA sequence common to most SRB was used to design a PCR primer for their selective amplification. When labeled with a fluorescent dye, this oligonucleotide also served as a hybridization probe to directly visualize putative SRB on coverslips colonized by the biofilm community, using epifluorescence microscopy. Comparison of two recovered sequences to those of characterized SRB revealed that they were close relatives of *Desulfovibrio vulgaris* and *Desulfuromonas acetoxidans*. Comparative sequence analysis was again used to design a fluorescently labeled oligonucleotide hybridization probe which targeted each population. When SRB populations within the biofilm were visualized by using the fluorescently labeled group- and population-specific probes, the major morphological type that hybridized to the group-specific probe was apparently not identified by either of the population-specific probes. This result again serves as caution to the possible biases of PCR amplification and the value of corroborating techniques.

Genetic fingerprints of community structure

We use the term "genetic fingerprint" to refer to techniques that provide a picture, or pattern, of community diversity based upon physical separation of unique nucleic acid species. Separation of individual species is usually accomplished by agarose or polyacrylamide gel electrophoresis, but has also included chromatographic techniques (Stahl et al. 1985). Discussion here is restricted to two recently described techniques for the study of relatively complex communities. The first uses small RNA species (tRNA and 5S rRNA) directly extracted from environmental samples to assess community diversity and changes in diversity through time or following perturbation (Hofle 1990). A recent study of particular note used this method to study the population dynamics of freshwater mesocosms following inoculation with large numbers of *E. coli* and *Pseudomonas putida* or their culture medium (Hofle 1993). The second approach combines general PCR amplification rRNA gene sequences (as described above) with denaturing gradient gel electrophoresis to resolve unique amplification products (unique rRNA gene sequences). The clear advantage of the PCR-based technique it that the sequences of individual amplification products can be determined. However, possible biases of PCR amplification must also be considered. The latter technique is addressed in a separate contribution to the volume and will not be further elaborated here (Muyzer et al. 1993 and this volume). These techniques are attractive in that they offer rapid access to general information about community diversity.

Activity of populations and single cells

Process measurements of microbial communities are well established among environmental microbiologists. More classical measures include microelectrodes, radiotracers, and isotope fractionation. These are covered in other contributions to this volume and are not the subject of this contribution. Discussion here is restricted to genetic and molecular techniques now being applied to environmental systems.

Measurement of transcriptional activity

Although the presence of specific genes or microorganisms in the environment now can be inferred by using direct molecular measures (e.g. DNA or RNA probes), such measures provide little information of activity of individual microorganisms or populations or their response to changing biotic and abiotic conditions. The development of more direct measures of single cell identity and activity should, therefore, offer a powerful complement to existing gene detection methods. One of the most exciting recent developments in this area has been the development of a bioluminescent reporter system based on fusion of lux genes to specific regulated genes (Shaw and Kado 1986; King et al. 1990). This has provided a format for non-destructive, on-line monitoring of specific gene transcription in the environment. Initial open environmental studies examined a lux gene cassette inserted into a plasmid containing the genes for naphthalene catabolism in *Pseudomonas fluorescens*. Bioluminescent light production from luciferase (lux) was demonstrated to coincide with naphthalene biodegradation in pure culture and following addition of the organism to soil slurries containing complex communities. More recent work has used this format to detect bioavailable mercury in the environment (Selifonova et al. 1993). Although not yet fully evaluated, this approach when combined with microscopy and image analysis should bring even single cells to this level of scrutiny. An important limitation of approach would appear to be the requirement of bioluminescence for oxygen, restricting the application of on line detection to aerobic settings.

Another approach to measuring transcriptional activity in the environment is to quantify mRNA abundance in nucleic acid extracted from natural material. Although the isolation of mRNA has become a standard tool in studies of prokaryotic and eukaryotic gene regulation, the approach is only now being applied to environmental microbiology. A recent report by Pichard and Paul (1991) compared sense and antisense RNA probes in measuring expression of ribulose-1,5-bisphosphate carboxylase/oxygenase in *Synechococcus* cultures and diel variation in natural phytoplankton populations. The combined use of sense and antisense probes as control for nonspecific hybridization. The conserved ribulose-1,5-bisphosphate carboxylase/oxygenase sequence offered a target common to the environmental variety of populations but introduced difficulties of quantitative interpretation. Another approach is the use of nuclease protection to measure specific message abundance, a standard technique in microbial genetics (Cai and Winkler 1993). A recent study by Fleming et al. (1993) used the nuclease protection format to detect specific message synthesis of a gene for naphthalene catabolism (NAH7 naphthalenedioxygenase). In this study the measured mRNA transcript level was positively correlated with naphthalene mineralization rates, soil naphthalene concentration, and gene frequency.

A very exciting recent publication by Hahn et al. (1993) describes the monitoring of single cell expression of a message for antibiotic synthesis (thiostrepton resistance gene) using digoxigenin labeled probes. Although this was a high copy message, the feasibility of measuring message expression in individual prokaryotic cells was established.

The use of ribosome abundance as a measure of single cell activity

In addition to the direct measures of gene expression discussed above, a more general assessment of individual organism activity is provided by measuring cellular ribosome abundance. The correlation between cellular ribosome content and growth rate was one of the earliest and most fundamental observations in microbial physiology. The relative ratio of RNA/DNA has been used as an indicator of both growth rate and biomass in planktonic marine organisms (Dortch et al. 1983). More recently, the fluorescent probes earlier described to visualize individual cells have been used to estimate cellular ribosome abundance (DeLong et al. 1989; Poulsen et al. 1993). We recently used this technique to infer the growth rate of a specific population of sulfate-reducing bacteria during the development of a multispecies biofilm on a fresh surface. This technique revealed a dramatic decrease in the activity/growth rate of this organism in a more established biofilm community (Figure 2).

Potential activity

Process versus gene abundance

We earlier discussed the need for tools to distinguish between presence (of a gene or organism) and activity. In addition to the methods already described, another approach is to relate gene or organism abundance to community processes. One good example of such a study is that of Smith and Tiedje (1992) in which nitrate reductase gene probe data were correlated with an enzyme activity assay and MPN estimates of denitrifying bacteria aquifer samples. The sequence for the dissimilatory heme-type nitrite reductase gene (nir) from *Pseudomonas stutzeri* strain JM300 was determined and compared to those of *P. aeruginosa* and a second *P. stutzeri* strain. Potential gene probes were derived from restriction fragments from within the gene. A probe generated from the C-terminus region of the nir gene hybridized with members of five different denitrifying genera, but not to non-denitrifying representatives spanning six different genera. A probe complimentary to an N-terminal 0.9Kb region failed to detect a *P. fluorescens* strain that was identified by the C-terminus probe. In anaerobic aquifer samples, gene probe hybridization signal, denitrifying activity and denitrifier MPN estimates all increased in response to the addition of nitrate or (to a lesser extent) nitrite. However, because of the format used (Southern blot analysis) it was difficult to correlate hybridization (band) intensity with population size. In another study, four gene probes targeting four different mer operons were used to measure the abundance of mer genes (upon which bacterial mercury resistance via organomercurial degradation and inorganic mercury reduction is based) in mercury contaminated waters (Barkay et al. 1991). The relative abundance of mer operons was highest in chronically exposed environments, and this corresponded with the ability of indigenous microbes to immediately initiate volatilization of added mercury. More recent studies by this same author used lux fusion constructs to assess mercury bioavailability (Selifonova et al. 1993).

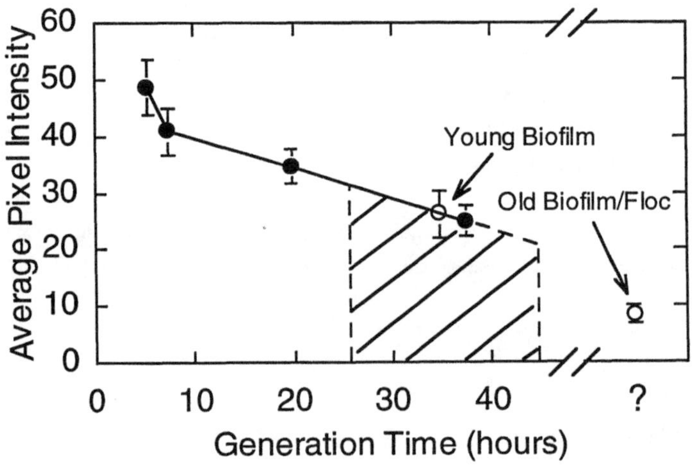

Fig. 2. Growth rates of sulfate reducing bacteria biofilm populations inferred through the use of *in-situ* hybridization with fluorescently labelled probes in conjunction with digital microscopy. The fluorescent signals (mean pixel intensity from images of hybridized cells) of cells from young versus old biofilm (○) were compared to signals from cells grown in pure culture at different growth rates (●). The cross hatching shows the range of growth rates spanned by the standard deviation derived from 50 cells in the young biofilm. The standard deviation for the old biofilm is based on nine cells, while the standard deviations for the pure culture cells are each based on 20 to 25 cells (data from Poulsen et al. 1993).

Still other examples of using gene probes to inspect natural populations within a functional context include studies of acetogenic bacteria (via formyltetrahydrofolate synthase gene probes; Lovell and Hui 1991) members of the genus *Desulfovibrio* (via uptake hydrogenase gene probes; Voordouw et al. 1990), and methanotrophs (via methane monoooxygenase gene probes; Tsien and Hanson 1992).

Can phylogenetic affiliation be used to predict physiology?

We take it as a given that the major lines of bacterial descent have evolved to fill different roles in the biosphere. However, there is considerable confusion concerning the ability to infer physiology from phylogenetic affiliation. There are many examples of close phylogenetic relationship between organisms that have remarkably different physiological attributes, for example close relationships between phototrophs, autotrophs and heterotrophs have been observed (for examples see Lane et al. 1992). But on the other hand there are examples of phylogenetically-defined groups that are remarkably coherent in physiological characteristics. The *Desulfovibrio* and methanogenic groups are notable examples of phylogenetic and physiological coherence. Other examples include ammonia and methane oxidizing bacteria. We are inclined to focus on the later as the more relevant example and would suggest that lack of apparent relationship is simply a consequence of our failure to recognize the defining attributes of different natural assemblages. The defining attributes are those that define both the evolution and the ecology of the assemblage. It is our estimation that as we learn more about the actual activities of different microbial populations by directly observing them in the environment, the phylogenetically cohesive characters will be closely aligned with their ecology.

Acknowledgements

Some of the research described in this contribution was supported by research grants from ONR and NSF to D.A.S.

References

Alvarez-Cohen L , McCarty PL , Bougygina E, Hanson RS, Brusseau GA, Tsien HC (1992) Characterization of a methane-utilizing bacterium from a bacterial consortium that rapidly degrades trichloroethylene and chloroform Appl Environ Microbiol 58: 1886-1893

Akkermans ADL , Hahn D, Mirza MS (1991) Molecular ecology of *Frankia:* Advantages and disadvantages of the use of DNA probes. Plant and Soil 137: 49-54

Amann RI , Binder BJ, Olson RJ, Chishom SW, Devereux R, Stahl DA (1990) Combination of 16S rRNA-targeted oligonucleotide probes with flow cytometry for analyzing mixed microbial populations. Appl Environ Microbiol 56: 1919-1925

Amann RI , Stromley J, Devereux R, Stahl DA (1992)a Molecular and microscopic identification of sulfate-reducing bacteria in multispecies biofilms . Appl Environ Microbiol 58: 614-623

Amann RI, Zarda B, Stahl DA, Schleifer K-H (1992b) Identification of individual prokaryotic cells by using enzyme-labeled, rRNA-targeted oligonucleotide probes. Appl Environ Microbiol 58: 3007-3011

Barkay T, Turner RR, Vandenbrook A, Liebert C (1991) The relationships of Hg(II) volatilization from a freshwater pond to the abundance of mer genes in the gene pool of the indigenous microbial community. Microb Ecol 21: 151-261

Britshgi TB, Giovannoni SJ (1991) Phylogenetic analysis of a natural marine bacterioplankton population by rRNA gene cloning and sequencing. Appl Environ Microbiol 57: 1707-1713

Bull AT, Hardman DJ (1991) Microbial diversity Curr Opinion Biotechnol 2: 421-428

Cai J , Winkler HH (1993) Identifiation of tlc and gltA mRNAs and determination of *in situ* RNA half-life in *Rickettsia prowazekii.* J Bacteriol 175: 5725-5727

Delong EF (1992) Archaea in coastal marine environments. PNAS 89: 5685-5689

Delong EF, Wickham GS, Pace NR (1989) Phylogenetic stains: Ribosomal RNA-based probes for the identification of single cells. Science 243: 1360-1363

Devereux R, Kane MD, Winfrey J, Stahl DA (1992) 16S rRNA hybridization probes to describe natural communities of sulfate reducing bacteria. Syst Appl Microbiol 15: 601-609

Dortch Q , Roberts TL, Clayton JR , Ahmed SI (1983) RNA/DNA ratios and DNA concentrations as indicators of growth rate and biomasss in planktonic marine organisms Mar Ecol Prog Ser 13: 61-71

Fleming JT, Sanserverino J, Sayler GS (1993) Quantitative relationship between naphthalene catabolic gene frequency and expression in predicting PAH degradation in soils at town gas manufacturing sites. Environ Sci Technol 27: 1068-1074

Hahn D, Amann RI, Zeyer J (1993) Detection of mRNA in *Streptomyces* by whole-cell hybridization with digoxigenin-labelled probes. Appl Environ Micribol 59: 2753-2757

Hofle MG (1990) Transfer RNAs as genotypic fingerprints of eubacteria. Arch Microbiol 153: 299-304

Hofle MG (1993) Bacterioplankton community structure and dynamics after large-scale release of nonindigenous bacteria as revealed by low-molecular weight RNA analysis. Appl Environ Microbiol 58: 3387-3394

Kane MD, Poulsen LK, Stahl DA (1993) Monitoring the enrichment and isolation of sulfate-reducing bacteria by using oligonucleotide hybridization probes designed from environmentally derived 16S rRNA sequences. Appl Environ Microbiol 59: 682-686

King JMH, DiGrazia PM, Applegate B, Burlage R, Sanseverino J, Dunbar P, Larimer F, Sayler GS (1990) Rapid, sensitive bioluminescent reporter technology for napthalene exposure and biodegradation. Science 249: 778-781

Lane DJ , Harrison AP, Stahl D, Pace B, Giovannoni SJ, Olsen GJ, Pace NR (1992) Evolutionary relationships among sulfur- and iron-oxidizing euacteria. J Bacteriol 174: 269-278

Larsen N, Olsen GJ, Maidak BL, McCaughey MJ, Overbeek R, Macke TJ, Marsh, TL, Woese CR (1993) The ribosomal RNA database project. Nucleic Acid Res 21: 3021-3023

Liesack W , Weyland N, Stackebrandt E (1991) Potential risks of gene amplification by PCR as determined by 16S rDNA analysis of a mixed-culture of strict barophillic bacteria. Microb Ecol 21: 191-198

Lim EL, Amaral LA, Caron DA, DeLong EF (1993) Application of rRNA-based probes for observing marine nanoplankton protists. Appl Environ Microbiol 59: 1647-1655

Lovell CR , Hui Y (1991) Design and testing of a functional group-specific DNA probe for the study of natural populations of acetogenic bacteria. Appl Environ Microbiol 57: 2602-2609

Manz W, Ulrich S, Ericksson P, Amann R, Schleifer, K-H, Stenstrom T-A (1993) In situ identification of bacteria in drinking water and adjoining biofilms by hybridization with 16S and 23S rRNA-directed fluorescent oligonucleotide probes. Appl Environ Microbiol 59: 2293-2298

Muyzer G, De Waal EC, Uitterlinden AG (1993) Profiling of complex microbial populations by denaturing gradient gel electrophoresis analysis of polymerase chain reaction-amplified genes coding for 16S rRNA. Appl Environ Microbiol 59: 695-700

Olsen BH (1991) Tracking and using genes in the environment. Environ Sci Technol 25: 604-611

Pichard SI, Paul JH (1991) Detection of gene expression in genetically engineered microorganisms and natural phytoplankton populations in the marine environment by mRNA analysis. Appl Environ Microbiol 57: 1721-1727

Poulsen LK, Ballard G, Stahl DA (1993) Use of rRNA fluorescence *in situ* hybridization for measuring the activity of single cells in young and established biofilms. Appl Environ Microbiol 59: 1354-1360

Putz J, Meinert F, Wyss U, Ehlers R-U, Stackebrandt E (1990) Development and application of oligonucleotide probes for molecular identification of *Xenorhabdus* species. Appl Environ Microbiol 56: 181-186

Sayler GS, Layton AC (1990) Environmental application of nucleic acid hybridization. Ann Rev Microbiol 44: 625-648

Schmidt TM , DeLong EF, Pace NR (1991) Analysis of marine picoplankton community by 16S rRNA gene cloning and sequencing. J Bacteriol 173: 4371-4378

Selifonova O, Burlage R, Barkay T (1993) Bioluminescent sensors for detection of bioavailable Hg(II) in the environment. Appl Environ Microbiol 59: 3063-3090

Shaw JJ , Kado CI (1986) Development of a *Vibrio* bioluminescence gene-set to monitor phytopathogenic bacteria during the ongoing disease process in a non-disruptive manner Bio/Technology 4: 560-564

Smith GB, Tiedje JM (1992) Isolation and characterization of a nitrate reductase gene and its use as a probe for denitrifying bacteria. Appl Environ Microbiol 58: 376-384

Spring S, Amann R, Ludwig W, Schleifer K-H, Peterson N (1992) Phylogenetic diversity and identification of non-culturable magnetotactic bacteria. System Appl Microbiol 15: 116-122

Stahl DA, Lane DJ, Olsen GJ, Pace NR (1985) Characterization of a Yellowstone hot spring microbial community by 5S rRNA sequences. Appl Environ Microbiol 49: 1379-1384

Stahl DA, Amann R (1991) Development and application of nucleic acid probes. In: Stackebrandt E, Goodfellow M (eds) Nucleic Acid Techniques in Bacterial Systematics. John Wiley & Sons Ltd

Steffan RJ, Atlas RM (1991) Polymerase chain reaction: applications in environmental microbiology. Ann Rev Microbiol 45: 137-161

Tsien HC, Hanson RS (1992) Soluble methane monooxygenase component-B gene probe for identification of methanotrophs that rapidly degrade trichloroethylene. Appl Environ Microbiol 58: 53-960

Tsien HC, Bratina BJ, Tsuji K, Hanson RS (1990) Use of oligonucleotide signiture probes for identification of physical groups of methylotrophic bacteria. Appl Environ Microbiol 56: 2858-2865

Voordouw G, Niviere V, Ferris FG, Fedorak PM, Westlake DWS (1990) Distribution of hydrogenase genes in *Desulfovibrio* spp and their use in identification of species from the oil field environment. Appl Environ Microbiol 56: 3748-3754

Ward DM, Bateson MM, Weller R, Ruff-Roberts AL (1992) Ribosomal RNA Analysis of Microorganisms as They Occur in Nature. In: Marshall KC (ed) Advances in Microbial Ecology, Vol 12. Plenum Press, New York, pp 219-286

Weller R, Weller JW, Ward DM (1991) 16S rRNA sequences of uncultivated hot spring cyanobacterial mat inhabitants retrieved as randomly primed cDNA. Appl Environ Microbiol 57: 1146-1151

Zarda B, Amann R, Wallner G, Schleifer K-H (1991) Identification of single bacterial cells using digoxigenin-labelled, rRNA-targeted oligonucleotides. J Gen Microbiol 137: 2823-2830

Determination of the genetic diversity of microbial communities using DGGE analysis of PCR-amplified 16S rDNA

Gerard Muyzer and Ellen C. de Waal[1]

Max-Planck-Institute for Marine Microbiology, Molecular Ecology Unit
Fahrenheitstraße 1, 28359 Bremen, Germany

Introduction

Microbial ecosystems contain a large diversity of bacterial species. They are dominated by complex interactions between the different microorganisms, whereby each of the individual species has a specific role in the maintenance of the system. The active communities can efficiently scavenge nutrients from the environment and eliminate toxic compounds. However, not all of the species are active; most of them are dormant until environmental conditions change to favour their growth. Because of the great metabolic diversity and flexible organisation microbial ecosystems can be found nearly everywhere. Examples are the bacterial biofilms in waste water treatment reactors, on ship walls, or the microbial mats, found in hypersaline environments, tidal sediments and hot springs.

Study of microbial communities has raised questions about the composition, structure and stability of these communities and about the activity and function of the individual inhabitants. Traditional microbiological techniques and conventional microscopy are insufficient means to answer these questions. Most of the bacteria in natural samples cannot be detected with conventional microscopy, because they adhere to soil and sediment particles and therefore remain "invisible". Fluorescent dyes, such as DAPI or acridine orange, have improved this, but yield no information as to which species are present. Activity measurements of bacteria in sediments have been performed, but they lack the specificity to discriminate between the action of different species. Physiological experiments have been used with great success to characterise isolated species. However, it is now widely recognised among microbiologists that less than 20% of the naturally occurring bacteria have been isolated and characterised so far (Wayne et al. 1987; Giovannoni et al. 1990; Torsvik et al. 1990). Selective enrichment cultures fail to mimic the conditions that particular micro-organisms require for proliferation in their natural habitat. However, the application of molecular biological techniques offer new opportunities for the analysis of structure and species composition of microbial communities.

---

[1] Geobiochemistry Unit, Leiden University, The Netherlands

NATO ASI Series, Vol. G 35
Microbial Mats
Edited by Lucas J. Stal and Pierre Caumette

Genetic diversity analysis

Different molecular approaches have been used to determine the genetic diversity of microbial communities. An overview of these approaches has been given by Pedrós-Alió (1993). Nucleic acids, DNA or RNA, are extracted from a mixed microbial population and are used to determine the complexity of the population and to identify its members. Direct analysis of the DNA using DNA reassociation experiments (Torsvisk et al. 1990) or DNA/DNA hybridisation (Lee and Fuhrman 1990, 1991) have revealed important information on the number of different genomes in natural samples, but these approaches are not suited to identify different *phylotypes*[2]. Different cloning strategies have also been performed to analyse genetic diversity: (1) "shotgun" cloning of DNA fragments into bacteriophage lambda (Schmidt et al. 1991), (2) cloning of rcDNA fragments, obtained after transcription of the 16S rRNA into cDNA (Ward et al. 1990; Weller et al. 1991), and (3) cloning of PCR-amplified 16S rDNA fragments (i.e., Giovannoni et al. 1990; Fuhrman et al. 1992; Stackebrandt et al. 1993). All these different cloning strategies result in libraries of recombinants from which the inserted 16S rDNA fragments can be sequenced. Using these cloning approaches, different bacteria were identified which were never found before. From these results it was concluded that the existing culture techniques were insufficient to isolate these species and that the genetic diversity of bacteria in nature is far greater than was ever thought.

A third strategy for diversity analysis is based on the electrophoresis of low-molecular-weight (LMW) RNA (5S rRNA and tRNA) in high-resolution polyacrylamide gels (Höfle 1992) or of PCR-amplified 16S rDNA fragments in polyacrylamide gels containing a linear gradient of DNA denaturants (Muyzer et al. 1993). Both approaches give a direct visualisation of the genetic diversity of the microbial populations, as the number of bands should be proportional to the number of species constituting the populations. Electrophoresis of LMW-RNA has been applied to study the structure and dynamics of a freshwater bacterioplankton community (Höfle 1992).

DGGE analyses of PCR-amplified 16S rDNA fragments have been used to study the genetic diversity of microbial communities from a range of different environments, such as from experimental bioreactors and intertidal sediments (Muyzer et al. 1993), and from hot springs and hydrothermal vents (unpublished results). After sequencing of the separated bands, the phylogenetic affiliation of the community members can be reconstructed, although, the phylogenetic resolution for LMW-RNA is limited, due to their small size (approx. 120 nucleotides for 5S rRNA and 85 nucleotides for tRNA). Figure 1 shows 3 different strategies to analyse the genetic diversity of mixed microbial populations and to identify the population members. Although most results yet have been obtained from the cloning strategies, they have several disadvantages: (1) it is time-consuming, laborious and expensive to sequence a large number of different colonies, which could have the same inserted 16S rDNA fragment, (2)PCR

---

[2] The term *phylotype* refers to a nucleotide sequence which has been obtained from an environmental sample and has phylogenetic affiliation with sequences from known species (after Pace).

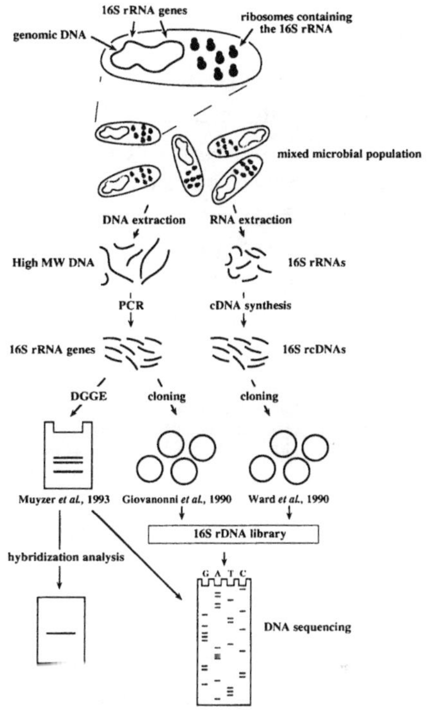

Fig. 1. Three different strategies to analyse the genetic diversity of mixed microbial populations (modified after Ward 1989).

fragments with identical ends might have different cloning efficiencies (Rudi Amann, pers. commun.), and therefore some species might not be represented in the library, (3) only qualitative information can be obtained, and (4) PCR errors might be sequenced (Giovannoni 1991). The DGGE procedure lacks these disadvantages. It is a fast procedure, it allows multiple sample analysis on the same gel, and because separated bands are sequenced directly, PCR errors are not detected. But, the analysis might be limited to partial 16S rDNA fragments only. Furthermore, the separation of many rDNA fragments might be difficult, although the resolution of separation can be enhanced be choosing a more narrow gradient of denaturants. However, DGGE analysis provides a direct display of the community composition in both a qualitative and semi-quantitative way. This makes it a excellent approach not only for analysing the genetic diversity of mixed microbial communities, but also for studying population dynamics, which is practically impossible to perform with the cloning strategies.

Denaturing gradient gel electrophoresis

In denaturing gradient gel electrophoresis (DGGE) DNA fragments of the same length but with different base-pair sequences can be separated. Separation in DGGE is based on the electrophoretic mobility of a partially melted double stranded DNA (dsDNA) molecule in polyacrylamide gels, which is decreased compared with that of the completely helical form of the molecule. The melting of the fragments proceeds in discrete so-called melting domains: stretches of base pairs with an identical melting temperature.

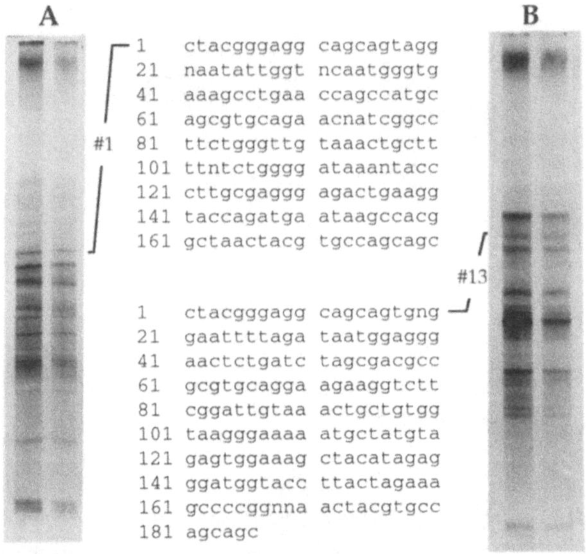

Fig. 2. Negative image of an ethidium-bromide stained DGGE separation pattern of 16S rDNA fragments obtained after enzymatic amplification of genomic DNA from a bacterial biofilm grown under aerobic conditions (A), and a biofilm grown under anaerobic conditions (B). Two different concentrations of the same sample have been applied to the gel. The sequence of one band from each sample has been determined .

Once the melting domain with the lowest melting temperature reaches this temperature at a particular position in the denaturant gradient gel (DGG), a transition of helical to partially melted molecules occurs, and migration of the molecule will practically halt. Sequence variation within such domains causes their melting temperature to differ. Sequence variants of particular fragments will therefore stop migrating at different positions in the denaturing gradient and hence can be separated effectively by DGGE (Lerman et al. 1984).

DGGE analysis of PCR-amplified 16S rDNA fragments from different microbial communities, i.e., microbial mats and experimental bioreactors, demonstrated the presence of several distinguishable bands in the separation pattern, which were most likely derived from the predominant species within these populations. Using a nucleotide probe specific for sulphate reducing bacteria (Amann et al. 1990) more detailed information was obtained about the species composition (Muyzer et al. 1993).

In this paper, we present the phylogenetic affiliation of some of the community members after sequencing of the individual bands and comparing the sequences to 16S rRNA sequences of know bacterial species.

Genomic DNA, extracted from bacterial biofilms of two different fluidized bed reactors, was used as target DNA for the amplification of the 16S rRNA genes as described in Muyzer et al. (1993). The PCR products were analysed by DGGE (Muyzer et al. 1993). Individual bands were excised from the gel and re-amplified. Some of the PCR products were sequenced directly as has been described by Corstjens and Muyzer (1993). The sequences were compared with sequences from the Ribosomal Database Project (RDP; Olsen et al. 1991) and the EMBL nucleotide sequence database. A bootstrap majority-rule consensus tree was constructed from the aligned sequences using the maximum parsimony algorithm implemented in the software program PAUP developed by Swofford (1991).

Fig. 2 shows the DGGE patterns of the PCR-amplified 16S rDNA fragments of a bacterial biofilm grown under aerobic conditions (Fig. 2A) and another, which was grown under anaerobic conditions (Fig. 2B). Up to 10 bands can be distinguished in the separation patterns, which are most likely derived from as many different species constituting the biofilms. The phylogenetic affiliation of 2 community members was determined after re-amplification and sequencing of the excised bands (Fig. 2). After comparison of these sequences with sequences from the RDP and the EMBL nucleotide sequence database, the highest similarity scores for the sequence of band #1 (from the aerobic biofilm) were found with members of the Flexibacter-Cytophaga-Bacteroides phylum, such as *Microscilla furvescens*, *Runella slithyformis*, and *Cytophaga hutchinsonii*. The similarity ranks for band #13 (from the anaerobic biofilm) were phylogenetically less consistent; high scores were found with *Thermosipho africanus*, which belongs to the Thermotogales, but also with *Eubacterium uniforme*, *Clostidium aminovalericum* and *Syntrophomonas wolfei*, which all belong to the Clostridia and relatives (Gram Positive phylum, High G+C subdivision).

Fig. 3 shows the affiliation of both sequences with known sequences taken from the RDP. The sequence of band #1 (*phylotype* #1) groups with *Runella slithyformis*, a member of the Flexibacter-Cytophaga-Bacteroides phylum. The sequence of band #13 (*phylotype* #13) groups with *Thermosipho africanus*, and not with the Clostridiae. This result is highly suprising, because *T. africanus* belongs to a unique group of extremely thermophilic micro-organisms, the Thermotogales, which are phylogenetically distant from all other eubacteria. However, *T. africanus* is able to grow at much lower temperatures, such as 37°C (Huber and Stetter 1992), the temperature at which the fluidized bed reactor was kept. The sequences were

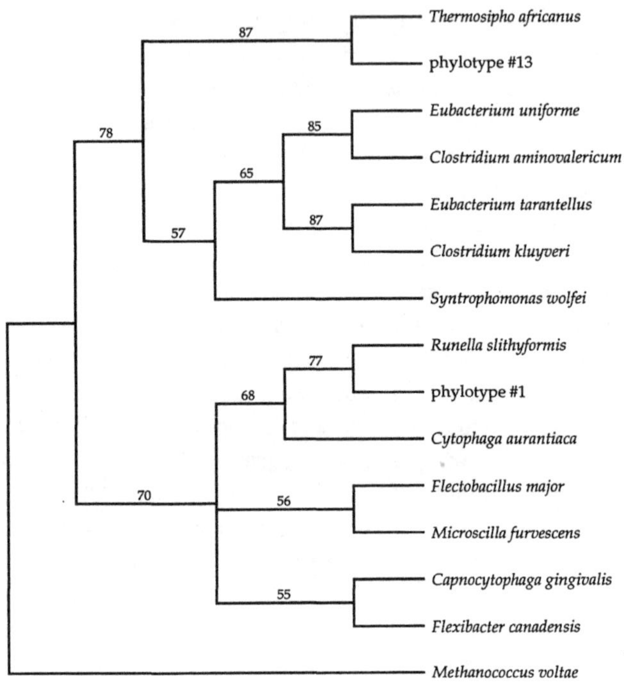

Fig. 3. Bootstrap consensus tree of 16S rDNA sequence data showing the affiliation of *phylotypes* #1 and # 13 with sequences from the Ribosomal Database Project (Olsen et al. 1991). The numbers on the branches indicate the percentage of the bootstrap replications that support the group descending from that branch. *M. voltae* is used as an outgroup.

also checked for a chimeric origin (Liesack et al. 1991) using the *Check Chimera* program implemented in the RDP (Olsen et al. 1991). A break point was found for both sequences at position 120, indicating the possibility of being chimeric molecules. However, after phylogenetic analysis of each individual chimeric fragment, no difference was found in tree topology for the two chimeric fragments (results not shown) and the full lenght fragment (Fig. 3). The indication that both fragments are chimeric molecules might be due to the small size of the fragment (about 200 bp) as well as the presence of a variable region, the V3-region (Neefs et al. 1990), which starts approximately at position 120 and continues to position 150. The small size of the fragment might also be the reason that the *T. africanus*/phylotype #13-cluster is grouping with the Clostridia, which is not correct. We are therefore now focussing on DGGE analysis of larger fragments of the 16S rDNA (about 600 bp) to improve the phylogenetic resolution of this approach, and to obtain further evidence of the presence of Thermotogales-like organisms in anaerobic bioreactors.

Acknowledgements

We are greatly indebted to Andre Uitterlinden (INGENY, Leiden, the Netherlands) for advice and Andreas Teske (MPI for Marine Microbiology, Bremen, Germany) for technical assistance. Luuk Tijhuis and Astrid van de Graaf (Delft University of Technology, Delft, the Netherlands) are thanked for providing us with bacterial samples. Niels Ramsing (MPI for Marine Microbiology, Bremen, Germany) and Gijs Kuenen (Delft University of Technology, Delft, the Netherlands) are acknowledged for their helpful discussions.

References

Amann RI, Binder B. Chisholm SW, Olsen R, Devereux R, Stahl DA (1990) Combination of 16S rRNA-targeted oligonucleotide probes with flow cytometry for analyzing mixed microbial populations. Appl Environ Microbiol 56: 1919-1925

Corstjens P, Muyzer G (1993) Phylogenetic analysis of the metal-oxidizing bacteria *Leptothrix discophora* and *Sphaerotilus natans* using 16S rDNA sequencing data. System Appl Microbiol 16: 219-223

Furhman JA, McCallum K, Davis AA (1992) Novel major archaebacterial group from marine plankton. Nature 356: 148- 149

Giovannoni SJ (1991) The polymerase chain reaction. In: Stackebrandt E, Goodfellow M (eds) Nucleic acid techniques in bacterial systematics. John Wiley and sons Ltd, Chichester, pp 177-203

Giovannoni SJ, DeLong EF, Olsen GJ, Pace NR (1990) Genetic diversity in Sargasso Sea bacterioplankton. Nature 345: 60-63

Höfle M (1992) Bacterioplankton community structure and dynamics after large-scale release of nonindignous bacteria as revealed by low-molecular-weight-RNA analysis. Appl Environ Microbiol 58

Huber R, Stetter KO (1992) The order Thermotogales. In: Balows A, Trüper HG, Dworkin M, Harder W, Schleiffer K-H (eds) The Prokaryotes 2nd edition, pp 3809-3815

Lee S, Fuhrman JA (1990) DNA hybridization to compare species compositions of natural bacterioplankton assemblages. Appl Environ Microbiol 56: 739-746

Lee S, Furhman JA (1991) Spatial and temporal variation of natural bacterioplankton assemblages studied by total genomic DNA cross-hybridization. Limnol Oceanogr 36: 1277-1287

Lerman LS, Fischer SG, Hurley I, Silverstein K, Lumelsky N (1984) Sequence-determined DNA separations. Annu Rev Biophys Bioeng 13: 399-423

Liesack W, Weyland H, Stackebrandt E (1991) Potential risks of gene amplification by PCR as determined by 16S rDNA analysis of a mixed-culture of strict barophilic bacteria. Microb Ecol 21: 191-198

Muyzer G, de Waal EC, Uitterlinden AG (1993) Profiling of complex microbial populations by denaturing gradient gel electrophoresis analysis of polymerase chain reaction-amplified genes coding for 16S rRNA. Appl Environ Microbiol 59: 695-700

Neefs J-M, van de Peer Y, Hendriks L, de Wachter R (1990) Compilation of small ribisomal subunit RNA sequences. Nucleic Acid Res 18: 2237-2317

Olsen GJ, Larses N, Woese CR (1991) The ribosomal RNA database project. Nucl Acids Res 19: 2017-2021

Pedrós-Alió C (1993) Diversity of bacterioplankton. TREE 8: 86-90

Schmidt TM, DeLong EF, Pace NR (1991) Analysis of a marine picoplankton community by 16S rRNA gene cloning and sequencing. J Bacteriol 173: 4371-4378

Stackebrandt E, Liesack W, Goebel BM (1993) Bacterial diversity in a soil sample from a subtropical Australian environment as determined by 16S rDNA analysis. FASEB J 7: 232-236

Swofford DL (1991) PAUP: Phylogenetic Analysis Using Parsimony version 3.1, Computer program distibuted by the Illinois Natural History Survey Champaign, Illinois

Torsvik V, Goksøyr J, Daae FL (1990) High Diversity in DNA of Soil Bacteria. Appl Environ Microbiol 56: 782-787

Ward DM (1989) Molecular probes for analysis of microbial communities. In: Characklis WG, Wilderer PA (eds) Structure and function of biofilms. John Wiley & Sons Ltd, pp 145-163

Ward DM, Weller R, Bateson MM (1990) 16S rRNA sequences reveal numerous uncultivated microorganisms in a natural environment. Nature 345: 63-65

Wayne LG, Brenner DJ, Colwell RR, Grimont PAD, Kandler O, Krichevsky MI, Moore C Moore WEC, Murray RGE, Stackebrandt E, Starr MP, Trüper HG (1987) Report of the ad hoc committee on reconciliation of approches to bacterial systematics. Int J Syst Bacteriol 37: 463-464

Weller R, Weller JW, Ward DM (1991) 16S rRNA Sequences of uncultivated Hot Spring cyanobacterial mat inhabitants retrieved as randomly primed cDNA. Appl Environ Microbiol 57: 1146-1151

Exopolymers in microbial mats: Assessing their adaptive roles

Alan W. Decho

Department of Environmental Sciences, School of Public Health, University of South Carolina, Columbia, SC. 29208 USA

Introduction

Microbial mat environments are characterized by thin-layered stratifications of microorganisms (Cohen 1989; Krumbein and Stal 1991) and geochemical gradients over small (μm) spatial scales (Jørgensen et al. 1983; Revsbech 1983). Sharp gradients in the concentrations of a variety of ions and molecules, such as O_2, CO_2, SO_4^{2-}, and PO_4^{3-}, have been observed within the layers of the mat. These gradients largely originate from the concentrated activities of specific groups of microorganisms within the mat layers (Jørgensen and Des Marais 1990). The relative stability of geochemical gradients (i.e. preventing rapid fluctuations) is important to maintain the efficient interactions between the different physiological groups of microorganisms contained within the mat layers.

Resident microbial flora of microbial mats release high molecular-weight mucous secretions called exopolymers (EPS). The exopolymer matrix can: 1) slow the diffusion of ions; 2) bind and store nutrients; 3) reduce desiccation; 4) protect cells against toxic compounds; and 5) enhance the cohesiveness and macro-physical stability of the mat and maintain the microspatial organization of non-motile microbial cells. Cohesive microbial exopolymers have likely existed since the earliest recorded life forms (i.e. precambrian stromatolites). It will be discussed how the intrinsic chemical properties of extracellular polymers can function to stabilize or even sharpen geochemical gradients, and constitute an important adaptive feature for mat microorganisms.

Diffusion-slowing properties

The movement of ions within a mat can be potentially reduced by the diffusion-slowing properties of exopolymers. Early studies showed that the diffusion of some ions and small molecules is significantly reduced when passing through an exopolymer matrix (compared to the diffusion of similar molecules in water) (Joyce and Dugan 1970; Mateson and Characklis 1976; LaMotta 1976). Rates of diffusion through exopolymers can vary depending on the localized concentration of the ion, the composition of the polymer (specifically the abundances of functional groups), and the concentrations of competing anions and cations present. Studies

NATO ASI Series, Vol. G 35
Microbial Mats
Edited by Lucas J. Stal and Pierre Caumette

of biofilm exopolymers (Mateson and Characklis 1976) have shown that a film of only 10-20 micrometers can reduce the diffusion of ions by 2-98%. Small monomeric molecules, such as glucose, acetate and individual amino acids have been shown to be similarly retarded when passing through an exopolymer matrix (Joyce and Dugan 1970).

It is important to understand what functions such secretions may have in maintaining short term geochemical gradients within mat microenvironments. For example, molecular oxygen can be toxic to certain biochemical processes, such as nitrogen fixation. Therefore, slowing the diffusive movement of molecular oxygen or converting the oxygen to less toxic forms can be favorable to the efficiency of such processes. Oxygen microelectrodes have been used to accurately and precisely measure oxygen profiles, and to determine consumption rates at the sediment-water interface and within microbial mat layers (Jørgensen and Des Marais 1990). These studies have shown that diffusion constitutes an important rate limiting step in the oxygen uptake of the surface sediment from the overlying water. Concentrated heterotrophic activities can result in hypoxic or anoxic microzones. Studies of water-column aggregates enveloped by an exopolymer matrix have shown that anoxic microzones exist, and can even allow methanogenic microbial activities to occur. Similarly, autotrophic production of oxygen (via photosynthesis) can result in supersaturated oxygenic microzones. Detailed measurements of the ion diffusion rates through exopolymer matrices of known compositions need to be made to begin assessing their relative importances to mat-forming cells.

The chemical properties of an exopolymer matrix may slow the diffusion of ions in several ways. The movement of an ion may be slowed by direct binding or electrostatic interactions with exopolymer molecules. The abundant functional groups on exopolymers, such as carboxyl and hydroxyl groups, may form anion-slowing or cation-binding barriers, depending upon the specific composition and conformation of the exopolymer (Geesey and Jang 1989).

The form of an ion may be changed as it moves through the matrix, thus altering its geochemical (and biological) effects. For example, molecular oxygen can be indirectly changed to a less toxic state. Molecular oxygen can directly interact with transition metals. Exopolymers readily bind transition metals, such as Mn, Fe, Cd, Pb, etc., and when in low oxidation states they are easily oxidized by molecular oxygen. Transition metals in high oxidation states can be reduced by organic molecules (Ghiorse 1986). Specific organic oxidizer molecules, localized within the exopolymer matrix (e.g. oxidizing proteins, thiols, phenols, or quinones) can react with reduced transition metals (such as Mn). Both processes can occur concurrently. Changes in pH due to photosynthesis can allow molecular oxygen to form complexes with transition metals removing them from the toxic molecular state.

Nutrient binding and storage

The exopolymer matrix binds and stores a range of molecules which can be utilized by heterotrophic and autotrophic cells (see Decho 1990 for review). Large molecules associate with exopolymers through a combination of ionic, covalent and Van der Waals interactions. The physical trapping of colloidal particles by the matrix can also occur. A variety of molecules and organic colloids can be stored by the matrix and later hydrolyzed by extracellular and cell-surface enzymes to smaller units or monomers before uptake by cells. Since this often requires extracellular enzymes, the binding of monomers to the exopolymer matrix slows their diffusive losses away from the cell once hydrolysis has occurred.

The localization of individual amino acids in proximity to phototrophic cells potentially can be important in sequestering nitrogen. Studies of phytoplankton cells by Palenik and Morel (1990) showed that cell surface enzymes oxidize amino acids and primary amines, to produce hydrogen peroxide, ammonia and alpha-keto acids (from amino acids) and aldehydes (from primary amines). Ammonia is taken up for growth, the remainder released. The peroxide can be used to reduce certain transition metals. Within microbial mats it is not known whether such oxidase enzymes play a determinative role in N-cycling. It is known that diatoms directly take up amino acids using transport systems, rather than utilizing them by oxidases (Flynn and Butler 1986). Some cyanobacteria possess internal oxidases, but may also possess cell surface oxidases (Meyer and Pistorious 1987). Since oxidase enzymes are activated during N-limiting conditions which occur periodically within mat microenvironments, they often require bacteria to hydrolyze ambient proteins or peptides to amino acids. Close associations of heterotrophic bacteria with autotrophic cells often occur in mat environments. Gene probes methodologies must be used to assay for oxidase enzyme capabilitics since the absence of direct enzyme activity may simply represent an artifact of the N-enriched culture methods needed to culture the organisms (Palenik and Morel 1990). Such studies will be important in determining how specific nutrients are localized, stored and cycled within mat microenvironments.

Protection against desiccation and toxic compounds

Mobile intertidal diatoms (Paterson 1989), which are intermittently exposed to air, also secrete exopolymers. The exopolymer matrix acts as a highly hydrated water storage reservoir immediately surrounding cells. Many exopolymers, when in a fully hydrated state, are greater than 99% water by weight. The water content of the exopolymer matrix can be lowered by the competitive binding of certain metal ions or molecules by dehydration reactions (Geesey and Jang 1989). The water content of isolated exopolymers can be grossly determined by simply measuring the wet weight vs. the dry weight. More precise and accurate estimations of water content can be determined by tritium exchange reactions which measure numbers of available sites where water can be bound by the matrix.

The exopolymer matrix can protect cells against potentially toxic compounds by binding these within the matrix and thus reducing the concentrations of the toxins actually encountered by the cells. Exopolymers competitively bind transition metals (such as Cu, Cd, Pb, Ag, Fe, Co, Ni) many of which can be toxic to cells. The presence of these metals, often triggers excessive exopolymer secretion. Since the binding of transition metals results in higher numbers of coordination bonds (than Ca or Mg), it also results in significant losses of water and condenses the tertiary structure of the exopolymer.

Physical stabilization of mats

Studies of mats have shown that their physical stability appears to be greatly enhanced by the presence of cohesive exopolymers. Several parameters affect the cohesiveness of exopolymers. The chemical composition is the most important factor which affects the stability, cohesiveness and physical state of an exopolymer. This determines the types of functional groups and hence the ability of an individual exopolymer molecules to interact with themselves and/or other molecules and ions. Subtle changes in composition can have significant effects on other properties of the exopolymer, such as: 1) tertiary state; 2) the types of bridging between adjacent exopolymer molecules; 3) adsorption of metals (transition metals vs. alkali earth metals); and 4) enzymatic interactions occurring within the exopolymer matrix.

The question of other potential roles of EPS (bio-optical properties, gene flow, enzyme activity)

In phototrophic microbial mats, light is a most important factor. With increasing depths, the light intensities decrease (or increase (see Kühl et al. this volume)) and the spectral composition of light changes due to selective absorption and backscatter. Exopolymers are an abundant component of the interstitial "translucent" portion of microbial mats. It is known that some cyanobacteria produce a light-induced extracellular pigment called scytonemin within their sheath (Golubic 1992). The complex branching and occasional ring structures found on some exopolymer molecules can potentially influence light absorption and refraction. It is not known how these high-MW exopolymers affect light absorption and scattering, or whether the spectral composition is altered when passing through an exopolymer matrix.

A second area involves the role of exopolymers in gene exchange (via plasmids) between adjacent microbial cells within a mat layer. Plasmids can contain genetic information for specific physiological capabilities. A third potential area of research involves the exopolymer matrix in maintaining a compatible microenvironment for the plethora of extracellular enzyme activities which occur in proximity to cells. How (and for how long) enzyme activities are maintained, once outside the cell is a question of great interest.

Exopolymers are high molecular weight secretions released by microbial flora within mats. Their abundances in mats and roles in other environments suggests they have important adaptive functions closely aligned with the resident mat microflora. They appear to influence a variety of both biological and geochemical processes. Pointed investigations must be conducted to specifically quantify their roles and adaptive significance to mat microorganisms and mat processes.

References

Cohen Y (1989) Photosynthesis in cyanobacterial mats and its relation to the sulfur cycle: a model for microbial sulfur interactions. In Cohen Y, Rosenberg E (eds) Microbial mats - physiological ecology of benthic microbial communities. ASM, Washington pp 26-36

Decho AW (1990) Microbial exopolymer secretions in ocean environments: their role(s) in food webs and marine processes. Oceanogr Mar Biol Annu Rev 28: 73-153

Flynn KJ, Butler I (1986) Nitrogen sources for growth of microalgae: role of dissolved free amino acids. Mar Ecol Prog Ser 34: 281-304

Geesey GG, Jang L (1989) Interactions between metal ions and capsular polymers. In: Beveridge TJ, Doyle RJ (eds) Metal ions and bacteria. John-Wiley and Sons, New York, pp 325-358

Ghiorse WC (1986) Microbial reduction of manganese and iron. In: Zehnder AJB (ed) Environmental microbiology of anaerobes. Wiley, New York

Golubic S (1992) Microbial mats of Abu Dhabi. In: Margulis L, Olendzenski L (eds) Environmental Evolution. MIT Press, Cambridge Mass, pp 102-130

Jørgensen BB, Des Marais DJ (1990) The diffusive boundary layer of sediments: Oxygen microgradients over a microbial mat. Limnol Oceanogr 35: 1343-1355

Jørgensen BB, Revsbech NP, Cohen Y (1983) Photosynthesis and structure of benthic microbial mats: Microelectrode and SEM studies of four cyanobacterial communities. Limnol Oceanogr 28: 1075-1093

Joyce GH, Dugan PR (1970) The role of floc-forming bacteria in BOD removal from waste water. Devel Ind Microbiol 11: 377-386

Krumbein WE, Stal LJ (1991) The geophysiology of marine cyanobacterial mats and biofilms. Kieler Meeresforsch Sonderheft 8: 137-145

LaMotta EJ (1976) Internal diffusion and reaction in biological films. Environ Sci Technol 10: 765-769

Mateson JV, Characklis WG (1976) Diffusion into microbial aggregates. Water Res 10: 877-881

Meyer R, Pistorius EK (1987) Some properties of photosystem II preparations from the cyanobacterium *Synechococcus* sp.: the presence of an L-amino acid oxidase. Biochim Biophys Acta 893: 426-433

Palenik B, Morel FMM (1990) Comparison of cell-surface L-amino acid oxidases from several marine phytoplankton. Mar Ecol Prog Ser 59:195-201

Paterson DM (1989) Short-term changes in the erodability of intertidal cohesive sediment related to the migratory behavior of epipelic diatoms. Limnol Oceanogr 34: 223-234

Revsbech NP (1983) *In situ* measurement of oxygen profiles of sediments by use of oxygen microelectrodes. In: Gnaiger E, Forster (eds) Polarographic oxygen sensors. Springer-Verlag, Berlin pp 265-273

The challenge to analyse extracellular polymers in biofilms

Thomas R. Neu

Laboratory for Materia Technica, University of Groningen.
Antonius Deusinglaan 1, 9713 AV Groningen, The Netherlands

Extracellular polymeric substances (EPS) represent a major fraction of microbial biofilms. The EPS within biofilms form a 3-dimensional network which is called biofilm matrix. The EPS immobilize the microbial cells within the biofilm and at the interface (Neu and Marshall 1990). The matrix is highly hydrated as it consists of more than 95% water. For this reason the biofilm matrix can be regarded as a layer of immobilized water (Cooksey 1992). So far the presence of the biofilm matrix has been demonstrated by various electron microscopic techniques which have been adapted from studies on microbial cell surfaces (Hancock and Poxton 1988; Mozes et al. 1991). The significance of EPS in biofilms has been elaborated in several review articles (Characklis 1973a; Characklis 1973b; Christensen 1989; Christensen and Characklis 1990; Geesey 1982; Neu 1992a). Nevertheless, the important compositional aspects of this matrix, mainly polysaccharides but also proteins, have been neglected in many studies on biofilms.

Generally, microbial polysaccharides have been studied for other reasons. They represent a structural component of the microbial cell surface. Due to this they are important as antigenic determinants and are involved in lectin-type interactions. They are also produced industrially to be used as viscosifiers and emulsifiers. Nearly all the knowledge on microbial polysaccharides has originated from these studies. The microorganisms used for these investigations however have been cultured under non-interfacial conditions, e. g., in liquid culture and in nutrient rich media. Apart from the term "EPS" a second term has to be introduced as it defines similar compounds. The term "glycocalyx" has been proposed for polysaccharide-containing structures lying outside the integral elements of the outer membrane of Gram-negative bacteria and the peptidoglycan of Gram-positive bacteria (Costerton et al. 1981). In the subsequent discussion however, the term EPS has been preferred since it refers not only to polysaccharide-containing structures of bacteria, but also to other polymers of the biofilm matrix produced by any biofilm microorganism (Cooksey 1992).

The common procedure to isolate cell surface bound microbial polysaccharides (capsule) is to extract or release them from the cell surface. Then they are precipitated from the aqueous phase. The released polymers (slime) are precipitated directly from the culture supernatant. In the following purification steps the polysaccharides may be dialysed to get rid of low molecular weight compounds. In addition, the polysaccharides are enzymatically digested to remove nucleic acids and proteins. Finally, they may be separated via column chromatography into different fractions and lyophilized. The result is a fluffy white substance which then can be

NATO ASI Series, Vol. G 35
Microbial Mats
Edited by Lucas J. Stal and Pierre Caumette

examined by the classical methods of carbohydrate analysis. These methods include quantitative analysis of the polysaccharide components, determination of the linkage type, sequencing of the repeating unit, elucidating the anomeric configuration and finally measuring of the physical properties of the polysaccharides (Aspinall 1982).

An important point regarding EPS composition analysis is that the ecological conditions at an interface are rather different compared to those of the water phase (Marshall 1976). As EPS are produced by microorganisms in response to their environment, the polymers synthesized at an interface or within a biofilm may be different to those isolated from the culture supernatant or from the cell surface. In comparison to the general procedure of polysaccharide analysis, there are several problems in analyzing the EPS of biofilms. Firstly, the biofilm matrix is built up by a mixed consortium of microorganisms. Therefore, the polymers in this matrix originate from different microorganisms. Secondly, microorganisms have the potential to produce more than one polymer. That means, during the development of the biofilm and due to the supply of nutrients, a single microbial species may produce different polymers. Thirdly, these polymers may not be a pure polysaccharide of one molecular size, but may be associated with proteins or other high molecular weight substances. The identity and role of these polysaccharide-associated proteins within a biofilm matrix has not been investigated at all. Fourthly, based on these points, it becomes obvious that taking a representative sample of this heterogenous matrix for analysis is a rather difficult task.

By looking up methologies of how to study microorganisms at interfaces one will find techniques covering various topics such as, chemical analysis of cell surface components, microscopic examination, determination of biomass, measurement of metabolic activity and specific interface-associated activities (Costerton 1980; Costerton et al. 1986; Marshall 1986; Hancock and Poxton 1988; Wardell 1988; Ladd and Costerton 1990; Fletcher 1990; Geesey and White 1990; Mozes et al. 1991). However, nearly all the techniques used so far provide only indirect evidence of the presence of biofilm matrix polymers. There are just two methods revealing information on these polymers: microscopy for visualization and determination of total polysaccharide for quantification. For electron and light microscopy several techniques have been developed to detect EPS. They include specific stains, lectins and antibodies (Allison and Sutherland 1984; Beesley and Dougan 1991; Handley 1991; Neu and Marshall 1991).

Some groups have tried to investigate the adhesive and biofilm matrix polymers by isolating them for chemical analysis or employing them in adsorption experiments (Allison and Sutherland 1987; Beech et al. 1991; Christensen et al. 1985; Corpe 1970; Neu et al. 1992; Platt et al. 1985; Pringle and Fletcher 1986; Read and Costerton 1987; Sutherland 1980; Uhlinger and White 1983; Wrangstadh et al. 1990). Nevertheless, the identity of the isolated compound as the true adhesive or biofilm matrix polymer remains unclear. Usually the EPS have been produced by bacteria growing in liquid culture in a nutrient-rich medium. Therefore, in all these studies the interpretation of the results has to proceed with some caution as the EPS examined may not be identical to the ones produced at the proximity of an interface.

The analysis of EPS is closely linked to the question of the true adhesive polymers of microorganisms. In the last few years new approaches to analyze adhesive and matrix EPS have been developed.

A biochemical analysis scheme to investigate microbial consortia was designed by White. He used different signature components of microorganisms to identify specific microbial groups, to measure biomass, microbial activity and the nutritional status of microorganisms. With this method the presence of EPS can also be determined due to typical carbohydrate components of EPS e.g., uronic acids (Fazio et al. 1982; White 1983).

Chemical treatment of biofilms has been used by Fletcher to study the response of the adhered bacteria and the biofilm matrix polymers. It was found that electrolytes contracted, whereas Tween 20 expanded the biofilm polymers. It was concluded that the polymers have acidic groups able to interact electrostatically as well as hydrophobic groups for hydrophobic interactions. Thus these chemicals influence finally the separation distance of bacteria at interfaces (Fletcher 1988; Fletcher et al. 1991; Marshall et al. 1989).

Another approach to analyze the adhesive polymers of biofilms via so-called microbial "footprints" has been employed by Neu and Marshall. These "footprints" can be found at the interface after artificially removing the microorganisms. The "footprint" material was characterized by electron microscopy and by using lectins. It could be shown that the lectins bind to the "footprints" thereby proving the presence of carbohydrates (Neu and Marshall 1991). More recently "footprints" have been discussed in terms of a microbial "information" the bacteria leave at interfaces after they naturally detach themselves and it was suggested to expand the term "footprint" to other molecules by which microorganisms may label interfaces (Neu 1992b).

To overcome the problem of culturing the microorganisms in suspension researchers have tried to grow the cells on membranes. Schneider et al. could show for *Pseudomonas* a dependency of EPS production on the molecular weight cutoff of the membrane. It has been speculated that diffusable compounds may act as a signal for EPS production (Schneider et al. 1990). The group of Colwell used comparable conditions referred to as semi-solid media. They grew marine bacteria on agar overlaid with a dialysis membrane. The extracellular polymer isolated from these cultures showed an unusual composition with major components such as calcium, sulphur, carbohydrate and lipid (Abu et al. 1991). This approach shows that by employing unconventional techniques new types of microbial polymers can be discovered.

A non-destructive on-line technique to study the interaction of exopolymers with interfaces is available by attenuated total reflection/Fourier transform infrared spectroscopy (ATR/FT-IR). This technique has already been used to investigate the adsorption of isolated polysaccharides, the development of biofilms as well as the biocorrosion due to bacteria and their exopolymers (Bremer and Geesey 1991a; Bremer and Geesey 1991b; Geesey et al. 1987; Jolley et al. 1989; Nichols et al. 1985; Nivens et al. 1993; Schmitt et al. 1992). The technique allows one to detect the appearance of certain characteristic chemical groups of polymers at interfaces. This application allows a depth of analysis to about 0.5 μm.

From several results it became obvious that the different physicochemical conditions at an interface influence not only the phenotypic expression (Alberti and Harshey 1990; Belas

1992; McCarter and Silverman 1990) but also the biochemical composition of microorganisms (Pitta et al. 1993; Valeur et al. 1988). By using reporter gene technology the activation/ inactivation of genes triggered by the presence of an interface was studied (Dagostino et al. 1991; Marshall and Goodman 1993). It could be shown for *Pseudomonas aeruginosa* that the gene expression for alginate biosynthesis is activated by the substratum. This is the first report on the activation of an EPS gene as a result of cell attachment (Davies et al. 1993).

As a strategy to elucidate the secrets of biofilm matrix polymers the following approaches may be succesful: 1) Isolation of the single true matrix polymers out of mixed biofilms to study the properties of the single EPS as well as the mixed EPS under defined conditions. 2) Application of probes to sense for specific single types of EPS within the biofilm matrix to reveal their distribution. 3) Development of online systems to follow the dynamics of EPS within the biofilm matrix and investigate temporal and spatial changes. 4) Exploitation of the experience gained from the application of laboratory biofilm reactors to study the EPS of the biofilm matrix. Nevertheless, the complete analysis of biofilm matrix polymers still awaits new methods and approaches to clarify the identity, property and function of this still, as Keith Cooksey called it, "mystic" product of microbial films.

References

Abu GO, Weiner RM, Rice J, Colwell RR (1991) Properties of an extracellular adhesive polymer from the marine bacterium *Shewanella colwelliana*. Biofouling 3: 69-84

Alberti L, Harshey RM (1990) Differentiation of *Serratia marcescens* 274 into swimmer and swarmer cells. Journal of Bacteriolgy 172: 4322-4328

Allison DG, Sutherland IW (1984) A staining technique for attached bacteria and its correlation to extracellular carbohydrate production. Journal of Microbiological Methods 2: 93-99

Allison DG, Sutherland IW (1987) The role of exopolysaccharides in adhesion of freshwater bacteria. Journal of General Microbiology 133: 1319-1327

Aspinall GO (1982) Chemical characterization and structure determination of polysaccharides. In: Aspinall GO (ed) The polysaccharides, Volume 1. Academic Press, New York, pp 35-131

Beech IB, Gaylarde CC, Smith JJ, Geesey GG (1991) Extracellular polysaccharides from *Desulfovibrio desulfuricans* and *Pseudomonas fluorescens* in the presence of mild steel and stainless steel. Applied Microbiology and Biotechnology 35: 65-71

Beesley JE, Dougan G (1991) Detection of surface antigens by immunogold labeling. In: Mozes N, Handley PS, Busscher HJ, Rouxhet PG (eds) Microbial cell surface analysis. VCH, Weinheim, pp 151-169

Belas R (1992) The swarming phenomenon of *Proteus mirabilis*. American Society for Microbiology News 58: 15-22

Bremer PJ, Geesey GG (1991a) Laboratory-based model of microbiologically induced corrosion of copper. Applied and Environmental Microbiology 57: 1956-1962

Bremer PJ, Geesey GG (1991b) An evaluation of biofilm development utilizing non-destructive attenuated total reflectance fourier transform infrared spectroscopy. Biofouling 3: 89-100

Characklis WG (1973a) Attached microbial growth - I. Attachment and growth. Water Research 7: 1113-1127

Characklis WG (1973b) Attached microbial growth - II. Frictional resistance due to microbial slime. Water Research 7: 1249-1258

Christensen BE, Kjosbakken J, Smidrod O (1985) Partial chemical and physical characterization of two extracellular polysaccharides produced by marine, periphytic *Pseudomonas sp.* strain NCMB 2021. Applied and Environmental Microbiology 50: 837-845

Christensen BE (1989) The role of extracellular polysaccharides in biofilms. Journal of Biotechnology 10: 181-202

Christensen BE, Characklis WG (1990) Physical and chemical properties of biofilms. In: Characklis WG, Marshall KC (eds) Biofilms. John Wiley & Sons, New York, pp 93-130

Cooksey KE (1992) Extracellular polymers in biofilms. In: Melo LF, Bott TR, Fletcher M, Capdeville B (eds) Biofilms - science and technology. NATO ASI Series E: Applied Sciences - Vol 223. Kluwer Academic Publishers, Dordrecht, pp 137-147

Corpe WA (1970) An acid polysaccharide produced by a primary film-forming marine bacterium. Developments in Industrial Microbiology 11: 402-412

Costerton JW (1980) Some techniques involved in study of adsorption of microorganisms to surfaces. In: Bitton G, Marshall KC (eds) Adsorption of microorganisms to surfaces. John Wiley & Sons, New York, pp 403-432

Costerton JW, Irwin RT, Cheng K-J (1981) The bacterial glycocalyx in nature and disease. Annual Review of Microbiology 35: 299-324

Costerton JW, Nickel JC, Ladd TI (1986) Suitable methods for the comparative study of free-living and surface associated populations. In: Pointdexter JS, Leadbetter ER (eds) Bacteria in Nature. Volume 2. Plenum Publishing, New York, pp 49-84

Dagostino L, Goodman AE, Marshall KC (1991) Physiological responses induced in bacteria adhering to surfaces. Biofouling 4: 113-119

Davis DG, Chakrabarty AM, Geesey GG (1993) Exopolysaccharide production in biofilms: Substratum activation of alginate gene expression by *Pseudomonas aeruginosa*. Applied and Environmental Microbiology 59: 1181-1186

Fazio SA, Uhlinger DJ, Parker JH, White DC (1982) Estimations of uronic acids as quantitative measures of extracellular and cell wall polysaccharide polymers from environmental samples. Applied and Environmental Microbiology 43: 1151-1159

Fletcher M (1988) Attachment of *Pseudomonas fluorescens* to glass and influence of electrolytes on bacterium - substratum separation distance. Journal of Bacteriology 170: 2027-2030

Fletcher M (1990) Methods for studying adhesion and attachment to surfaces. Methods in Microbiology 22: 251-283

Fletcher M, Lessmann JM, Loeb GI (1991) Bacterial surface adhesives and biofilm matrix polymers of marine and freshwater bacteria. Biofouling 4: 129-140

Geesey GG (1982) Microbial exopolymers: Ecological and economic considerations. ASM News 48: 9-14

Geesey GG, Iwaoka T, Griffith PR (1987) Characterization of interfacial phenomena occuring during exposure of a thin copper film to an aqueous suspension of an acidic polysaccharide. Journal of Colloid and Interface Science 120: 370-376

Geesey GG, White DC (1990) Determination of bacterial growth and activity at solid-liquid interfaces. Annual Review in Microbiology 44: 579-602

Hancock I, Poxton I (eds) (1988) Bacterial cell surface techniques. John Wiley & Sons, Chichester

Handley PS (1991) Detection of cell surface carbohydrate components. In: Mozes N, Handley PS, Busscher HJ, Rouxhet PG (eds) Microbial cell surface analysis. VCH, Weinheim, pp 87-107

Jolley JG, Geesey GG, Hankins MR, Wright RB, Wichlacz PL (1989) In situ, real-time FT-IR/CIR/ATR study of the biocorrosion of copper by gum arabic, alginic acid, bacterial culture supernatant and *Pseudomonas atlantica* exopolymer. Applied Spectroscopy 43: 1062-1067

Ladd TI, Costerton JW (1990) Methods for studying biofilm bacteria. Methods in Microbiology 22: 285-307

Marshall KC (ed) (1976) Interfaces in microbial ecology. Harvard University Press, Cambridge

Marshall KC (1986) Microscopic methods for the study of bacterial behaviour at inert surfaces. Journal of Microbiological Methods 4: 217-227

Marshall KC, Goodman AE (1993) Effects of adhesion on microbial cell physiology. Conference Bioadhesion II, Louvain-la-Neuve, Belgium

Marshall PA, Loeb GI, Cowan MM, Fletcher M (1989) Response of microbial adhesives and biofilm matrix polymers to chemical treatments as determined by interference reflection microscopy and light section microscopy. Applied and Environmental Microbiology 55: 2827-2831

McCarter L, Silverman M (1990) Surface-induced swarmer cell differentiation of *Vibrio parahaemolyticus*. Molecular Microbiology 4: 1057-1062

Mozes N, Handley PS, Busscher HJ, Rouxhet PG (eds) (1991) Microbial cell surface analysis. VCH, Weinheim

Neu TR, Marshall KC (1990) Bacterial Polymers: Physicochemical aspects of their interactions at interfaces. Journal of Biomaterials Applications 5: 107-133

Neu TR, Marshall KC (1991) Microbial "footprints" - a new approach to adhesive polymers. Biofouling 3: 107-133

Neu TR (1992a) Polysaccharide in Biofilmen. In: Präve P, Schlingmannn M, Crueger W, Esser K, Thauer R, Wagner F (eds) Jahrbuch Biotechnologie Band 4. Carl Hanser Verlag, München, pp 73-101

Neu TR (1992a) Microbial "footprints" and the general ability of microorganisms to label interfaces. Canadian Journal of Microbiology 38: 1005-1008

Neu TR, Dengler T, Jann B, Poralla K (1992) Structural studies of an emulsion-stabilizing exopolysaccharide produced by an adhesive, hydrophobic *Rhodococcus* strain. Journal of General Microbiology 138: 2531-2537

Nichols PD, Henson JM, Guckert JB, Nivens DE, White DC (1985) Fourier transform-infrared spectroscopic methods for microbial ecology: Analysis of bacteria, bacteria-polymer mixtures and biofilms. Journal of Microbiological Methods 4: 79-94

Nivens DE, Chambers JQ, Anderson TR, Tunlid A, Smit J, White DC (1993) Monitoring microbial adhesion and biofilm formation by attenuated total reflection/Fourier transform infrared spectroscopy. Journal of Microbiological Methods 17: 199-213

Pitta T, Godchaux III W, Leadbetter ER (1993) Protein content of peptidoglycan of liquid-grown cells differs from that of surface-grown, gliding *Cytophaga johnsonae*. Archives of Microbiology 160: 214-221

Platt RM, Geesey GG, Davies JD, White DC (1985) Isolation and partial chemical analysis of firmly bound exopolysaccharide from adherent cells of a freshwater sediment bacterium. Canadian Journal of Microbiology 31: 675-680

Pringle JH, Fletcher M (1986) Adsorption of bacterial surface polymers to attachment substrata. Journal of General Microbiology 132: 743-749
Read RR, Costerton JW (1987) Purification and characterization of adhesive exopolysaccharides from *Pseudomonas putida* and *Pseudomonas fluorescens*. Canadian Journal of Microbiology 33: 1080-1090
Schmitt J, Nivens D, Flemming H-C, White DC (1992) In-situ monitoring der Entwicklung von Biofilmen mit Hilfe der FT-IR/ATR Spektroskopie und die Rolle der EPS für die Sorptionseigenschaften von Biofilmen. Jahrestagung GdCH Fachgruppe Wasserchemie, Dresden, Germany
Schneider H, Ross N, Levitan R (1990) Surface associated factors in the formation of biofilms: Effects of surface porosity on slime formation by *Pseudomonas aeruginosa*, *Staphylococcus aureus* and *Staphylococcus epidermis*. In: Bont JAM de, Visser J, Mattiasson B, Tramper J (eds) Physiology of immobilized cells. Elsevier, Amsterdam, pp 99-102
Sutherland IW (1980) Polysaccharides in the adhesion of marine and freshwater bacteria. In: Berkeley RCW, Lynch JM, Melling J, Rutter PR, Vincent B (eds) Microbial adhesion to surfaces. Ellis Horwood, Chichester, pp 329-338
Uhlinger DJ, White DC (1983) Relationship between physiological status and formation of extracellular polysaccharide glycocalyx in *Pseudomonas atlantica*. Applied and Environmental Microbiology 45: 64-70
Valeur A, Tunlid A, Odham G (1988) Differences in lipid composition between free-living and initially adhered cells of a gram-negative bacterium. Archives of Microbiology 149: 521-526
Wardell JN (1988) Methods for the study of bacterial attachment. In: Austin B (ed) Methods in aquatic bacteriology. John Wiley & Sons, Chichester, pp 389-415
White DC (1983) Analysis of microorganisms in terms of quantity and activity in natural environments. In: Slater JH, Whittenbury R, Wimpenny JWT (eds) Microbes in their environments. 34. Symposium of the Society for General Microbiology. Cambridge University Press, Cambridge, pp 37-66
Wrangstadh M, Szewzyk U, Östling J, Kjelleberg S (1990) Starvation-specific formation of a peripheral exopolysaccharide by a marine *Pseudomonas* sp. strain S9. Applied and Environmental Microbiology 56: 2065-2072

New cultivation techniques and laboratory model systems for investigating the growth of stratified microbial communities

Julian Wimpenny, Sarah Kinniment, Liese Ganderton and David Stickler

School of Pure and Applied Biology, University of Wales, Cardiff, CF1 3TL, Wales

Introduction

Algal mats are one example of a large range of microbial ecosystems which are spatially heterogeneous and dominated by physico-chemical gradients. Such gradients can be expressed over a huge range of physical dimensions. pH gradients at the nm level occur around clay crystal domains Bacterial colonies show oxygen gradients from fully saturated to anoxic over 20 to 30 μm whilst biofilms and algal mat communities range from tens of μm to mm in depth. Gradients over soil profiles are expressed in the cm range whilst stratified water bodies incorporate gradients that range from meters to hundreds of metres. As these examples indicate, probably the majority of microbial systems are spatially heterogeneous.

Algal mat communities can be considered as one form of biofilm, that is a microbial community which is predominantly a two dimensional layer on a surface. The biofilm attaches to a solid surface whilst the algal mat generally forms above previous layers as for example the mat community in Solar Lake. Biofilm growing on surfaces exposed to light, for example that growing on the surface of solid substrata in a trickling filter or many epilithon communities growing on rocks, contain phototrophic species. These systems show temporal as well as spatial changes in concentrations of solutes including oxygen, sulphide etc. associated with dark-light cycles.

Whilst gradient systems have been investigated *in situ* by many workers we feel that there is a place for laboratory growth models to investigate specific aspects of the biology of such systems under controlled conditions in the laboratory.

A number of experimental models were developed by my group in Cardiff. These include:

1. Gradostats or bidirectionally linked multistage continuous culture devices which generate opposing steady state solute gradients.

2. Gel stabilised systems. Closed systems in which cells proliferate in opposing gradients of for example oxygen and a carbon source.

3. The bacterial colony as model. This is a 'natural' system representing a simple model for morphogenesis.

NATO ASI Series, Vol. G 35
Microbial Mats
Edited by Lucas J. Stal and Pierre Caumette

4. Biofilm models. The importance of biofilm suggested the need for a laboratory device generating reproducible biofilm (Wimpenny 1992).

These models have been discussed before on several occasions. See for example (Wimpenny 1985, 1988, 1989, 1991). Of these only the film fermenter will be discussed here.

The Constant Depth Film Fermenter

The Constant Depth Film Fermenter (CDFF) was developed by Coombe et al. (1984) and modified to its current design by Peters and Wimpenny (1988a,b). It consists of a rotating stainless steel disc in which are located 15 PTFE film pans each containing 6 4.7 mm diameter removable film plugs. These are recessed a measured amount, commonly 300 μm. Film develops in these recesses. The disc and its pans rotate beneath a PTFE scraper blade which, once the film pans are full, removes any extra growth that protrudes above the surface. Under these circumstances a quasi steady-state film forms in the pans.

The whole assembly is located inside a borosilicate glass tubing section sealed at both ends with PTFE gaskets and stainless steel head and base plates. Pans can be removed and replaced aseptically through a port in the head plate. Fresh medium enters through an inlet and drips over the film pans where it is carried round as a thin layer by the scraper blade. Waste medium leaves the system via a second port in the base plate and the latter also houses a bearing and drive assembly for the motor drive to the mother disc.

After setting up and sterilizing the system, medium inoculated with culture is pumped into the CDFF. This is recycled via peristaltic pumps for a period of generally 24 hours allowing time for the cells to attach to the plug surface. After this period the system is switched to fresh medium for the duration of the experiment. During growth samples were taken at selected time intervals. Each pan contained six film plugs allowing assays to be performed sufficiently often to be statistically significant.

Pseudomonas aeruginosa from metal-working fluid

Metal working fluids are needed for the lubrication and cooling of machine tools and represent a family of materials used by industry. They are often emulsions which are quite susceptible to microbial contamination when the emulsions tend to break down. Biofilm of these organisms accumulates on the walls of reservoirs and associated equipment.

A number of isolates derived from contaminated metalworking fluids were investigated using the CDFF. Of these one strain of *Pseudomonas aeruginosa* was investigated more fully. It was grown in a medium containing a mixture of carboxylic acids and triethanolamine representing some of the components found in metalworking fluids. High concentrations of this medium led to slimy growth rather than biofilm. Diluting the medium solved this problem leading to a proper biofilm which filled the pans. Growth was determined as an increase in protein and in viable count. Repeat experiments indicated that film production was reproducible.

Electron microscopy

Scanning electron microscopy of the biofilm surface revealed a dense mass of apparently healthy cells. Transmission electron microscopy revealed that cells appeared healthy over the first 150 μm into the film. Below this level empty lysed cells became progressively more common suggesting that below 150 μm the cells were dying.

Microelectrode measurements

Microelectrodes were used to determine oxygen tension above and within the film and to determine what pH gradients, if any, were present. Oxygen disappears in the film at approximately the point where TEM's show a change to lysing cells suggesting that cells near the base of the film were anoxic. Since *Ps. aeruginosa* is a strict aerobe in the absence of nitrate, it is assumed that death is due to anoxia. Only shallow pH gradients were observed in these biofilms.

Cryosectioning biofilm

Very few investigations into biofilm physiology have attempted to examine the distribution of activities across its profile. Exceptions include Ritz (1969) who cryosectioned dental plaque and reported that *Neisseria* were located in the upper and *Veillonella* in the lower layers of the structure. Banks and Bryers (1990) examined the distribution of binary pairs of microbes across a biofilm when grown under different nutritional conditions.
We have developed methods of cryosectioning biofilm in 12 μm slices to investigate the distribution of viability and of adenylates across the film profile.

Distribution of viability

Early attempts at measuring viability gave highly erratic results and a large drop in viable count compared with total counts for the unfrozen biofilm. This was solved by preincubating the film with a cryoprotectant dextran solution, which was shown in separate experiments to penetrate to the base of the film. Viable counts rose to a maximum in films of *Ps. aeruginosa* just below the surface. They then fell near the base of the film to about 1% of the maximum count.

Distribution of adenylates

Cryosectioning followed by quickly dropping sections into ice cold perchloric acid allowed assays of the adenylates ATP, ADP and AMP (Kinniment and Wimpenny 1992). From these data total adenylates and the adenylate energy charge could be calculated. Broadly, AMP fell, ATP rose and ADP stayed approximately the same from the base to the surface of the film. Total adenylates showed a peak at the same position at which viable counts were at a maximum. AEC values were low (about 0.25) at the base rising to 0.45 nearer the surface. Such values are low compared to values found in normal planktonic cultures but probably represent nutrient limitation within the biofilm structure. They also confirm values reported for an oral biofilm by Wimpenny et al. (1988).

The effects of biocides on biofilm

It is now accepted that organisms growing within a biofilm are often much more resistant to the effects of biocides, antibiotics and preservatives. The film fermenter has been used to determine the effects of a number of biocides used in the metal working industry. These include formaldehyde, chlorocresol and the isothazolinone biocide Kathon 886 MW. In each case liquid shake cultures were more sensitive to the agent than were the film grown organisms. For example planktonic cells are 50 times more sensitive than are their biofilm counterparts to formaldehyde.

A nine-membered oral biofilm

Marsh and his co-workers have constructed a community of nine oral bacteria characteristic of the dental plaque ecosystem. These have been grown in homogeneous chemostat culture (McKee et al. 1988; Bradshaw et al. 1989, 1990). This community has been grown in Cardiff in a chemostat and the effluent from the chemostat used to inoculate a CDFF. When grown on BM medium a stable biofilm community developed., however the balance between the different species changed. The dominant species in the chemostat (which was anaerobic) was *Fusobacterium nucleatum* whilst the aerobic *Neisseria subflava* was the most prevalent organism in the film fermenter. The latter was run with aeration in order to replicate conditions found in a dental plaque system growing in the mouth. Transmission electron micrographs have been recorded across the entire 300 μm film. It was concluded, based on simple morphology, that the *Neisseria* was present in the upper layers of the film.

Films from infected indwelling catheters

Indwelling bladder catheters are used in enormous numbers (some 50 million annually world wide) to deal with urinary retention and incontinence in elderly and chronically debilitated patients. In all patients undergoing long term catheterisation ($>$28 days) urinary tract infection occurs. A major complication of these infections is the formation of encrustations which can obstruct the lumen of the catheter and block the drainage of urine from the bladder. Such a blockage can induce urinary retention with its associated pain and distress and can culminate in fever, sepsis and shock. Cox et al. (1989) proposed that encrustation is initiated by the colonisation of the catheter with urease producing bacteria. The urease in these catheter biofilms hydrolyses urea to produce alkaline conditions that bring about crystallization of calcium and magnesium salts. The biofilm matrix probably also serves to stabilize the developing encrustation. We have examined encrusted catheters taken from patients and confirmed that urease producing organisms, particularly *Proteus mirabilis*, are commonly found in the colonising biofilms, sometimes as the sole organism but often in mixed culture.

Such a crystal forming mixed species catheter biofilm has been reconstructed in the CDFF. Strains of *Pr. mirabilis* and *Pseudomonas aeruginosa* that had been isolated from catheters, were grown in the CDFF in artificial urine. The biofilms reach a maximum cell density by 24h and have a standard composition of some 10^9 *Ps. aeruginosa* and 10^7 *Pr. mirabilis* cfu cm^{-2}. Calcium and magnesium salts start to accumulate in the biofilm from 30h and the film reaches the maximum thickness of 300 μm by 42-48h. The pH of the biofilm examined at 48h ranges from 6.3 at the surface to 6.7 at the base. This model system is now being used to examine ways of inhibiting formation of the encrustations.

Conclusions

Though not directly related to algal mats, the work described in this paper indicates the sort of approaches that can be taken in investigating substantially flat stratified microbial communities.

References

Bradshaw DJ, McKee AS, Marsh P (1989) The use of defined inocula stored in liquid nitrogen for mixed culture chemostat studies. Journal of Microbiological Methods 9:123-128

Bradshaw DJ, McKee AS, Marsh P (1990) Prevention of population shifts in oral microbial communities *in vitro* by low fluoride concentrations. Journal of Dental Research 69: 436-441

Bryers, JD Banks MK (1990) Assessment of biofilm ecodynamics. In: De Bont JAM, Visser J, Mattiasson B, Tramper J (eds) Physiology of Immobilized Cells. Elsevier, Amsterdam, pp 49-62

Coombe RA, Tatevossian A, Wimpenny JWT (1982). An *in vitro* model for dental plaque. Paper given at the 30th Meeting of the International Association for Dental research, British Division, University of Edinburgh, Scotland

Cox AJ, Hukins DWL and Sutton, TM (1989) Infection of catheterised patients: bacterial colonisation of encrusted Foley catheters shown by scanning electron microscopy. Urological Research 17: 349-352

Kinniment SL, Wimpenny JWT (1992) Measurements of the distribution of adenylate concentrations and adenylate energy charge across *Pseudomonas aeruginosa* biofilms. Applied and Environmental Microbiology 58: 1629-1635

McKee AS, McDermid AS, Ellwood DC, Marsh P (1988) The establishment of reproducible complex communities of oral bacteria in the chemostat using defined media. Journal of Applied Bacteriology 55: 263-275

Peters AC, Wimpenny JWT (1988a) A constant depth laboratory model film fermentor. Biotechnology and Bioengineering 32: 263-270

Peters AC, Wimpenny, JWT (1988b) A constant depth laboratory film fermenter. In: Wimpenny JWT (ed) Handbook of Laboratory Model Systems for Microbial Ecosystems, vol 1. CRC Press Boca Raton, Florida, pp 175-195

Ritz, HL (1969) Fluorescent antibody staining of *Neisseria, Streptococcus* and *Veillonella* in frozen sections of human dental plaque. Archives of Oral *Biology* 14: 1073-1083

Wimpenny JWT (1985) Novel growth systems. Microbiological Sciences 2: 53-60

Wimpenny JWT (1988) Handbook of Laboratory Model Systems for Microbial Ecosystems. CRC Press, Boca Raton, Florida

Wimpenny JWT (1989) Laboratory model systems for the experimental investigation of gradient communities. In: Cohen Y, Rosenberg E (eds) Microbial Mats. Physiological Ecology of Benthic Microbial Communities. American Society for Microbiology, Washington DC, pp 366-383

Wimpenny JWT (1991) Growth systems in microbiology. Science Progress 75: 205-225

Wimpenny JWT (1992) Microbial systems: patterns in time and space. Advances in Microbial Ecology 12: 469-522

Panel discussion: New methods in microbial mat research

Lee Prufert-Bebout (rapporteur) and Yehuda Cohen[1] (chairman)

Max-Planck-Institut für marine Mikrobiologie, Fahrenheitstraße 1, D-28359 Bremen, Germany

Introduction

The chair set forth two broad topics to begin this session: 1) What limitations exist to the use of new technologies and what are the prospects for overcoming them? 2) What are the promising aspects for use of these technologies, especially with regard to possibilities of using them in concert? These themes were touched on repeatedly as participants discussed a variety of research questions and approaches. This summary is divided into three sections corresponding to those areas discussed at most length. Those being: 1) Promising aspects, limitations and future directions in molecular biological research. 2) How the various inter-disciplinary views or paradigms that we hold of microbial mats are influenced by, and in turn affect, our data interpretation and future research goals. 3) What challenges and benefits exist with regards to the designing of laboratory and computer-based microbial mat models. This session became a welcome opportunity for the participants to step back from the "technologies" and discuss more directly and philosophically what questions we need to answer, and in turn, what directions should we be taking to develop our technologies to allow us to do so.

Directions in molecular biological research

Microbiologists have long recognized that due to our biases towards studying those species that we can either see or cultivate, we run the risk of developing incorrect assumptions about environments based on our limited observations of only a small proportion of those organisms naturally occurring. One of the most promising aspects of the exploitation of the rapidly growing field of molecular biological techniques is the possibility of generating more accurate pictures of the true biodiversity and distribution of organisms in microbial mat systems. Oligonucleotide probes in conjunction with gross core-sectioning or fine resolution image analysis systems, can give previously unobtainable detailed information about the distribution of specific groups of organisms. Nucleotide sequencing (most notably 16S rRNA) either

[1] The Moshe Shilo Center for Marine Biogeochemistry. The Institute of Life Sciences, The Hebrew University of Jerusalem. Jerusalem 91904, Israel.

NATO ASI Series, Vol. G 35
Microbial Mats
Edited by Lucas J. Stal and Pierre Caumette

directly or following polymerase chain reaction (PCR) amplification, in addition to new techniques such as denaturing gradient gel electrophoresis (DGGE), are allowing researchers to get closer to being able to directly from the field assess diversity, distribution and relative abundance of microbial populations (Stahl et al. 1985; Olson et al. 1986; Ward et al. 1992; Muyzer et al. 1993). Species which are as yet uncultivated, present in low numbers, or which appear identical to other species and hence were previously not detected, may now be identified. This may in turn lead us to modify and improve our standard culture techniques. In at least one case, such analyses and detection of "new strains" prompted renewed cultivation attempts using alternate (infinite dilution) methods with good success (Ward et al. this volume). On a smaller scale use of labeled-immunological probes to specific enzymes in conjunction with transmission electron microscopy allow for the localization of those enzymes and hence of specific processes in individual cells (Stal and Bergmann 1990). The ability to distinguish between identical appearing bacterial strains (for example two sulfate-reducing bacteria) using these methods may allow researchers to conduct the type of co-culture system competitive studies that were previously possible for only macroscopically distinguishable species (Van Gemerden and Beeftink 1978).

Though this area of research holds much promise, participants in this discussion were uniform in their recommendation that much more methodological work needed to be done in order to avoid misleading conclusions. One concern was the question of whether our current techniques are capable of uniformly extracting all of the relevant nucleotides from an environmental sample. It was concurred and emphasized that more information was needed to understand how nucleotide extraction efficiency or bias is affected by factors such as sediment composition, chemical environment and extracellular polysaccharide matrix characteristics. Additionally, for techniques requiring an amplification step, recent findings that PCR techniques may generate false chimeric sequences should be kept in mind (Liesack et al. 1991; Kopczynski et al. 1994). The possibility that method limitations may be variable depending on environment should also be further assessed. At this point it was agreed that stringent tests and book-keeping should be done in order to allow us to develop the confidence needed for data interpretation. Utilizing several of the techniques available in a comparative manner in a single environment was suggested as one beneficial approach.

Due to the popularity, relative ease of methodology, and recent research funding trends, we are experiencing a rapid explosion of molecular biological data, which due to the concerns stated above, may be difficult to adequately interpret and compare. The need for good technical information availability cannot be overstressed. New data base services such as RDP (Larsen et al. 1993) and EMBL (Higgins et al. 1992) which provide means for centralization of information are timely and will become increasingly important tools in the advancement of this field. In conjunction with this, there is a strong parallel need for adequate ground truthing in pure cultures. Traditional microbiological and physiological studies were recognized as being absolutely essential to, and non-replaceable by molecular research. So called "unculturable bacteria" still await innovative microbiologists using novel approaches such as fluctuating chemostats to be brought into culture. It is by the combined use of the old and new technologies that researchers see the most significant advances to be made in future years.

Microbial mat paradigms

Key to the formulation of new research questions are our pre-existing concepts of the system. As a consequence of the ability to sequence RNA and DNA, and hence quickly assign newly detected strains into a phylogenetic tree, more researchers are coming to view microbial systems in evolutionary terms; visualizing the emergence of new niches, functional groups and physiological pathways, in response to one another and to both small localized and larger planetary-scale geochemical environmental changes. In making estimations of biodiversity, it was cautioned that we must be careful to not mix relevant time scales by extrapolating back from data collected over a small time frame to a larger scale interpretation. Likewise, we should remember that phylogenetic trees are inferences of history, and that the length and timing of branching events may be very different in real time. To improve this situation the search continues in the geologic record for biomarkers that can be clearly linked to the emergence of new pathways or species so that real time constraints can be assigned to phylogeny. Due to the problems of organic material preservation, the most promising areas for such advances are in the discovery of isotopic signatures linked with specific biological processes, and in learning to better recognize impact signatures of the biosphere and lithosphere on one another.

Many researchers felt that there had not been an extended observational or natural history phase of study of microbial mats preceeding the development and application of the powerful technologies now available. A problematic consequence of this is that the largest barrier to improving our understanding of how microbial mat systems work may not be a need for the development of better technologies, but rather a need for the development of better questions. Though it was put forth that relevant questions should be easily adapted from other areas of microbiology, most felt that microbial mat systems were unique enough to warrant their own process of observation, and question formulation followed by analytical study.

Currently expanding appreciation of multiple possible physiological pathways in a wide range of strains of common microbial mat organisms is fostering yet another intriguing and fundamentally different view of the traditional organization of mats into layers of predictable species doing particular roles. In the presence of high degrees of physiological diversity it appears possible that different physiological roles may be played by a large variety of organisms which may in effect act as "functional guilds" for maintaining specific chemical transformations in fluctuating mat environments (Ward et al. this volume).

It was suggested that some useful parallels could be drawn by examining analogies between mats and tissues, both of which have components intricately and intimately situated to optimize interactions. Such efficient exchange and cohesiveness which typify microbial mats, continue to present formidable technical challenges in order to measure the chemical environments seen by the biota on the scale relevant to the organisms. Though the ever increasing use of various microsensors (O_2, pH, S^{2-}, N_2O) has greatly expanded our knowledge of the fluctuations of some key parameters in these systems, other parameters (e.g., DIC) are still problematic.

A major unknown factor in these environments, is how the chemical and physical nature of the abundant varieties of extracellular polysaccharides (EPS) and proteinacious sheath materials affect these systems. A recurring question is whether or not microzone environments of distinctly different redox and/or chemical composition from the ambient matrix can exist and be maintained in part by permeability barriers created by EPS. Such microzones have been theorized as sites for oxygen inhibited processes such as nitrogen-fixation and sulfate-reduction occurring in aerobic portions of mats, however consensus on interpretation of the data available (microsensor oxygen profiles and tetrazolium salt reduction) and theoretical diffusion calculations and modeling to assess this possibility has not been reached. Answers to difficult questions of composition, diffusion permeability to various chemical species, as well as other basic properties are needed to help resolve these issues. Indeed a clear understanding of what benefits these sheath, mucilage and capsule materials confer to the organisms secreting them is far from complete and is one area where great contributions to our understanding of many facets of microbial mat ecology stand to be made.

Model systems

It was suggested that reducing inherent complexity by constructing simplified model microbial mat systems in the laboratory could provide valuable information. Under controlled conditions large replicate arrays could be generated and subjected to different incubation regimes. Community responses could then be quantified by a variety of methods (e.g., microsensors, molecular probes, microscopy). One problem with this approach is that in simplified systems one cannot assess fully what key organisms and chemical pathways are left out making it problematic to draw conclusions pertinent to real systems. However, it is possible that observing the disparity between simplified systems and natural microbial mats, and by stepwise varying conditions or community compositions to try to approach stable mat-like behaviour is exactly where their usefulness may lie. An important factor to consider in system design is the degree to which it is open to natural type exchange-interactions with air and water, as well as with meio- and macrofaunal organisms.

Computer modeling was mentioned as an alternative method of studying microbial mats. Computer simulations could allow for rapid visualization of effects of varying parameters of community composition or physiological rates. Additionally, when known physiological parameters are put into such a model and run the points were models generate conditions not observed in natural communities may be of use in pointing out where gaps in our conceptualization of the system and/or future research efforts should lie. The existing geochemical model of H. Blackburn for sediments was mentioned as a resource that microbial mat research could adapt and take advantage of (Blackburn 1993). A model specifically taking into account typically observed rate data in microbial mats which simulates interactions between some of the major groups has also recently been designed (De Wit personal communication)

Summary

In summary it was agreed that interdisciplinary efforts will prove the most fruitful in answering remaining riddles in many areas of microbial-mat research. Molecular biological techniques are rapidly advancing and yielding exciting finds as well as stimulating new ways of viewing microbial mat systems. Though enthusiasm for new techniques abounds, consensus agreed that new technologies will not advance our knowledge far without being linked to classical microbiological, physiological, biochemical and genetic studies. Much basic work in these areas needs to be done, and participants expressed hope that researchers and funding bodies alike will keep this in mind in designing and evaluating research proposals over the coming years. High quality classical observational studies should also not be neglected in our eagerness to focus on "high-tech" approaches since such classical evaluation of natural history may well direct us to the most interesting and relevent questions where new techniques can be utilized to the greatest benefit. There is a continuing need for greater understanding of the composition and characteristics of the mucilaginous matrix of these systems without which many intriguing questions must remain unanswered. Implementation of new laboratory and computer model systems may well prove valuable to our understanding by virtue of their disciplined approach of moving from the simple to the complex.

References

Blackburn TH, Blackburn ND (1993) A reaction diffusion model of C-N-S-O species in a stratified sediment. FEMS Microbiol Ecol 102: 207-215

Higgins DG, Fuchs R, Stoehr P, Cameron N (1992) The EMBL Data Library. Nucleic Acids Research 21: supplement 2071-2074

Kopczynski ED, Bateson MM, Ward DM (1994) Recognition of chimeric SSU rRNA sequences composed of genes from uncultivated microorganisms. Appl Environ Microbiol in press

Larsen N, Olsen GJ, Maidak BL, McCaughey MJ, Overbeek R, Macke TJ, Marsh TL, Woese CR (1993) The ribosomal database project. Nucleic acids research 21: supplement 3021-3023

Liesack W, Weyland H, Stackebrandt E (1991) Potential risks of gene amplification by PCR as determined by 16S rDNA analyses of a mixed-culture of strict barophilic bacteria. Microb Ecol 21: 191-198

Muyzer G, De Waal EC, Uitterlinden AG (1993) Profiling of complex microbial populations by denaturing gradient gel electrophoresis analysis of polymerase chain reaction-amplified genes coding for 16S rRNA. Appl Environ Microbiol 59: 695-700

Olson GJ, Lane DJ, Giovannoni SJ, Pace NP, Stahl DA (1986) Microbial ecology and evolution: A ribosomal RNA approach. Annual Review of Microbiol. 40: 337-365

Stahl DA, Lane DJ, Olsen GJ, Pace NR (1985) Characterization of a Yellowstone hot spring microbial community by 5S rRNA sequences. Appl Environ Microbiol 49: 1379-1384

Stal LJ, Bergman B (1990) Immunological characterization of nitrogenase in the filamentous non-heterocystous cyanobacterium *Oscillatoria limosa*. Planta 182: 287-291

Van Gemerden H, Beeftink HH (1978) Specific rates of substrate oxidation and product formation in autotrophically growing *Chromatioum vinosum* cultures. Arch Microbiol 119: 135-143.

Ward DM, Bateson MM, Weller R, Ruff-Roberts AL (1992) Ribosomal analysis of microorgansims as they occur in nature. Adv Microbial Ecol 12: 219-286

III. Microscale interactions in microbial mats

Diffusion processes and boundary layers in microbial mats

Bo Barker Jørgensen

Max Planck Institute for Marine Microbiology, Fahrenheitstraße 1, D-28359 Bremen, Germany

On a bacterial scale, the interface between a microbial mat and the overflowing water is a remarkable environment governed by low Reynolds number hydrodynamics, by diffusional solute transport, and by exposure to extreme chemical fluctuations and gradients. The thin surface layer of benthic phototrophic mats, in which all photosynthesis and most of the respiration of the mat community takes place, is generally about a mm thick. Yet, these mats have a productivity and organic turn-over of a similar magnitude as planktonic ecosystems. Since the euphotic zone of mats is typically 10^3-10^5-fold thinner than that of the water column, the microbial activity per unit volume is correspondingly 10^3-10^5-fold higher. As a consequence, there is a dynamic balance between the rapid production and consumption of oxygen, the concentration of which may fluctuate between >1 atm. partial pressure and total depletion within minutes during shifting light conditions. The microorganisms living at the mat surface are physiologically adapted to these chemical extremes. Many also have a highly developed motility and tactic response to the environmental factors. The responses are often simple for the individual cells, but together they may lead to complex behavioral patterns of the whole populations. This paper will review some of the microbiologically important properties of the interface and how some bacteria are adapted to this environment.

A viscous world

The bulk-flow of water above a microbial mat is physically governed by inertial forces which drive the vertical transport of dissolved compounds by convection and eddy diffusion. Internal friction of the water has little effect on the large scale flow, but in microbial dimensions it becomes the dominating force. It changes the physical environment into a very different world in which the water appears very viscous and "sticky". The vertical flow component becomes strongly impeded within a few mm from the mat surface, i.e. within the viscous sublayer, and the eddy diffusivity, E, steeply decreases. At a certain distance above the mat, within a fraction of a mm, E falls below the molecular diffusion coefficient, D, and molecular diffusion becomes the fastest transport mechanism for dissolved compounds. This thin layer of the water column is termed the *Diffusive Boundary Layer*, *DBL*. Although diffusion is often considered a slow transport mechanism, the DBL is generally not rate-limiting for fluxes of dissolved compounds

NATO ASI Series, Vol. G 35
Microbial Mats
Edited by Lucas J. Stal and Pierre Caumette

across the sediment-water interface. For some chemical species such as oxygen, however, microbial mats appear to be exceptions to this, as shown below.

Table 1. Characteristic diffusion times, *t*, of O_2 in water at 20°C over different dimensions, *z*, ($D = 2.0 \times 10^{-5}$ cm^2 s^{-1}). Diffusion times in a microbial mat would be about twice that long due to the lower diffusion coefficient.

Distance (*z*)	Time (*t*)
1 μm	0.3 ms
10 μm	30 ms
100 μm	3 s
1 mm	4 min
1 cm	7 h
10 cm	25 d
1 m	8 yr

The diffusion time, *t*, for dissolved molecules decreases with the square of the distance, z, ($t = z^2/2D$, where *z* is the standard deviation of molecules from their starting position rather than their mean deviation). The diffusion time of an oxygen molecule over the dimension of a bacterium is thus only about one millisecond (Table 1). The characteristic diffusion times over the dimensions of the euphotic mat zone, however, are in the order of minutes, or hours.

The Diffusive Boundary Layer

Although hydrodynamicists have been aware of diffusive boundary layers for many years (e.g., Boudreau and Guinasso 1982), the physical reality and significance of the DBL became particularly apparent during our early measurements of oxygen microgradients in sediments (Jørgensen and Revsbech 1985). It was only by carefully checking the exact position of the - nearly invisible - microelectrode tip relative to the solid-water interface, however, that we originally discovered that steep oxygen gradients from the sediment continued several hundred micrometer up into the water phase. These results have since been theoretically supported by hydrodynamic models of the boundary layer flow (Rahm and Svensson 1989; Dade 1993). Fig. 1A shows an example of oxygen microgradients in the 0.16-0.6 mm thick DBL of a cyanobacterial mat, measured at different velocities of water flow over the mat. Data show that an increase in flow velocity caused A) a thinner DBL, B) steeper gradients, C) higher diffusion fluxes through the DBL, D) a higher oxygen concentration at the mat surface and, E) a deeper oxygen penetration into the mat.

When the benthic microbial community has a sufficiently high respiration potential, oxygen is used up just below the mat surface, and nearly the whole oxygen gradient is found within the DBL. This appears to be the case in the thin, white mats formed by filamentous, sulfide-oxidizing bacteria, *Beggiatoa* spp., on highly sulfidic sediments (Fig. 1B; Jørgensen and Revsbech, 1983). In this case, the respiration rate of the entire bacterial community is limited and regulated by the DBL diffusion flux of oxygen, which in turn is governed by the DBL thickness and thus by the oxygen concentration and flow velocity of the overlying water.

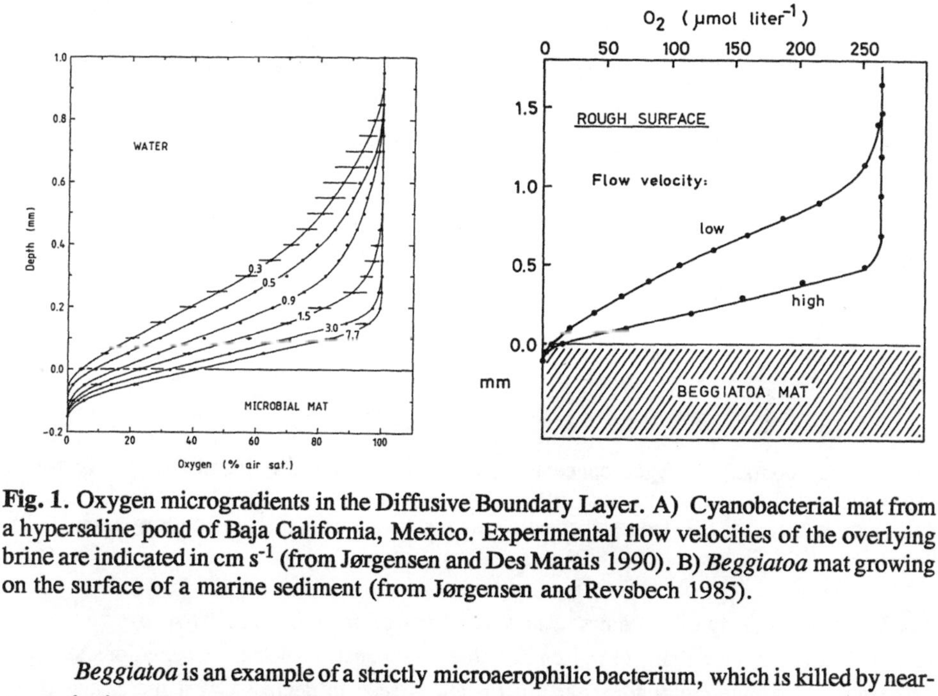

Fig. 1. Oxygen microgradients in the Diffusive Boundary Layer. A) Cyanobacterial mat from a hypersaline pond of Baja California, Mexico. Experimental flow velocities of the overlying brine are indicated in cm s^{-1} (from Jørgensen and Des Marais 1990). B) *Beggiatoa* mat growing on the surface of a marine sediment (from Jørgensen and Revsbech 1985).

Beggiatoa is an example of a strictly microaerophilic bacterium, which is killed by near-atmospheric oxygen concentrations, but which at the same time is dependent on oxygen for respiration. Only because the total community has such a high respiration rate are the individual organisms protected from the harmful effects of high oxygen concentrations and can grow as aerobes in a nearly anoxic environment. A chemotactic response in *Beggiatoa*, i.e. a step-up phobic reaction to oxygen, enables these long filaments to efficiently position themselves and curl up in a thin mat just at the oxygen-sulfide interface (Møller et al. 1985). Many other oxygen-sensitive bacteria, such as several purple sulfur bacteria, which occasionally grow densely on coastal sediment surfaces exposed to oxic water, probably depend on a similar protective function of the DBL.

The unstable oxygen concentrations in the DBL had initially discouraged more detailed microelectrode studies of its distribution but, upon closer examination, the stochastic fluctuations of oxygen encountered within this thin water film provided important information on the microscale hydrodynamics, information which has yet to be interpreted and incorporated into hydrodynamic models of the DBL. The fluctuations are apparent from the time-course of oxygen concentrations at different positions within a DBL as shown in Fig. 2.

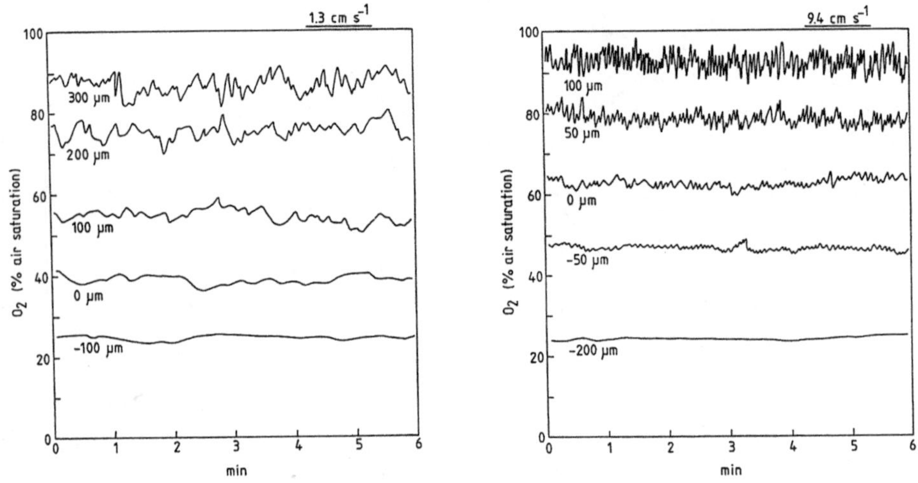

Fig. 2. Time course of oxygen concentration at different levels (indicated in μm) above or below the sediment-water interface at two flow velocities. The "noise" is caused by eddies constantly hitting the DBL from the overflowing water (from Gundersen and Jørgensen 1990).

At a low flow velocity of 1.3 cm s^{-1}, slow varations of oxygen are apparent over the 6 min period. With flow increasing to 9.4 cm s^{-1}, the time constant of these fluctuations strongly decreased. Also the amplitude of the fluctuation is seen to decrease with increasing flow, whereas it increased with distance from the mat out towards the turbulent layer. The smooth oxygen gradients presented in Fig. 1 are thus based on time-averaged measurements at each position. The data presented in Fig. 1 and 2 were all obtained using small-scale flow incubations in the laboratory. Similar results are, however, also found by *in situ* microelectrode measurements with a benthic lander instrument in the DBL directly on the sea-floor (Gundersen and Jørgensen 1990).

It is remarkable that, although the boundary layer is inherently unstable, the DBL is sharply defined by the oxygen microgradients. It forms a thin water film or blanket over the solid surface, the thickness of which can be defined by the transition between the constant water concentration and the DBL gradient of oxygen. Over the 3-dimensional topography of a mat surface, the DBL is seen to closely follow structures of mm-size dimensions with a mean thickness of 0.5-0.6 mm (Fig. 3). As water flows over this mountainous microlandscape, the DBL is even distinctly thinner on the upstream side of the small mounds than on their lee side.

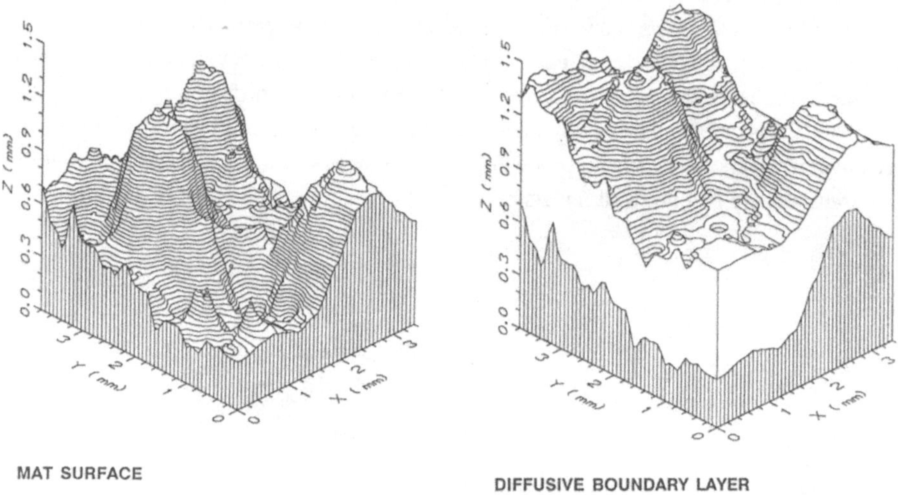

Fig. 3. 3-dimensional plot of the topography of a microbial mat surface and of the upper boundary of the DBL in the flowing water above. Structures of 0.5-1.0 mm hight, which protrude over the 13 mm^2 large mat area, caused slightly smoothened deflections of the DBL (from Jørgensen and Des Marais 1990).

The small thickness, ca. 5 μm, of the microelectrode tip combined with the observation, that mat surface structures of up to 100-200 μm size had no significant effect on the topography of the DBL, led us to believe that the presence of microelectrodes did not affect the DBL. Glud et al. (in press) recently checked this assumption by measuring oxygen microgradients in the DBL of a marine sediment simultaneously with two microelectrodes, one coming from above and one protruding up from below through the sediment itself. As the latter was known not to affect the DBL, it could be used to check the effect of the former.

The results showed that the upper microelectrode caused a 25-45% thinning of the DBL just below the tip (Fig. 4). An effect was observed even when the upper electrode was positioned 1 mm above, and thus outside, the DBL. On the downstream side the DBL was relatively thicker. This microelectrode artifact is probably caused by hydrostatic pressure variations due to acceleration and deceleration of the water flow as it passes the thin microelectrode, but the hydrodynamic effect has yet to be analyzed more carefully. In most sediments, this effect on the DBL thickness does not influence the oxygen fluxes significantly, but in microbial mats, with their extremely high oxygen uptake rate, the calculated flux will be overestimated.

Also the surface structures themselves cause accelerations and decelerations of the water flow and thus lateral pressure variations which could drive a vertical convective flow of water and solutes into or out of the mat. Such a mechanism has been demonstrated in porous marine sediments where macroscopic surface structures cause a considerable, passive pore-water flow (Hüttel and Gust 1992). Mats, however, are presumably rather impermeable to such a flow due to particle agglutination by extracellular polysaccharides. There are several other mechanisms which tend to enhance the solute exchange over that of molecular diffusion (cf. Hofman et al. 1991). The diffusive boundary layer is thus physically complex, and the interactions of flow-driven hydrodynamics and bioconvection due to micro- or macrofauna in relation to molecular solute transport are still not well understood.

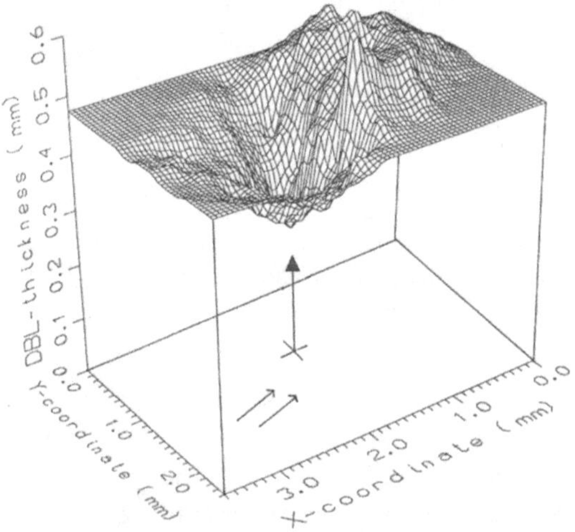

Fig. 4. Compression and expansion of the upper boundary of the DBL caused by the vertical introduction of a microelectrode from above to a position (vertical arrow) at 0.3 mm above the sediment surface. The flow direction of the overlying water is indicated by arrows (From Glud et al. in press).

The diffusive boundary layer constitutes a partial barrier to the flux of solutes across the mat-water interface (cf. Santschi et al. 1983). It delays the transfer of chemical signals from the water to the microorganisms living on the mat surface, a delay corresponding to the diffusion time for molecules across the DBL. For oxygen penetrating through a 500 μm thick DBL, the delay is about one minute, cf. Table 1. Assuming a pure molecular diffusion through the DBL, chemical microgradients measured in this water film can be used to calculate the total flux of oxygen to or from the mat. Fig. 5 shows the one-dimensional steady-state diffusion equations which can be used to calculate this flux.

Modeling of bacterial metabolism in mats

Flux calculations from DBL gradients are preferable to similar calculations from gradients within the mat because the latter require knowledge of the porosity and the effective diffusion coefficient in the mat, both of which are difficult to determine at sufficient spatial resolution.

In a diatom biofilm, or in the trickling filter biofilm from a sewage treatment plant, effective diffusion coefficients of oxygen were surprising low, only 55-60% of the molecular diffusion coefficient (Revsbech 1989; R. Glud and K. Jensen, pers. comm.). A similar reduction to 55% of the diffusion coefficient of glucose has been found in dense suspensions of yeast cells in agar (Cronenberg and van den Heuvel 1991). For hydrogen sulfide and sulfate, values determined by radiotracers in cyanobacterial mats were even lower, 34% and 32% of the diffusion coefficients in water, respectively (Jørgensen et al. 1979).

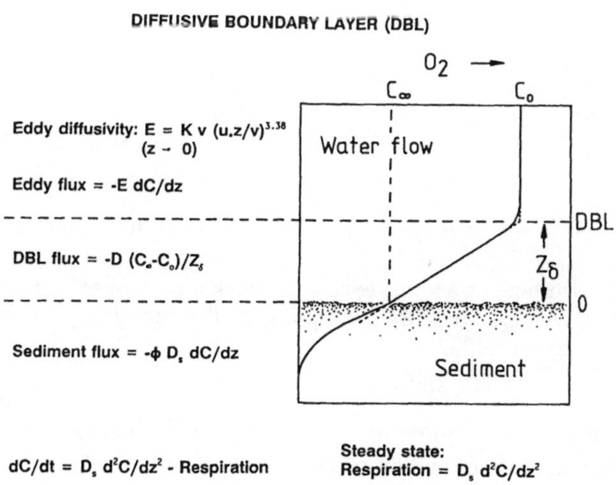

Fig. 5. Diagram of oxygen distribution through the water, DBL and sediment with equations used for the calculation of oxygen fluxes and consumption rates.

As shown in Fig. 5, a simple microbial process such as dark respiration can be directly calculated at steady-state from Fick's second law, using the shape of oxygen microgradients in the mat (cf. Crank 1983). Respiration at each depth is equal to the diffusion coefficient multiplied by the second derivative of concentration versus depth, i.e. to the curvature of the oxygen gradient. The stronger the curvature, the higher the respiration rate. The application of such an approach to the oxygen uptake of coastal marine sediments showed close agreement with concurrent DBL flux calculations (Rasmussen and Jørgensen 1992). Bioconvection and other transport factors accounted for an additional 25-30% of the total oxygen uptake of the sediment. The calculated oxygen consumption rate was roughly constant throughout the oxic zone during winter, while during summer it was highest just at the oxic-anoxic interface (Fig. 6). At this interface, oxygen respiring bacteria were probably receiving abundant hydrogen sulfide and other energy-rich substrates diffusing up from deeper, anoxic sediment layers.

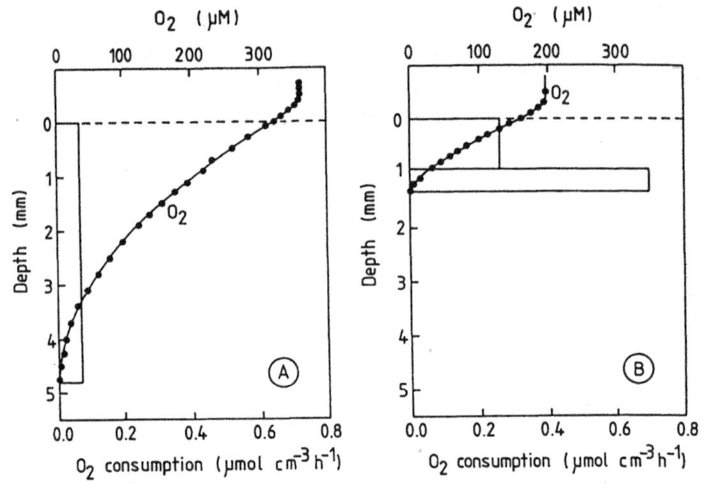

Fig. 6. Oxygen distributions and modeled consumption rates in sediment from Aarhus Bay, Denmark, before (A) and after (B) the deposition of a spring phytoplankton bloom (from Rasmussen and Jørgensen 1992).

Similar to these calculations of oxygen consumption, microgradients of sulfide can be used to calculate the rate of sulfide oxidation and of its formation from bacterial sulfate reduction. This was done at high spatial resolution for a heterotrophic biofilm growing on a rock under flowing, oxygenated water (Kühl and Jørgensen 1992). Sulfide microelectrode measurements taken at 50 μm depth intervals showed that sulfate reduction occurred only below the oxygen-nitrate zone at 1.1 mm depth and stopped at 2.1 mm depth due to sulfate depletion (Fig. 7).

There was a rapid 100% recycling of the produced hydrogen sulfide within the mat. Sulfide diffused up to the oxic zone where it was completely oxidized in a layer of 0.3 mm thickness and at a turn-over time of a few seconds. Oxygen respiration was maximal just at the mat-water interface but had a secondary maximum at the oxygen-sulfide interface where the intensive sulfide oxidation consumed 50% of all oxygen diffusing into the biofilm. This example shows how strongly the microbial zonation and respiration processes are governed by diffusion in such a biofilm, and how the oxidation of organic matter inside the biofilm, which surficially would appear to be oxic, can be dominated by a closed sulfur cycle.

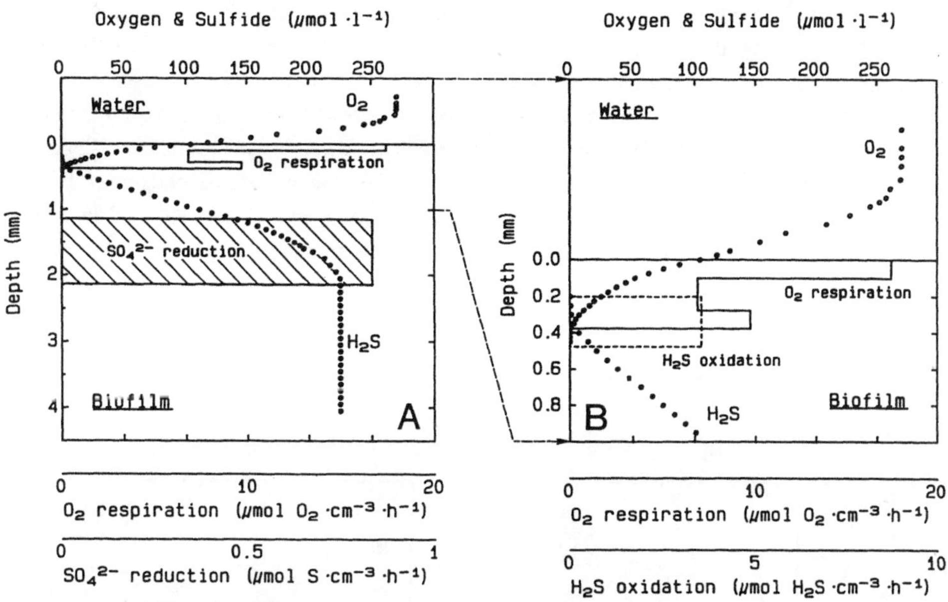

Fig. 7. Distributions of oxygen respiration, sulfate reduction, and sulfide oxidation in a 4.5 mm thick biofilm as modeled from measured steady-state gradients of O_2 and H_2S (from Kühl and Jørgensen 1992).

Beggiatoa mats at deep-sea hydrothermal vents

A final example is given here to illustrate how well adapted mat-forming bacteria such as *Beggiatoa* may be to a life at the oxygen-sulfide interface. In Fig. 1B it was shown how the community respiration of a *Beggiatoa* mat on a sediment surface may be limited by oxygen diffusion through the DBL. This regulation limits the actual thickness of such mats to 0.5-1 mm. It was therefore a surprising discovery, during a dive with the deep-sea submersible, ALVIN, to the hydrothermal vent area at 2,000 m depth in Guaymas Basin, Gulf of California, that 1-2 cm thick mats of giant *Beggiatoa* occurred in patches on the sediment surrounding fields of hot pore-water emanation (Jannasch et al. 1989). By *in situ* microsensor measurements of oxygen, sulfide, pH and temperature directly on the bacterial mats, Gundersen et al. (1992) discovered a small-scale hydrothermal fluid circulation around and partly through these mats (Fig. 8).

The circulation causes a pulsatory seawater flow into the coarse mesh of the 40-120 μm thick filaments and thereby enhances the supply of oxygen and sulfide to the bacteria. The bacteria thus exploit the hydrothermal convection and thereby overcome the diffusion limitation by the DBL. At the same time, they must resist the shear force of the vertical water flow, which may explain why such an unusual filament size has been developed in this particular environment. In contrast to the normal coastal habitats of *Beggiatoa*, the Guaymas Basin mats are not harmed by oxygen because its concentration in the bottom water is only 8% of atmospheric saturation. Had thermal convection not overcome the diffusion resistence of the DBL, the community would have been under 10-fold stronger oxygen limitation than their coastal relatives due to a 10-fold weaker oxygen gradient. The unusual size of these *Beggiatoa* filaments, however, raises another question of whether the individual cells may become diffusion limited in their uptake of energy substrates (cf. Koch and Wang 1982). The answer to this awaits progress in the ability to culture, or at least to maintain, these giant bacteria from the deep sea.

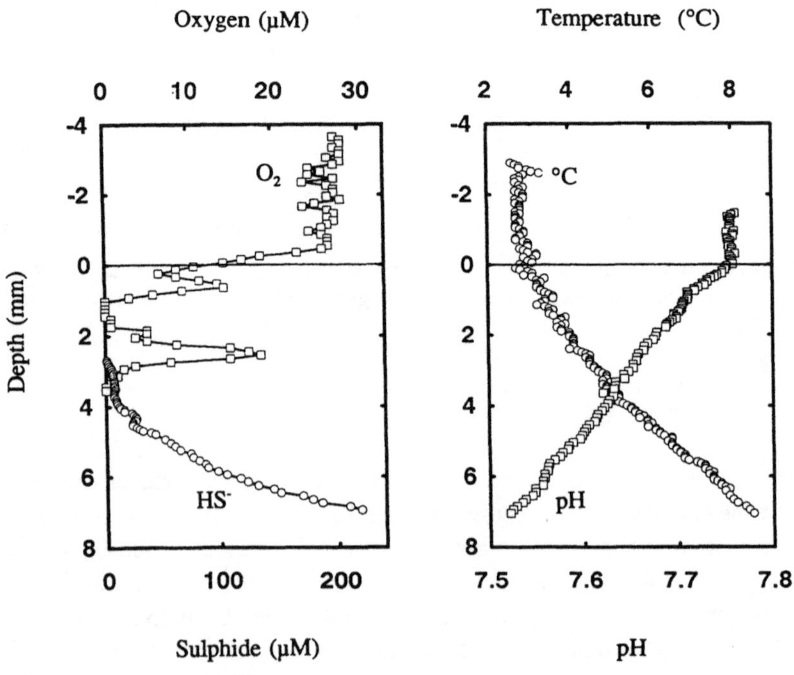

Fig. 8. Distributions of oxygen, sulfide, pH and temperature measured in situ in a mat of giant *Beggiatoa* from the hydrothermal vent area of Guaymas Basin (from Gundersen et al. 1992).

References

Boudreau BP, Guinasso Jr NL (1982) The influence of a diffusive sublayer on accretion, dissolution, and diagenesis at the sea floor. In: Fanning KA, Manheim FT (eds) The dynamic environment of the ocean floor. Lexington, Mass, pp 115-145

Crank J (1983) The mathematics of diffusion. Clarendon Press, Oxford, pp 414

Cronenberg CCH, van den Heuvel JC (1991) Determination of glucose diffusion coefficients in biofilms with micro-electrodes. Biosensors Bioelectronics 6: 255-262

Dade WB (1993) Near-bed turbulence and hydrodynamic control of diffusional mass transfer at the sea floor. Limnol Oceanogr 38: 52-69

Glud RN, Gundersen JK, Revsbech NP, Jørgensen BB. Effects on the benthic diffusive boundary layer imposed by microelectrodes. Limnol. Oceanogr., in press

Gundersen JK, Jørgensen BB (1990) Microstructure of diffusive boundary layers and the oxygen uptake of the sea floor. Nature 345: 604-607

Gundersen JK, Jørgensen BB, Larsen E, Jannasch HW (1992) Mats of giant sulphur bacteria on deep-sea sediments due to fluctuating hydrothermal flow. Nature 360: 454-455

Hofman PAG, de Jong SA, Wangenvoort EJ, Sandee AJJ (1991) Apparent sediment diffusion coefficients for oxygen and oxygen consumption rates measured with microelectrodes and bell jars: applications to oxygen budgets in estuarine intertidal sediments (Oosterschelde, SW Netherlands). Mar Ecol Prog Ser 69: 261-272

Hüttel M, Gust G (1993) Impact of bioroughness on interfacial solute exchange in permeable sediments. Mar Ecol Prog Ser 89: 253-267

Jannasch HW, Nelson DC, Wirsen CO (1989) Massive natural occurrence of unusually large bacteria (*Beggiatoa* sp.) at a hydrothermal deep-sea vent site. Nature 342: 834-836

Jørgensen BB, Des Marais, DJ (1990) The diffusive boundary layer of sediments: Oxygen microgradients over a microbial mat. Limnol Oceanogr 35: 1343-1355

Jørgensen BB, Revsbech NP (1983) Colorless sulfur bacteria, *Beggiatoa* spp. and *Thiovulum* spp., in O_2 and H_2S microgradients. Appl Environ Microbiol 45: 1261-1270

Jørgensen BB, Revsbech NP (1985) Diffusive boundary layers and the oxygen uptake of sediments and detritus. Limnol Oceanogr 30: 111-122

Jørgensen BB, Revsbech NP, Blackburn TH, Cohen Y (1979) Diurnal cycle of oxygen and sulfide microgradients and microbial photosynthesis in a cyanobacterial mat sediment. Appl Environ Microbiol 38: 46-58

Koch AL, Wang CH (1982) How close to the theoretical diffusion limit do bacterial uptake systems function? Arch Microbiol 131: 36-42

Kühl M, Jørgensen BB (1992) Microsensor measurements of sulfate reduction and sulfide oxidation in compact microbial communities of aerobic biofilms. Appl Environ Microbiol 58: 1164-1174

Møller MM, Nielsen LP, Jørgensen BB (1985) Oxygen responses and mat formation by *Beggiatoa* spp. Appl Environ Microbiol 50: 373-382

Rahm L, Svensson U (1989) On the mass transfer properties of the benthic boundary layer with an application to oxygen fluxes. Neth J Sea Res 24: 27-35

Rasmussen H, Jørgensen BB (1992) Microelectrode studies of seasonal oxygen uptake in a coastal sediment: role of molecular diffusion. Mar Ecol Prog Ser 81: 289-303

Revsbech NP (1989) Diffusion characteristics of microbial communities determined by use of oxygen microsensors. J Microbiol Meth 9: 111-122

Santschi PH, Bower P, Nyffeler UP, Azevedo A, Broecker WS (1983) Estimates of the resistance to chemical transport posed by the deep-sea boundary layer. Limnol Oceanogr 28: 899-912

Cycling of carbon, sulfur, oxygen and nutrients in a microbial mat

Donald E. Canfield[1] and David J. Des Marais[2]

Max Planck Institute for Marine Microbiology, Fahrenheitstraße 1, 28359 Bremen, Germany

Microbial mats are among the most productive aquatic ecosystems on Earth, yet, in many cases, the waters from which they grow are depleted in the basic nutrient elements. How, then, are nutrients cycled to allow for such high rates, and what ultimately controls these rates? To begin to address these issues, the cycling of carbon, oxygen, sulfur and nutrients has been explored over several years in *Microcoleus chthonoplastes*-dominated cyanobacterial mats from the hypersaline salt ponds in Guerrero Negro, Baja California Sur, Mexico (D'Amelio et al. 1989; Canfield and Des Marais 1993).

The processes regulating the cycling of the biogenic elements, and a generalized cartoon of the mat environment, are shown in Fig. 1. During the day, O_2 is produced by oxygenic photosynthesis, with, typically, a distinct O_2 maximum at some depth below the mat-brine interface. Whereas photosynthesis is the only source for O_2, potential sinks include sulfide oxidation, O_2 respiration, and diffusion from the mat into the overlying brine. The following equation may be used to describe the O_2 cycle, provided that the O_2 sources are in balance with the sinks:

$$\text{Photosynthesis} = \text{Diffusion out} + \text{S-oxidation} + O_2 \text{ respiration} \qquad 1)$$

The cycling of carbon during the day is rather opposite to that of O_2. Whereas photosynthesis is a source of O_2, it is a sink for dissolved inorganic carbon (DIC). Anoxygenic photosynthesis, utilizing H_2S as an electron donor, and producing S^0 or SO_4^{2-} is a further sink for DIC. Bacterial production also provides a potential sink for DIC. Sources of DIC include respiration within the photic zone (both by O_2 and sulfate, see Canfield and Des Marais 1991; 1993), and diffusion from the brine, as well as the underlying anaerobic portion of the mat. Assuming that sulfate reduction is the most important anaerobic carbon oxidation process, the sources of carbon from within the mat are completely provided by sulfate reduction and O_2 respiration.

---

[1] On leave from: School of Earth and Atmospheric Sciences, Georgia Institute of Technology, Atlanta, GA 30332-0340 USA
[2] Ames Research Center, Moffett Field, CA 94035-1000 USA

NATO ASI Series, Vol. G 35
Microbial Mats
Edited by Lucas J. Stal and Pierre Caumette

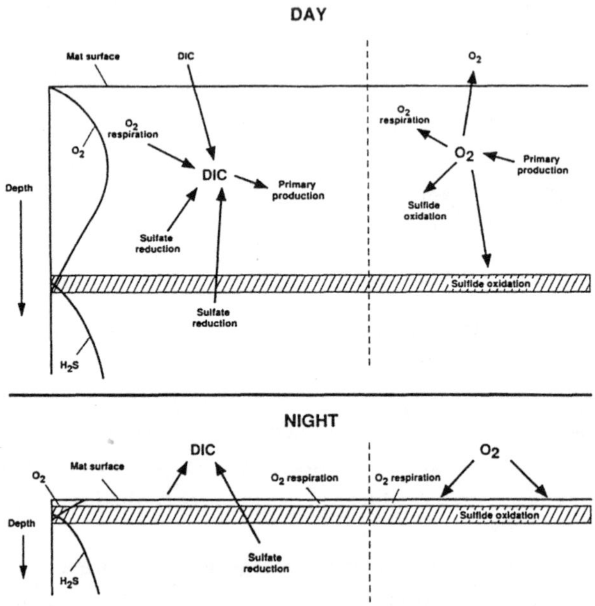

Fig. 1. Cartoon depicting the principles of carbon and O_2 cycling, day and night, in cyanobacterial mats. See text for details.

A general equation describing the DIC balance may be written, focusing on either the sources (eqn 2) or sinks (eqn 3) of carbon within the mat:

$$\text{Carbon fixed} = \text{Diffused in (brine)} + \text{Sulfate reduction} + O_2 \text{ respiration} \qquad 2)$$

$$\text{Carbon fixed} = \text{Oxygenic photosynthesis} + \text{Anoxygenic photosynthesis} + \text{Bacterial production} \qquad 3)$$

At night, photosynthesis does not occur, and without O_2 production, the O_2-H_2S interface moves very close to the mat surface (Fig. 1; Revsbech et al. 1983). The mat acts as a sink for O_2, which is potentially used in both sulfide oxidation and O_2 respiration. Sources of DIC include sulfate reduction and O_2 respiration. Of the DIC produced, some may be used in chemoautotrophic carbon production, with the remaining DIC free to diffuse from the mat. Equations summarizing these processes may be written:

$$O_2 \text{ Diffusion in} = \text{S-oxidation} + O_2 \text{ respiration} \qquad 4)$$

$$\text{DIC out} = \text{Sulfate reduction} + O_2 \text{ respiration} - \text{Bacterial production} \qquad 5)$$

A number of different techniques have been employed to quantify rates of the above mentioned processes. These have been described in detail in Canfield and Des Marais (1993) and are summarized in Table 1. In addition, we report here on mat-brine ammonia fluxes, and pore water concentrations obtained using the FIA (flow injection analysis) system of Hall and Aller (1992), and pore water extraction techniques as described by Bebout (1992).

Table 1. Parameters measured or calculated to constrain O_2 and DIC budgets.

Parameter	Method
Oxygenic photosynthesis	O_2 microelectrodes
Anoxygenic photosynthesis	Estimate from sulfur budget
Sulfate reduction rate	Radiotracer
O_2 respiration	O_2 mass balance
Mat-brine exchange	Flux chambers

The C, O_2 and S budgets discussed here have been obtained from field trips in April 1990 (April 90) and November 1990 (Nov 90). During these two different sampling times, somewhat different mat environments were encountered, which affected the cycling of the biogenic elements. In Nov 90, the mat was similar to earlier reports (D'Amelio et al. 1989) with *Microcoleus chtholoplastes* dominating the phototrophic community and with *Beggiatoa* spp. the most abundant sulfide oxidizing organism. This circumstance is expressed chemically with the daytime O_2-H_2S interface well below the deepest extent of measurable oxygenic photosynthesis (Jørgensen and Des Marais, 1986; Canfield and Des Marais, 1993). This implied that most of the available light for photosynthesis was absorbed high in the mat and not available to fuel an active anoxygenic photosynthetic community. By contrast, in April 90, though *Microcoleus* was still the dominant oxygenic phototrophic organism, *Beggiatoa* was nearly absent, and *Chloroflexus* sp. was abundant at the O_2-H_2S interface. Also, active oxygenic photosynthesis occurred to the O_2-H_2S interface implying that just below this region light was still available to fuel anoxygenic photosynthesis by *Chloroflexus*.

A summary of the data used to construct the chemical budgets is given in Table 2. In addition to these data, for April 90, the O_2 budget was further constrained by the diel distribution of S^o, where roughly 57% of the daytime sulfide production accumulated as S^o in the surface 2mm of the mat and was quickly lost from this region during the night (see Canfield and Des Marais 1993). The S^o loss mechanism(s) is not known and could include oxidation, reduction or disproportionation (Thamdrup et al. 1993; Finster and Bak 1993). It is significant that S^o accumulation during the day reduces the demand for O_2 in sulfide oxidation. If sulfide oxidation is completely accomplished by photoautotrophs then no O_2 will be used in daytime sulfide oxidation. This is a possible scenario for the April 90 mat, but not likely during the Nov

90 trip. Since S^o was not monitored during Nov 90, and anoxygenic photosynthesis was probably not significant, it will be assumed that all of the daytime sulfate reduction was oxidized with O_2 to sulfate during this time.

Table 2. Data used to construct O_2 and DIC Budgets[a]:

	April 90 17°C	April 90 30°C	Nov 90 20°C
DIC flux-Day[b]	-34.5	-98.0	-69.0
O_2 flux-Day	15.3	55.0	51.0
DIC flux-Night	32.4	118.0	75.0
O_2 flux-Night	-13.9	-60.0	-61.0
O_2 production	75.7	236.0	84.2
Sulf. Red.-Day[c]	5.5	22.5	20.8
Sulf. Red.-Night	6.9	30.0	26.5

[a] All values reported as mmoles m^{-2}, calculated over the time period of the flux chamber deployments. See Canfield and Des Marais (1993) for details.
[b] Negative values are fluxes into the mat, and positive values fluxes out of the mat.
[c] Presented as rates of sulfate reduction. April rates were calculated from nighttime flux chamber results and O_2 balance. See Canfield and Des Marais (1993) for details.

With the above considerations, an overall O_2 budget can be constructed. During the day (eqn. 1), rates of O_2 production by photosynthesis (Table 2) and diffusion of O_2 from the mats were measured directly. Rates of O_2 utilization in sulfide oxidation have been obtained from sulfate reduction rates as described above. Minimum and maximum rates of O_2 uptake by sulfide oxidation were obtained for April 90, assuming, first, that all sulfide oxidation was caused by anoxygenic phototrophs (no O_2 is used in sulfide oxidation) and, second, that all sulfide oxidation was nonphotosynthetic using O_2. The only unknown parameter is O_2 respiration which is obtained by difference. At night (eqn. 4), O_2 diffusion rates were measured directly and sulfide oxidation rates have been obtained from sulfate reduction rate measurements assuming complete oxidation to sulfate. Again, O_2 respiration is obtained by difference. The results are shown in Fig. 2.

It is clear that, at night, sulfide oxidation is the most important sink for O_2 in the mat. This is perhaps not too surprising as, at night, O_2 may penetrate only 100 μm or less into the mat (Revsbech et al. 1983; Canfield and Des Marais 1993) leaving only a very thin zone for O_2 respiration to occur. During the day, the fate of O_2 within the mat varies depending on when the mat was sampled. During April 90, most of the O_2 produced was recycled within the mat and did not escape to the overlying brine. Of the various sinks, O_2 respiration was by far the most important. It is not clear whether O_2 was respired in photorespiration or some aerobic heterotrophic pathway.

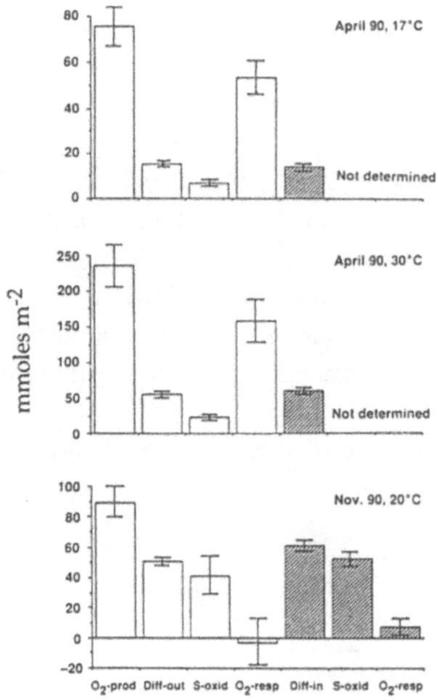

Fig. 2. Histogram showing fate of O_2 day (open area) and night (shaded area) in the mat. Bars show one standard deviation errors. In April 90, S-oxid is zero if oxidation is by anoxygenic phototrophs, and estimate of O_2 respiration increases by a corresponding amount (see text for details).

Lastly, although rates of processes were strongly influenced by temperature, the relative significance of the different O_2 sinks did not change. Hence, all metabolic activity seemed to scale with temperature by roughly the same amount.

During Nov 90, a significant proportion of the O_2 produced by photosynthesis was also recycled within the mat, although sulfide oxidation was far more important than in April. Also, there was no indication for significant O_2 respiration in Nov 90, quite in contrast to April of the same year. We know that the mat microbial population and light fields were quite different between April 90 and the Nov 90 trip, but the exact link between these parameters and the fate of O_2 within the mat remains unresolved.

A budget for DIC can be constructed in much the same way as for O_2. As mentioned earlier, budgets may be constructed from both sources and sinks of DIC. Considering the sources (eqn. 2), the DIC diffused into the mat was measured directly, DIC from sulfate reduction was calculated assuming two DIC for each sulfate reduced, and the DIC from O_2 respiration comes from the O_2 budget (Fig. 2) assuming one carbon for each O_2. From the DIC sinks (eqn. 3), rates of carbon fixation from oxygenic photosynthesis are from O_2 production rates (Table 2) assuming one carbon fixed for each O_2 produced. This stoichiometry requires that ammonia is the preferred source of N for the cyanobacteria, which seems to be the case (Bebout et al. this volume). Anoxygenic photosynthesis is assumed to be significant only during April 90, with rates calculated from the sulfur budget (see above), assuming that all daytime sulfide oxidation is accomplished by anoxygenic phototrophs. Although we recognize the potential of DIC fixation by bacteria (eqn. 3), too little is known of these rates for inclusion in our budget. A summary of the DIC budget is given in Table 3. A few conclusions are clear. First, even though a dense *Chloroflexus* layer indicates a significant role for anoxygenic photosynthesis during April 90, the sulfur budget limits maximum rates to less than 10% of the total C-fixation. Also, from the DIC sources, the mat itself, as quantified by the sum of sulfate reduction and O_2 respiration, is always a significant source of DIC for photosynthesis. Lastly, as with O_2, temperature affects the relative rates of microbial processes by roughly the same amount.

Table 3. Daytime DIC Budget.

	April 90 17°C		April 90 30 °C		Nov 90 20 °C	
DIC Sinks:						
Oxygenic photosynthesis	75.7	(8.3)[a]	236	(26)	84.2	(9.8)
Anoxygenic Photosynthesis	5.6	(1.4)	23	(5.8)	0	
Total	81.3	(8.4)	259	(27)	84.2	(9.8)
DIC sources:						
Diffusion from overlying water	34.5	(1.8)	98	(4.9)	69	(3.5)
Sulfate Reduction	10.9	(1.3)	45	(5.4)	41.6	(13)
Oxic Respiration	53.5	(8.4)	158	(27)	-3.2	(16)
Total	98.9	(8.7)	301	(28)	107.4	(21)

[a] one standard deviation

A puzzling aspect of the daytime DIC budget is that the total amount of carbon fixed is always greater when calculated from DIC sources than from DIC sinks. A similar discrepancy is also observed at night where more DIC diffuses from the mat than is apparently made available by the sum of sulfate reduction and O_2 respiration (eqn. 5; data not shown, see Canfield and Des Marais 1993). These discrepancies ultimately arise from a mismatch between the fluxes of DIC and O_2 across the mat brine interface (Table 2). Whereas over a diel cycle the O_2 and DIC budgets balance independently (Table 2), there is always more DIC diffusing into the mat than O_2 diffusing out during the day, and more DIC out of the mat than O_2 in during the night. Hence, during the day, carbon is incorporated into the mat with an oxidation level greater than zero (see Canfield and Des Marais 1993), and a similarly high oxidation level carbon is liberated from the mat during the night. This implies that the relatively oxidized carbon is a transient storage product within cyanobacterial cells, although the chemical nature of this carbon is unknown.

Our results document the cycling of DIC, O_2 and S over a full diel cycle for cyanobacterial mats at near steady-state; that is, that show little net growth as evidenced by the close match between day and night fluxes of DIC and O_2 (Table 2; subsequent field trips have confirmed this similarity of fluxes). We find that temperature affects both autotrophic and heterotrophic processes by the same magnitude, and that much, and sometimes most, of the carbon for photosynthesis is derived from within the mat itself. This leads us to speculate that the mat is a closely coupled system where high rates of carbon fixation by photosynthesis both fuel and require high rates of carbon oxidation. In principle, since this mat is at steady-state, enough nutrients are generated by oxidative processes over 24 hours to fuel primary production during the day, with a small additional external source to be accreted to balance the small amount of carbon buried in the mats. This proposition becomes compelling if the mats can retain the nutrients liberated over a 24 hour cycle, to utilize them as required during photosynthesis. In fact, the mats themselves seem to be the most probable source of nutrients since the ponds in which the mats grow are very nutrient depleted (Javor 1983); thus, only a limited supply of nutrients from the ponds seems likely.

To test whether the mats can in fact retain nutrients liberated during the night, when photosynthetic carbon fixation does not occur, the pore water concentrations of ΣCO_2 [$=CO_2(aq) + H_2CO_3 + HCO_3^- + CO_3^{2-}$] and ammonia were monitored both day and night during May 1993 (Fig. 3). The results show a modest ΣCO_2 draw-down during the day (due to the pore water extraction method used, this draw-down was probably underestimated) presumably due to ΣCO_2 uptake by photosynthesis, and a pronounced ΣCO_2 increase in the upper mat at night. This increase results from very rapid rates of organic carbon oxidation in this region (Canfield and Des Marais 1991; 1993), producing DIC, without the ΣCO_2 demand of photosynthesis. For ammonia, no significant nighttime increase is observed, implying that carbon, and not nitrogen, is mineralized and escapes the mat at night. This is consistent with the contention that mats effectively retain nutrients for potential use in biomass production as needed (see also Bebout 1992).

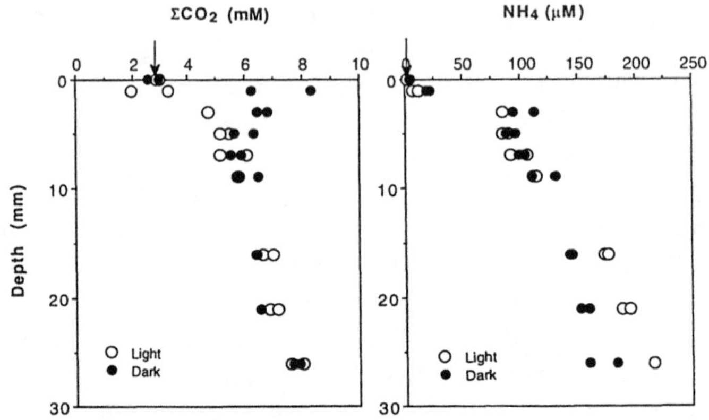

Fig. 3. Pore water concentrations of ΣCO_2 and ammonia, day and night, for *Microcoleous* mat collected May 1993.

It is possible (though not likely) that the low surface ammonia values in Fig. 3 could have resulted from ammonia oxidation in this region of the mat. We tested for this possibility by monitoring ΣCO_2 and ammonia in a flux chamber experiment in the dark where the water was fully aerobic at first, then became completely anoxic as the dissolved O_2 was consumed by the mat (Fig. 4). Under anoxic conditions, ammonia oxidation should not occur. The results showed a continuous liberation of ΣCO_2 from the mat during both oxic and anoxic incubation.

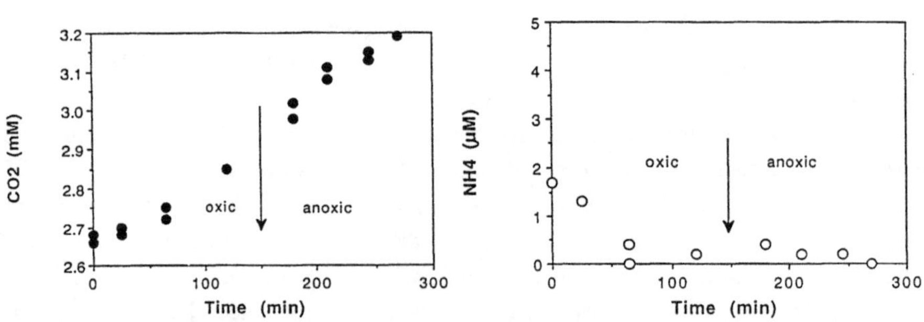

Fig. 4. Results of experiment where ammonia and ΣCO_2 were monitored over time in a flux chamber that started with O_2 at air saturation. O_2 was depleted after about 2 hours of uptake by the sediment. Although continuous ΣCO_2 production was observed, no ammonia was liberated to the water under either aerobic or anaerobic conditions.

Yet, ammonia was retained under all circumstances. Hence, ammonia retention is further demonstrated for these mats during the night, when active heterotrophic carbon mineralization occurs. Overall, it seems that our earlier suggestion is correct (Canfield and Des Marais 1993), that rates of organic carbon breakdown and carbon fixation by photosynthesis are tightly coupled, with the rate of one substantially influencing the rate of the other.

Acknowledgements

We thank Lucas Stal for his kind invitation to a very enjoyable conference, and to the Exportadora de Sal in Baja California, Sur, Mexico, for their support. Work was supported by a grant from NASA's Exobiology Program.

References

Bebout BM (1992) Interactions of Nitrogen and Carbon Cycling in Microbial Mats and Stromatolites, PhD Diss, Univ. N. Carolina, Chapel Hill, 187 pp

Canfield DE, Des Marais DJ (1991) Aerobic sulfate reduction in microbial mats. Science 251: 1471-1473

Canfield DE, Des Marais DJ (1993) Biogeochemical cycles of carbon, sulfur, and free oxygen in a microbial mat. Geochim Cosmochim Acta 57: 3971-3984

D'Amelio ED, Cohen Y, Des Marais DJ (1989) Comparative functional ultrastructure of two hypersaline submerged cyanobacterial mats: Guerrero Negro, Baja California Sur, Mexico, and Solar Lake, Sinai, Egypt. In: Cohen Y, Rosenberg E (eds) Microbial Mats, Physiological Ecology of Benthic Microbial Communities. American Society for Microbiology, Washington, pp 97-113

Finster K, Bak F (1993) Complete oxidation of propionate, valerate, succinate, and other organic compounds by newly isolated types of marine, anaerobic, mesophilic, Gram negative, sulfur-reducing eubacteria (1993) Appl Envir Microbiol 59: 1452-1460

Hall POJ, Aller RC (1992) Rapid, small volume, flow injection analysis for ΣCO_2 and NH_4^+ in marine and freshwaters. Limnol Oceanog 37: 1113-1119

Javor BJ (1983) Planktonic standing crop and nutrients in a saltern ecosystem. Limnol Oceanogr 28: 153-159

Jørgensen BB, Des Marais DJ (1986) Competition for sulfide among colorless and purple sulfur bacteria in cyanobacterial mats. FEMS Microbiol Ecol 38: 179-186

Revsbech NP, Jørgensen BB, Blackburn TH (1983) Microelectrode studies of the photosynthesis and O_2, H_2S and pH profiles of a microbial mat. Limnol Oceanogr 28: 1062-1074

Thamdrup B, Finster K, Hansen JW, Bak F (1993) Bacterial disproportionation of elemental sulfur coupled to chemical reduction of iron or manganese. Appl Envir Microbiol 59: 101-108

Nitrogen cycling in microbial mat communities: The quantitative importance of N-fixation and other sources of N for primary productivity

Brad M. Bebout, Hans W. Paerl[1], James E. Bauer[2], Donald E. Canfield and David J. Des Marais[3]

Max-Planck-Institut für marine Mikrobiologie, Fahrenheitstraße 1, D-28359 Bremen, Germany

Introduction

Although N-fixation has been assumed to contribute substantially to the overall N requirements of primary production in benthic microbial mat communities, few quantitative studies have been undertaken. Several factors have contributed to the lack of quantitative data including: 1) pronounced temporal variability in rates of N-fixation (which necessitate rate measurements over entire diel cycles), 2) difficulties in relating rates of nitrogenase activity (acetylene reduction) to actual N fixed (due to possible natural variations in this ratio, documented in other systems; Bothe, 1982), and 3) differences in methodologies used by various authors to assay N-fixation. These restraints on the acquisition of quantitative N-fixation data, as well as comparable restraints on obtaining reliable estimates of carbon fixation in these communities (Revsbech et al. 1981), result in some uncertainty as to the importance of N-fixation in supporting primary productivity in microbial mats.

Compounding the difficulty in determining the relative importance of N-fixation is the fact that other sources of N to microbial mats are seldom quantified. Other potential sources of N include: 1) dissolved inorganic nitrogen (DIN) species (NO_3^- and NH_4^+) present in the water column, 2) DIN (generally NH_4^+) made available by the degradative activities of bacteria in the anaerobic deeper layers of the mats, 3) dissolved organic nitrogen (DON) present in the water column and in porewaters due to degradation of organic matter in the deeper layers of the mats, and 4) recycling (production and subsequent consumption of N) within the photic zone of the mat community. Of these sources, N-fixation, and uptake of N from the water column can be considered "new" in that they originate from outside of the photosynthetic zone. Degradation of previously-produced biomass in the mat also results in the liberation of N which

[1] Institute of Marine Sciences, University of North Carolina at Chapel Hill, Morehead City, NC, USA
[2] Department of Oceanography, Florida State University, Tallahassee, FL, USA
[3] NASA Ames Research Center, Moffet Field, CA, USA

NATO ASI Series, Vol. G 35
Microbial Mats
Edited by Lucas J. Stal and Pierre Caumette

could be considered to be either new, or recycled, depending primarily on the relative age of the organic matter being degraded and its proximity to the photosynthetically active community (where the N requirement is generated). "New" sources of N must be available to support net photoautotrophic growth by the community. We report here on the relative importance of various sources of N to two different microbial mat systems having different growth rates (and therefore "new" N requirements).

Materials and Methods

The microbial mats examined in this study were located on the Baja California peninsula, near the town of Guerrero Negro, Baja California Sur, Mexico. These mats have been previously described (Javor and Castenholz 1984). We compared mats located in an intertidal area on the edge of Laguna Ojo de Liebre (composed primarily of the cyanobacterium *Lyngbya* sp.) with submerged *Microcoleus chthonoplastes*-dominated mats growing in a hypersaline ($\sim$8.5 %) pond managed for the production of salt by the company Exportadora del Sal, C.A. N-fixation was estimated using the acetylene reduction technique, and the theoretical ratio of actylene reduced to N fixed (3 mol C_2H_4:2 mol N). A discussion of the effects of variations in this theoretical ratio in these mats will be given elsewhere (Bebout et al. in prep.). The flux of nutrients across the mat-water interface in both mats was measured using glass or polycarbonate flux chambers (Canfield and Des Marais 1993). Concentrations of nitrogenous nutrients were assayed using standard analyses in porewaters extracted from microbial mats using techniques described in detail by Bebout (1992).

Results and Discussion

Measurements of rates of N transformations and N fluxes were obtained in November 1989, April 1990, November 1990, and November 1991. Dissolved inorganic carbon (DIC) diffused into both mats during the day owing to the carbon demands of photosynthesis, and out of both mats at night due to respiration (Canfield and Des Marais 1993). Over a diel cycle, however, there was net retention of DIC (growth) in the *Lyngbya* mat (45 mmol $Cm^{-2}d^{-1}$) but no detectable net retention of DIC in the *Microcoleus* mat. Both the *Lyngbya*- and *Microcoleus*-dominated microbial mat communities were composed of organic material with an elemental composition (C and N) very similar to the Redfield ratio, and this elemental composition was preserved with depth. There were slightly higher molar C/N ratios and greater variability in the ratio measured in the *Lyngbya* mats (8.4 $\pm$ 0.9 in *Lyngbya*, 6.3 $\pm$ 0.2 in *Microcoleus*). The DIC which is retained by the mats over a diel cycle must be combined with N also originating from outside the mat in order to maintain the observed elemental composition of the mat organic matter. There is a large flux of DIC into the mats during the daytime (hereafter referred to as "daytime DIC uptake") which must be either stored in C rich compounds such as glycogen, or combined with nitrogen (and other elements) into new organic matter. Since there is a net flux of N as well as C into the mat during the daytime, and since there are sources of N at depth which would flux out of the mat if they were not actively retained (see below), we can disregard the possibility that the mat is storing only C during the day. In the following discussion, we will express the N flux data as percentages of that needed to produce organic matter at the measured rates of daytime DIC uptake and at the observed elemental ratios. This will allow us to directly compare the relative importance of the different sources of N to the two different microbial mats. Values for the *Microcoleus* mats are given as mean ± standard deviation from all four sampling times; values for the *Lyngbya* mats are given as the means ± ranges from two samplings (Nov 1989 and Nov 1991).

Pronounced diel variability in N-fixation was observed in both mats, with highest rates occuring at night (Fig. 1). There were large differences in the integrated rates of N-fixation between the two communities despite similar rates of daytime DIC uptake. N-fixation accounted for 16.6 ± 7.3% of the N required to support daytime DIC uptake in the *Lyngbya* mat, and 1.8 ± 1.3% in the *Microcoleus* mat. The amount of N brought in through N-fixation alone therefore leaves a large deficit in the amount of N needed to support daytime DIC uptake in these mats. This daily N requirement must be provided through sources of N not usually considered in these systems.

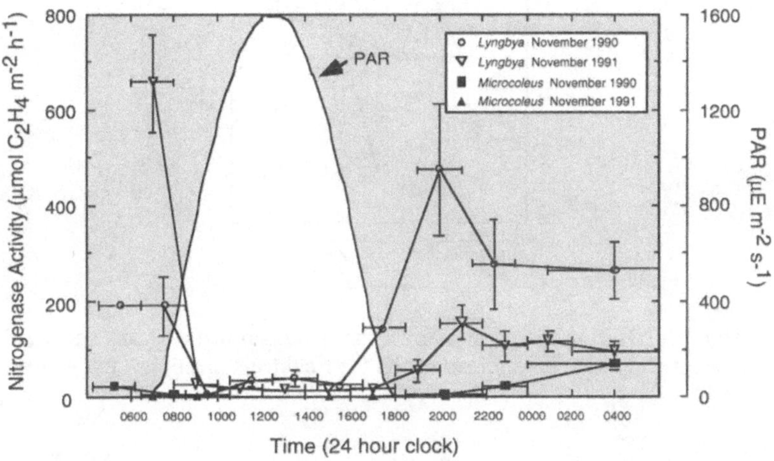

Fig. 1. Rates of N-fixation (acetylene reduction) in the Baja California microbial mats. Horizontal error bars show incubation intervals, vertical error bars show standard deviations associated with three replicate rate measurements.

A net flux of DIN into the microbial mats from the water was observed in both mats. Because of depletion of NH_4^+ in the flux chambers over the incubation periods, and because separate experiments revealed that both mats were able to take up NH_4^+ from the water at high rates (data not shown), we have chosen to use a calculated maximum value based on uptake across a 500 μm thick diffusive boundary layer at a maximum water column NH_4^+ concentration. Further consideration of this calculation will be given elsewhere (Bebout et al. in prep.). The estimate given by the calculation is 1.5 mmol N m^{-2} d^{-1} for each mat or 10.6% of the N requirement based on daytime DIC uptake for the *Lyngbya* mat and 14.4% of the requirement in the *Microcoleus* mat.

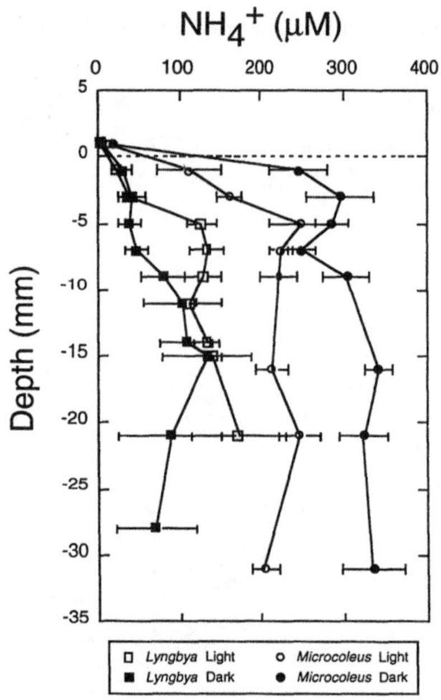

Fig. 2. Profiles of NH_4^+ in the porewaters of the microbial mats. Means and standard error bars are shown for replicate determinations on the field trips mentioned in the text.

Profiles of dissolved NH_4^+ in the porewaters of the mats revealed that there were large concentrations of potentially available NH_4^+ below the photic zone (Fig. 2). NO_3^- concentrations were at or near the limits of detection (less than ~1 μM, data not shown), and are omitted in the budget presented here. The NH_4^+ profiles were modeled using Fick's first law in order to calculate the flux of N into the surface layers of each mat. Since NH_4^+ export was not observed in the flux measurements, and assuming rates of nitrification-coupled denitrification in these mats are low (Bebout, et al. in prep), we assume here that the entire diffusive flux of NH_4^+ from the deeper layers is retained by the surface communities of both mats. The diffusive fluxes into the photoautotrophic layer from depth were calculated to be 14.8 ± 14.8% of the N requirement based on daytime DIC uptake in the *Lyngbya* mat and 73.1 ± 64.4% in the *Microcoleus* mat.

After summing our estimates of the potential sources of N, we find that there is a shortfall in the amount of N needed to produce organic matter from the quantities of DIC taken up during the day in the *Lyngbya* mat. It appears likely from these measurements that the *Lyngbya* mats are storing very high C/N ratio compounds during the day. Our measurements of near Redfield ratio organic matter in the mats (made at 2 mm resolution) argues against the storage of material significantly different from this ratio, but we cannot ignore the possibility that the thin surface layer of photosynthetically active cells has an elemental ratio different from that of the (integrated) top 2 mm of each mat (the resolution of our sampling technique). Storage of photosynthetically-derived energy (presumably C-rich photosynthate) for nighttime N-fixation has recently been demonstrated for another microbial mat containing a *Lyngbya* sp. (Bebout et al. 1993). Although our estimates of the flux of NH_4^+ from depth in the *Microcoleus* mat have rather large uncertainties associated with them, it appears possible that in this mat the flux of N from depth could supply the entire N requirement of daytime DIC uptake. Therefore, this mat appears to be able to derive its entire N demand through the recycling of previously produced biomass.

Absent from our current budget are DON flux measurements. Although DOC fluxes in these mats account for only a small fraction of the total C flux (Canfield and Des Marais 1993), the possibility remains that DON may be a relatively more important component of the N flux. Microbial mats in other locations have been shown to actively incorporate DON from the water column (Paerl et al. in press). The quantitative importance of this process in the Baja mats is currently under investigation. Also absent from our current budget are measurements of N loss due to denitrification. Measurements of denitrification made using the acetylene block technique have revealed undetectable rates of this process in these mats (data not shown) but problems associated with the use of this technique in sulfidic environments make these measurements suspect. Efforts are underway to determine the significance of denitrification in these mats using newly described methodologies (LP Nielsen, this volume).

A summary of the magnitudes of the various sources of N to both microbial mats is shown in Fig. 3. The DIC flux data of Canfield and Des Marais (1993) indicates that there is effectively no net growth over a diel cycle (retention of DIC from the water column by the mat) occuring in the *Microcoleus* mat. By contrast, the *Lyngbya* mat was shown to be an actively growing mat and the C retained must be combined with enough N also originating from outside the mat to create organic matter at the observed elemental ratios. Of the sources of N considered here, N-fixation, and uptake of N from the water column can be considered to be "new", i.e. they originate from outside the zone of production of biomass. The amount of N brought in through N-fixation is 43.7 ± 19.1% of that required for the production of organic matter from the carbon retained over a diel cycle (growth) by the *Lyngbya* community. The addition of N fluxing into the mat from the water column to the N gained from N-fixation results in a total of 71.5 $\pm$ 19.1% of the amount of nitrogen necessary to produce organic matter from the DIC retained by this microbial mat. Since the flux of N produced at depth is a combination of the contributions of NH_4^+ production at depth and NH_4^+ production and uptake in the photic zone, we cannot be sure how much of this source can actually be considered "new" N. The large flux of N attributable to the sum of these processes, is, however, testimony to the importance of recycling in supporting daytime DIC uptake in these mats (particularly in the *Microcoleus* mats).

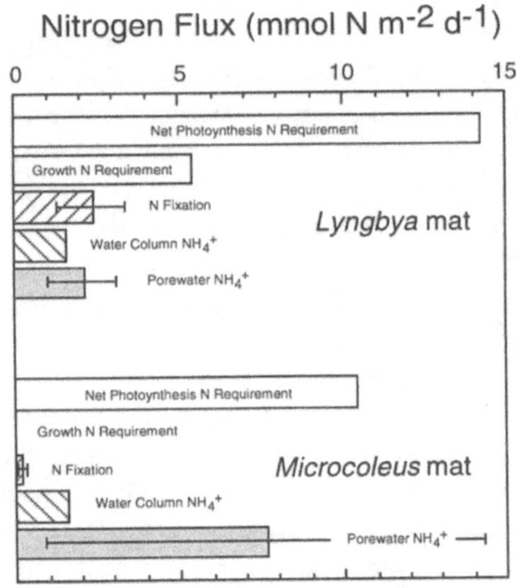

Fig. 3. Fluxes of N into the two microbial mats. Error bars depict the standard deviation of measurements made on 3 field trips in the *Microcoleus* mat, and the range of measurements made on two field trips in the *Lyngbya* mat.

Our findings of high rates of recycling of N within some microbial mat communities have important implications for the relationship of these benthic communities to the nutrient poor environments in which they occur.

The Lyngbya-dominated microbial mat used in this study occurs in an environment subject to frequent physical disruption. This mat, then, is often in the "pioneering" stages of habitat colonization and exhibits DIC retention as well as higher rates of N-fixation. The *Microcoleus*-dominated mat, in contrast, is not subjected to physical stresses which remove it from its substrate. The lack of DIC retention, and low rates of N-fixation are indicative of a community nearing a steady state production vs. degradation of biomass (perhaps due to limitation of primary productivity by another element). Higher rates of N-fixation have been previously reported in "colonizing" vs. "established" communities (Villbrandt et al. 1990). Finally, very little or no N leaves these communities, once it has been brought in via either the process of N-fixation or by uptake from the water column. Microbial mats may not be sources of N to the overlying water column, but rather sinks. It seems unlikely that the N-fixation which occurs in mats directly supports productivity in the overlying water column due to the small fluxes of N out of these systems.

Acknowledgements

Financial support from the Max Planck Institute for Marine Microbiology (to BMB), US National Science Foundation grants OCE 901246 and 9115706 (to HWP), NASA grant NAGW-3117 (to DEC), and from the NASA Exobiology Program (to DJDM) is gratefully acknowledged. We appreciate a review by N.P. Revsbech which greatly improved the manuscript.

References

Bebout BM (1992) Interactions of nitrogen and carbon cycling in microbial mats and stromatolites. Ph.D. Dissertation, University of North Carolina at Chapel Hill

Bebout BM , Fitzpatrick MW, Paerl HW (1993) Identification of the sources of energy for nitrogen fixation and physiological characterization of nitrogen-fixing members of a marine microbial mat community. Appl Environ Microbiol 59: 1495-1503

Bothe H (1982) Nitrogen fixation. In: Carr NG, Whitton BA (eds) The Biology of the Cyanobacteria. Blackwell Scientific Publications, Oxford, pp 87-104

Canfield DE, Des Marais DJ (1993) Biogeochemical cycles of carbon, sulfur and free oxygen in a microbial mat. Geochim Cosmochim Acta 57: 3971-3984

Javor BJ, Castenholz RW (1980) Laminated microbial mats, Laguna Guerrero Negro, Mexico. Geomicrobiol J 2: 237-273

Paerl HW, Bebout BM, Joye SB, Des Marais DJ (1993) Microscale characterization of dissolved organic matter production and uptake in marine microbial mat communities. Limnol Oceanogr 38: 1150-1161

Revsbech NP, Jørgensen BB, Brix O (1981) Primary production of microalgae in sediments measured by oxygen microprofile, $H^{14}CO_3^-$ fixation, and oxygen evolution methods. Limnol Oceanogr 26: 717-730

Villbrandt M, Stal LJ, Krumbein WE (1990) Interactions between nitrogen fixation and oxygenic photosynthesis in a marine microbial mat. FEMS Microbiol Ecol 74: 59-72

Anaerobic dark energy generation in the mat-building cyanobacterium *Microcoleus chthonoplastes*

Roy Moezelaar and Lucas J. Stal

Department of Microbiology, University of Amsterdam
Nieuwe Achtergracht 127, 1018 WS Amsterdam, The Netherlands

Introduction

Microbial mats are characterized by strong diel fluctuations in oxygen concentration which to a large extent can be attributed to the physiology of the cyanobacteria (Revsbech et al. 1983). In the light these organisms carry out oxygenic photosynthesis resulting in oxygen supersaturation of the mat. In the dark the primary mode of energy generation is respiration of endogenous glycogen (Smith 1982). However, in well-established microbial mats diffusion of oxygen in the mat will normally be insufficient to cover the demands and, as a result, the mat will turn anoxic. Evidently, the cyanobacteria have to switch to another way of energy generation in order to survive. Organisms isolated from such habitats were investigated for their mechanisms of anaerobic dark energy generation and turned out to be capable of fermentation. For instance, *Oscillatoria limnetica*, which thrives in the sulfide-rich hypolimnion of Solar Lake (Sinai) displays a homolactic fermentation using endogenous storage glucan as the substrate (Oren and Shilo 1979). Alternatively, this organism is capable of anaerobic respiration using elemental sulfur as the electron acceptor. Another mat-building cyanobacterium, *Oscillatoria limosa*, carries out a heterolactic and homoacetic fermentation simultaneously (Heyer et al. 1989). In the presence of elemental sulfur this organism stops producing hydrogen and sulfide is produced instead. *Oscillatoria terebriformis* was isolated from a hot-spring microbial mat and produces lactate form exogenous glucose or fructose (Richardson and Castenholz 1987).

The cosmopolitan filamentous cyanobacterium *Microcoleus chthonoplastes* is often the dominant organism in marine and hypersaline microbial mats. Its capacity to cope with dark anoxic conditions therefore can be predicted. We have set up a study to investigate the mechanisms of anaerobic dark energy generation in this organism.

Fermentation

The capability of *M. chthonoplastes* to obtain energy from the fermentation of glycogen (storage polyglucose) was tested with photoautotrophically pre-grown cells incubated in anoxic buffers (25 mM Hepes/KOH, pH 8 containing 3% (w/v) NaCl) in the dark (Moezelaar et al. 1992). When cells were transferred to anoxic dark conditions a decrease of the glycogen content

NATO ASI Series, Vol. G 35
Microbial Mats
Edited by Lucas J. Stal and Pierre Caumette

of the cells was observed (Fig. 1). Concomitantly, acetate, ethanol, and formate were excreted in the medium. It was concluded that they were the products of glycogen fermentation because the kinetics of production correlated inversely with the decrease of glycogen. Characteristically, about one mol of both acetate and ethanol were produced for every mol of glycogen-glucose degraded. In addition to the fermentation products mentioned above, hydrogen and some lactate were formed as well (Table 1). Succinate was not found. Carbon dioxide was assumed to equal the amount of hydrogen. Although the carbon recovery was satisfactory (95 %) the O/R balance was not (55 %). This indicated that not all of the electrons had been recovered. A possible explanation for the low O/R balance is offered by the observation that *M. chthonoplastes* accumulates ferric iron in its sheath. Formate may be oxidized non-enzymatically to carbon dioxide by ferric iron, the latter being reduced to ferrous iron (Ehrlich 1990). A higher recovery of formate would increase the O/R balance considerably whereas the carbon recovery would increase only slightly.

Table 1. Fermentation of glycogen in *Microcoleus chthonoplastes* and the influence of elemental sulfur on fermentation. Amounts of substrate and products in μmol.

		fermentation	fermentation with S^0
substrate	glucose	2.32 ± 0.22	2.41 ± 0.26
products	acetate	2.52 ± 0.09	4.10 ± 0.09
	ethanol	2.69 ± 0.10	0.22 ± 0.02
	formate	1.93 ± 0.04	0.55 ± 0.05
	hydrogen	0.71 ± 0.02	0
	lactate	0.07	0.05
	sulfide	-	12.1

-, not measured

The osmoprotectant in *M. chthonoplastes*, glucosylglycerol, was not used as a substrate during dark anoxic incubations. This was checked because in *Oscillatoria limosa* the osmoprotectant trehalose was degraded via a homoacetic fermentation (Heyer et al. 1989).

Fermentation in *M. chthonoplastes* was not inhibited in the presence of chloramphenicol. This indicated that switching to fermentation did not depend on *de novo* protein synthesis. Clearly, all the enzymes involved in fermentation metabolism are synthesized constitutively during photoautotrophic growth.

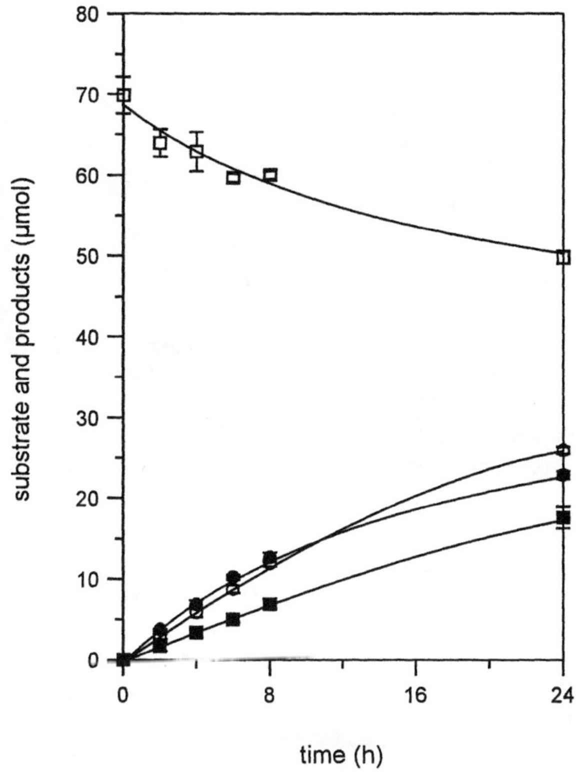

Fig. 1 Fermentation of glycogen by *Microcoleus chthonoplastes*. Symbols: □, glycogen-glucose; ■, ethanol; ○, formate; ●, acetate.

Pathways of fermentative metabolism

The pathway of fermentative metabolism in *M. chthonoplastes* was investigated by assaying cell-free extracts of photoautotrophically grown cells for enzyme activities (Table 2). Aerobically, in cyanobacteria glycogen is degraded via the oxidative pentose phosphate pathway, but its involvement during fermentation is unlikely. The production of equimolar amounts of acetate and ethanol during fermentation by cell suspensions rather suggested the involvement of either the Embden-Meyerhof glycolytic pathway or the Entner-Doudoroff pathway, but of neither of these routes the key enzymes have been found thusfar. However, the enzymes glyceraldehyde-3-phosphate dehydrogenase and pyruvate kinase were present. These enzymes play a role in both pathways. Remarkably, glyceraldehyde-3-phosphate dehydrogenase was active only with NAD as a cofactor.

The production of formate hints to the involvement of pyruvate:formate lyase, but this enzyme has not been found thusfar. Most likely, the production of hydrogen results from the cleavage of formate by formate:hydrogen lyase. The soluble hydrogenase which is a component of this enzyme complex, was found to be present in photoautotrophically grown cells. In addition, the enzymes catalyzing the formation of the end products ethanol (alcohol dehydrogenase), acetate (acetate kinase), and lactate (lactate dehydrogenase) were all present. Again the dehydrogenases showed activity with NAD only and not with NADP. In contrast, the enzymes glucose-6-phosphate dehydrogenase and 6-phosphogluconate dehydrogenase which are part of the oxidative pentose phosphate cycle used only NADP as a cofactor.

Table 2 Specific activities of enzymes [in U.(mg of protein)$^{-1}$] in cell free extracts of photoautotrophically pregrown *Microcoleus chthonoplastes*.

Enzyme	Cofactor	Specific activity
glucose-6-phosphate dehydrogenase	NADP	0.089
6-phosphogluconate dehydrogenase	NADP	0.021
glyceraldehyde-3-phosphate dehydrogenase	NAD	0.029
pyruvate kinase		0.059
hydrogenase		0.019
alcohol dehydrogenase	NAD	0.005
acetate kinase		0.013
lactate dehydrogenase	NAD	0.032

Influence of elemental sulfur on fermentation

Under anoxic conditions in the dark *M. chthonoplastes* used elemental sulfur as an electron sink reducing it to sulfide when incubated anoxically in the dark (Stal and Krumbein 1986). The presence of elemental sulfur influenced the fermentation pattern in two ways (Moezelaar and Stal, unpublished). Whereas the rate of glycogen degradation remained the same, the production of acetate was greatly increased at the expense of ethanol. This would result in a higher yield of ATP. Also, very little formate was produced and the evolution of hydrogen ceased completely. The production of sulfide indicated that the reducing equivalents otherwise used for the release of ethanol and hydrogen had been transferred to elemental sulfur. If all reducing equivalents were transferred to elemental sulfur then one mol of glucose would be transformed to two moles of both acetate and carbon dioxide, and 4 mol of sulfide would be produced:

$$\text{glucose} \rightarrow 2\ \text{acetate} + 2\ CO_2 + 8\ [H]$$
$$8\ [H] + 4\ S \rightarrow 4\ H_2S$$

However, the amount of sulfide produced and glycogen degraded were higher than would be expected on the basis of the amount of fermentation products. It is therefore conceivable that a small fraction of the glycogen was oxidized completely to carbon dioxide via the oxidative pentose phosphate pathway and that the electrons released as NADPH were transferred to elemental sulfur. In that case for each mol of glucose 12 mol of sulfide are expected. The oxidative pentose phosphate pathway does not have any site of substrate-level phosphorylation. Hence, under anoxic conditions this pathway would only make sense if electron transfer from NADPH to elemental sulfur did involve the generation of a proton motive force over the cytoplasmic membrane (sulfur respiration).

Ecological implications

M. chthonoplastes is very well adapted to the fluctuating oxygen concentrations in microbial mats. This is in the first place because of the metabolic versatility of the organism. The possession of a fermentative metabolism enables the organism to generate metabolic energy under anoxic conditions in the dark and thus to maintain viability. In addition, elemental sulfur can be used as an electron acceptor, which may give rise to higher ATP yields. All the enzymes involved in this metabolism are synthesized constitutively. The switch from photoautotrophic metabolism to fermentation therefore does not require any induction of protein synthesis. This reactivity is of paramount importance for survival in the mat where oxygen depletion in the dark may occur within minutes.

The excretion of fermentation products by *M. chthonoplastes* is very important for the ecology of the mat. It can be calculated that in *Microcoleus*-dominated mats fermentation products constitute the bulk of photosynthetically fixed carbon that is excreted. Acetate, ethanol, and formate are all substrates for sulphate-reducing bacteria which eventually mineralize them completely.

References

Ehrlich HE (1990) Geomicrobiology (2nd ed) Marcel Dekker Inc, New York and Basel, pp 283-346

Heyer H, Stal LJ, Krumbein WE (1989) Simultaneous heterolactic and acetate fermentation in the marine cyanobacterium *Oscillatoria limosa* incubated anaerobically in the dark. Arch Microbiol 151:558-564

Moezelaar R, Bijvank SM, Kuiper F, Stal LJ (1992) Anaerobic dark energy generation in cyanobacteria. Abstract 6th International Symposium on Microbial Ecology, Barcelona, Spain

Oren A, Shilo M (1979) Anaerobic heterotrophic dark metabolism in the cyanobacterium *Oscillatoria limnetica*: sulfur respiration and lactate fermentation. Arch Microbiol 122:77-84

Revsbech NP, Jørgensen BB, Blackburn TH, Cohen Y (1983) Microelectrode studies of the photosynthesis and O_2, H_2S and pH profiles of a microbial mat. Limn Oceanogr 28:1062-1074

Richardson LL, Castenholz RW (1987) Enhanced survival of the cyanobacterium *Oscillatoria terebriformis* in darkness under anaerobic conditions. Appl Environm Microbiol 53:2151-2158

Smith AJ (1982) Modes of cyanobacterial carbon metabolism. In: Carr NG, Whitton BA (eds) The biology of cyanobacteria. Blackwell Scientific Publications, Oxford, pp 47-85

Stal LJ, Krumbein WE (1986) Metabolism of cyanobacteria in anaerobic sediments. Actes de Colloques 3 (2ième Colloque International de Bacteriologie marine, Brest, 1-5 Octobre 1984), pp 301-309

Production and consumption of volatile organosulfur compounds in microbial mats

Pieter T. Visscher[1] and Ronald P. Kiene[2]

Rosenstiel School of Marine and Atmospheric Science,
Miami, FL 33149 USA

Introduction

Marine microbial mats typically support high rates of sulfate reduction. As a result, sulfur cycling is one of the dominant biogeochemical processes in these ecosystems. Detailed studies on inorganc sulfur cycling are available, but little is known about organic sulfur transformations. Sulfur-containing amino acids and dimethylsulfoniopropionate (DMSP) are the major biogenic precursors of volatile organosulfur compounds. DMSP, which functions as an osmolyte, is cleaved to acrylate and dimethylsulfide (DMS; White 1982). Production of DMS and methane thiol (CH_3SH) from decaying microbial mats in Yellowstone was reported by Zinder et al. (1977). Upon hydrolysis with NaOH, 200 μmol DMS l^{-1} sediment was retrieved from intact marsh sediments (Kiene 1988), and DMSP, DMS, CH_3SH and dimethyldisulfide (DMDS) were found in slurried samples of a marine microbial mat (Visscher et al. 1991; Visscher and Van Gemerden 1993). Kiene and Visscher (1987) found production of CH_3SH from methionine and DMS from DMSP in anoxic salt marsh sediments. Visscher and Van Gemerden (1991b) reported rapid production of DMS, CH_4 and traces of CH_3SH in oxic and anoxic mat slurries when DMSP was added. In addition to cleavage, DMSP also undergoes demethylation, during which 3-methiolpropionate (MMPA) and 3-mercaptopropionate (MPA) are formed successively (Mopper and Taylor 1986; Kiene and Taylor 1989). MPA was a dominant thiol in anoxic sediments of Biscayne Bay, and DMSP addition resulted in increasing MPA production in slurries (Kiene and Taylor 1989). Sulfur-containing amino acids added to coastal sediments generated a variety of thiols, including CH_3SH and MPA (Kiene et al. 1990). DMS is formed by microbial reduction of dimethyl-sulfoxide (DMSO) (Kiene and Capone 1988), DMS and CH_3SH are produced during metabolism of methoxylated aromatic compounds (Finster et al. 1990), and CH_3SH is generated by demethiolation of MMPA (Taylor and Gilchrist 1991). Abiotically, H_2S and polysulfides might act as methyl acceptor and react with unsaturated carbon bonds to form organosulfur compounds (Aizenshtat et al. 1983; Vairavamurthy et al. 1992). MPA, CH_3SH, DMS and

[1] Present address: Department of Marine Sciences, The University of Connecticut, 1084 Shennecosset Road, Groton, CT 06370, USA
[2] Department of Marine Sciences, University of South Alabama, Mobile, AL 36688 USA

NATO ASI Series, Vol. G 35
Microbial Mats
Edited by Lucas J. Stal and Pierre Caumette

acrylate are further metabolized by various marine bacteria (De Zwart and Kuenen 1992; Taylor 1993). The present study examined cycling of DMSP and potential microbial populations involved in DMSP consumption.

Experimental

Sediments samples were taken from a developing microbial mat located within a salt marsh on Sapelo Island, GA. The salinity was 90-130 and 70-80 ppt in the surface layer and at 15 mm depth, respectively. Sampling, slurry preparation and enumeration of viable DMSP and DMS utilizing populations and anoxygenic phototrophs have been described previously (Visscher et al. 1991; Visscher and Van Gemerden 1993). All sediment experiments were replicated. Most Probable Number (MPN) incubations with DMSP and DMS (600 μM) under anoxic conditions contained 5 mM nitrate. DMS and CH_3SH were measured by headspace analysis on a gas chromatograph with FID or FPD (Visscher et al. 1991). MPA was measured after treatment with 1 mM tributylphosphine (Mopper and Taylor 1986) using the OPA-HPLC procedure described in Kiene and Taylor (1989).

Results

Amended DMSP was degraded rapidly in oxic microbial mat slurries, during which DMS and MPA were formed (Fig 1). MPA accumulated more rapidly and reached a higher transient

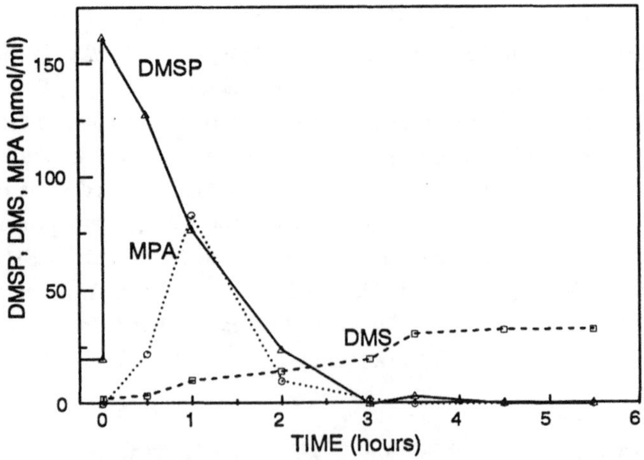

Fig. 1. Consumption of amended DMSP (165 μM) and production of MPA and DMS in an oxic microbial mat slurry from Sapelo Island, GA.

concentration. Traces of CH_3SH were observed occasionally (not shown). Endogenous DMS was consumed under oxic and anoxic conditions (Table 1). Addition of nitrate enhanced the rate of anoxic DMS consumption (Fig 2). Populations of aerobic DMSP utilizers were $2 \cdot 10^7$, $3 \cdot 10^6$, and $4 \cdot 10^4$ cells.cm^{-3}, at 0-5, 5-10, and 10-20 mm depth horizon, respectively, and anaerobic DMSP-utilizers were present at $1.1 \cdot 10^7$, $1.1 \cdot 10^7$, and $2 \cdot 10^6$ cells.cm^{-3} sediment. From these same sampling intervals, aerobic DMS utilizing populations were $4 \cdot 10^6$, $1 \cdot 10^5$, and $2.3 \cdot 10^4$ cells.cm^{-3} sediment and anaerobic DMS utilizers $1.1 \cdot 10^6$, $2.3 \cdot 10^6$, and $9 \cdot 10^5$ cells.cm^{-3}. Anoxygenic phototrophs were only found in the 0-5 and 5-10 mm depth horizons ($2 \cdot 10^5$ and $8 \cdot 10^4$ cells.cm^{-3} sediment, respectively).

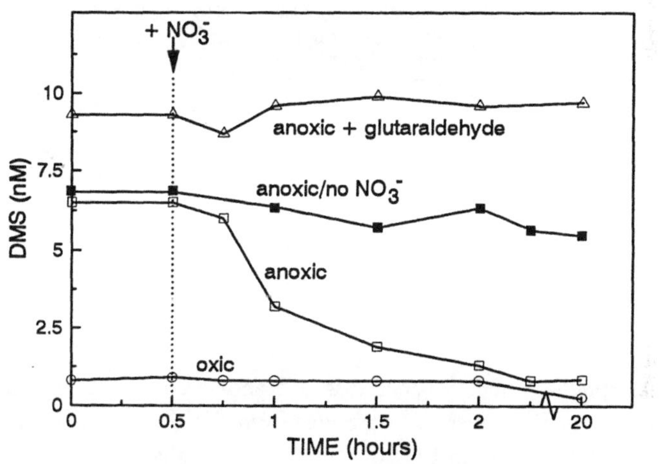

Fig. 2. Time course of DMS consumption in a slurried microbial mat of Sapelo Island, GA. Open symbols indicate treatments which received NO_3^-.

Discussion

Volatile organosulfur compounds are commonly found in marine sediments, including microbial mats (Aizenshtat et al. 1983; Visscher et al. 1991). DMSP was present in pure cultures of the cyanobacterium *Microcoleus chthonoplastes* (Visscher and Van Gemerden 1991b), which often is found in mature mats. Brackish water and marine strains of *Oscillatoria agardhii* and *M. lyngbyaceus* also contain DMSP (Granroth and Hattula 1976; White 1982). The cyanobacterial layer of the microbial mat of Sapelo Island contained *Microcoleus*, *Oscillatoria* and *Spirulina* spp. Formation of DMS after addition of NaOH and HPLC analysis confirmed the presence of DMSP in the upper 10 mm of the mat (R.P. Kiene, unpubl. data). DMSP was released by *M. chthonoplastes* at 12 nmol.mg^{-1} protein.h^{-1} upon a decrease in salinity of the medium (Visscher and Van Gemerden 1991b). Similarly, DMSP was released when distilled water was added to microbial mats samples associated with mangroves which contained *Microcoleus* sp. (Visscher unpubl. data). Simultaneous cleavage and demethylation of DMSP generates DMS, MPA, and traces of CH_3SH, as found in this study.

Table 1. Concentration and potential consumption rate of DMSP and DMS in the top 10 mm of various microbial mats. Rates shown under oxic (anoxic) conditions; conc = concentration, spec. rate = specific consumption rate (consumption in slurry divided by cell number obtained from MPN), nd = not determined, * = estimate

location	Chl*a* ($\mu g.cm^{-3}$)	DMSP			DMS		
		conc (μM)	consumption (nmol. $cm^{-3}.h^{-1}$)	spec rate (fmol. $cell^{-1}. h^{-1}$)	conc (μM)	consumption (nmol. $cm^{-3}.h^{-1}$)	spec rate (fmol. $cell^{-1}.h^{-1}$)
Sapelo	40	16	55	4.6	12	18 (15)	9 (36)
Biscayne Bay[1]	12	4	11 (23)	17	8	3 (6)	4 (27)
Texel[2]	310	230	20 (66)	nd	220	6*	24

[1] Visscher et al, unpubl. data
[2] Visscher and Van Gemerden 1991

DMS was rapidly consumed under oxic and anoxic conditions. Similarly, DMSP addition to a microbial mat slurry resulted in transient formation of DMS, which was consumed rapidly (Visscher and Van Gemerden 1991b). MPN's revealed high populations of DMSP and DMS consumers in the present study. Previously, high populations of (potential) DMS(P) utilizers were detected in a microbial mat, including anoxygenic phototrophs and thiobacilli (Visscher et al. 1991; Visscher and Van Gemerden 1991b). In this study, higher numbers of nitrate-respiring DMS utilizers were found in deeper strata, which is in agreement with the observation that nitrate addition increased the rate of DMS consumption in anoxic slurries. Recently, a facultative anaerobic, nitrate respiring *Thiobacillus* sp. was isolated on DMS from the microbial mat on Sapelo (Visscher and Taylor 1993). In addition, methanogens catabolize DMS and CH_3SH (Kiene et al. 1986; Finster et al. 1992) and phototrophs oxidize DMS to DMSO coupled to CO_2 fixation (Visscher and Van Gemerden 1991a). Under oxic conditions marine thiobacilli catabolize CH_3SH, DMS, DMDS and a range of alkyl sulfides (Smith and Kelly 1988; Visscher et al. 1991; Visscher and Taylor 1993). Inhibitor studies suggested involvement of methanogens and sulfate reducers in DMS consumption in microbial mats (Visscher and Van Gemerden 1991b). In addition, DMS-metabolizing hyphomicrobia, pseudomonads and anoxygenic phototrophs which have been isolated from non-marine environments (see De Zwart and Kuenen 1992; Taylor 1993 for reviews) might contribute to DMS consumption in coastal sediments as well. Studies with different sediments show rapid consumption by microbes (Table 1), which is supported by pure culture studies. For example, specific consumption rate for DMS under anoxic conditions was 680 $nmol.mg^{-1}$ $protein.min^{-1}$ for *Thiocapsa roseopersicina* (Visscher and Van Gemerden 1991a), 49 $nmol.mg^{-1}.min^{-1}$ for *Thiobacillus* strain ASN-1 and 48 $nmol.mg^{-1}$ $protein.min^{-1}$ for a *Methanosarcina* sp. (Finster et al. 1992). DMS consumption rates under oxic conditions were 47 for *T. thioparus* strain T5 and 65 $nmol.mg^{-1}$ $protein.min^{-1}$ for *Thiobacillus* strain ASN-1. All of these isolates except for the methanogen, were obtained from microbial mats. In conclusion, both sources and sinks for DMS(P) and CH_3SH exist in microbial mats and rapid consumption of these compounds in slurries suggests these ecosystems could be a sink for organosulfur compounds.

Acknowledgements

This work was supported by National Science Foundation grants OCE 90-12157 and OCE 92-03728 and by The Netherlands Organization for Scientific Research. PTV thanks the UGA Marine Insitute for providing housing.

References

Aizenshtat Z, Stoler A, Cohen Y, Nielsen H (1983) The geochemical sulphur enrichment of recent organic matter by polysulfides in the Solar-Lake. In: Bjoroy M et al. (eds) Advances in Organic Geochemistry. Wiley & Sons, Chichester UK, pp 279-288

De Zwart J, Kuenen JG (1992) C_1-cycle of sulfur compounds. Biodegr 3: 37-59

Finster K, King GM, Bak F (1990) Formation of methylmercaptan and dimethylsulfide from methoxylated aromatic compounds in anoxic marine and fresh water sediments. FEMS Microbiol Ecol 74: 295-302

Finster K, Tanimoto Y, Bak F (1992) Fermentation of methanethiol and dimethylsulfide by a newly isolated methanogenic bacterium. Arch Microbiol 157: 425-430

Granroth B, Hattula T (1976) Formation of dimethyl sulfide by brackish water algae and its possible implication for the flavor of Baltic Herring. Finn Chem Lett 6: 148-150

Kiene RP (1988) Dimethyl sulfide metabolism in salt marsh sediments. FEMS Microbiol Ecol 53: 71-78

Kiene RP, Oremland RS, Catena A, Miller LG, Capone DG (1986) Metabolism of reduced sulfur compounds in anaerobic sediments and by a pure culture of an estuarine methanogen. Appl Environ Microbiol 52: 1037-1052

Kiene RP, Visscher PT (1987) Production and fate of methylated sulfur compounds from methionine and dimethylsulfoniopropionate in anoxic salt marsh sediments. Appl Environ Microbiol 53: 2426-2434

Kiene RP, Capone DG (1988) Microbial transformations of methylated sulfur compounds in anoxic salt marsh sediments. Microb Ecol 15: 275-291

Kiene RP, Taylor BF (1989) Demethylation of dimethylsulfoniopropionate and production of thiols in anoxic marine sediments. Appl Environ Microbiol 54: 2208-2212

Kiene RP, Malloy KD, Taylor BF (1990) Sulfur-containing amino acids as precursors of thiols in anoxic coastal sediments. Appl Environ Microbiol 56: 156-161

Mopper K, Taylor BF (1986) Biogeochemical cycling of sulfur: thiols in marine sediments. In:

Sohn M (ed) Organic Marine Chemistry. Am Chem Soc, Washington DC, pp 324-339

Smith NA, Kelly DP (1988) Isolation and physiological characterization of autotrophic sulphur bacteria oxidizing dimethylsulfide as sole source of energy. J Gen Microbiol 134: 1407-1417

Taylor BF (1993) Bacterial transformations of organic sulfur compounds in marine environments. In: Oremland RS (ed) Biogeochemistry of global change. Chapman & Hall, New York NY, pp 745-781

Taylor BF, Gilchrist DC (1991) New routes for aerobic biodegradation of dimethylsulfoniopropionate. Appl Environ Microbiol 57: 3581-3584

Vairavamurthy A, Mopper K, Taylor BF (1992) Occurence of particle bound polysulfides and their significance of their reaction with organic matter in marine sediments. Geophys Res Lett 19: 2043-2046

Visscher PT, Quist P, Van Gemerden H (1991) Methylated sulfur compounds in microbial mats: in situ concentrations and metabolism by a colorless sulfur bacterium. Appl Environ Microbiol 57: 1758-1763

Visscher PT, Van Gemerden H (1991a) Photo-autotrophic growth of *Thiocapsa roseopersicina* on dimethyl sulfide. FEMS Microbiol Lett 81: 247-250

Visscher PT, Van Gemerden H (1991b) Production and consumption of dimethylsulfoniopropionate in marine microbial mats. Appl Environ Microbiol 57: 3237-3242

Visscher PT, Van Gemerden H (1993) Sulfur cycling in laminated marine microbial ecosystems. In: Oremland RS (ed) Biogeochemistry of global change. Chapman & Hall, New York NY, pp 672-690

Visscher PT, Taylor BF (1993) A new mechanism for the aerobic catabolism of dimethyl sulfide. Appl Environ Microbiol 59: 3784-3789

White RH (1982) Analysis of dimethyl sulfonium compounds in marine algae. J Mar Res 4: 173-181

Zinder SH, Doemel WN, Brock TD (1977) Production of volatile sulfur compounds during the decomposition of algal mats. Appl Environ Microbiol 34: 859-860

Panel discussion:
Microscale interactions in microbial mats

Carsten Lassen (rapporteur) and Hans W. Paerl[1] (chairman)

Institute of Ecology, University of Aarhus, Ny Munkegade, 8000 Aarhus, Denmark

The main part of the panel discussion concentrated on the fate of carbon in mats of Guerrero Negro, Baja California, Mexico. In mats of *Microcoleus chthonoplastes* growing at a salinity of 100‰ the carbon is recycled in the mat and the consumption of carbon is much higher than the flux from the overlaying water. Bicarbonate constituted the most significant part of the carbon flux whereas flux of dissolved organic carbon (DOC) from planktonic production does not seem to be significant in these mats. ^{14}C uptake experiments carried out in other mats showed that carbon is trapped in the euphotic zone during day. There is a mismatch in the dissolved inorganic carbon (DIC) budget of the mats arising from an imbalance between the DIC and O_2 fluxes across the mat/water interface. During the day, more DIC diffuses into the mat than O_2 is released and at night more DIC is lost than O_2 is consumed. There are two possible mechanisms to explain this observation. High oxidation state carbon e.g. aspartic acid, is produced during day and used for growth during night or glycolate is formed as s result of photorespiration. It was noted that glycolate and to some extent acetate fuels sulfate reduction in mats. If formation of glycolate explains the observed flux imbalance it must be stored during day and oxidized at night.

Experiments have shown that the products of photosynthesis have an isotopic composition nearly identical to the water, but there was a discrepancy with the isotopic composition of the microbial mat biomass as a whole. It was asked why recycling of the carbon would not result in decreasing $\delta^{13}C$. It was concluded that all the carbon is trapped in the mat during day and there is no possibility for an isotopic discrimination of the carbon supplied from either remineralization or diffusion from the water. The negative $\delta^{13}C$ of the mat may be due to carbon isotope discrimination of sulfide oxidizing bacteria that are active during night. It was noted that the isotopic composition of the CO_2 flux from the water to the sediment during night will be different from the composition of the remineralized carbon. Due to the isotope discrimination, the gradient of ^{13}C between the mat and the water will be steeper than the gradient of ^{12}C. A simple model was proposed to divide the net carbon flux into a downward and upward flux of different isotopic composition. By use of isotope probes and mass spectrometry it may be possible to distinguish between different carbon sources for photosynthesis.

---

[1] Institute of Marine Sciences, University of North Carolina at Chapel Hill, Morehead City, North Carolina 28557, USA

NATO ASI Series, Vol. G 35
Microbial Mats
Edited by Lucas J. Stal and Pierre Caumette

Another aspect of the Guerrero Negro mats that was discussed during the meeting were the fluxes of NH_4^+. A model was presented proposing that the *Microcoleus* mat could be regarded as a 'steady state' community (with respect to N) with a flux of NH_4^+ from deeper layers into the productive layer. With an accretion of 8 mm of mat per year it was proposed that burial may be regarded as a downward flux on a different time scale. It was noted that 'steady state' referred to the whole mat where the accretion of 8 mm was balanced by complete mineralization.

IV. Diel and spatial variations of physico-chemical parameters and processes in microbial mats

The carbon isotope biogeochemistry of microbial mats

David J. Des Marais and Donald E. Canfield[1]

Ames Research Center,
Moffett Field, CA 94035-1000, USA

Studies of the distribution of stable carbon isotopes within ecosystems frequently offer useful insights into the structure and function of those ecosystems. The isotopic composition of the organic matter in a community such as a microbial mat is controlled by those enzymes which create and subsequently alter organic compounds. The extent to which these enzymes can affect the isotopic composition of organics is influenced by the relative fluxes of carbon, both within a community and between the community and its environment. Because the isotopic composition of organic matter in sedimentary rocks resists thermal alteration better than organic molecular structures, isotopic studies of ancient rocks have been important for Precambrian paleobiology (Hayes et al. 1992; Des Marais et al. 1992a).

Carbon isotopic compositions are reported as permil (parts per thousand) differences between a sample and a standard, as follows:

$$\delta_{sam} = ((^{13}R_{sam}/^{13}R_{PDB}) - 1)1000 \qquad (1$$

where $^{13}R_{sam}$ and $^{13}R_{PDB}$ represent the $^{13}C/^{12}C$ values for the sample and the Peedee Belemnite calcite standard, respectively. PDB is very similar isotopically to dissolved inorganic carbon (DIC) in seawater. Because biota fix ^{12}C preferentially, δ values of organic matter are usually negative.

Most isotopic studies of mats have been surveys of the carbon isotopic compositions of the various mat types in a range of environments. A major contrast exists (Table 1) between modern microbial mats in marine hypersaline environments (Behrens and Frishman 1971; Aizenshtat et al. 1984; Schidlowski et al. 1984) and their ancient stromatolite counterparts in marine carbonates (e.g., Eichmann and Schidlowski 1975; Strauss et al. 1992). It is useful to explore this isotopic contrast; it might reflect the evolution either of the biota or of their environment. A key first step is to examine modern mats.

The δ_{mat} value reflects the net isotopic discrimination by the processes which sustain the network of carbon pathways (Fig. 1). If two competing reactions at a branch point in this network differ in their isotopic discrimination, then the $\delta^{13}C$ values of their respective products also differ (see Monson and Hayes 1980).

---

[1] Present address: Max Planck Institute for Marine Microbiology, Fahrenheitstrasse 1, D 28359 Bremen, Germany

NATO ASI Series, Vol. G 35
Microbial Mats
Edited by Lucas J. Stal and Pierre Caumette

Table 1. δ values of organic matter in microbial mats and stromatolites (data summarized by Des Marais et al. 1992b; Des Marais 1993).

Locality and dominant flora	δ_{mat}	Stromatolite age, b.y.	$\delta^{13}C$
Solar Lake, cyanobacteria	-4 to -16	0.58 - 0.95	-31.8
Gavish Sabkha, cyanobacteria	-6 to -17	1.05 - 1.44	-34.4
Guerrero Negro, cyanobact.	-10 to -14	1.50 - 2.10	-38.1
Shark Bay, diatoms	-10 to -13	2.50 - 2.70	-44.1
Shark Bay, cyanobacteria	-7 to -19		

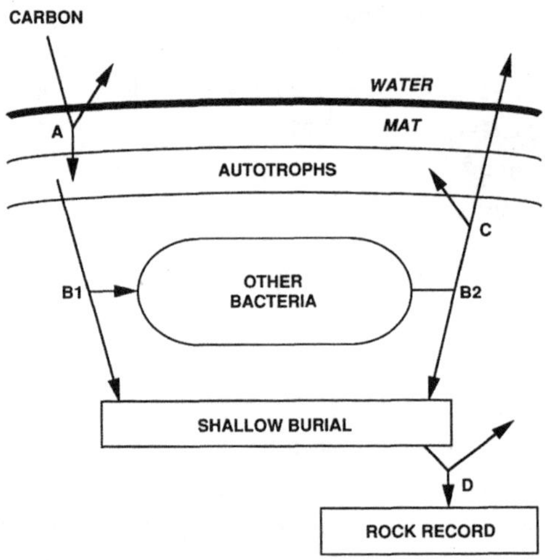

Fig. 1. Carbon pathways in a microbial mat. Labels are defined in the text.

The above concept is illustrated by data from cultures of *Anabaena variabilis* grown in a medium sparged with mixtures of air and CO_2. The following equation (Hayes 1983) relates isotope discrimination to the relative rates of diffusion versus fixation reactions:

$$\delta_P = \delta_{CO2} - (1 - f_P)(\alpha_{fix} - 1)10^3 \qquad (2$$

where δ_P and δ_{CO2} correspond to newly-formed photosynthate and dissolved CO_2 in the culture medium, respectively; and f_P is that fraction of CO_2 (on a scale from 0 to 1) which is converted to photosynthate. Unreacted CO_2 diffuses back into the medium. The term α_{fix} is the ratio of the rate at which $^{12}CO_2$ is assimilated, relative to the rate for $^{13}CO_2$, under conditions of unlimited CO_2 supply. Fixation of CO_2 by cyanobacteria, diatoms and purple sulfur bacteria occurs principally via the enzyme ribulose 1,5 bisphosphate carboxylase/oxygenase ("RuBisCO," e.g., see Schidlowski et al. 1983). For green plant RuBisCO, $\alpha = 1.029$ (Roeske and O'Leary 1984). If CO_2 in the medium is not depleted, and if its exchange between the medium and RuBisCO is rapid, then f_P approaches zero and δ_P becomes more negative. If the CO_2 uptake becomes diffusion-limited, then f_P approaches unity and δ_P becomes more positive and approaches δ_{CO2}.

The term f_P was varied by manipulating the CO_2 levels and cell density (Table 2). Values of f_P were estimated using the concentrations of CO_2 entering and leaving the culture medium (see Table 2 footnote c). Using Equation 2, values of δ_P were calculated and compared to δ_P values actually measured for cells of *Anabaena* from the cultures. A good correspondence was observed between calculated and measured values (Table 2).

Table 2. Isotopic discrimination by cultures of *Anabaena variabilis* [a]

Percent CO_2 in gas			δ_{cells} [b]	
Into culture	Out	f_P [c]	calculated[d]	culture
0.14	0.05	0.61	-11.3	-11.4
0.15	0.01	0.93	-2.0	-3.5
0.29	0.21	0.28	-20.9	-16.1
1.0	0.48	0.52	-13.9	-14.5
11.0	8.9	0.19	-23.5	-24.7
11.7	9.4	0.20	-23.2	-25.6

[a] Grown in ASN III medium at 30°C, pH in range 7 to 8, illuminated at 200 $\mu E\ m^{-2}\ s^{-1}$ using cool white fluorescent lamps.
[b] Values expressed relative to δ_{CO2} of zero permil.
[c] $f_P = 1 - ((\%\ CO_2 \text{ out of culture})/(\%\ CO_2 \text{ into culture}))$
[d] α is assumed to equal 1.029 (Roeske and O'Leary 1984).

In a photosynthesizing microbial mat, an important branch point accompanies the uptake of dissolved inorganic carbon (DIC) from the overlying water column (process A in Fig. 1). DIC entering the mat is either fixed into biological material or it diffuses back out. The factors controlling isotopic discrimination in microbial mats are, of course, more complex than Equation 2 implies. Mat organisms in addition to cyanobacteria fix CO_2. Also, some of the fixed DIC has been recycled by organic decomposition (Canfield and Des Marais 1993). Variations in CO_2 levels affect the degree to which cyanobacteria actively transport HCO_3^- during fixation (e.g., Badger and Andrews 1982). Isotopic discrimination during active HCO_3^- transport in algae is much less than with direct uptake of CO_2 (Sharkey and Berry 1985).

In *Chlorogloeopsis* mats in a New Zealand thermal spring (Table 3), isotopic discrimination varies markedly with changes in CO_2 concentrations (Des Marais et al. 1992b), whereas changes in total DIC concentrations (principally HCO_3^- are too modest to cause the change in fractionation. This is consistent with the interpretation that, as CO_2 levels decline, active HCO_3^- transport becomes important and suppresses isotopic discrimination.

Table 3. Discrimination by *Chlorogloeopsis* mats growing in a thermal spring, Orakei Korako, New Zealand (Des Marais et al. 1992b).

	Concentration, mM		
T,°C	[DIC]	[CO_2]	δ_{mat} [a]
60	2.5	0.39	-20.3
53	2.3	0.12	-12.5
47	2.0	0.03	-5.5
42	1.9	0.02	-4.8

[a] Values are relative to δ_{CO2} value of zero permil for dissolved CO_2.

Mat δ_P values (Table 1) indicate that discrimination is much less than the maximum value expected from RuBisCO. Most of these reported values were not accompanied by δ_{DIC} values, therefore discrimination cannot be accurately estimated. However, if these δ_{DIC} values were within a few permil of zero, which is typical of marine DIC, then discrimination was low, indicating that DIC uptake in these mats was largely diffusion-limited.

It has been proposed that the high salinities and high temperatures often associated with hypersaline mats can lower the solubility of CO_2 and create diffusion limitation and thus suppress isotopic discrimination (Schidlowski et al. 1984). However, in a study at Guerrero Negro, Baja California, Mexico, discrimination was nearly constant in mats growing in the salinity range 70 to >120‰ (Des Marais et al. 1989). We hypothesize that increasing salinity lowered rates of CO_2 fixation (Des Marais et al. 1992c) as it lowered CO_2 solubility. Therefore the net effect upon f_P was small (see Equation 2) and isotopic discrimination remained relatively constant.

To assess environmental effects upon isotopic discrimination, field measurements were made in the tidal flats and seawater evaporation ponds of Exportadora de Sal, S.A., Guerrero Negro, Baja California Sur, Mexico. Mats dominated by *Microcoleus* and *Lyngbya* cyanobacteria were selected to represent both subtidal and also intertidal to supratidal mats, respectively. The *Microcoleus* mats grew in the range 90-108‰ salinity, at a water depth of approximately 70 cm. Seasonal temperatures varied from 12°C (morning, January) to 30°C (late afternoon, June). The *Lyngbya* mats were emergent and inactive most of the time, being either partially dried or in shallow, highly saline (>120‰) pools. Photosynthesis occurred principally when these mats were flooded with lagoon water with a salinity in the range 38 to 50‰. Sections of mat were incubated using this lagoon water in aquaria, wherein benthic chambers were placed over the mats for a full diel cycle. Changes in the chemical composition of dissolved species were monitored (for details, see Canfield and Des Marais 1993). The observations were made in four time intervals over diel cycles. We examined those environmental parameters (water turbulence, temperature, pH and DIC concentration) which might alter discrimination by affecting upon either the CO_2 supply or photosynthesis.

Each benthic chamber was constructed from a pyrex beaker which was sawed off and placed upside-down over the mats (see Canfield and Des Marais 1993, Fig. 2). The chamber was fitted with a glass stirring bar which was rotated at a speed sufficient to create a reproducible diffusion boundary layer over the mats. This was apparently achieved by stirrer rotation rates at or above 12 rpm, which created water currents in the chambers of about 5 cm/sec or greater. At the slower stir rate of 4.5 rpm, the larger diffusion boundary layer suppressed carbon uptake rates to about half of those observed at 12 rpm. This more restricted diffusion of inorganic carbon also suppressed isotopic discrimination during the daytime by *Lyngbya* mats (Table 4). In *Microcoleus* mats, essentially no isotopic discrimination was observed during the daytime, even at the higher stir rate. Thus, in all but the experiment described above, stirring rates were maintained at 12 rpm.

The range of α values measured for photosynthesizing *Lyngbya* mats is consistent with δ_{mat}. For example, an α value of 1.007 would be consistent with a δ_{mat} value of -7. This observation implies that other factors such as assimilation of dissolved or particulate organic carbon by the mat, or processes occurring at night, etc. probably did not greatly affect δ_{mat}. Indeed, the δ_{DIC} value measured for carbon leaving the mat at night is isotopically very similar to δ_{mat}. Therefore, loss of this carbon has a minimal effect.

In contrast, the α values measured for the *Microcoleus* mats indicate that essentially no isotopic discrimination occurred during carbon fixation (α = 1.000 in the stirring rate experiment, Table 4), therefore it cannot explain the δ_{mat} value observed (-7 relative to δ_{DIC}, Table 4).

It is useful to explore how the range of conditions experienced by the mat can affect discrimination. For example, rates of DIC diffusion in and out of the mat are not as sensitive to temperature as are carbon fixation rates. Therefore, at lower temperatures, the rate of carbon supply increases, relative to carbon fixation, a trend which favors greater isotopic discrimination. This parameter was explored over the range of temperatures (15 to 30°C) typically experienced by subtidal *Microcoleus* mats. However, no significant variation in α was observed with temperature (Table 4), even though the carbon uptake rate varied by a factor of 2.8 over this range (Canfield and Des Marais 1993).

Table 4. Isotope discrimination in intertidal and subtidal hypersaline mats incubated during the daytime (7 a.m. to 6 p.m.).

Mat type	Temp., °C	Miscellaneous	α[a]	δ_{mat}[b]
		Stirring rate, rpm		
Lyngbya	20	4.5	1.001	-6.0
Lyngbya	20	12	1.006 to 1.011	-6.0 to -7.1
Microcoleus	20	4.5	1.000	-6.5
Microcoleus	20	12	1.000	-6.3
		Brine [DIC], pH		
Microcoleus	17	2.5mM, 8.6	0.999	-7.1
Microcoleus	20	2.5mM, 8.6	1.002	-6.5
Microcoleus	30	2.5mM, 8.7	1.001	-6.9
		Brine [DIC], pH		
Microcoleus	15	2.4mM, 7.1	1.002	-7.5
Microcoleus	20	2.5mM, 8.7	1.000	-7.1
Microcoleus	20	11.0mM, 8.0	1.004	-7.4

[a] α equals the rate of ^{12}C uptake divided by the rate for ^{13}C (see Mariotti et al. 1981).
[b] Values are expressed relative to a value for δ_{DIC} of zero permil.

As mentioned earlier, discrimination is affected by CO_2 concentrations. The effect of CO_2 was explored in two ways. First, the pH of the brine was lowered from 8.5 to 7.1 to increase the ratio of $[CO_2]/[HCO_3^-]$. This pH change had no effect upon the rate of carbon uptake, therefore the rate of CO_2 diffusion into the mat should have increased, relative to its rate of uptake by the mat. However, as Table 4 shows, little or no increase in discrimination was observed. In the second approach, the rate of DIC supply was increased by increasing [DIC] in the brine by more than a factor of four, while keeping pH relatively constant (Table 4). The resultant increase in α from values in the range 0.999 to 1.002 to a value of 1.004 is small but probably real. However, the high DIC concentrations required to give this modest increase in fractionation are never experienced by the natural mat population under study.

In summary, over the range of environmental conditions typically experienced by the *Microcoleus* mats, isotopic discrimination which is directly associated with photosynthesis cannot account for the 7 permil depletion observed for δ_{mat} relative to δ_{DIC}. Other factors must be responsible.

Excretion, respiration and fermentation are important for processing organic matter in mats (see B1 and B2 in Fig. 1), yet they apparently alter δ_{mat} values minimally. Dissolved organic carbon (DOC) in the porewaters near the mat surface has a δ value similar to δ_{mat} (Bauer et al. 1991), suggesting that little discrimination occurs during the decomposition of cellular material to DOC. Organic matter buried at depth in mat sediments is isotopically similar to that near the mat surface (Des Marais et al. 1992c; Aizenshtat et al. 1984), similarly indicating little diagenetic alteration of δ_{mat}. This might be due in part to the abundant mucilage produced by mat phototrophs; this mucilage is relatively resistant to decomposition, and thus it preserves the isotopic signature created by the near-surface community.

One aspect of the *Microcoleus* mat which merits attention is the fate of organic carbon which is remineralized to DIC. This DIC source is important for photoautotrophs (Canfield and Des Marais 1993), supplying as much as half or more of the carbon fixed during photosynthesis (process C in Fig. 1). Because rates of photosynthesis exceed rates of remineralization (Canfield and Des Marais 1993), this reassimilation is expected to be very efficient, with little of the remineralized DIC reentering the overlying water column during the daytime. Because this reassimilation is essentially 100 percent efficient during the day, carbon isotopes cannot be fractionated. However the uptake of this isotopically light carbon (δ = -7) demonstrates that δ_{mat} values are at least partially maintained by internal recycling of DIC. At night, the mats do lose DIC to the overlying water (Table 5). If some of the remineralized DIC was recovered by mat chemoautotrophs near the mat/brine interface, selective uptake of ^{12}C might occur and would be evidenced by ^{13}C-enriched unused DIC escaping into the water column. Indeed, *Beggiatoa* has been observed to migrate to the mat/water interface in order to maintain its position at the sulfide/O_2 boundary (Jørgensen and Des Marais 1986). Chemoautotrophs can fractionate DIC substantially during carbon uptake (Ruby et al. 1987).

Both the DIC flux and δ_{DIC} were monitored over a full diel cycle. In simultaneous nighttime incubations, replicate sections of mat were incubated as follows: in one benthic chamber, dissolved [O_2] was maintained at between 90 and 100% air saturation; in the other chamber [O_2] was allowed to decline to 0% air saturation over a 3 hour period; the incubation was terminated after 4.5 hours. The results of these experiments are given in Table 5.

Isotopic discrimination was observed throughout the diel cycle. As stated earlier, the photosynthetic uptake of DIC during daytime (α = 1.002) could not account for δ_{mat} values near -7. However, DIC was released from the mat at night which is several permil enriched in ^{13}C, relative to its presumed source, namely decomposed mat organic matter (δ = -7). This ^{13}C enrichment might indicate that a fraction of the remineralized carbon has been

Table 5. Incubation of a *Microcoleus* mat over a full diel cycle, salinity = 110%, temperature = 20°C.

Time interval	DIC flux,[a] $mM.m^{-2}$	δ_{DIC}[b] of flux	Mean $[O_2]$,[c] % air sat.
6:45AM to Noon	-24	-1.8	220
12:30PM to 6:00PM	-17	-1.3	180
7:15PM to Midnight	15	+1.4	95
12:30AM to 6:15AM	25	+1.6	90
7:15PM to Midnight	10	+3.4	35

[a] Negative values signify uptake of DIC into the mat (loss from chamber).
[b] δ values of DIC which crosses the mat/brine interface; expressed relative to d value of DIC in chamber, which is set equal to zero permil.
[c] Mean value of $[O_2]$ in the benthic chamber during the incubation interval.

refixed by isotopically-selective chemoautotrophic bacteria. It should be possible to estimate the flux and isotopic composition of the carbon which is refixed, using the following equation (Hayes 1983):

$$\delta_{out} = \delta_{remin.} + (f_{refix.})(\beta - 1)10^3 \qquad (3$$

where δ_{out} corresponds to the DIC diffusing out of the mat (given in the third column of Table 5); $\delta_{remin.}$ is the δ value of the total DIC component released as HCO_3^- during organic decomposition by sulfate reducing bacteria (assumed to equal -7, the δ_{mat} value); and f_{refix} is the fraction of this DIC which is refixed by chemoautotrophs. The term β represents the overall discrimination associated with the conversion of DIC (mostly HCO_3^- in the pH range 8 to 9) to photosynthate. For chemoautotrophs, Ruby et al. measured a value of 1.025 (Ruby et al. 1987). The nighttime DIC fluxes can be estimated using equation 3 and the values in Table 5 for the experiment under oxygenated conditions (90 to 95% air saturation). About 61 mM m^{-2} DIC was produced by organic decomposition. Chemoautotrophs refixed about 21 mM m^{-2} of this DIC, and 40 mM m^{-2} was released to the benthic chamber. By isotope mass balance, the δ value of the refixed carbon was -23. To estimate the total daily contribution to the mat organic carbon inventory and, thus, to δ_{mat}, the δ value of the total DIC fixed per day is calculated by adding the DIC component imported from the overlying water to the DIC which is refixed from sources within the mat. It was assumed that the total internal DIC source during the day equals 61 mM m^{-2}, the value measured during the night (see above). Thus, the carbon fixed during the day equals the flux from the overlying water (41 mM m^{-2}, δ = -1.6, Table 5) plus the carbon refixed from within (61 mM m^{-2}, δ = -7). At night, DIC is fixed by the chemoautotrophic bacteria (21 mM m^{-2}, δ = -23). By mass balance, the δ value of DIC fixed over a full 24 hour period equals -7.9, which is more negative than the δ_{mat} value of -7.1. Therefore chemoautotrophs contribute isotopically light carbon which balances the heavier carbon which is assimilated from the overlying water during the day. The δ_{mat} value of the *Microcoleus* mat apparently is maintained by a close interaction between the carbon and sulfur cycles.

In summary, δ_{mat} is controlled by variations in the rate of DIC supply, relative to its rate of uptake by autotrophs. Discrimination correlates with environmental CO_2 concentrations. The trend of increasing $\delta_{organic}$ values in Proterozoic stromatolitic carbonates (Table 1) might reflect a decline in CO_2 and DIC levels. In today's marine hypersaline environment, discrimination during photosynthesis is minimal due to diffusion-limited DIC uptake and the importance of active HCO_3^- transport. Internal DIC recycling and chemautotrophy, perhaps by sulfide-oxidizing bacteria, maintain the small depletion in δ_{mat}, relative to δ_{DIC}, which is observed in subtidal *Microcoleus* mats. The depletion in the δ_{mat} value of *Lyngbya* mats is due principally to discrimination during photosynthesis.

Acknowledgements

We thank Exportadora de Sal, S.A for their support of the field work. Anne Tharpe provided excellent laboratory analyses. This work was supported by a grant from NASA's Exobiology Program.

References

Aizenshtat Z, Lipiner G Cohen Y (1984) Biogeochemistry of carbon and sulfur cycle in the microbial mats of the Solar Lake (Sinai). Microbial Mats: Stromatolites. Alan R Liss, Inc., New York, pp 281-312

Badger MR, Andrews TJ (1982) Photosynthesis and inorganic carbon usage by the marine cyanobacterium *Synechococcus sp.* Plant Physiology 70: 517-523

Bauer JE , Haddad RI, Des Marais DJ (1991) Method for determining stable isotope ratios of dissolved organic carbon in interstitial and other natural marine waters. Mar Chem 33: 335-351

Behrens EW, Frishman SA (1971) Stable carbon isotopes in blue-green algal mats. Geology 79: 94-100

Canfield DE, Des Marais DJ (1993) Biogeochemical cycles of carbon, sulfur and free oxygen in a microbial mat. Geochim Cosmochim Acta 57: 3971-3984

Des Marais DJ (1993) Tectonic Control of the Crustal Organic Carbon Reservoir during the Precambrian. Chemical Geology : in press

Des Marais D J, Cohen Y, Nguyen H, Cheatham M, Cheatham T, Munoz E (1989) Carbon isotopic trends in the hypersaline ponds and microbial mats at Guerrero Negro, Baja California Sur, Mexico: Implications for Precambrian stomatolites. Microbial Mats: Physiologial Ecology of Benthic Microbial Communities. American Society for Microbiology, Washington pp 191-203

Des Marais DJ, Bauld J, Palmisano AC, Summons RE, Ward DM (1992b) The Biogeochemistry of Carbon in Modern Microbial Mats. The Proterozoic Biosphere. Cambridge University Press, New York pp 299-308

Des Marais DJ, D'Amilio E, Farmer JD, Jørgensen BB, Palmisano AC, Pierson BK (1992c) Case Study of a Modern Microbial Mat-Building Community: the Submerged Cyanobacterial Mats of Guerrero Negro, Baja California Sur, Mexico. The Proterozoic Biosphere. Cambridge University Press, New York.

Des Marais DJ, Strauss H, Summons RE, Hayes JM (1992a) Carbon isotope evidence for the stepwise oxidation of the Proterozoic environment. Nature 359: 605-609

Eichmann R, Schidlowski M (1975) Isotopic fractionation between coexisting organic carbon-carbonate pairs in Precambrian sediments. Geochim Cosmochim Acta 39: 585-595

Hayes JM (1983) Practic and Principles of Isotopic Measurements in Organic Geochemistry. Organic Geochemistry of Contemporaneous and Ancient Sediments. Society of Economic Paleontologists and Mineralogists, Bloomington 5-1 - 5-31

Hayes JM, Des Marais DJ, Lambert IB, Strauss H, Summons RE (1992). Proterozoic Biogeochemistry. The Proterozoic Biosphere. Cambridge University Press, New York, pp 83 - 134

Jørgensen BB, Des Marais DJ (1986) Competition for sulfide among colorless and purple sulfur bacteria in cyanobacterial mats. FEMS Microbiol Ecol 38: 179-186

Mariotti AJ, Germon C, Hubert P, Kaiser P., Loetolle R, Tardieux A, Tardieux P (1981) Experimental determination of nitrogen kinetic isotope fractionation: some principles; illustration for the denitrification and nitrification processes. Pl Soil 62: 413-430

Monson KD, Hayes JM (1980) Biosynthetic control of the natural abundance of carbon 13 at specific positions within fatty acids in *Escherichia coli*, evidence regarding the coupling of fatty acid and phospholipid synthesis. J Biol Chem 255: 11435-11441

Roeske CA, O'Leary M (1984) Carbon isotope effects on the enzyme-catalyzed carboxylation of ribulose bisphosphate. Biochemistry 1984: 6275-6284

Ruby EG, Jannasch HW, Deuser WG (1987) Fractionation of stable carbon isotopes during chemoautotrophic growth of sulfur-oxidizing bacteria. Appl Environ Microbiol 53: 1940-1943

Schidlowski M, Hayes JM, Kaplan IR (1983). Isotopic inferences of ancient biochemistries: carbon, sulfur, hydrogen, and nitrogen. Earth's Earliest Biosphere. Princeton University Press, Princeton pp 149-186

Schidlowski M, Matzigkeit U, Krumbein WE (1984) Super-heavy organic carbon from hypersaline microbial mats. Naturwissenschaften 71: 303-308

Sharkey T D Berry JA (1985). Carbon isotope fractionation of algae as influenced by an inducible CO_2 concentrating mechanism. Inorganic Carbon Uptake by Aquatic Photosynthetic Organisms. American Society of Plant Physiologists, Davis, pp 389-401

Strauss H, Des Marais DJ, Hayes JM, Summons RE (1992). The Carbon-Isotopic Record. The Proterozoic Biosphere. Cambridge University Press, New York, pp 117-127

The fluxes of inorganic carbon and CO_2-dependent genes involved in the cyanobacterial inorganic carbon-concentrating mechanism: A view on some of the open questions

Rakefet Schwarz, Judy Lieman-Hurwitz, Michal Ronen-Tarazi, Chana Gabai, Miriam Hassidim, Leonora Reinhold and Aaron Kaplan

Department of Botany, The Hebrew University of Jerusalem, 91904 Jerusalem, Israel.

In this manuscript we briefly discuss some of the open questions in relation to the response of photosynthetic microorganisms to changes in the ambient concentration of CO_2. We focus on the regulation and activity of the inorganic carbon-concentrating mechanism (CCM) in cyanobacteria and its contribution to the fractionation of stable carbon isotopes ($\delta^{13}C$).

The $\delta^{13}C$ serves as an important natural tracer to study past and present changes in biological and geochemical carbon cycling. (see Jasper and Hayes 1990; Rau et al. 1989; Raven 1991; Rundel et al. 1989; and citations therein). The fractionation is thought to result in large part from the activity of the universal CO_2 fixing enzyme, rubisco, which discriminates against the heavier isotope (see Guy et al. 1993). In consequence, the $\delta^{13}C$ of the photosynthetic products is lower than that in the bulk CO_2. Changes in the isotopic composition of atmospheric CO_2 since the industrial revolution (due to consumption of fossil fuel originating from photosynthetic activity and hence containing lighter carbon) are recorded in various natural materials such as tree rings, coral skeletons and deep sea sediments. The use of carbon isotopes as a major tool in the study of the global carbon cycle is, however, hampered by our poor understanding of biological carbon isotope fractionation. There are noticeable variations in the $\delta^{13}C$ values observed in different latitudes; also in phytoplankton vs benthic organisms and organisms representing different taxonomic groups. Studies by Rau et al. (1989) indicated strong correlation between the concentration of dissolved CO_2, as affected by temperature, and the $\delta^{13}C$ values. The mechanisms leading to these variations, however, are poorly understood. Furthermore, it has been widely accepted that the primary oceanic productivity is not limited by the concentration of inorganic carbon (Ci) but recent studies clearly indicated that this might not be the case (Riebesell et al. 1993).

Many photosynthetic microorganisms are capable of adapting to a wide range of CO_2 concentrations. This is largely accomplished by regulating the degree of expression of an inorganic carbon-concentrating mechanism (CCM) operating in these organisms (see Miller

NATO ASI Series, Vol. G 35
Microbial Mats
Edited by Lucas J. Stal and Pierre Caumette

et al. 1990; Coleman 1991; Kaplan et al. 1991, 1993; Raven 1991; Badger and Price 1992 for recent reviews and literature citations). The activity of the CCM increases dramatically following exposure to low (air level) CO_2 and decreases after a treatment with high (0.1%-5%) CO_2. It results in a large intracellular inorganic carbon (Ci) pool (about 50 mM in cyanobacteria) and, consequently, of CO_2 at the site of rubisco. This mechanism enables aquatic organisms to perform efficient photosynthesis even though the concentration of dissolved CO_2 at equilibrium with air is 5 to 20-fold lower than the affinity (Km) of their RuBisCO for CO_2.

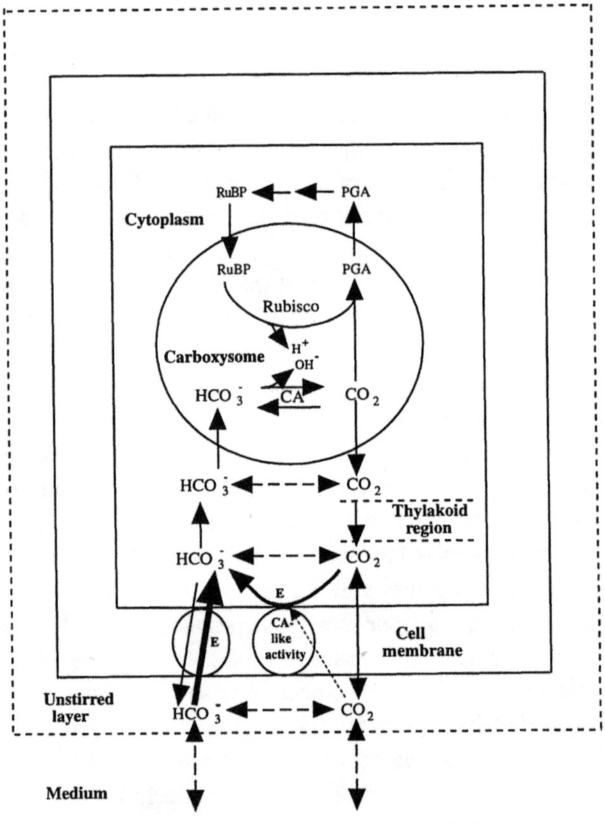

Fig. 1. A schematic representation of the Ci fluxes associated with the operation of the CCM in cyanobacteria. Energy-dependent Ci fluxes are designated (E) and presented as heavy lines. The dashed arrow indicates that the Ci species are not at chemical equilibrium in the cytoplasm and the unstirred layer.

The presence of the CCM has been documented in a number of freshwater and marine photosynthesizing microorganisms and it is most probably widely spread (Raven 1991; but see Bloye et al. 1992).

The scheme in Fig. 1 presents a "working hypothesis" of the major components and Ci fluxes associated with the CCM in cyanobacteria. Briefly, Ci enter the cells from the surrounding unstirred layer and accumulates in the cytoplasm via energy-dependent system(s). Regardless of whether bicarbonate or CO_2 is supplied, the former is the species which proponderates in the cytoplasm and the Ci species are not at chemical equilibrium here. Bicarbonate passes through the thylakoid region and enters the carboxysomes where localized carbonic anhydrase (CA) catalyzes the formation of CO_2. Part of the CO_2 generated is fixed at the closely adjacent carboxylation sites of RuBisCO (which is mainly located in these polyhedral bodies). Part of the CO_2 may leak from the carboxysomes to the cytoplasm. Here, a plasmalemma-located CA-like activity converts it back to bicarbonate in an energy-dependent process and thus minimizes the wasteful leak of CO_2 from the cells.

The activity of the CCM may well contribute to the observed $\delta^{13}C$ due to the effect of its associated Ci fluxes on the isotopic composition of the internal Ci pool which is relatively closed. It has been noted that the higher the activity of the CCM, the smaller is the discrimination against the heavier isotope (Berry, in Rundel et al. 1989). It is likely (though yet to be experimentally established) that the natural variations in $\delta^{13}C$ reflect variations in the flux balance between Ci influx, efflux and photosynthesis. Every cellular or environmental parameter which affects this balance should have a considerable effect on the $\delta^{13}C$ of the organic matter produced.

Models which quantitatively describe the unidirectional fluxes of Ci and photosynthesis in cyanobacteria, as a function of the ambient concentration of Ci, have been developed (Reinhold et al. 1991). These models might be used to predict the expected $\delta^{13}C$ under various conditions. The kinetic parameters used in these models, however, were derived from experiments where the cells were allowed to use the ambient Ci followed by the addition of different Ci concentrations for a short period. It is necessary to test whether the results obtained are applicable under steady state conditions. Preliminary results of experiments performed with Drs. Silverman and Tu (not shown) indicated that, in the presence of 2 mM Ci, the net efflux of CO_2 is considerably smaller than that predicted by the models. Moreover, the data suggest complex regulation of the Ci fluxes by ambient CO_2. The nature and role of the putative control site are yet to be established. Elucidation of the mechanisms involved should have significant impact on our understanding of the regulation of the Ci fluxes associated with the CCM in cyanobacteria.

Some of the genes involved in the ability of cyanobacteria to adapt to various levels of CO_2 have been identified with the aid of CO_2-dependent mutants. In *Synechococcus* PCC 7942, transcripts originated from *purK* (encoding subunit II of AIR carboxylase involved in purine biosynthesis, Schwarz et al. 1992) and *cmpA* (encoding the cytoplasmic-membrane-located 42 kDa polypeptide) were only observed when the cells were exposed to low CO_2. An elevated level of transcripts following exposure to low CO_2 has been noted in the cases of *ndhB* (encoding subunit II of NADH dehydrogenase, Marco et al. 1993), *rbc* and *ccmO* (located immediately upstream of *rbc*, unpublished). The observed responses to ambient CO_2 stimulated the search for CO_2-dependent promoters. Several such promoters have been detected in *Synechococcus* PCC 7942 with the aid of *LacZ* as a reporter gene (Scanlan et al. 1990). The elements involved in conferring responsiveness to CO_2 and the mechanisms by which the concentration of CO_2 affects the expression of genes in this and in other organisms are yet to be established.

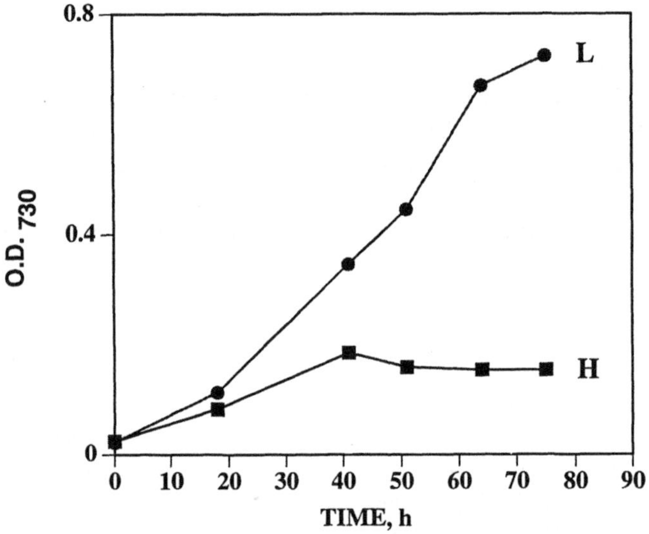

Fig. 2. Growth curves of a *Synechococcus* mutant in which a promoter less cat was fused to the promoter region of *cmpA*. The cells were grown in thepresence of chloramphenicol and high (H) or low (L) CO_2.

Fusion of a promoter-less cat (encoding chloramphenicol resistance) within the 5' coding sequence of *cmpA* in *Synechococcus* PCC 7942 resulted in a mutant which possessed chloramphenicol acetyl transferase activity and was thus capable of growing in the presence of chloramphenicol, but only under low CO_2 conditions (Fig. 2). These data provide supporting evidence that the promoter region of *cmpA* contains elements involved in the response to ambient CO_2. Analysis of the minimal promoter region and the sequences responding to ambient CO_2 is in progress.

Insertion of a cartridge encoding kanamycin resistance within a *SalI* site located 9.5 kbp upstream of *rbc* (the region where many mutations leading to dependence on CO_2 for growth in *Synechococcus* PCC 7942 were mapped, suggesting a cluster of genes involved in this ability (Kaplan et al. 1991, 1993) resulted in a mutant capable of growing in the presence of kanamycin under low, but not high CO_2. Sequence analysis (Accession No. X72391) revealed the presence of an open reading frame of 839 amino acids. Search in the gene bank indicated high homology between the first 700 amino acids of ORF839 and *topA* encoding topoisomerase I in *E. coli*. The role of DNA topology in the response to ambient CO_2 is being investigated. Analysis of the C-terminus of ORF839 revealed the presence of a clear motif of histidine kinase (Fig. 3). Over the last few years there is a growing body of evidence for the involvement of histidine kinase in a "two component" signal transduction (Stock et al. 1989). It remains to be seen whether ORF839 participate in the perception of the "CO_2 concentration" signal.

```
                  I                II                                       III
CheA     RAAHSIKGGAG  PEKRLEAGKNVVGNL  VTDDGAGLNR  [41]  SGRGVGMDVV
NRII     GLAHEIKNPLG  PDQIEQVLLNIVRNA  VEDNGPGIPS  [15]  GGTGLGLSIA
PhoR     NVSHELRTPLT  EDQLRSAISNLVYNA  VEDNGPGIAP  [21]  GGSGLGLAIV
EnvZ     GVSHDLRTPLT  PLSIKRAVANMVVNA  VEDDGPGIAP  [19]  SGTGLGLAIV
SpoII    GIAHEIRNPLT  QNQLKQVFINLIKNA  VKDEGEGIPE  [15]  KGTGLGLMVT
DctB     GVAHEINQPVA  RIRLEQVLINLLQNA  VADNGPGIPT  [15]  SGLGLGLVIS
PgtB     TLAHEINQPLN  EVRIQQVLVNVLANA  IADNGPGWPV  [15]  VGLGIGLSIS
CpxA     DISHELRTPLT  PNALESALENIVRNA  VDDDGPGVSP  [21]  GGTGLGLAIV
PhoM     ALTHELKSPLA  PALLEQALGNLLDNA  VLDTGSGIPD  [19]  KSSGLGLAFV
VirA     GIAHEFNNILG  PLELQQVLINICKNA  ISDNGGGIPE  [17]  GGTGLGLASV
PhoO     DLTHSLKTALA  QNDFVEVMGNVLDNA  VEDDGPGIPH  [17]  PGQGVGLAVA
UhpB     RELHDDIGQTI  RVTLFRVCQEGLNNI  IEDDGSGLPA  [13]  TRNGAGWHIT
FixL     TLAHELNQPLS  RIQVQQVLINLMRNA  VEDTGGGIPE  [15]  SGMGIGLSIS
LegS     REIHDGPAQML  EVALFRLAQEAVSNA  IKDNGKGFDL  [30]  SKIGLGTFIM

ORF839   RVEHYCGKKP   RFTCLRVPTVPTSNL  VCDLGGGEKD  [26]  KSRGRGKEPI
```

Fig. 3. Alignment of various histidine kinases (after Stock et al. 1989) with the relevant region in ORF839 of *Synechococcus*. The H, N, DXGXG and GXG are boxed.

Acknowledgement

This research was supported by a grant from the USA-Israel Binational Science Foundation (BSF), Jerusalem, Israel; and The Ministry of Science of Niedersachsen, Germany.

References

Badger MR, Price GD (1992) The CO_2 concentrating mechanism of cyanobacteria and green algae. Physiol Plant 84: 606-615

Bloye SA, Karagouni AD, Carr NG (1992) A continous culture approach to the question of inorganic carbon concentration by *Synechococcus* species. FEMS Microbiol Lett 99: 79-84

Coleman JR (1991) The molecular and biochemical analyses of CO_2-concentrating mechanisms of cyanobacteria and green algae. Plant Cell Environ 14: 861-867

Guy RD, Fogel ML, Berry JA (1993) Photosynthetic fractionation of the stable isotopes of oxygen and carbon. Plant Physiol 101: 37-47

Jasper JP, Hayes JM (1990) A carbon isotope record of CO_2 levels during the late quaternary. Nature 347: 462-464

Kaplan A, Schwarz R, Lieman-Hurwitz J, Reinhold L (1991) Physiological and molecular aspects of the inorganic carbon concentrating mechanism in cyanobacteria. Plant Physiol 97: 851-855

Kaplan A, Schwarz R, Lieman-Hurwitz J, Reinhold L (1993) Physiological and molecular studies on the response of cyanobacteria to changes in the ambient inorganic carbon concentration. In: Bryant D (ed.) The molecular biology of the cyanobacteria. Kluwer Academic Pub, Dordrecht, The Netherlands (in press)

Marco M, Ohad N, Schwarz R, Lieman-Hurwitz J, Gabay C, Kaplan A (1993) High CO_2 concentration alleviates the block in photosynthetic electron transport in an *ndhB*-inactivated mutant of *Synechococcus* sp. PCC 7942. Plant Physiol 101: 1047-1053

Miller AG, Espie GS, Canvin DT (1990) Physiological aspects of CO_2 and HCO_3 transport by cyanobacteria: a review. Can J Bot 68: 1291-1302

Rau GH, Takahashi T, Des Marais DJ (1989) Latitudinal variations in plankton $\delta^{13}C$: implications for CO_2 and productivity in past oceans. Nature 341: 516-518

Raven JA (1991) Implications of inorganic C utilization: ecology, evolution and geo-chemistry. Can J Bot 69: 908-924

Riebesell U, Wolf-Gladrow DA, Smetacek V (1993) Carbon dioxide limitation of marine phytoplankton growth rates. Nature 361: 249-251

Reinhold L, Kosloff R, Kaplan A (1991) A model for inorganic carbon fluxes and photosynthesis in cyanobacterial carboxysomes. Can J Bot 69: 984-988

Rundel PW, Ehleringer JR, Nagy KA (eds) (1989) Stable Isotopes in Ecological Research. Springer-Verlag New York

Scanlan DJ, Bloye SA, Mann NH, Hodgson DA, Carr NG (1990) Construction of *lacZ* promoter probe vectors for use in *Synechococcus*: application to the identification of CO_2- regulated promoters. Gene 90: 43-49

Schwarz R, Lieman-Hurwitz J, Hassidim M, Kaplan A (1992) Phenotypic complementation of high-CO_2-requiring mutants of the cyanobacterium *Synechococcus* sp. strain PCC 7942 by inosine 5'-monophosphate. Plant Physiol 100: 1987-1993

Stock JB, Ninfa AJ, Stock AM (1989) Protein phosphorylation and regulation of adaptive responses in bacteria. Microbiol Rev 53: 450-490

Oxygenic photosynthesis and light distribution in marine microbial mats

Carsten Lassen, Helle Ploug, Michael Kühl[1], Bo Barker Jørgensen[1] and Niels Peter Revsbech

Institute of Biological Sciences, Department of Microbial Ecology, University of Aarhus D-8000 Aarhus C, Denmark

Introduction

Marine intertidal sediments are often colonized by dense populations of phototrophic microorganisms forming stratified communities with diatoms at the very surface and an underlying population of cyanobacteria (Stal et al. 1985). Underneath the layers of oxygenic phototrophs, purple and green sulfur bacteria frequently form additional colored bands (Nicholson et al. 1987). Microalgae in the top layers shade the underlying sediment of those regions of the light spectrum which they preferentially absorb (Jørgensen et al. 1987; Pierson et al 1987; Lassen et al. 1992b; Ploug et al. 1993). The distinct stratification, which is often observed for different types of phototrophic organisms, may thus be strongly influenced by their complementary utilization of the light spectrum. Below the layers of the oxygenic phototrophs, scalar irradiance in the visible spectrum (400-700 nm, PAR) is depleted 10-100 times more than light in the near infrared spectrum (NIR) (Pierson et al. 1990; Lassen et al. 1992b; Kühl and Jørgensen 1992). The NIR absorption bands of the bacteriochlorophyll-protein complexes of the anoxygenic phototrophs are essential for the presence of the dense populations of these organisms underneath the oxygenic phototrophs.

It is not known to what extent complementary spectral utilization of the incident light has an ecological significance for the zonation of the oxygenic phototrophs. The extension of the euphotic zone of oxygenic photosynthesis in microbial mats is often only about 1 mm. The aim of the present study was to quantify the efficiency by which available light in sediments is used for photosynthesis, to measure action spectra for photosynthesis and from these data to estimate the extent to which changes in the spectral light composition was of competitive advantage to cyanobacteria over diatoms.

[1] Present address: Max Planck Institute for Marine Microbiology, Fahrenheitstraße 1, D-28357 Bremen, Germany

NATO ASI Series, Vol. G 35
Microbial Mats
Edited by Lucas J. Stal and Pierre Caumette

Photosynthetic efficiency

The investigated microbial mats were collected in August 1990 and 1991 at sheltered sandy sediments of Limfjorden, Denmark, in areas periodically desiccated. Light gradients in the sediments were measured by 70 μm wide fiber-optic scalar irradiance (4π) sensors (Lassen et al. 1992a). Gross oxygenic photosynthesis was measured by the light-dark-shift technique (Revsbech and Jørgensen 1983) using oxygen microelectrodes with a 90% response time of 0.3–0.4 s (Revsbech 1989).

When illuminated at high irradiance, the depth distribution of photosynthesis exhibited a maximum at the very surface and a secondary maximum at 1.0 mm, reflecting the zonation of the dominant oxygenic phototrophs (Fig. 1). At the depth of the secondary maximum, PAR was only 12 μE $m^{-2}s^{-1}$ i.e. less than 2% of incident irradiance. The efficiency by which the available light was utilized for oxygenic photosynthesis could be calculated by dividing the photosynthetic rates for each depth interval by the scalar irradiance (PAR) at the top of the interval (Fig. 2). At the secondary maximum, the efficiency was 10-fold higher than the efficiency in the upper 0.0-0.6 mm of the sediment when the sediment was illuminated at high irradiance (60% of *in situ* irradiance on a clear day).

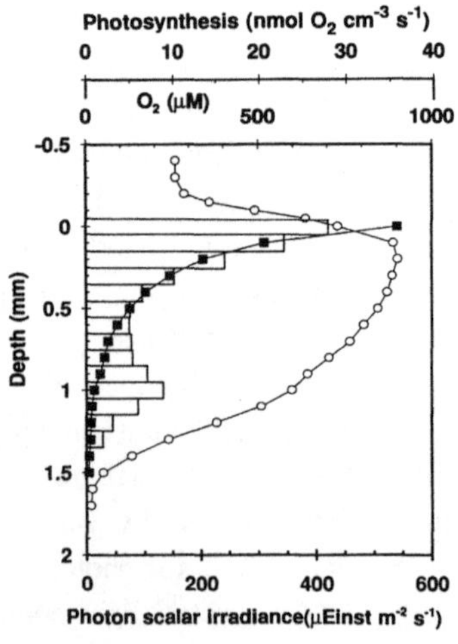

Fig. 1. Depth profiles of photosynthesis (bars), O_2 concentration (○) and photon scalar irradiance (■) in a microbial mat with a dense surface layer of pennate diatoms and an underlying layer of *Oscillatoria* spp. Incident irradiance was 740 μE m^{-2} s^{-1}. Reproduced from *Limnology and Oceanography* (Lassen et al. 1992 b).

The efficiency as calculated here is an ecological parameter and will both depend on the concentration of phototrophes and the efficiency by which the absorbed light is used for photosynthesis. The latter depends on the intensity of incident light and will decrease as the intensity exceeds the saturation onset parameter, E_k, of the population of phototrophes (4π). At lower irradiances the efficiency will thus increase. The local maximum of photosynthesis and the maximum photosynthetic efficiency at 12 μE m^{-2} s^{-1} were probably an effect of the observed accumulation of cyanobacteria at this depth.

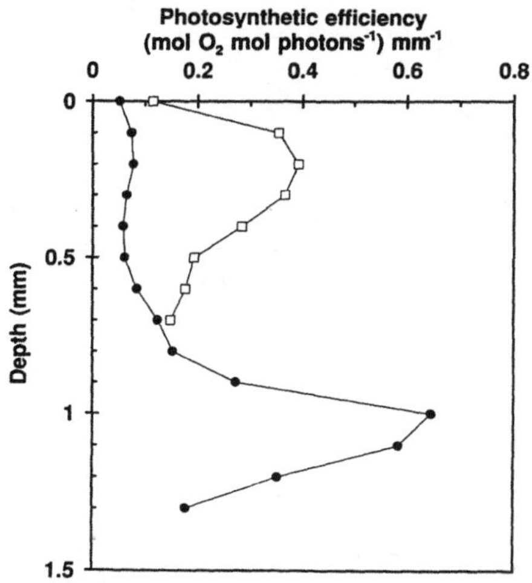

Fig. 2. Photosynthetic efficiency at 740 (●) and 55 μEm^{-2}s^{-1} (□). Data (●) calculated from Fig. 1. Reproduced from (Lassen et al. 1992 b) with permission from the publisher.

Light quality

To analyze the ecological significance of measured differences in spectral photosynthesis for the zonation of diatoms and cyanobacteria, the relative quality of the ambient light was calculated for the two populations as a function of depth (Ploug et al. 1993). The relative light quality was calculated from measured action spectra for photosynthesis for the two populations and the spectral distribution of scalar irradiance in the sediment. At the top of the sediment, the spectral distribution of scalar irradiance changed dramatically due to absorption by diatom photopigments, and the spectral window of 550-650 nm constituted an increasing part of the available light with depth (Fig. 3). Light in this part of the spectrum was efficiently used for photosynthesis by the cyanobacteria. Relative to the quality of PAR incident at the surface, the light quality for the diatoms decreased to 85% at 1 mm depth whereas the relative light quality for the cyanobacteria increased to 107% at this depth (Fig 4). In other words, the cyanobacteria at the lower boundary of the euphotic zone had a competitive advantage of 26% over the diatoms due to their relatively more efficient utilization of the light spectrum available at this depth.

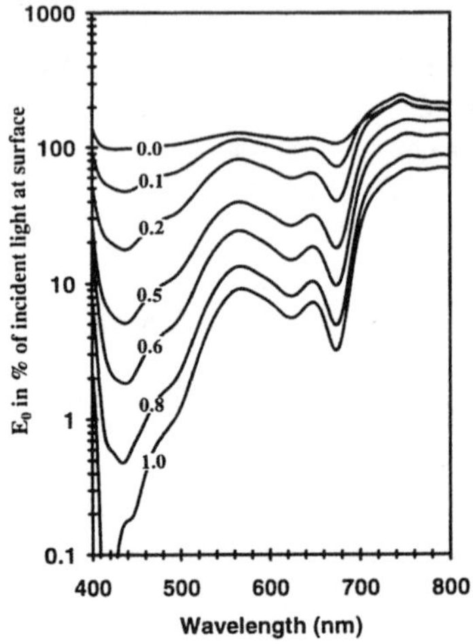

Fig. 3. Depth distribution of scalar irradiance, E_0, normalized to the incident light at the sediment surface, E_{0d}. The numbers on the curves indicate the depth in the sediment in mm. Reproduced from FEMS Microbiology Ecology (Ploug et al. 1993) with permission from the publisher.

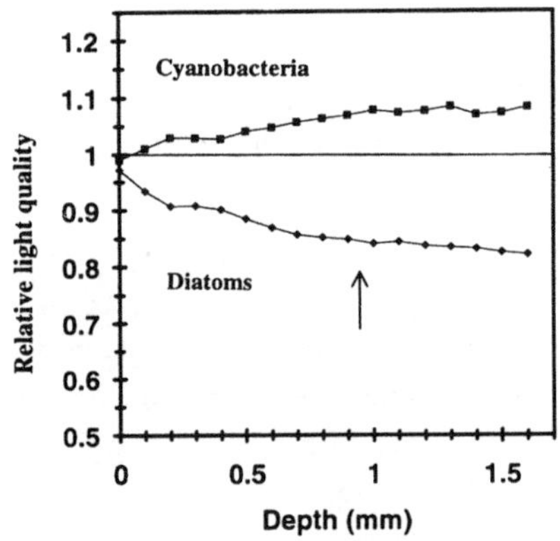

Fig. 4. Changes in the relative light quality with depth calculated from the depth distribution of spectral scalar irradiance (400-700 nm) and the action spectra for photosynthesis of the two phototrophic populations. The value 1 equals no change in relative light quality. The arrow indicates the lower boundary of the euphotic zone. Reproduced from FEMS Microbiology Ecology (Ploug et al. 1993).

The results indicate that complementary utilization of the spectral light can only be one of several factors explaining the zonation af the oxygenic phototrophs. The zonation of the cyanobacteria is probably determined by the combined effect of light and chemical microgradients.

High resolution photosynthesis measurements

Gross photosynthesis was in these studies measured by the light-dark-shift method using a strip chart recorder. The extension of the zone of oxygenic photosynthesis in these microbial mats was only about 1 mm resulting in steep gradients of O_2 between the most active regions and the surroundings. After the onset of darkness, the diffusional fluxes of O_2 will therefore change very fast. By computer simulation it has been shown that the photosynthesis profile as measured by this method will underestimate the photosynthetic rates in the most active layers and overestimate the rates in adjacent layers (Revsbech et al. 1986; Glud et. al 1992). For this reason, an improved method was developed where initial rates of oxygen decrease after darkening were estimated from 3rd order polynomial fits to the oxygen curve within the first 0.3-4.0 s after darkening recorded by a personal computer (Lassen et al., in preparation). The initial rate of oxygen decrease was calculated as the slope of the polynomial in 0.3 s (90% response time of the electrode).

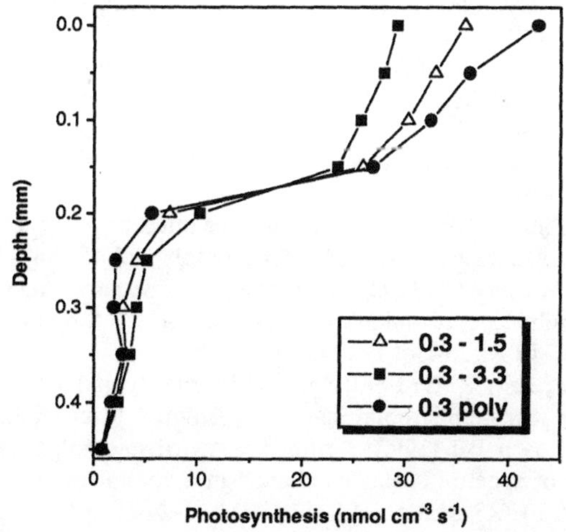

Fig. 5. Gross photosynthesis in cyanobacterial mat from Limfjorden, Denmark. The photosynthetic rates were calculated from 3rd order polynomial fit to the interval 0.3-3.0 s after darkening (0.3 poly) or from linear regression to the indicated intervals (s). The water flow above the sediment was 4-5 cm s^{-1} Adapted from Lassen et al, manuscript in preparation

In a microbial mat with a dense surface layer of cyanobacteria, maximum photosynthetic rates calculated from the polynomial were 18% higher than the rates calculated from linear regression to the interval 0.3-1.5 s (Fig 5). Just below the highly photosynthesizing layer, photosynthesis calculated from the linear regression was in contrast 98% higher than the rates calculated from the polynomial. These differences reflect the different spatial resolutions of the two methods, and the underestimation of the rates in one layer using linear regression is to some extent counterbalanced by an overestimation in other layers. But due to diffusion of O_2 from the sediment to the water, the depth integrated photosynthesis will systematically be underestimated by the linear regression method. In these measurements the depth integrated photosynthesis calculated from the polynomial was 5% higher than calculated from linear regression. For detailed studies of light-photosynthesis properties in microbial mats, the higher resolution of the photosynthesis measurements when polynomial fits are used, can be essential.

References

Glud RN, Ramsing NB, Revsbech NP (1992) Photosynthesis and photosynthesis-coupled respiration in natural biofilms quantified by use of oxygen microsensors J Phycol 26: 51-60

Jørgensen BB, Cohen Y, Des Marais DJ (1987) Photosynthetic action spectra and adaptation to spectral light distribution in a benthic cyanobacterial mat Appl Environ Microbiol 53: 879-886

Kühl M, Jørgensen BB (1992) Spectral light measurements in microbenthic phototrophic communities with a fiber-optic microprobe coupled to a sensitive diode array detector. Limnol Oceanogr 37: 1813-1823

Lassen C, Ploug H, Jørgensen BB (1992a) A fibre-optic scalar irradiance microsensor: application for spectral light measurements in sediments. FEMS Microbiol Ecol 86: 247-254

Lassen C, Ploug H, Jørgensen BB (1992b) Microalgal photosynthesis and spectral scalar irradiance in coastal marine sediments of Limfjorden, Denmark. Limnol Oceanogr 37: 760-772

Nicholson AM, Stolz JF, Pierson BK (1987) Structure of a microbial mat at Great Sippewissett March, Cape Cod, Massachusetts. FEMS Microbiol Ecol 45: 343-364

Pierson BA, A Oesterle, Murphy GL (1987) Pigments, light penetration, and photosynthetic activity in the multi-layered microbial mats of Great Sippewissett Salt March, Massachusetts. FEMS Microbiol Lett 45: 365-376

Pierson BK, Sands VM, Frederick JL (1990) Spectral irradiance and distribution of pigments in a highly layered marine microbial mat. Appl Environ Microbiol 56: 2327-2340

Ploug H, Lassen C, Jørgensen BB (1993) Action spectra of microalgal photosynthesis and depth distribution of spectral scalar irradiance in a coastal marine sediment of Limfjorden, Denmark. FEMS Microbiol Ecol 102: 261-270

Revsbech NP (1989) An oxygen microelectrode with a guard cathode. Limnol Oceanogr 34: 474-478

Revsbech NP, Jørgensen BB (1983) Photosynthesis of benthic microflora measured with high spatial resolution by the oxygen microprofile method : Capabilities and limitations of the method. Limnol Oceanogr 28: 749-756

Revsbech NP, Madsen B, Jørgensen BB (1986) Oxygen production and consumption in sediments determined at high spatial resolution by computer simulation of oxygen microelectrode data. Limnol Oceanogr 31: 293-304

Stal LJ, Van Gemerden H, Krumbein WE (1985) Structure and development of a benthic marine microbial mat. FEMS Microbiol Ecol 31: 111-125

The effects of irradiance, temperature and desiccation on cyanobacterial photosynthesis: A possible explanation for the diurnal changes in surface waterblooms

Bas W. Ibelings and Ben de Winder[1]

School of Biological Sciences, University of Bristol, Bristol BS 8 1 UG, United Kingdom

Introduction

During surface waterbloom formation a population of buoyant cyanobacteria becomes telescoped to the lake surface in the absence of mixing (Reynolds and Walsby 1975). Bloom formation involves an abrupt change in environmental conditions, most notably the exposure to full sunlight of cells in the top-layer of the bloom. In contrast, cells in deeper layers have to cope with anoxic, dark conditions (Ibelings and Mur 1992). Abeliovich and Shilo (1972) already observed photooxidative death of cyanobacteria in surface blooms. In addition to high irradiance stress, cells in the surface bloom are exposed to elevated temperature, and formation of a dry, desiccated crust has also been observed (Zohary and Pais Madeira 1990). The presence of additional stresses enhances the likelihood of photoinhibition (Demmig-Adams and Adams 1992). Photoinhibition is the decrease in quantum yield of photosynthesis induced by exposure to an irradiance, higher than that which can be used with maximum quantum efficiency (Powles, 1984). In this study we further investigated the impact of such conditions in a bloom on photosynthesis of the cyanobacterium *Microcystis*. Our investigations concentrated on the performance of Photosystem II, which is not only sensitive to photoinhibition, but to other stress factors as well (Havaux 1992).

Materials and methods

Organisms

Colonies of a recently isolated *Microcystis* strain from Lake Nieuwe Meer, The Netherlands, were grown in continuous culture, and received a 12:12 light/dark regime, with an irradiance of 100 μmol photons m^{-2} s^{-1}. Freshly harvested samples from a

---

[1] Laboratory for Microbiology, University of Amsterdam, 1018 WS Amsterdam, The Netherlands.

NATO ASI Series, Vol. G 35
Microbial Mats
Edited by Lucas J. Stal and Pierre Caumette

Microcystis spp. community in Lake Nieuwe Meer were used to form surface waterblooms in the laboratory.

Microelectrode measurements

Microcystis was left to float up and form a surface waterbloom in an 0.8 m deep column. The column was kept at room temperature. Illumination was provided from above by a Schott KL 1500 halogen light source, equipped with an infrared filter. Irradiance was varied to give a "stepped" sinusoidal light regime, mimicking that at a lake surface on a cloudless day. Irradiance was raised from 50 - 350 - 650 - 950 - 1250 μmol photons m^{-2} s^{-1}, and reduced in the reverse order. Cells were illuminated for 60 minutes at each irradiance before profiles of oxygen and gross rates of photosynthesis were measured in duplicate. Clark-type oxygen microelectrodes were made according to the specifications given by Revsbech and Ward (1983). Rates of photosynthesis were measured as the initial decrease in oxygen concentration, during the first two seconds after darkening of the cells (Revsbech et al. 1981).

Fluorescence measurements

A Walz PAM 101 fluorometer was used to measure fluorescence of the surface layer of the bloom *in situ*. The minimal- and maximal fluorescence yield (F_o and F_m) were monitored as described by van Kooten and Snel (1990). Pulses of 12000 μmol photons m^{-2} s^{-1} and 0.9 s duration were applied every 3 minutes. From this, variable fluorescence F_v (= F_m- F_o), and the ratio of F_v /F_m were calculated. F_v /F_m has been shown in several studies to be closely related to the maximum quantum yield of PS II (see Demmig-Adams and Adams 1992). Fluorescence vs. temperature curves were made using cells from the *Microcystis* culture, incubated in the dark in a Walz KS-101 suspension cuvette. Temperature was controlled within 0.1 °C by circulating water from a Lauda waterbath, and was raised by 0.4 °C $minute^{-1}$.

Desiccation studies

Microcystis cells were taken from the culture, immobilized on nylon filters, and washed in 10-fold diluted BG-11 medium buffered with TAPS (N-tris[hydroxymethyl]-3-amino-propanesulfonic acid) of pH 8.2. Each filter contained approximately 3.5 μg Chl (determined spectroscopically at 665 nm after extraction in dimethyl-formamide). Controlled desiccation was accomplished by exposing a series of the filters to a stream of dry air, leading to fast desiccation of the filters at the start of the sequence, but slower towards the end. After 2 hours desiccation $^{14}CO_2$ was injected. Illumination was provided during a 20 minutes period by an Osram 500 W halogen lamp at an irradiance of 50 or 1500 μmol photons m^{-2} s^{-1}; the water content of the filters was then determined by weighing (see De Winder et al. [1989] for details of the method used).

Pigments

Pigments were extracted in 90 % acetone, and analyzed using reverse phased ion-pair chromatography on an Isco model 2350 HPLC, at a wavelength of 455 nm.

Results

Both maximum and integrated rates of gross photosynthesis in a *Microcystis* bloom increased initially during the 10 h sinusoidal light regime (Table 1). When irradiance was raised to 950 μmol photons m^{-2} s^{-1}, however, rates of photosynthesis and oxygen concentrations in the bloom started to decline, and the maximum rate of photosynthesis was shifted to deeper layers. In the second half of the light period the decline became more pronounced; photosynthetic activity was virtually absent in the top 800 μm of the bloom. The inhibition of photosynthesis, induced by exposure to high irradiance, did not recover when irradiance was reduced. The occurrence of photoinhibition was confirmed by a 50% drop in F_v /F_m. The ratio decreased as result of quenching of F_m, not as result of a rise in F_o (Figure 1). Increases in F_o have been related to damage of PS II resulting in slow

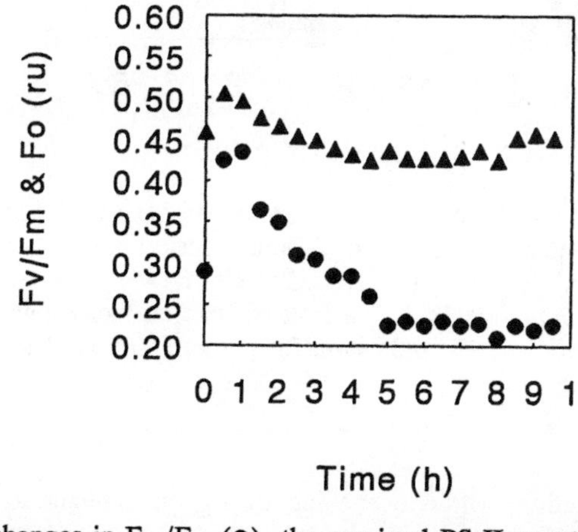

Fig. 1. Diurnal changes in F_v /F_m (●), the maximal PS II quantum yield, and F_o (▲), the minimal fluorescence yield (relative units) of the *Microcystis* bloom of Table 1.

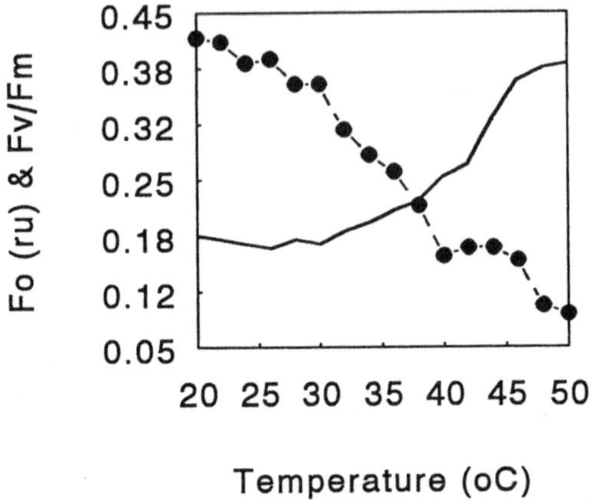

Fig. 2. Increase in F_o (relative units; solid line) of *Microcystis* cells during a 0.4 °C minute^{-1} rise in temperature, during which F_v/F_m (•) decreased.

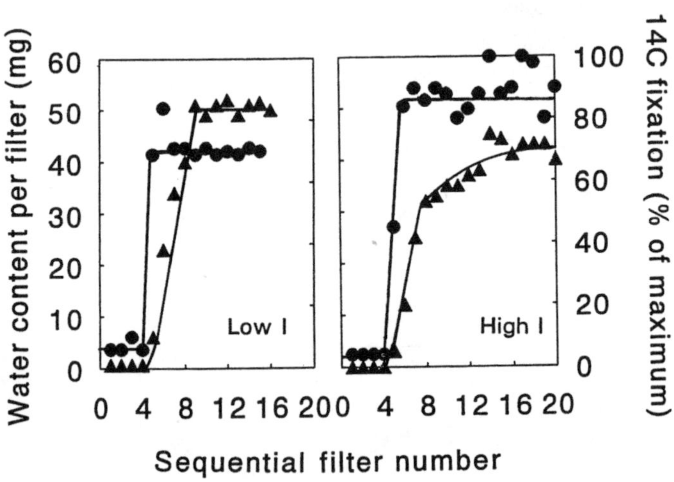

Fig. 3. Impact of desiccation on rates of ^{14}C fixation of *Microcystis* on filter. Left axis shows remaining water content of filters after 2 hours desiccation (▲), right axis rates of ^{14}C fixation (% of the maximum rate; normalized to Chl a (•)). Incubation at low or high irradiance (50 or 1500 μmol photons m^{-2} s^{-1}). Rates of carbon fixation were only hampered by virtually complete desiccation. No indication for photoinhibition of photosynthesis were found.

recovery from photoinhibition, while quenching of F_m, is thought to be indicative of increased thermal energy dissipation (Demmig-Adams and Adams, 1992). The absence of any recovery of F_v /F_m would nevertheless suggest that considerable damage to PS II occurred.

F_v /F_m of *Microcystis* cells decreased in response to a gradual rise in temperature (Figure 2); F_o started to rise from 30 °C onwards, the rate of increase gradually accelerated and saturated at 50°C. At 30 °C PS II began to loose its stability (T_c *sensu* Schreiber and Berry 1977); its disintegration was complete at 50 °C (T_p).

Table 1. Changes in maxium rate of gross photosynthesis (P), depth where this maxiumum rate was reached, integrated rates of gross photosynthesis, and integrated oxygen contents of a *Microcystis* bloom receiving a "stepped" sinusoidal light regime. Downwelling irradiance as indicated in the first column of the table

Irradiance /μmol photons m^{-2} s^{-1}	Maximum rate of gross P /mM O_2 h^{-1}	Depth of maximum rate of P /μm	Integrated gross rates of P /mmol O_2 m^{-2} h^{-1}	Integrated O_2 content /mmol m^{-2}
50	13.5	100	4.5	0.25
350	18	100	14.4	0.83
650	26	400	18.3	0.82
950	17	400	14.2	0.77
1250	12	1200	7.8	0.58
950	10	800	6.9	0.10
650	9	500	5.2	0.03
350	3.5	500	2	0.02
50	2	0	0.2	0.02

It was found that photosynthesis in *Microcystis* was resistant to desiccation. Rates of ^{14}C fixation were equal on fully wet and partially desiccated filters (Figure 3). Carbon fixation dropped only on filters where almost no water was detected (filter numbers 1-4). No indication of photoinhibition or an interaction between high irradiance and desiccation stress was found. The rates of ^{14}C fixation after the 1500 μmol photons m^{-2} s^{-1} incubation were similar to the values found at 50 μmol photons m^{-2} s^{-1} over the whole range of filters. Photoinhibition, however, is a time dependent process. It is possible that the 20 minutes incubation at high irradiance were insufficient to induce severe photoinhibition. Monitoring of variable fluorescence during the desiccation process also implied that PS II was desiccation resistant (Figure 4). F_v initially rose when filters started to desiccate. The 50 % higher value of F_v/F_m would suggest that desiccation initially improved PS II quantum yield. Fv rose due to an increase in F_m, which was only partially alleviated by a (small) rise in F_0. F_v declined abruptly when desiccation was almost complete. This decline occurred within the 2 hour desiccation period that was used in the ^{14}C fixation experiments on filter positions 1-4 only; cells on these filter-positions showed no subsequent carbon fixation.

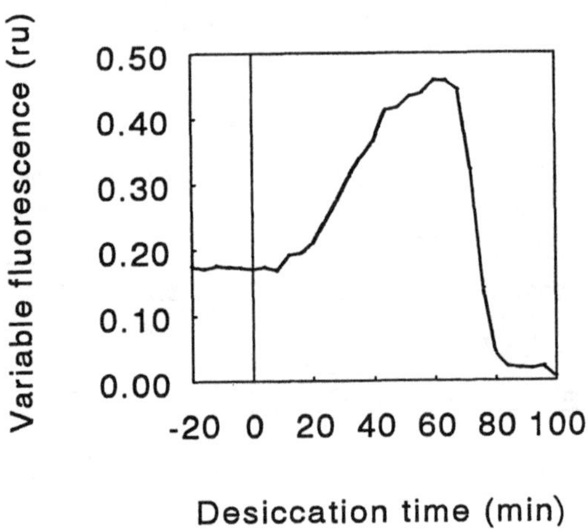

Fig. 4. Changes in variable fluorescence (F_v) during fast desiccation of *Microcystis* cells. Graph displays pattern for filter position 1 (see Fig.4). An initial rise of 50 % in F_v/F_m, resulting from the increase in F_v would suggest an improved functioning of PS II of cells on partially desiccated filters. When desiccation was nearly complete, F_v collapsed.

Discussion

In this paper the effects of high irradiance, elevated temperature, and desiccation on photosynthesis of the bloom forming cyanobacterium *Microcystis* were studied. When algae are exposed to an irradiance above that which can be used with maximum quantum efficiency of photosynthesis, excessive energy will be dissipated by PS II, either in the antenna or in the reaction centre itself (Krause and Weis 1991). When these mechanisms are saturated, net photoinhibition will occur. Under conditions where additional constraints on photosynthesis are absent, cells are expected to recover readily from photoinhibition (Demmig-Adams and Adams 1992). The extensive photoinhibition, measured as a reduction of rates of gross photosynthesis, and quenching of variable fluorescence (Table 1 and Figure 1), and the lack of recovery would suggest that stress factors other than high irradiance were present in the bloom. Phycobilisomes of cyanobacteria in lichens became dissociated from PS II during desiccation, resulting in an increase in F_o and drop in F_v (Bilger et al. 1989). Our results show that only virtually complete desiccation induced the decline in F_v (Figures 3 and 4), partial desiccation did not affect photosynthetic activity. It is likely that the slime layer surrounding the cells in the *Microcystis* colonies acted as a buffer during desiccation, as suggested by Shepard (1987).The initial rise of F_v and F_v/F_m during desiccation would suggest an increase in the quantum yield of PS II, although care must be taken to draw conclusions solely based on fluorescence measurements. Desiccation possibly induced a tighter interaction between PS II proteins and the lipids of the thylakoid membrane (see Havaux 1992).

The rise of F_o with an increase in temperature (Figure 2) has been described for higher plants. It was suggested that the fluorescence rise is the result of the uncoupling of the antenna from the reaction centre, although a change in the redox state of Q_A has also been mentioned (Bukhov et al. 1990). The moderate increase in temperature of the bloom presumably affected the PS II quantum yield of *Microcystis* cells in the bloom (as judged from the decrease of F_v/F_m with temperature), which in turn would increase the likelihood of photoinhibition. It seems improbable that an interaction between temperature and high irradiance alone can explain the severe photoinhibition. Both desiccation and temperature can, however, be more extreme under natural conditions (Zohary and Pais Madeira 1990) and the presence of additional stresses may well enhance photoinhibition and photooxidative death in lakes.

Alternatively, the extensive photoinhibition could be explained if the *Microcystis* cells that made up the bloom lacked sufficient mechanisms for dissipation of excess excitation energy. The xanthophyll zeaxanthin has been proposed to be a central mediator of the dissipation process (see Demmig-Adams and Adams 1992). Cyanobacteria synthesise zeaxanthin when grown under high irradiance (e.g. Ibelings et al. submitted), although they lack the eukaryotic xanthophyll cycle in which zeaxanthin can be formed rapidly. The *Microcystis* cells had a zeaxanthin/chlorophyll-a ratio of 0.03; which is comparable to the value found for a *Microcystis* culture receiving a light regime simulating deep mixing in lakes, but seven times lower than the value for cells receiving a sinusoidal light regime (Ibelings et al. submitted). This indicates that the *Microcystis* cells used to form the bloom had been subjected to low average light conditions in the lake, prior to sampling. In contrast Paerl et al. (1983) studied a population of *Microcystis* that was resistant to photoinhibition. The cells were found to contain a high zeaxanthin content, presumably because the population stayed within the top 0.5 m of the watercolumn throughout most of the summer. The *Microcystis* cells of our study were adapted to deep mixing, and were subsequently unable to cope with high irradiance as encountered in surface waterblooms, and suffered photoinhibitory damage as a result. Under such conditions surface waterblooms should be considered as a loss factor for the population.

Acknowledgements

This work was supported by a grant from the U.K. National Environment Research Council. We are grateful for discussions with F. Garcia-Pichel and A.E. Walsby.

References

Abeliovich A, Shilo M (1972) Photooxidative death in blue green algae. J Bact 111: 682-689

Bilger W, Rimke S, Schreiber U, Lange OL (1989) Inhibition of energy transfer to photosystem II in lichens by dehydration: Different properties of reversibility with green and blue-green phycobionts. J Plant Physiol 134: 261-268

Bukhov NG, Sabat SC, Mohanty P (1990) Analysis of chlorophyll a fluorescence changes in weak light in heat treated *Amaranthus* chloroplasts. Photosyn Res 23: 81-87

Demmig-Adams B, Adams III WW (1992) Photoprotection and other responses of plants to high light stress. Ann Rev Plant Phys Plant Molec Biol 433: 599-626

Havaux M (1992) Stress tolerance of Photosystem II *in vivo*. Plant Physiol 100: 424-432

Ibelings BW, Mur LR (1992) Microprofiles of photosynthesis and oxygen concentration in *Microcystis* scums. FEMS Microbiol Ecol 86: 195-202

Kooten van O, Snel JFH (1990) The use of chlorophyll fluorescence in plant stress physiology. Photosyn Res 25: 147-150

Krause GH, Weis E (1991) Chlorophyll fluorescence and photosynthesis; The basics. Ann Rev Plant Physiol Plant Molec Biol 42: 313-349

Paerl HW, Tucker J, Bland PT (1983) Carotenoid enhancement and its role in maintaining blue-green algal (*Microcystis aeruginosa*) surface blooms. Limnol Oceanogr 28: 847-857

Powles SB (1984) Photoinhibition of photosynthesis induced by visible light. Ann Rev Plant Physiol 35: 15-44

Revsbech NP, Ward DM (1983) Microelectrode studies of interstitial water chemistry and photosynthetic activity in a hot spring microbial mat. Appl Environm Microbiol 48: 270-275

Revsbech NP, Jørgensen BB, Brix O (1981) Primary production of microalgae in sediments measured by oxygen microprofile $H^{14}CO_3^-$ fixation and oxygen exchange methods. Limnol Oceanogr 25: 403-411

Reynolds CS, Walsby AE (1975) Waterblooms. Biolog Rev 50: 437—481

Schreiber U, Berry JA (1977) Heat induced changes of chlorophyll fluorescence in intact leaves correlated with damage of the photosynthetic apparatus. Planta 136: 233-238

Shepard KL (1987) Evaporation of water from the mucilage of a gelatinous algal community Br Phycol J 22: 181-185

Winder de B, Matthijs HCP, Mur LR (1989) The role of water containing substrate on the photosynthetic response of three drought tolerant phototrophic microorganisms isolated from a terrestrial habitat. Arch Microbiol 152: 458-462

Zohary T, Pais Madeira AM (1990) Structural physical and chemical characteristics of *Microcystis aeruginosa* hyperscums from a hypertrophic lake. Freshw Biol 23: 339-352

Denitrification, nitrification and nitrogen assimilation in photosynthetic microbial mats

Lars Peter Nielsen and Niels Peter Sloth

Department of Microbial Ecology, Institute of Biological Sciences, Bd. 540, University of Aarhus, DK-8000 Århus C, Denmark

Introduction

The mat building photosynthetic microbes need nitrogen to produce biomass. The actual availability of combined nitrogen depends on the balance between sources and sinks in the mat. Sources are dissolved nitrogen in the overlying water, nitrogen from mineralization processes, and nitrogen fixation, and sinks are burial of nitrogen, efflux of dissolved nitrogen, and denitrification.

In this paper we will discuss the ability of a mat community to minimize nitrogen loss by denitrification when nitrogen is scarce. The discussion is based on data from an experimental study of mixed cyanobacteria-diatom mats in marine mesocosms using the ^{15}N isotope pairing assay for denitrification.

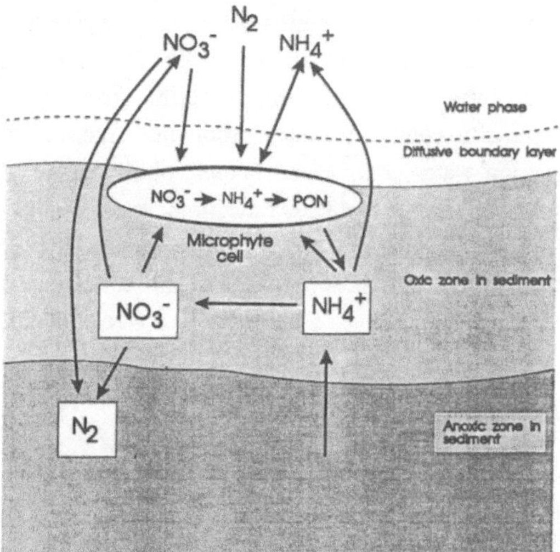

Fig. 1. Nitrogen cycling in a stratified system with benthic, photosynthetic microbes. See text for explanation.

NATO ASI Series, Vol. G 35
Microbial Mats
Edited by Lucas J. Stal and Pierre Caumette

First of all it is important to recognize that nitrogen processes like most other biological processes in mats are not running in a stable homogenous environment, but show marked spatial and diel variations linked to steep concentration gradients and light gradients. The general nitrogen cycle in a photosynthetic mat community is outlined in Fig. 1. Photosynthesis is restricted to the upper photic zone where O_2 producing microphytes assimilate combined nitrogen from the sediments below or from the overlying water. Ammonia is directly incorporated into organic compounds while NO_3^- and NO_2^- first have to be reduced to ammonia by assimilatory processes. The oxic zone where the nitrifying bacteria are active extends below the photic zone. The nitrification process is fueled by ammonia from mineralization processes in the oxic and anoxic zones or diffusing from the overlying water. Denitrification activity is restricted to the anoxic zone and must depend on the diffusion of NO_3^- or NO_2^- from the overlying water column or from the nitrification zone. The oxic zone becomes much more shallow during the night when no oxygen is produced in the sediment. Assimilation of nitrogen may proceed during the night, however.

Mesocosm experiment

Mesocosms enclosing 1.8 m^2 of sediment with the overlying water column of 4 m were set up in May 1991 in Knebel Vig, Denmark. Knebel Vig is a sheltered lagoon exchanging water with the Bay of Århus (average salinity 23 ‰). The sediment here is covered by a dense mat of cyanobacteria and diatoms (500 mg chlorophyll m^{-2}) which is highly resistant to resuspension (Sloth, Nielsen Riemann, subm.) and seemingly with no significant macrofauna grazing. Planktonic production is mostly nitrogen limited, Secci depth is 8 m on the average and concentrations of NO_3^- and NH_4^+ are always low ($< 2\ \mu M$) from April to September (County of Aarhus 1992). At the site benthic primary production exceeds the integrated primary production in the overlying water column and benthic phototrophic microbes are generally considered to be of prime importance to nutrient dynamics and O_2 balances in the coastal areas less than 15 m deep (Nielsen, Riemann & Sloth, subm.; Sundbäck 1986).

Nitrate was added to replicate mesocosms as an initial pulse of 30 μM (PULSE) or by multiple additions amounting 30 μM over a period of two weeks (MULTIPLE). A third pair of mesocosms was not given any addition (CONTROL). Sediment cores (10 cm^2) were retrieved from the bottom and incubated with and without light at *in situ* conditions in the laboratory. Sediment-water fluxes of O_2, CO_2, NO_3^- and NH_4^+ were determined from concentration changes in the overlying water phase during 1 hour incubations as described by Nielsen et al. (subm.) and Sloth et al. (subm.). Subsequently, $^{15}NO_3^-$ (30 μM) was added to the water phase and the rates of denitrification were determined from the production of single and double labelled dinitrogen according to the isotope pairing technique (Nielsen 1992). This technique allows the calculation of denitrification of both NO_3^- from the overlying water and NO_3^- from nitrification within the sediment.

The measurements presented here were all made one week after the initiation of the experiment. Benthic primary production was evident from high rates of O_2 release and CO_2 uptake in the light. No significant differences in respiration and production could be seen among the experimental mesocosms and all data are therefore pooled in Fig. 2.

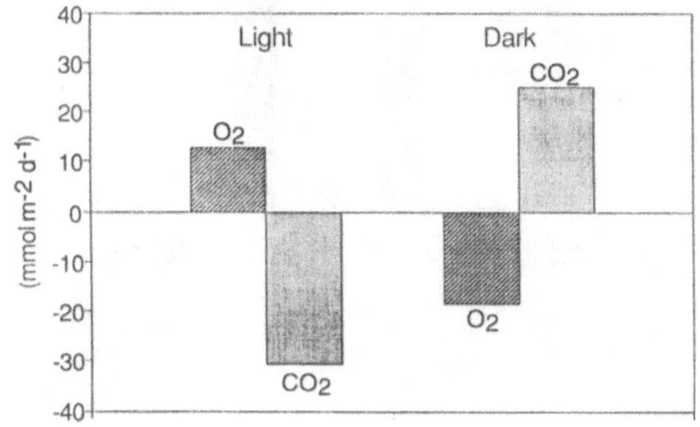

Fig. 2. Sediment-water fluxes of O_2 and CO_2 in light and darkness. Positive numbers denote net release from the sediment.

The NO_3^- concentration in the PULSE mesocosms had dropped to 2.6 μM at the time of sampling while the concentration in the MULTIPLE mesocosms was somewhat higher (4.4 μM) as the two-day dose was added shortly before sampling. Nitrate and NH_4^+ were generally taken up by the mat at high rates in the light and with a marked positive correlation with the concentration (Fig. 3). In the dark the fluxes were more scattered and NH_4^+ was generally released to the water column.

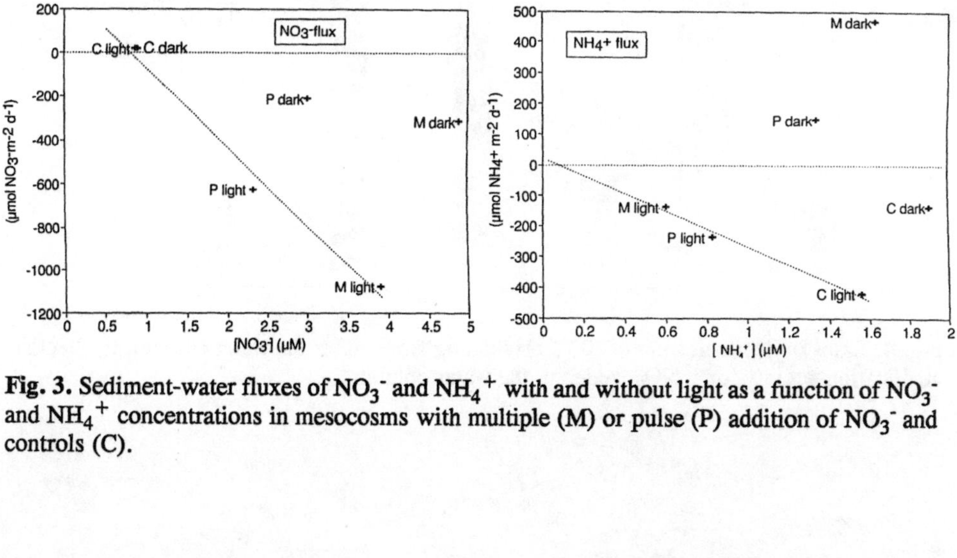

Fig. 3. Sediment-water fluxes of NO_3^- and NH_4^+ with and without light as a function of NO_3^- and NH_4^+ concentrations in mesocosms with multiple (M) or pulse (P) addition of NO_3^- and controls (C).

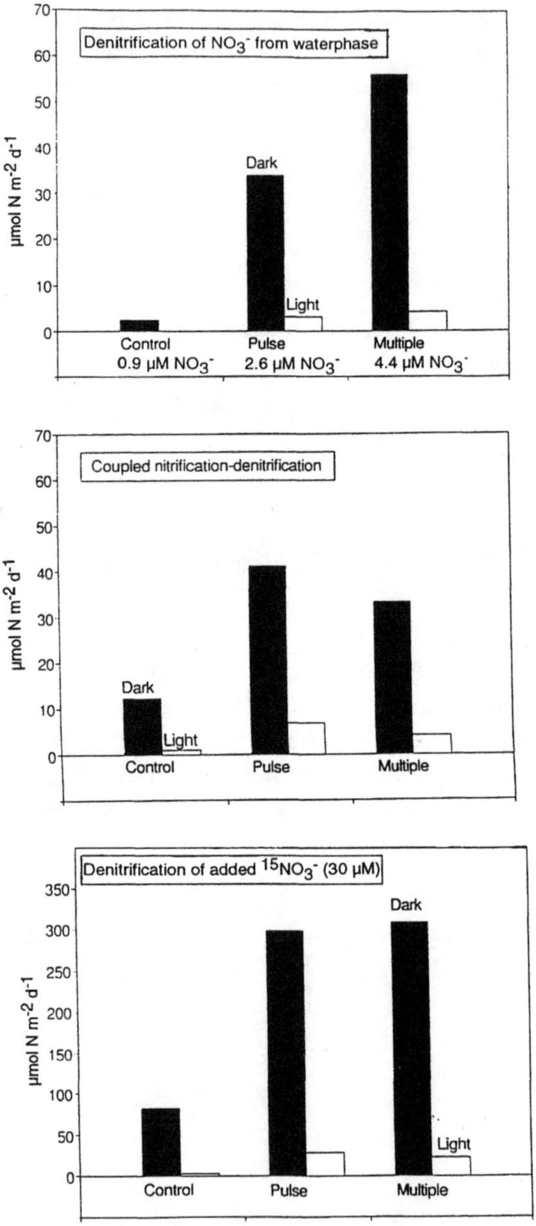

Fig. 4. Rates of denitrification of NO_3^- originating from either the water column, the benthic nitrification activity, or $^{15}NO_3^-$ added to the water column.

Denitrification rates were very small compared to the measured fluxes of O_2, CO_2, NO_3^- and NH_4^+ (Fig. 4). Both NO_3^- from nitrification in the mat and NO_3^- from the overlying water were important sources. The mesocosm with no NO_3^- added showed the lowest denitrification rates, and rates in the dark were generally 10 times higher than rates in light. Rates of denitrification of NO_3^- from the overlying water were dependent on the concentration of NO_3^-. The denitrification rate of the added $^{15}NO_3^-$ provided an independent measure of the "potential" for denitrification of NO_3^- from the overlying water (Fig. 4) and this showed a much higher potential in the mesocosms with NO_3^- added.

Discussion

The investigated mat community assimilated NO_3^- and NH_4^+ from the water column very efficiently as indicated by the high uptake rates in the light strictly correlated with the concentrations. The slopes for NO_3^- and NH_4^+ in the light were the same, and using Fick's first law of diffusion and the molecular diffusion coefficient of NO_3^- and NH_4^+ ($1.3\ 10^{-5}\ cm^2\ s^{-1}$; Li and Gregory 1974) it was calculated that the maximal diffusion path was 0.4 mm. This distance was the same as the thickness of the diffusive boundary layer in the stirred cores (Jens Gundersen, pers. comm.), thus indicating that the microphytes captured essentially every NO_3^- and NH_4^+ molecule touching the sediment. The net uptake was less in the dark, and for ammonia the release could actually exceed the uptake.

The rates of denitrification were generally small compared to rates measured in subphotic sediments with similar O_2 uptake out in the Bay of Aarhus (200-500 umol N $m^{-2}\ d^{-1}$; Blackburn & Henriksen 1983; Jensen et al. 1988; Nielsen in prep.). The presence of benthic microphytes apparently limited denitrification activity and the much higher NO_3^- uptake rate compared to the denitrification rate directly suggested a competition between assimilative and dissimilative nitrogen consumption processes. Looking at the stratification of the processes (Fig. 1) it is not surprising that NO_3^- assimilating microbes in the photic zone should reduce the flux of NO_3^- from the overlying water to the anoxic denitrification zone. The measurements of denitrification of added ^{15}N also showed that the mats in the NO_3^- enriched mesocosms had become less effective as NO_3^- filters, indicating partly nitrogen saturation of the microphytes. Light also had a negative effect on coupled nitrification-denitrification apparently due to increased nitrogen assimilation by the phototrophs in the light. This is supported by the observation that NO_3^- additions to the mesocosms enhanced coupled nitrification-denitrification. Other studies in a nitrogen rich stream, estuary and lake environments showed that nitrification and hence coupled nitrification-denitrification could be stimulated by benthic photosynthesis due to increased availability of O_2 (Risgaard-Petersen et al. 1994; Anette Pind unpublished). It is not obvious, however, why the rates of denitrification of NO_3^- from the nitrification zone were so much affected by the phototrophs in the present study. The coupling of nitrification and denitrification is associated with the oxic-anoxic interface below the photic zone, thus having a prior access to ammonia from the anoxic zone, and the efflux of ammonia in the dark and the high oxygen consumption rate indicated a significant production of ammonia in the sediment. A number of other studies in nitrogen poor mat environments have generally shown very low nitrification activity, indicating that benthic microphytes may limit the nitrification activity by other means than just reducing the availability of nitrogen (manuscripts in prep.). The actual interactions of benthic photosynthesis and nitrification might be resolved when NO_3^- and NH_4^+ microsensor are applied to microbial mats (Revsbech ibid).

Acknowledgements

This work was a part of the SAFE project (EEC MAST program contract 0020). We wish to thank Sten P. Andersen and Preben G. Sørensen for excellent analytical work and assistance in the field. Also thanks to Bo Riemann for organizing the mesocosm project.

References

Blackburn TH, Henriksen K (1983) Nitrogen cycling in different types of sediments from Danish waters. Limnol Oceanogr 28: 477-493

County of Århus (1992) Hydrography, nutrients and plankton in Aarhus Bay 1990-1991 (In Danish). Århus.

Jensen MH, Andersen TK, Sørensen J (1988) Denitrification in coastal bay sediment: regional and seasonal variation in Aarhus Bight, Denmark. Mar Ecol Prog Ser 48: 155-162

Li Y-H, Gregory S (1974) Diffusion of ions in sea water and in deep sea sediments. Geochim Cosmochim Acta 38: 703-714

Nielsen LP (1992) Denitrification in sediment determined from nitrogen isotope pairing. FEMS Microb Ecol 86: 357-362

Risgaard-Petersen N, Rysgaard S, Nielsen LP, Revsbech NP (1994) Diurnal variation of denitrification based on NO_3^- from overlying water and of coupled nitrification-denitrification in sediments colonized by benthic microphytes. Limnol Oceanogr In press

Sundbäck K (1986) What are the benthic microalgae doing on the bottom of Laholm Bay? Ophelia, suppl. 4: 273-286

Nitrogen fixation dynamics in microbial mats

Hans W. Paerl, Brad M. Bebout [1], Carolyn A. Currin, Matthew W. Fitzpatrick and James L. Pinckney

Institute of Marine Sciences, University of North Carolina at Chapel Hill, Morehead City, NC 28557 USA

Introduction

(N) limitation is a widespread feature of estuarine and oceanic waters (Dugdale 1967; Carpenter and Capone 1983). Despite this, benthic microbial mats frequently flourish in geographically-diverse intertidal and subtidal sand/mud flat, lagoon, reef, marsh and mangrove habitats characterizing such waters (Cohen et al. 1984; Cohen and Rosenberg 1989). Mats are particularly prolific in oligotrophic tropical and subtropical waters, where an appreciable portion of the primary production can be attributed to benthic microalgae (Whitton and Potts 1982; Bauld 1984). The remarkable success of mats in these N-limited environments has been attributed to the ability of specific groups of mat microorganisms (cyanobacteria, photosynthetic-, heterotrophic- and lithotrophic bacteria) to "fix" (reduce) atmospheric N (N_2), thereby providing biologically-available N (NH_3) to mat flora and fauna (Wiebe et al. 1975; Potts and Whitton 1977; Carpenter et al. 1978; Paerl et al. 1981; Stal et al. 1984). Nitrogen fixation is often the sole biologically-derived source of "new" N supporting primary production in these waters. In the face of N loss (as N_2) in mats (and other ecosystem components) via denitrification (Joye and Paerl 1993) and to a lesser extent ammonification, N_2 fixation is also of importance in regulating flux and mass balance of this important nutrient.

The rates, distribution, and environmental controls on N_2 fixation are critical determinants of mat primary and secondary production. The importance of N_2 fixation in mat N budgets and fluxes is addressed in a companion manuscript by Bebout et al. (This Volume). Here, we will examine the functional groups of microorganisms responsible for mat N_2 fixation in relation to environmental factors controlling this process at scales relevant to mat microhabitats. The objective of this overview is to examine the physiological ecology of mat diazotrophs as it pertains to mat biogeochemical gradients, carbon (energy) and nutrient cycling processes discussed elsewhere.

---

[1] Max Planck Institute for Marine Microbiology , Fahrenheitstraße 1, D-28359 Bremen, Germany.

N_2 fixers in microbial mats

Nitrogen fixation is a prokaryotic process, confined to specific eubacterial and cyanobacterial genera (Stewart and Gallon 1980; Bothe 1982; Paerl 1990) . The enzyme complex, nitrogenase, responsible for the conversion of N_2 to NH_3, only functions under anoxic conditions, forcing confinement of N_2 fixation to obligate anaerobes, microaerophiles, and, among oxygenic cyanobacteria in O_2-devoid cells (heterocysts) or intra-cellular regions supporting localized O_2 depletion (Stewart and Gallon 1980). In cell- free extracts, nitrogenase is irreversibly inactivated when exposed to O_2 (Yates 1977). In intact cells, nitrogenase exists in active and inactive forms (Stal and Bergman 1990; Zehr et al. 1993). Diazotrophs appear to rapidly alter these forms in response to ambient O_2 concentrations. In this manner, nitrogenase may be inactivated when unfavorably high O_2 conditions prevail, without destruction of the enzyme; conversely, under anoxic conditions previously-inactivated nitrogenase may be reactivated or new nitrogenase synthesized. The ability to interchange active/inactive forms of nitrogenase may be instrumental in ensuring rapid commencement of nitrogenase activity, once favorable conditions (e.g. low $[O_2]$) exist.

Given metabolic flexibility in the face of rapid shifts in vertical oxygen gradients (Jørgensen et al. 1979; Bebout et al. 1987; Paerl et al. 1989), mats provide diverse and dynamic ecological niches for diazotrophs. Indeed, microbiological (culturing and isolation), immuno-chemical (polyclonal antibodies specific for nitrogenase) and molecular (southern blotting, probing, polymerase chain amplification of N_2 fixing genes (*nif*H,D,K) have revealed a diverse assemblage of microbes capable of N_2 fixation (Guerinot and Colwell 1985; Currin et al. 1990; Kirshtein et al. 1991, 1993). The list of N_2-fixing taxa will undoubtedly grow as immunochemical, molecular probing, and characterization (sequencing) techniques are more widely deployed.

Environmental controls on N_2 fixation in mats

The above discussion suggests that pO_2 should be an effective regulator of N_2 fixation. Laboratory and field studies have shown this to be true (Stal et al. 1984; Paerl et al. 1989; Fay 1992). However, given rapid (on the order of minutes) changes in mat vertical pO_2 and redox gradients in conjunction with alterations in nutrient exchange with overlying water, additional environmental constraints on N_2 fixation are possible. These include; 1) nutrient limitation (most often mentioned are P, Fe, Mo, CO_2), 2) ammonium (end product) suppression of nitrogenase, 3) temperature (diel and seasonal) alterations, and 4) in the case of motile cyanobacteria and photosynthetic bacteria, irradiance. Ironically, while adequate irradiance is essential for supporting phototrophic N_2 fixation, it can also exert a strong control. For example, vertical attenuation of light confines certain phototrophs to depths where appropriate spectral components can be exploited using specialized photopigment systems (Kühl and Jørgensen 1992). Furthermore, the extent of light penetration often dictates the shape and depth of O_2 gradients and delineates oxic/anoxic regimes. Biotic interactions, including mutualistic exchange of metabolites, grazing, and symbioses may also impact mat N_2 fixation potential. Given close interactions among these variables, we will discuss the following environmental controls on N_2 fixation in an interrelated, synergistic manner.

Table 1. N_2-fixing functional microbial groups inhabiting mats.

Functional group	Representative genera
Photosynthetic	
Anoxygenic (Bacteria)	*Chlorobium, Chromatium*
Oxygenic (Cyanobacteria)	
- Heterocystous filamentous	*Calothrix, Scytonema*
- Nonheterocystous filamentous	*Oscillatoria, Lyngbya*
	Microcoleus, Plectonema
- Unicellular	*Gloeocapsa, Synechococcus*
Chemolithotrophic (Bacteria)	*Thiobacillus*
Heterotrophic (Bacteria)	
Aerobic	*Azospirillum, Azotobacter*
Facultative anaerobes	*Klebsiella, Vibrio*
Anaerobic	*Clostridium, Desulfovibrio*

Irradiance

Irradiance exerts a strong control on photosynthetic production of organic matter, which provides carbon (reductant) and energy (ATP) needed to support N_2 fixation. However, O_2 produced by oxygenic photosynthesis is a potent inhibitor of N_2 fixation (Yates 1977; Fay 1992). In nature, the positive (C supply) and negative (O_2 evolution) aspects of photosynthesis interact to control diel rates of N_2 fixation, as witnessed in mat systems dominated by non-heterocystous cyanobacteria (Stal et al. 1984; Bebout et al. 1987; Villbrandt et al. 1990). Photosynthetic production potentially exerts another negative control on N_2 fixation, by competing for similar essential nutrients, including P, Fe, and trace metals. There are also competitive interactions between electron carriers used for energy production and photosynthesis (Schrautemeier et al. 1984).

Temperature

Nitrogen fixation, which is an enzyme (nitrogenase) catalyzed process, can be regulated by ambient temperatures, on diel or seasonal bases. Rapid diel temperature shifts may influence photosynthetic (dark reaction), metabolic and biosynthetic rates among phototrophs and heterotrophs. Seasonally, temperature regimes influence diazotrophic community composition. In temperate regions, successional shifts within and among phototrophic, heterotrophic and chemolithotrophic communities frequently accompany seasonal temperature changes (Polimeni 1976; Bautista and Paerl 1985; Pinckney and Zingmark 1993; see Fig. 4). Here, temperature may play joint regulatory roles with irradiance (i.e. low vs. high sun angles, variable daylength), and salinity ("freezing out" of salts in winter and enhanced evaporation during summer) alterations.

Nutrient limitation

Structurally and functionally, N_2 fixation has specific elemental requirements. The energetic (ATP) demands for phosphorus (P) can be high when periods of enhanced production (and biomass accumulation) coincide with N limitation, creating relatively high demands for N_2 fixation. P limitation may, at times, be operative under these circumstances, as shown in short-term (1-5 day) nutrient addition experiments (Paerl et al. 1993b). P limitation is, however, infrequent, confined to specific habitats (e.g., high carbonate, high irradiance tropical reef and lagoonal systems), and yet to be substantiated in temperate mats (Paerl et al. 1993b). It appears that, at least in the short term (i.e. days to weeks), endogenous P stores and regeneration, as well as "new" external P inputs (water column sedimentation, runoff, ground water) are often sufficient for meeting mat P requirements.

As structural components (cofactors) of nitrogenase, iron (Fe) and molybdenum (Mo) are nutrients essential for N_2 fixation. These elements have received considerable scrutiny as a potential limiting factors of aquatic N_2 fixation (Howarth et al. 1988; Rueter et al. 1992). Adequate Fe supply may be important for sustaining optimal N_2 fixation among certain freshwater planktonic cyanobacterial bloom populations (Wurtzbaugh and Horne 1983). Recently, we (Paerl et al. submitted) found N_2 fixation and growth in the marine planktonic non-heterocystous bloom former *Trichodesmium* to periodically be controlled by Fe supply. Assessments of Fe limitation of N_2 fixation in benthic microbial mats encountered in similar waters, and other geographically-diverse regions (Baja California, Tomales Bay, Northern California, Exuma and San Salvador Islands, Bahamas) yielded uniformly negative results (Paerl et al. 1987; 1993b). Fe may be sequestered (from overlying waters) and recycled. Similarly, no evidence for Fe limitation of photosynthetic CO_2 fixation could be obtained.

Molybdenum (Mo) was suggested as a possible limiting nutrient for marine N_2 fixation by Howarth and Cole (1985). The evidence for Mo limitation is, however, indirect and based on the observation that seawater contains relatively high concentrations of SO_4^{2-} ($\approx$28 mM), a structural analog of the commonly available form of Mo, MoO_4^{2-} (Howarth and Cole 1985). Uptake competition has been observed between SO_4^{2-} and MoO_4^{2-}, to the extent that MoO_4^{2-} uptake may be reduced by this interaction in seawater (Cole et al. 1993). While Mo uptake may be negatively affected by high ambient SO_4^{2-}, there is no evidence that the rather small cellular requirements for Mo are compromised (Paerl et al. 1987; Paulsen et al. 1991) We have exhaustively searched for potential Mo limitation among diverse planktonic and benthic N_2-fixing assemblages, using short-(1 week) and long-term bioassays, with no support for Mo limitation. In addition, Collier (1985) and others found soluble Mo acted as a conservative nutrient in open ocean water, indicating no localized depletion of this nutrient.

Mats examined thus far generally fail to exhibit short-term (days to weeks) inorganic nutrient (P, metals) limitation of N_2 fixation (Paerl et al. 1993b). In contrast, additions of organic matter (DOM), in the form of sugars, sugar alcohols and organic acids, can readily and rapidly (within 24 h) stimulate mat N_2 fixation (Paerl et al. 1987; 1993b) (Fig. 1). Stimulation may be attributable to DOM assimilation for growth and energy needs among heterotrophs. Stimulation may also be indirect, since DOM enhances mat respiration (O_2 consumption), which in turn could relieve O_2 inhibition of nitrogenase. Both means of enhancing N_2 fixation may operate contemporaneously.

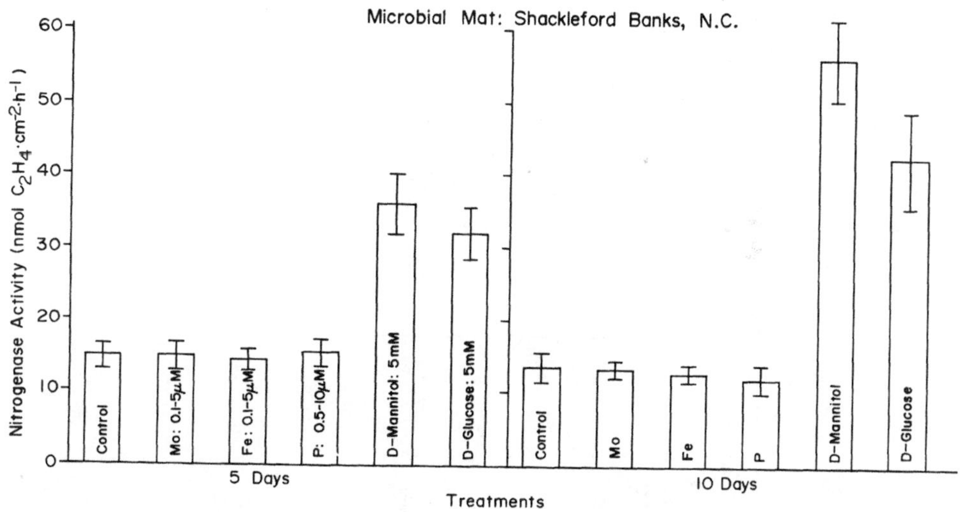

Fig. 1. Impacts of a variety of inorganic and organic nutrient additions (5 and 10 days after addition) on nitrogenase activity in an intertidal, cyanobacteria-dominated mat (Shackleford Banks, NC). Controls are untreated. Means and standard deviations of triplicate samples are shown.

Ammonia/Ammonium end product suppression of N_2 fixation

Nitrogenase synthesis and activity, are regulated by intracellular N concentrations (Gallon and Stal 1992). For gene expression (most thoroughly studied in *Klebsiella pneumoniae*), there is a complex interaction between several operons (*nif, ntr* and *pln*) with several sites of sensitivity to intracellular combined N which collectively provide a mechanism of induction during N deficiency and repression during N sufficiency. Similarly, N_2 fixation in sediments can be inhibited by as little as 50 μM (but generally higher) NH_4^+ (Howarth et al. 1988), although Bebout et al. (1987) observed high rates of N_2 fixation in mats containing >100 μM NH_4^+.

Oxygen inhibition as an environmental control of N_2 fixation

Mat primary production is dominated by oxygenic cyanobacteria and eukaryotic microalgae (diatoms, chlorophytes, euglenophytes and chrysophytes). N_2 fixing cyanobacteria are significant contributors to O_2 supersaturation observed during daylight hours (Jørgensen et al. 1979; Bebout et al. 1987; Paerl et al. 1989). Ironically, the cyanobacterial N_2-fixing community is often dominated by non-heterocystous filamentous and coccoid genera (*Oscillatoria, Microcoleus, Lyngbya, Schizothrix, Synechococcus, Gloeocapsa*). The frequent scarcity of heterocystous genera is mysterious, since it has been shown that daytime O_2 evolution inhibits nitrogenase activity in mats (Stal et al. 1984; Bebout et al. 1987; Paerl et al. 1989; Villbrandt et al. 1990). Diel assessments of mat N_2 fixation and photosynthesis illustrate the "O_2 problem". These studies, conducted on heterocystous and non-heterocystous cyanobacterial mats in the same saltmarsh, show contrasting patterns of activity (Fig. 2). While heterocystous (*Scytonema, Calothrix*) mats exhibit daytime maxima of CO_2 fixation (or O_2 evolution) and N_2 fixation, non-heterocystous mats reveal contrasting patterns, with N_2 fixation maxima confined to either early morning (Stal et al. 1984; Villbrandt et al. 1990), early evening or nighttime periods (Bebout et al. 1987; Paerl et al. 1989).

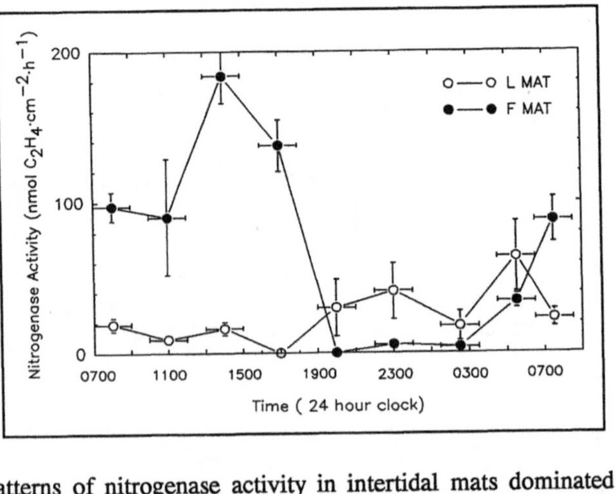

Fig. 2. Diel patterns of nitrogenase activity in intertidal mats dominated by filamentous heterocystous (F mat) vs. non-heterocystous (L; *Lyngbya* mat) cyanobacteria. Mats were located in a saltmarsh system near Morehead City, N.C. and assayed under *in situ* irradiance and temperature during May 1991.

Physiological and molecular evidence for *in situ* O_2 inhibition of mat nitrogenase activity also exists. The most obvious experimental evidence for this inhibition is the sudden and often dramatic increase in nitrogenase activity (NA) in previously-illuminated non-heterocystous cyanobacterial mats treated with the photosystem II (O_2 evolution) inhibitor 3-(3,4-dichlorophenyl)-1,1-dimethylurea (DCMU; 10^{-5}M) (Fig. 3). In the presence of DCMU, mat pO_2 drops dramatically because O_2 production stops while O_2 consumption continues. This drop is accompanied by NA stimulation. Longer (days) exposure to DCMU leads to a cessation of N_2 fixation, which has been attributed to eventual exhaustion of photoreductant (Bebout et al. 1993).

Immunochemical (western blots) probing studies of nitrogenase indicate the presence of active and inactive forms in mat diazotrophs (Stal and Bergman 1990). During daylight, the inactive form tends to increase relative to the active form. At night, a larger proportion of the total nitrogenase is active, resulting in enhanced N_2 fixation rates relative to daylight periods (Villbrandt et al. 1992).

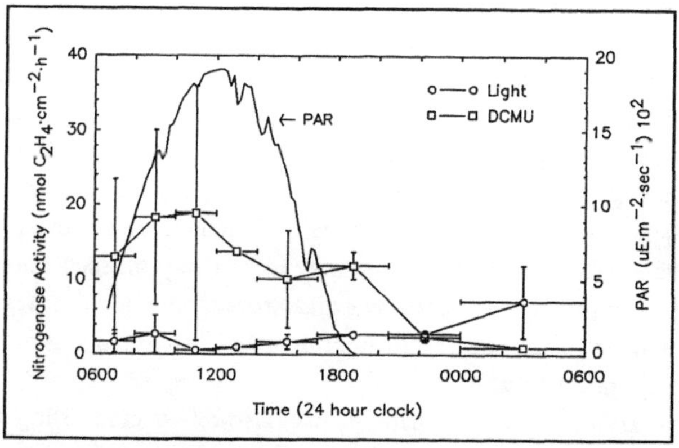

Fig. 3. Impacts of DCMU (10^{-5}M) (□) on nitrogenase activity in a non-heterocystous (*Microcoleus, Lyngbya*) cyanobacterial mat, located on Carrot Island, NC, USA. Untreated controls are (O). Daily PAR flux is shown. Standard deviations of triplicates and length of incubation periods are shown. Time is on a 24 h clock.

Evaluating N_2 fixation dynamics in mat communities

Uncertainty exists as to microorganisms responsible for observed diel and seasonal patterns of NA in microbial mats. There is little doubt that heterocystous and numerically-dominant non-heterocystous cyanobacterial genera such as *Microcoleus, Oscillatoria, Lyngbya,* and *Schizothrix* are significant contributors to fixed N inputs (Stal et al. 1984; Bebout et al. 1993). However, the fact that nighttime (or daytime dark-incubated, daytime DCMU- incubated) N_2

fixation frequently occurs in inter- and subtidal mats, suggests that microorganisms utilizing heterotrophic metabolism for supporting NA are an important component of mat N_2 fixation dynamics. Potentially-important heterotrophs include anaerobic and microaerophilic bacteria (*Klebsiella, Clostridium, Vibrio, Pseudomonas, Spirillum*) (Table 1).

The potentially-significant roles of associated eubacteria as diazotrophs should also not be ruled out. Culturing and molecular (*nif* H) probing studies (Paerl et al. 1989; Kirshtein et al. 1991, 1993; Zehr et al. in prep.) reveal diverse N_2-fixing eubacteria in mats. Immunofluorescence (using polyclonal antibodies to nitrogenase) and microautoradiographic studies show intimate relationships between mat cyanobacteria and eubacteria (Paerl et al. 1993a). In some instances N_2-fixing eubacteria were found tightly attached to N_2 fixing cyanobacteria. Microautoradiographs show bacteria associated with cyanobacteria (sometimes within cyanobacterial sheaths) assimilate sugars, organic acids and amino acids (Paerl et al. 1993a).

Recent studies (Paerl et al. 1989; Bebout et al. 1993) indicate that newly synthesized organic carbon compounds produced during daytime photosynthetic periods are a source of reductant supporting nighttime N_2 fixation in mats dominated by non-heterocystous cyanobacteria and a variety of eubacteria. This points to accumulated photosynthate as the energy source for day and nighttime N_2 fixation. Possibly transfer of metabolites (e.g., fixed C from cyanobacteria to eubacteria, which in turn utilize C to enhance O_2 consumption and support N_2 fixation) may in part reflect the observed dark-stimulated N_2 fixation in mats. Year-round studies in NC, USA lagoonal mats show seasonal differences in diel patterns magnitudes of light- vs. dark-mediated N_2 fixation (Fig. 4) and seasonal differences in the extent to which DOC enrichment stimulates NA (not shown). Both may in part rely on successional shifts among phototrophic and heterotrophic members of mat diazotrophic communities. Dark-dominated N_2 fixation typifies summer through fall, while light-dominated N_2 fixation is more common from winter through early spring (Fig. 4).

Seasonally-varying N_2 fixation patterns may also reflect contrasting photoperiods and irradiance intensities. Microscopic observations reveal maximal biomass of cyanobacteria during spring-summer months, while non-photosynthetic (presumably heterotrophic) bacteria and diatoms assume a greater share of total microbial biomass during winter. We are currently assessing phototrophic community shifts using photosynthetic accessory pigments (i.e. fucoxanthin, diatoxanthin, zeaxanthin) as chemotaxonomic markers (Millie et al. 1993).

With respect to potential heterotrophic contributors to mat N_2 fixation, a potentially large group of heterotrophic N_2 fixers are the SO_4^{2-} reducing bacteria (i.e. *Desulfovibrio*). Considering the proposed SO_4^{2-} : MoO_4^{2-} hypothesis of Howarth and Cole (1985), it would be ironic if SO_4^{2-} reducers proved to be significant N_2 fixers in mats. Given the recent availability of phylogenetic (16S RNA) and functional DNA/RNA probes for these heterotrophs, coupling of these probes to *nif* probes would prove a powerful combined tool for testing this possibility. The role of SO_4^{2-} reducers as potentially-significant N_2 fixers might be additionally surprising, since the anoxic regions of mats in which these microorganisms are found also contain high amounts of pore water NH_4^+, a known suppressor of NA in cultures. Spatial distributions (on the microscale) of NH_4^+ in the mat matrix may however be quite heterogeneous, allowing diazotrophy to coexist (in NH_4^+- devoid microzones) with relatively high bulk phase NH_4^+ levels.

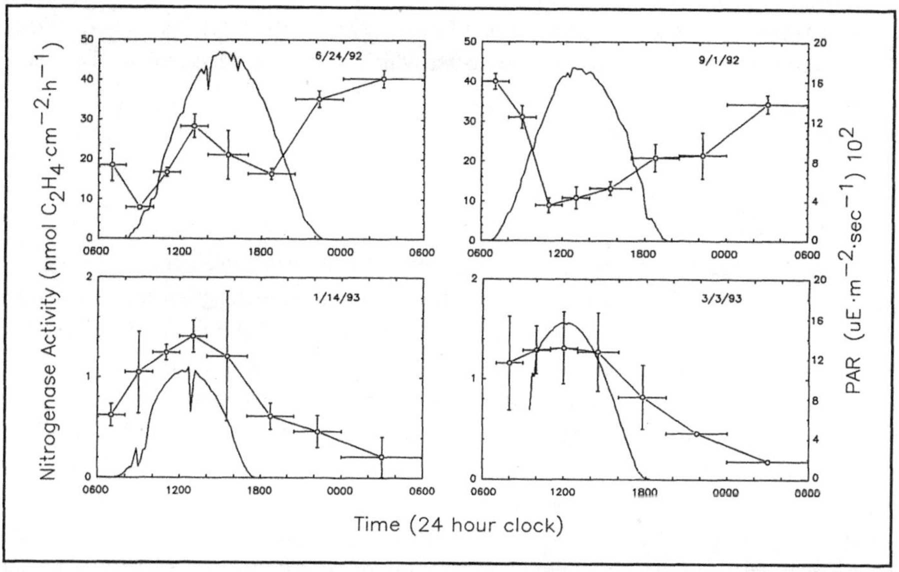

Fig. 4. Seasonal differences in diel patterns of nitrogenase activity in an intertidal cyanobacterial mat ystem, Carrot Island, NC, USA. Daily PAR curves, S.D. values and length of incubation periods (under natural conditions) are shown.

In attempting to identify all the "players" involved in mat N_2 fixation dynamics, anoxic phototrophs, which include the photosynthetic sulfur (H_2S) oxidizing bacteria (*Chromatium, Chlorobium, Rhodospirillum*) and chemolithotrophs should be considered. These eubacterial groups thrive in distinct biogeochemical zones, often only a few microns below the oxic-anoxic interface. While these diazotrophic groups are considered to be remnants of the O_2-devoid Precambrian era, when anoxygenic photosynthesis and chemolithotrophy were dominant means of primary production, modern-day mats, especially those exhibiting high rates of primary production, organic matter deposition and degradation provide a range of habitats for these autotrophic N_2 fixers along vertical O_2 gradients. The preferred location for anoxic phototrophs is the uppermost region of the anoxic zone, where adequate illumination exists for net photosynthetic production; often evident as a thin pink or green layer just above the sulfide-rich anoxic zone. Because they are frequent and numerically-significant inhabitants of such mats, photosynthetic bacteria may be significant N_2 fixers. They may also be responsible for observed DCMU-stimulated NA. Photosynthesis by phototrophic bacteria, which lack PS II, is insensitive to DCMU. The observed stimulation of NA might occur in response to increased anoxia throughout the mat following DCMU additions, thus enlarging the niche in which photosynthetic bacteria can thrive.

Photosynthetic bacteria, as well as many cyanobacteria, are also known to possess light-stimulated means of taking up dissolved organic matter. This DCMU-insensitive process, referred to as photoheterotrophy (Neilson and Lewin 1974; Paerl 1991; Paerl et al. 1993a), could enhance mat assimilation of organic matter which in turn could stimulate N_2 fixation potentials.

Concluding remarks

A suite of compelling, and in some ways, paradoxical ecological, evolutionary and biogeochemical questions emerge as we evaluate our current knowledge of mat N_2 fixation dynamics and their environmental controls. An obvious and pressing question is: given what appears to be an indispensible and persistent need for N_2 fixation as a source of "new" N for supporting mat production, combined with the strong environmental constraint that O_2 represents, why don't heterocystous cyanobacteria play a more dominant role as diazotrophs? Are there functional or structural constraint precluding heterocystous cyanobacterial dominance? These questions will be addressed by Stal et al. (this issue).

Steady progress is being made at identifying and characterizing (as to functional roles and environmental constraints) the entire suite of phototrophic, heterotrophic and chemolithotrophic microorganisms responsible for temporal and spatial patterns and overall rates of N_2 fixation in mats. Applications of immunochemical (species and taxon-specific antibodies coupled to fluorochromes and immunogold EM methods), molecular (species and group-specific DNA/RNA probes, polymerase chain reaction (PCR), coupled to degenerate primers, subsequent sequencing of amplified and cloned fragments of *nif* genes) (chapters by Stahl and Ward in this issue) and microscale physiological measurements (microelectrodes, redox indicators, tetrazolium salts, microautoradiography) (chapters by Jørgensen and Revsbech in this issue) will help characterize microbial N_2-fixing consortia and their environmental limitations/controls in the context of mat N cycling. A particularly powerful approach is to utilize these techniques in a contemporaneous, complimentary manner for addressing questions concerning the spatial (along physical-chemical gradients) and functional roles (contributions to community production, N_2 fixation and N cycling, mineralization, etc.) of mat microorganisms utilizing autotrophic and heterotrophic metabolisms.

The prospects for utilizing functional (process-related) and phylogenetic (16S RNA) nucleic acid probing and PCR characterization of resident microflora in parallel with microscale measurements (microelectrodes, microautoradiography) of biogeochemical zonation in mats are excellent (Stahl; this issue). A particularly exciting aspect of this approach will be the use of fluorescent or radiolabeled cell-specific probing, which will complement non-destructive physiological indicators such as microautoradiography. This approach will enable us to visualize "players" in their respective microhabitats. Relevant "target" communities for such efforts include sulfate reducers, methanogens, a variety of microheterotrophs, and cyanobacteria able to exploit both anoxygenic and oxygenic photosynthesis. Recent methodological refinements coupled to field applications of the abovementioned molecular, microsensor and microphysiological techniques have opened up numerous avenues for their deployment in natural mat communities. These technologies have become critically-important items in the biogeochemist's and ecologist's "toolbox", as functional and structural complexities characterizing microbial mats are elucidated and integrated in the context of production and nutrient cycling dynamics.

Acknowlegements

We thank L. Stal and others for critically reviewing this manuscript. Research support was provided by the US National Science Foundation, Projects OCE 9012496 and OCE 9115706 as well as NOAA-Coastal Ocean Program/Sea Grant Project R/EHP-1.

References

Bauld J (1984) Microbial mats in marginal marine environments: Shark Bay, Western Australia, and Spencer Gulf, South Australia. In: Cohen Y, Castenholz RW, Halvorson HO (eds) Microbial Mats: Stromatolites. Alan R Liss, Inc., New York, pp 39-58

Bautista MF, Paerl HW (1985) Diel N_2 fixation in an intertidal marine cyanobacterial mat community. Mar Chem 16: 369-377

Bebout BM, Paerl HW, Crocker KM, Prufert LE (1987) Diel interactions of oxygenic photosynthesis and N_2 fixation (acetylene reduction) in a marine microbial mat community. Appl Environ Microbiol 55: 369-384

Bebout BM, Fitzpatrick MW, Paerl HW (1993) Identification of the sources of energy for nitrogen fixation and physiological characterizations of a marine microbial mat community. Appl Envir Microbiol 59: 1495-1503

Bothe H (1982) Nitrogen fixation. In: Carr NG, Whitton BA (eds) The Biology of Cyanobacteria. Blackwell Scientific Publications Ltd , Oxford, pp 87-104

Carpenter EJ, Capone DG (1983) Nitrogen in the Marine Environment. Academic Press Inc, New York

Carpenter EJ, Van Raalte CD, Valiela I (1978) Nitrogen fixation by algae in a Massachusetts salt marsh. Limnol Oceanogr 23:318-327

Cohen Y, Castenholz RW, Halverson HO (eds) (1984) Microbial Mats: Stromatolites. Alan R Liss Inc. New York

Cohen Y, Rosenberg E (1989) Microbial Mats: Physiological Ecology of Benthic Microbial Communities. Americ Soc Microbiol Washington DC

Cole JJ, Lane JM, Marino R, Howarth RW (1993) Molybdenum assimilation by cyanobacteria and phytoplankton in freshwater and salt water. Limnol Oceanogr 38: 25-35

Collier RW (1985) Molybdenum availability in the Northeast Pacific Ocean. Limnol Oceanogr 30: 1351-1354

Currin CA, Paerl HW, Suba GR, Alberte RS (1990) Immunofluorescence detection and characterization of N_2 fixing microorganisms from aquatic environments. Limnol Oceanogr 35: 59-71

Dugdale RC (1967) Nutrient limitation in the sea: Dynamics, identification and significance Limnol Oceanogr 12: 685- 695

Fay P (1992) Oxygen relations of nitrogen fixation in cyanobacteria. Microbiol Rev 546: 340-373

Gallon JR, Stal LJ (1992) N_2 fixation in non-heterocystous cyanobacteria: an overview. In: Carpenter EJ, Capone DG, Rueter J (eds) Marine pelagic cyanobacteria: *Trichodesmium* and other diazotrophs. Kluwer Academic Publishers, Amsterdam, pp 115-139

Guerinot ML, Colwell RR (1985) Enumeration, isolation and characterization of N_2-fixing bacteria from seawater. Appl Envir Microbiol 50: 350-355

Howarth RW, Cole JJ (1985) Molybdenum availability, nitrogen limitation and phytoplankton growth in natural waters. Science 229: 653-655

Howarth RW, Marino R, Cole JJ (1988) Nitrogen fixation in freswater, estuarine and marine ecosystems. 2. Biogeochemical controls. Limnol Oceanogr 33: 669-697

Jørgensen BB, Revsbech NP, Blackburn TH, Cohen Y (1979) Diurnal cycle of oxygen and sulfide microgradients and microbial photosynthesis in a cyanobacterial mat sediment. Appl Environ Microbiol 38: 46-58

Joye SB, Paerl HW (1993) Contemporaneous N fixation and dentrification in marine microbial mats: rapid response to runoff events. Mar Ecol Progr Ser 94: 267-274

Kirshstein JD, Paerl HW, Zehr J (1991) Amplification, cloning, and sequencing of a nifH segment from aquatic microorganisms and natural communities. Appl Environ Microbiol 57: 2645-2650

Kirshstein JD, Zehr JP, Paerl HW (1993) Determination of N_2 fixation potential in the marine environment: application of the polmerase chain reaction. Mar Ecol Progr Ser 95: 305-309

Kühl M, Jørgensen BB (1992) Spectral light measurements in microbenthic phototrophic communities with a fiber-optic microprobe coupled to a sensitive diode array detector. Limnol Oceanogr 37: 1813-1823

Millie DF, Paerl HW, Hurley JP (1993) Microalgal pigment assessment using high-performance liquid chromatography: a synopsis of organismal and ecological applications. Can J Fish Res Aquat Sci (in press)

Neilson AH, Lewin RA (1974) The uptake and utilization of organic carbon by algae: an essay in comparative biochemistry. Phycologia 13: 227-264

Paerl HW (1990) Physiological ecology and regulation of N_2 fixation in natural waters. Adv Microb Ecol 11:305-344

Paerl HW (1991) Ecophysiological and trophic implications of light-stimulated amino acid utilization in marine picoplankton. Appl Environ Microbiol 57:473-479

Paerl HW, Webb KL, Baker J, Wiebe WJ (1981) Nitrogen fixation in waters. In: Broughton WJ (ed) Nitrogen Fixation, V . 1 Ecology, Clarendon, pp193-240

Paerl HW, Crocker KM, Prufert LE (1987) Limitation of N_2 fixation in coastal marine waters: relative importance of molybdenum, iron, phosphorus and organic matter availability. Limnol Oceanogr 32: 525-536

Paerl HW, Bebout BM, Prufert LE (1989) Naturally occurring patterns of oxygenic photosynthesis and N_2 fixation in a marine microbial mat: physiological and ecological ramifications. In: Cohen Y, Rosenberg E (eds) Microbial mats: physiological ecology of benthic communities. American Society for Microbiology, Washington DC, pp 326-341

Paerl, HW, BM Bebout, SB Joye and DJ Des Marais (1993a) Microscale characterization of dissolved organic matter production and uptake in marine microbial mat communities. Limnol Oceanogr 38:1150-1161

Paerl HW, Joye SB, Fitzpatrick M (1993b) An evaluation of nutrient limitation of CO_2 and N_2 fixation in marine microbial mats. Mar Ecol Progr Ser (in press)

Paulsen DM, Paerl HW, Bishop PE (1991) Evidence that molybdenum-dependent nitrogen fixation is not limited by high sulfate in marine environments. Limnol Oceanogr 36: 1325-1334

Pinckney JL, Zingmark R (1993) Biomass and production of benthic microalgal communities in five typical estuarine habitats. Estuaries (in press)

Polimeni C (1976) Seasonality and life history of a blue-green algal mat on Shackleford Banks, NC. M. Sc. Thesis, Univ. of North Carolina at Chapel Hill 129 pp

Potts M, Whitton BA (1977) Nitrogen fixation by blue-green algal communities in the intertidal zone of the Lagoon of Aldabra Atoll. Oecologia 27: 275-283

Rueter, JG, Hutchins DA, Smith RW, Unsworth NL (1992) Iron nutrition of *Trichodesmium*. In: Carpenter EJ, Capone DG, Reuter JG (eds) Marine Pelagic Cyanobacteria: *Trichodesmium* and other diazotrophs. Kluwer, Dordrecht, Netherlands pp 289-306

Schrautemeier B, Böhme H, Böger P (1984) *In vitro* studies on pathways and regulation of electron transport to nitrogenase with cell-free extracts from heterocysts of *Anabaena variabilis*. Arch Microbiol 137: 14-20

Stal LJ, Grossberger S, Krumbein WE (1984) Nitrogen fixation associated with the cyanobacterial mat of a marine laminated microbial ecosystem. Mar Biol 82: 217-224

Stal LJ, Bergman B (1990) Immunological characterization of nitrogenase in the filamentous non-heterocystous cyano X$3 bacterium *Oscillatoria limosa*. Planta 182: 287-291

Stewart WDP, Gallon JR (eds) (1980) Proc. of an Internat. Symp. on Nitrogen Fixation. Oxford Univ. Press, Oxford UK

Villbrandt M, Stal LJ, Krumbein WE (1990) Interactions between nitrogen fixations and oxygenic photosynthesis in a marine cyanobacterial mat. FEMS Microbiol Ecol 74: 59-72

Villbrandt M, Stal LJ, Bergman B, Krumbein WE (1992) Immunolocalization and western blot analysis of nitrogenase in *Oscillatoria limosa* during a light-dark cycle. Bot Acta 105: 90-96

Whitton BA, Potts M (1982) Marine littoral. In: Carr NG, Whitton BA (eds) The Biology of Cyanobacteria. Blackwell, pp 515-542

Wiebe WJ, Johannes RE, Webb KL (1975) Nitrogen fixation in a coral reef community. Science 188: 257-259

Wurtzbaugh WA , Horne AJ (1983) Iron in eutrophic Clear Lake, CA: Its importance for algal nitrogen fixation and growth. Can J Fish Aquat Sci 40: 1419-1429

Yates MG (1977) Physiological aspects of nitrogen fixation. In: Newton W, Postgate JR, Rodriguez-Barruero C (eds) Recent Developments in Nitrogen Fixation. Academic Press, New York, pp 219-270

Zehr JP, Wyman M, Miller V, Duguay L, Capone DG (1993) Modification of the Fe protein of nitrogenase in natural populations of *Trichodesmium thiebautii*. Appl Envir Microbiol 59: 669-676

Relationships between functional groups of organisms in microbial mats

Frank P. van den Ende and Hans van Gemerden

Department of Microbiology, University of Groningen
Kerklaan 30, 9751 NN Haren, The Netherlands

Introduction

Microbial mats develop under a wide range of environmental conditions, and can be found in hypersaline coastal lagoons, hot springs, alkaline lakes, and marine intertidal flats (Cohen 1984, 1989; Jørgensen and Cohen 1977; Javor and Castenholz 1981, 1984; Jørgensen et al. 1983; Bauld 1984; Stal et al. 1985; Nicholson et al. 1987; Pierson et al. 1987). These laminated ecosystems characteristically are dominated by only a few functional groups of microbes. The driving force of most microbial mats is photosynthesis by cyanobacteria (CyaB) and algae. Subsequently, dissimilatory sulfate-reducing bacteria (SRB), using excretion-, lysis-, and decomposition products of CyaB, produce sulfide. The sulfide can be reoxidized to sulfate by colorless sulfur bacteria (CSB) and purple sulfur bacteria (PSB). Aerobic heterotrophic organisms are functionally important as their activity leads to oxygen depletion, and fermentative organisms provide growth substrates for SRB. In microbial mats these metabolically different groups of microbes live together in a layer of 5-10 mm thickness. Their combined metabolic activities result in steep environmental microgradients, particularly of oxygen and sulfide. Sulfide is inhibitory for most oxygenic phototrophs. Sulfide production immediately underneath the layer of CyaB might inhibit their growth, and, consequently, that of the entire ecosystem. On the other hand, anaerobic PSB and SRB are hampered by oxygen.

This contribution focuses on microbial interactions in benthic marine microbial mats. These mats develop on sandy sediments in the upper part of the intertidal zone that are irregularly flooded. Understanding the properties of dominant representatives of the functional groups may help to understand how these peculiar microbial communities can function as a complete ecosystem on such a small scale.

Vertical distribution of sulfur bacteria in microbial mats

From the physiology of the different functional groups described above, one would expect that CyaB inhabit the top layers, underneath which layers of CSB are overlying layers of PSB, whereas the deeper layers are harbored by SRB. Due to differences in chlorophylls and accessory pigments of CyaB and PSB, as well as the blackening of deeper layers due to FeS

NATO ASI Series, Vol. G 35
Microbial Mats
Edited by Lucas J. Stal and Pierre Caumette

precipitation, microbial mats clearly have a laminated appearance when observed with the naked eye. However, not only are CSB named so for obvious reasons, also populations of SRB in itself are colorless, and PSB may lack photopigments resulting in little or no coloration. Therefore, the visible lamination of the mats may not reflect the actual distribution of the different groups of organisms.

Light absorption by CyaB is in part complementary to that of PSB, enabling the latter to develop underneath the oxygenic phototrophs. Also, reflection of light by the quartz surfaces ensures sufficient light, even in deeper layers (Jørgensen and Des Marais 1986; Jørgensen 1988,1989; Lassen et al. 1992). Analysis of the vertical distribution of photopigments on a (sub-) millimeter scale showed that the populations of CyaB and PSB overlap. However the maximum concentration of bacteriochlorophyll *a* (Bchl) is found 1-2 mm deeper in the sediment than the maximum concentration of chlorophyll *a* (Pierson et al. 1987; De Wit et al. 1989).

Most probable number (MPN) counts in a microbial mat on the Frisian Island of Texel (The Netherlands) showed that the maximum population densities of PSB and CSB are found in the top 5 mm of the mat. Also, the population density of SRB peaked in the top 5 mm (Visscher et al. 1992a). It was observed that sulfate reduction was most prominent in the top layer (Visscher et al. 1992a,b; Visscher and Van den Ende this volume), often coinciding with the presence of oxygen (Cohen 1984; Skyring 1984; Skyring et al. 1989; Canfield and Des Marais 1991). An increasing number of data have shown that (many) SRB no longer can be considered obligate anaerobes (Dilling and Cypionka 1990; Cypionka this volume). The vertical distribution of CSB and SRB has not been studied with the same resolution as that of the CyaB and PSB, so there might be, and probably is, some small-scale stratification. However it is clear that their distributions do overlap.

Experiments with isolated organisms

Experimental work with pure cultures of organisms isolated from microbial mats is highly relevant due to the dominance of a few functional groups of microbes, each with a low species diversity as judged from morphological and kinetic criteria. Some illustrative examples are given below. The organisms used were obtained either by direct isolation (*Microcoleus chthonoplastes*), or by isolation from the highest dilution of mat samples resulting in growth on selective media (*Thiocapsa roseopercisina* and *Thiobacillus thioparus*). These methods for isolation ensure a high probability that the organisms are numerically important in the mat.

Sulfide utilization by Microcoleus chthonoplastes

CyaB, having photosystems I and II, primarily are oxygenic phototrophs using H_2O as the electron donor. However, many species also are able to perform an anoxygenic photosynthesis in which sulfide acts as the electron donor (Cohen et al. 1975a,b; Garlick et al. 1977; De Wit and Van Gemerden 1987b; De Wit et al. 1988; Stal 1991). In *M. chthonoplastes*, sulfide concentrations < 0.15 mM stimulate growth, at higher concentrations the organism gradually shifts to anoxygenic photosynthesis (Cohen 1984). It takes about 3h before sulfide utilization starts, thereafter sulfide is stoichiometrically converted to thiosulfate (De Wit and Van Gemerden 1987b). *Microcoleus* is hampered in its growth by the absence of O_2. In the presence of 3(3,4-dichlorophenyl)1,1-dimethylurea (DCMU), which selectively inhibits photosystem II, growth on sulfide does not occur, unless oxygen is supplied externally (De Wit et al. 1988). *Microcoleus* thus is an obligately aerobic organism. Conceivably, the oxygen is required for the synthesis of polyunsaturated fatty acids (Padan 1979; De Wit et al. 1988). Growth of *Microcoleus* is severely inhibited by sulfide, the relation between the specific growth rate μ and the concentration of sulfide is shown in Fig. 1. Anoxygenic photosynthesis is not inhibited by elevated sulfide concentrations (to at least 1.3 mM sulfide), resulting in the intracellular storage of glycogen (De Wit and Van Gemerden 1987b). These reserve materials can be used for fermentation during the anoxic night periods.

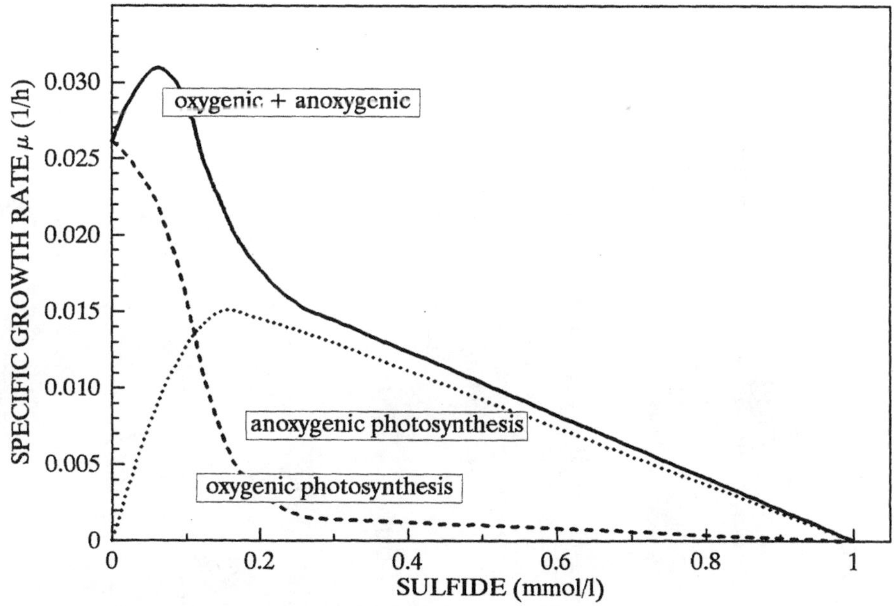

Fig. 1. Relation between the specific growth rate of *Microcoleus chthonoplastes* strain 11 and the concentration of sulfide, and the respective contribution of oxygenic and anoxygenic photosynthesis. Based on data of De Wit et al. 1988.

The impact of oxygen on growth and photopigment synthesis in *Thiocapsa roseopersicina*

PSB primarily are anoxygenic phototrophic organisms using sulfide and other reduced forms of sulfur as the electron donor for the reduction of CO_2 to cellular carbon. Because of the small vertical dimensions of microbial mats, oxygen, produced by the directly overlying layer of CyaB, penetrates into the layer of PSB.

T. roseopersicina, has a number of metabolic characteristics that enable it to thrive under rather harsh conditions. Important features are the abilities of phototrophic and chemotrophic growth (Bogorov 1974; Kämpf and Pfennig 1980; De Wit and Van Gemerden 1987a), the utilization of polysulfides (Visscher et al. 1990), and the assimilation of low molecular weight organic compounds. Chemotrophy, and the impact of fluctuations of oxygen and light are discussed below.

In *Thiocapsa* oxygen inhibits the synthesis of Bchl, but not growth. Prolonged culturing under oxic conditions results in colorless organisms, which then rely on chemotrophy. Chemotrophy inevitably results in lower yields compared to phototrophy, because not all reducing equivalents can be used for the reduction of CO_2 (compare regimen 24/0 and 0/24 in Fig. 2). The chemotrophic yield of *Thiocapsa* is similar to that found in thiobacilli (Kelly 1982).

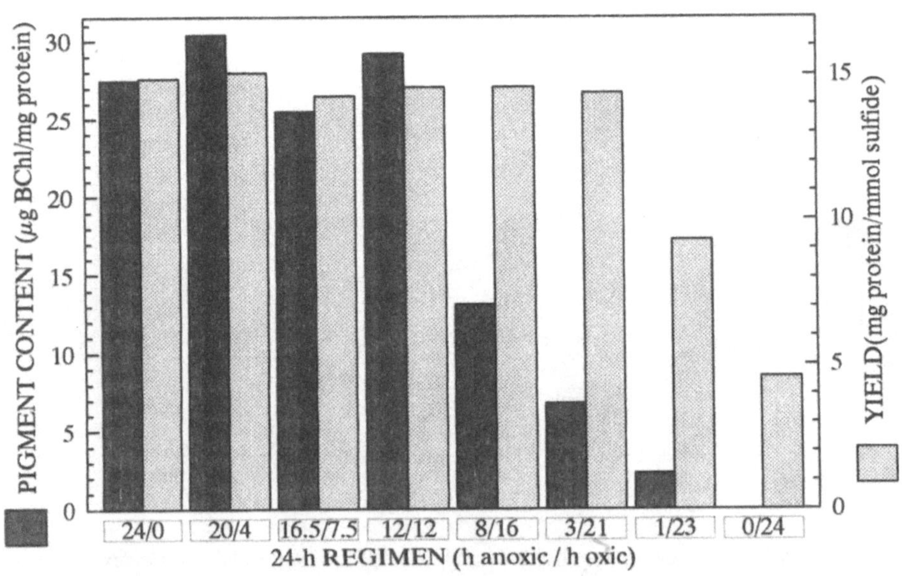

Fig. 2. Pigment content and yield on sulfide of *Thiocapsa roseopersicina* strain M1 grown on various anoxic/oxic regimen in the light. Data were collected at the end of the anoxic period after >5 volume changes. Based on De Wit and Van Gemerden, 1990a and Schaub and Van Gemerden 1993.

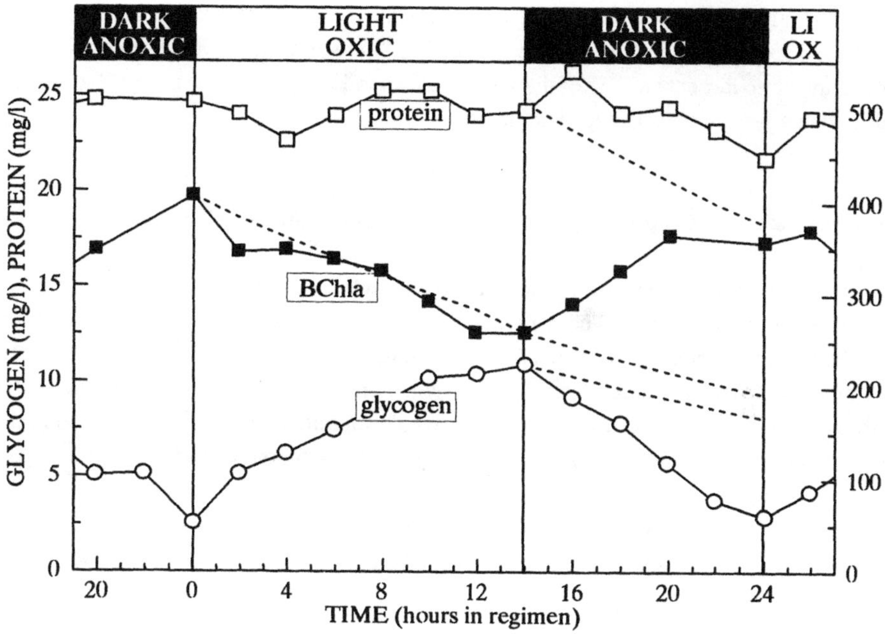

Fig. 3. Time course of the concentration of protein, bacteriochlorophyll *a*, and glycogen in a sulfide-limited continuous culture of *Thiocapsa roseopersicina* strain M1 subjected to a 14h oxic-light/10h anoxic-dark regime. Data were collected after 5 volume changes. Redrawn from De Wit and Van Gemerden, 1990b.

Incubation of *Thiocapsa* under a regimen of 21h oxic/3h anoxic conditions (continuous light) resulted in a similar yield as observed during continuously anoxic conditions, despite the fact that the specific Bchl content was only 20% of that reached during continuously anoxic conditions, showing that the organisms were growing fully phototrophically (Fig. 2).

Incubation of *Thiocapsa* at a regimen 23h oxic/1h anoxic in continuous light revealed a yield intermediate between values obtained during full phototrophic growth and full chemotrophic growth (Fig. 2). It can therefore be concluded that *Thiocapsa* is capable of simultaneously using its chemotrophic and phototrophic growth potential.

In microbial mats, oxygen depletion usually coincides with darkness, and at daytime the synthesis of photopigments is inhibited by oxygen. Mimicking these conditions by culturing *Thiocapsa* in a continuous culture under a 14h oxic-light/10h anoxic-dark regime, resulted in a time-dependent equilibrium in which the yield was the same as expected for full phototrophic growth. During the oxic-light periods, synthesis of photopigments was not observed. However, in the anoxic-dark periods, BChl synthesis was found to occur at a specific rate exceeding the dilution rate (Fig. 3). This phenomenon is explained by the fact that during the preceding oxic-light period, glycogen had been accumulated intracellularly. It was calculated that the synthesis of BChl and spirilloxanthin, the major carotenoid in *Thiocapsa*, could be accounted for by the degradation of no more than 17% of the glycogen. Degradation of the remaining glycogen thus could account for the observed growth in the dark.

The impact of oxygen limitation on sulfide oxidation by *Thiobacillus thioparus*

CSB are chemotrophic organisms primarily oxidizing sulfide and other reduced forms of sulfur with oxygen to obtain energy. The oxidation of reduced sulfur species also provides reducing equivalents for the reduction of carbon dioxide to cellular carbon. In microbial mats, maximum rates of sulfide oxidation take place at the oxygen/sulfide interface. In order to mimic these environmental conditions, chemostat cultures of *T. thioparus* were subjected to conditions ranging from severe sulfide limitation to severe oxygen limitation. Data are shown in Fig. 4.

Flushing the headspace with gas mixtures containing 100-70% air, at which progressively decreasing concentrations of oxygen in the culture were detected, resulted in virtually complete oxidation of sulfide to sulfate. Using gas mixtures containing 60-30% air,

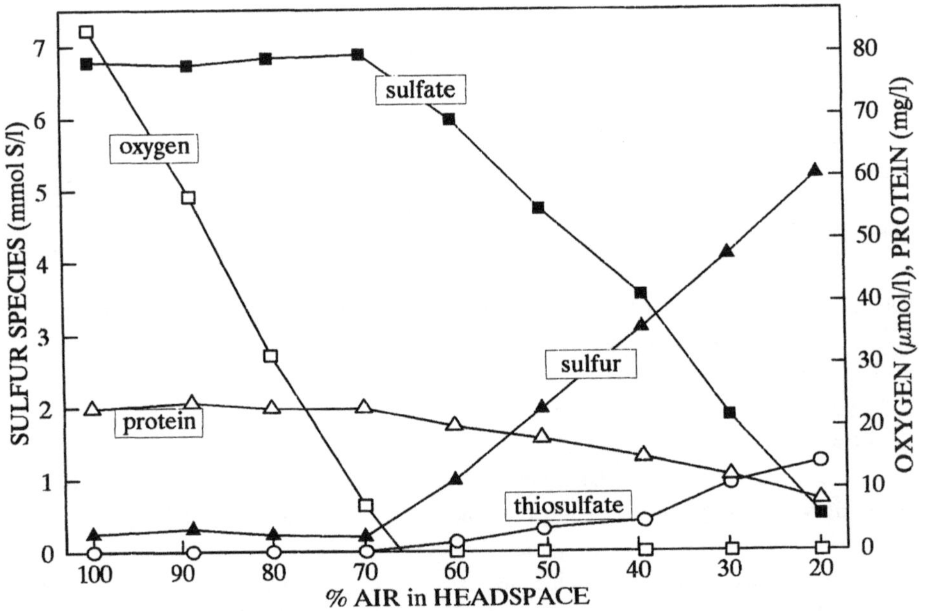

Fig. 4. Product formation by *Thiobacillus thioparus* strain T5 under various air/nitrogen mixtures in the headspace. The oxygen concentration in the culture is also shown. Redrawn from Van den Ende and Van Gemerden, 1993.

resulting in oxygen concentrations below the limit of detection in the culture, sulfide still could not be detected, but the concentration of other reduced sulfur species increased with decreasing oxygen availability.

It is anticipated that a similar partial sulfide oxidation occurs in microbial mats at the oxygen/sulfide interface.

Co-cultures of *Thiobacillus thioparus* and *Thiocapsa roseopersicina*

The sulfide affinity of *Thiocapsa* is insufficiently high to enable this organism to compete successfully with *Thiobacillus*, irrespective whether *Thiocapsa* is growing phototrophically or chemotrophically. However, as outlined above, oxygen limitation results in incomplete sulfide oxidation by *Thiobacillus* and the production of somewhat less reduced sulfur compounds by *Thiobacillus* can be expected to compensate for the lower sulfide affinity of *Thiocapsa*.

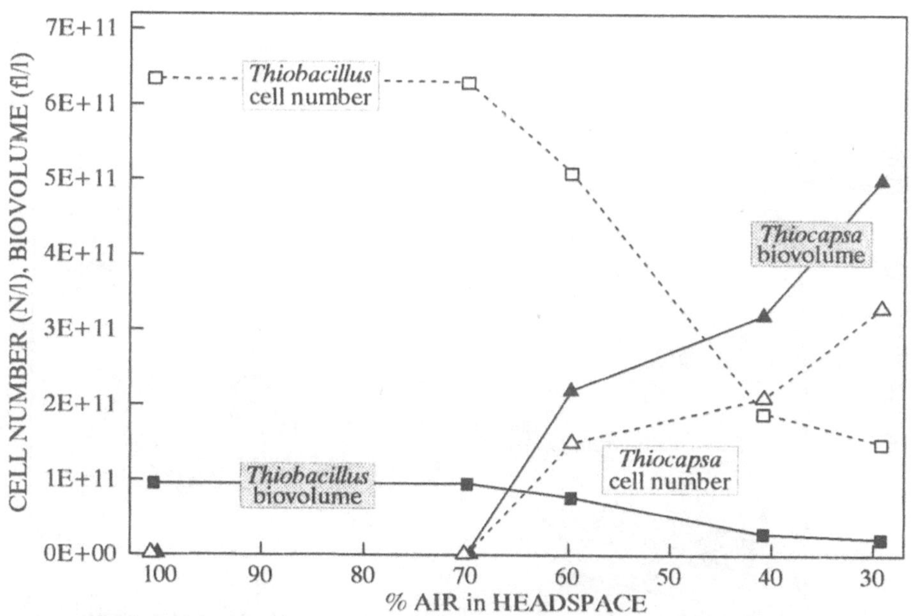

Fig. 5. Cell number and total biovolume in a co-culture of the colorless sulfur bacterium *Thiobacillus thioparus* strain T5 and the purple sulfur bacterium *Thiocapsa roseopersicina* strain M1 under various air/nitrogen mixtures in the headspace. Oxygen in the culture was below the limit of detection below 70% air in the headspace. Unpublished data of Anniet Laverman.

Attempts to co-culture *Thiobacillus* and *Thiocapsa* indeed resulted in stable coexistence once the supply of oxygen became limiting. As a result of the effective removal of oxygen by *Thiobacillus*, pigment synthesis in *Thiocapsa* is no longer inhibited, which allows it to grow phototrophically resulting in maximum yields: as shown in Fig. 5, the increase in *Thiocapsa* biomass with decreasing oxygen availability, is much higher than the decrease in *Thiobacillus* biomass. The result thus is co-existence between a chemotrophically growing colorless sulfur bacterium and a phototrophically growing purple sulfur bacterium.

Mat building: an interactive process

The mats discussed harbor few motile organisms like *Beggiatoa* and *Chromatium*, and the PSB and CSB present are confronted with the ambient fluctuations of oxygen and sulfide. Also SRB appear not to migrate to deeper layers during the day when the concentrations of O_2 are high (Visscher et al. 1992a).

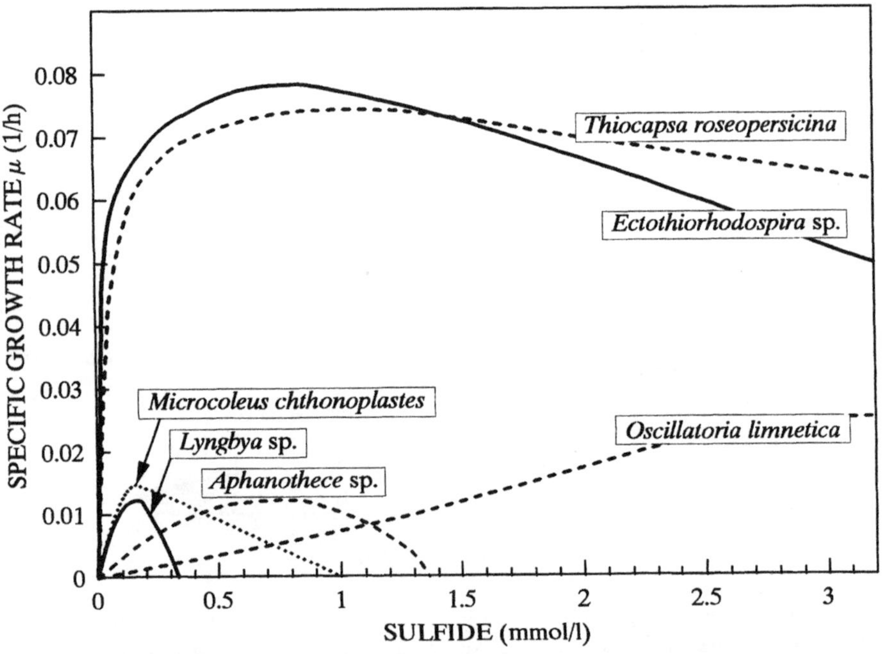

Fig. 6. Relation between the specific growth rates of phototrophic bacteria and the consumption of sulfide. Based on short turn CO_2 fixation data of Garlick et al. 1977, and continuous culture data of De Wit et al. 1988 and Visscher et al. 1990.

The pure culture studies discussed above have shown that the organisms are adapted to unfavorable high and fluctuating concentrations of oxygen and sulfide. The aerobic CyaB *M. chthonoplastes* can grow on sulfide, the anaerobic PSB *T. roseopersicina* can grow chemotrophically with oxygen, and some of the SRB can use oxygen. However in the microbial mat these properties are probably less important as a mode of growth, but rather reflect mechanisms that allow these organisms to survive in the microbial mat environment.

For the mat-building *Microcoleus* the ecological relevance of the ability to utilize sulfide as the electron donor in anoxygenic photosynthesis appears to be detoxification rather than growth. In the presence of sulfide, growth of CyaB is slow compared to that of PSB, moreover, sulfide utilization requires induction. More importantly, the affinity of CyaB for sulfide is very low compared to other sulfide-oxidizing microorganisms (Fig. 6). For growth, the CyaB rely on the powerful mechanism of oxygenic photosynthesis and not on the oxidation of sulfide.

The expected interactions in microbial mats between CyaB, CSB, PSB, SRB, aerobic heterotrophs and fermentative bacteria are visualized in Fig. 7. For clarity the cyanobacterial layer is separated from the layer with other bacteria, but in the microbial mat these layers show an overlap. Factors of crucial importance are the concentrations of oxygen and sulfide, but the

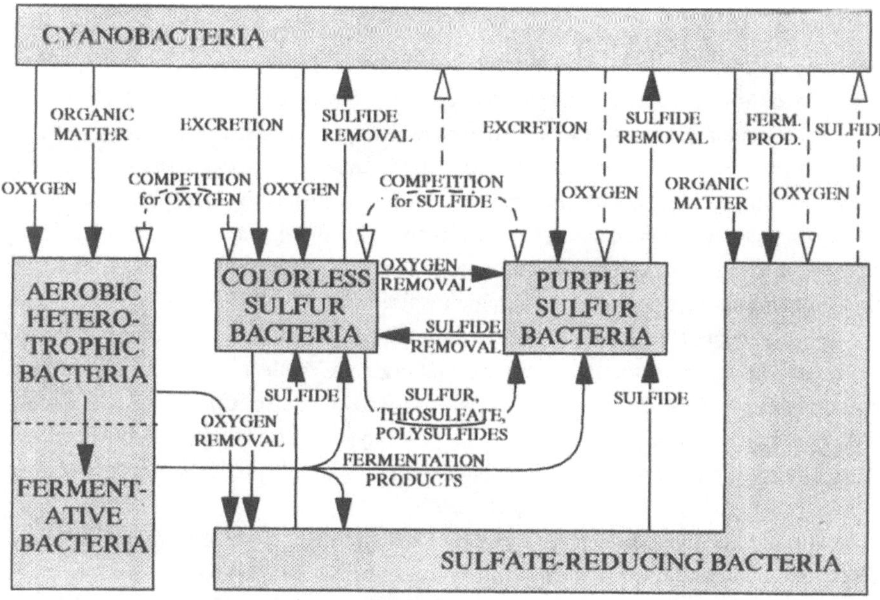

Fig. 7. Schematic representation of the interactions between cyanobacteria, aerobic heterotrophic bacteria, fermentative bacteria, colorless sulfur bacteria, purple sulfur bacteria and sulfate reducing bacteria in microbial mat. Black arrows indicate positive interactions, white arrows negative interactions.

presence of other sulfur compounds is relevant as well. Particularly thiosulfate, a potential substrate for PSB, CSB, and SRB, appears to be of importance in microbial mats (B.B. Jørgensen pers. comm.). SRB depend on CyaB for carbon- and energy sources but at the same time they are slowed down by the oxygen produced by the latter group. In turn, growth of the CyaB is inhibited by sulfide. CSB need oxygen and for that reason largely depend on the CyaB. In PSB, photopigment synthesis is inhibited by oxygen, and this may force these organisms to turn to chemotrophy resulting in much lower yields. On the other hand, both types of sulfide-oxidizing microbes are stimulated by excretion products of CyaB. Finally, colorless and PSB compete for mutual electron donors. Oxygenic photosynthesis carried out by the CyaB, is the driving force in most microbial mats, but the involvement of other microbial groups is by no means trivial.

In the resulting community, aerobic bacteria (CyaB and heterotrophs), predominantly anaerobic bacteria (SRB and PSB), and specialist oxic/anoxic interface bacteria (CSB), thrive in co-existence. Each of these functional groups is adapted to cope with alternating advantageous and hostile conditions. Yet, together they manage to reduce the hostility of the environment, resulting in a sustaining system with a higher productivity than would be possible without their collective effort.

Oxygenic photosynthesis and aerobic respiration are complementary processes. In theory, sufficient oxygen is produced during the day to allow for complete aerobic degradation of photosynthates and storage compounds during the night. However, due to the small vertical dimensions of microbial mats, a substantial amount of oxygen produced during the day, resulting in 200-400% super-saturation, will escape to the overlying water or the atmosphere. Therefore, microbial mats can be expected to have an oxygen deficit shortly after sunset, which is further increased as during the growth season nights are short and days are long. The organic remains, resulting from CyaB productivity, form an exquisite energy source for other microorganisms, and the presence of aerobically respiring organisms will shorten the time span during which oxygen is available. Therefore, in the dark CyaB must rely on fermentation or anaerobic respiration, rather than on aerobic respiration.

With respect to the availability of substrates for sulfate-reducing and other heterotrophic bacteria, conditions in the top layers are most favorable. Excellent growth substrates for SRB (lactate, acetate, ethanol, H_2) are produced by dark fermentation of CyaB (Stal et al. 1989; Stal 1991), resulting in sulfide production. The shallow oxygen penetration may slow down sulfate reduction in the top layer during the day, but continuous sulfide production in deeper layers is inevitable.

The bulk of the sulfide is oxidized by CSB and PSB. Their effective removal of the sulfide, as it diffuses upwards from deeper production sites or is produced *in situ*, not only supports their own growth, but is also beneficial to the overlying CyaB.

Apart from their collective effort in sulfide removal, CSB and PSB have little in common and will compete for this reduced sulfur compound. The extent to which H_2S is oxidized by either group very much depends on the availability of oxygen. With no oxygen available, sulfide will be exclusively oxidized by PSB, provided light is available. With excess oxygen, virtually all sulfide will be oxidized by CSB, despite the fact that PSB are capable of chemotrophic growth. This is explained by the fact that CSB have much higher affinities for sulfide than PSB, as is the case for thiosulfate (Visscher et al. 1992b). In microbial mats most of the sulfide is oxidized at the oxygen/sulfide interface at low oxygen concentrations. Under these conditions *Thiobacillus* will not be able to oxidize all sulfide completely to sulfate because of oxygen deficiency. The production of somewhat less reduced sulfur compounds by *Thiobacillus* does provide *Thiocapsa* with sufficient electron donors. Effective removal of oxygen by *Thiobacillus* prevents inhibition of pigment synthesis in *Thiocapsa*. The expected result is coexistence between chemotrophically growing *Thiobacillus* and phototrophically growing *Thiocapsa*, operating in concert to remove sulfide. Such a result was indeed observed in the laboratory, applying oxygen-limited chemostats.

High oxygen concentrations, although in itself advantageous to thiobacilli, would inhibit SRB, resulting in decreased rates of sulfide formation. Utilization of oxygen by thiobacilli thus not only promotes their own growth, but is also advantageous for sulfate reducers.

Thiocapsa thus benefits from the presence of *Thiobacillus*. It can be argued that *vice versa* CSB benefit from the presence of PSB. In the absence of the latter, sulfide removal by CSB would not be complete due to lack of oxygen. In the absence of PSB, sulfide would reach the layer of CyaB, possibly resulting in severe inhibition of growth of CyaB and cessation of oxygen production. Bearing in mind that the cyanobacterium *M. chthonoplastes* is the driving force of these microbial mats - also providing physical protection to organisms growing underneath - the result would be disastrous. This problem is elegantly solved by sulfide removal without the participation of oxygen, as carried out by PSB.

In summary, the driving force of microbial mats are the CyaB. However, their continued oxygenic photosynthesis is made possible by the beneficiary effects of CSB and PSB in the functional removal of the sulfide produced by SRB after oxygen-dependent heterotrophs and fermentative bacteria have degraded the organic compounds originating from lysis and excretion. CSB and PSB operate in concert because the former excrete electron donors for the PSB when oxygen is short in supply. The shortage of oxygen is the consequence of the small vertical dimension of microbial mats, resulting in oxygen escaping to the atmosphere during daytime. Mat building thus can be considered a joint venture of functionally different groups of microbes.

Acknowledgements

We thank Rutger de Wit, Pieter T. Visscher, Bartholomeus E.M. Schaub and Jan Beukema for their contribution.

References

Bauld J (1984) Microbial mats in marginal marine environments: Shark Bay, Western Australia, and Spencer Gulf, South Australia. In: Cohen Y, Castenholz RW, Halvorson HO (eds) Microbial Mats: Stromatolites. Liss, New York, pp 39-58

Bogorov LV (1974) About the properties of *Thiocapsa roseopersicina* strain BBS isolated from the estuary of the White Sea. Microbiology 43: 275-280

Canfield DE, Des Marais DJ (1991) Aerobic sulfate reduction in microbial mats. Science 251: 1471-1473

Cohen Y (1984) Oxygenic photosynthesis, anoxygenic photosynthesis, and sulfate reduction in cyanobacterial mats. In: Klug MJ, Reddy CA (eds) Current Perspectives in Microbial Ecology. ASM Washington DC, pp 435-441

Cohen Y (1989) Photosynthesis in cyanobacterial mats and its relation to the sulfur cycle: a model for microbial sulfur interactions. In: Cohen Y, Rosenberg E (eds) Microbial Mats. Physiological Ecology of Benthic Microbial Communities. ASM Washington DC, pp 22-36

Cohen Y, Jørgensen BB, Padan E, Shilo M (1975a) Sulfide dependent anoxygenic photosynthesis in the cyanobacterium *Oscillatoria limnetica*. Nature 257: 489-492

Cohen Y, Padan E, Shilo M (1975b) Facultative anoxygenic photosynthesis in the cyanobacterium *Oscillatoria limnetica*. J Bacteriol 123: 855-861

De Wit R, Van Gemerden H (1987a) Chemolithotrophic growth of the phototrophic sulfur bacterium *Thiocapsa roseopersicina*. FEMS Microbiol Ecol 45: 117-126

De Wit R, Van Gemerden H (1987b) Oxidation of sulfide to thiosulfate by *Microcoleus chthonoplastes*. FEMS Microbiol Ecol: 45: 7-13

De Wit R, Van Gemerden H (1990a) Growth of the phototrophic purple sulfur bacterium *Thiocapsa roseopersicina* under oxic/anoxic regimens in the light. FEMS Microbiol Ecol 73: 69-76

De Wit R, Van Gemerden H (1990b) Growth and metabolism of the purple sulfur bacterium *Thiocapsa roseopersicina* under combined light/dark and oxic/anoxic regimens. Arch Microbiol 154: 459-464

De Wit R, Jonkers HM, Van den Ende FP, Van Gemerden H (1989) *In situ* fluctuations of oxygen and sulphide in marine microbial sediment ecosystems. Neth J Sea Res 23: 271-281

De Wit R, Van Boekel WHM, Van Gemerden H (1988) Growth of the cyanobacterium *Microcoleus chthonoplastes* on sulfide. FEMS Microbiol Ecol 53: 203-209

Dilling W, Cypionka H (1990) Aerobic respiration in sulfate-reducing bacteria. FEMS Microbiol Lett 71: 123-128

Garlick S, Oren A, Padan E (1977) Occurrence of facultative anoxygenic photosynthesis among filamentous unicellular cyanobacteria. J Bacteriol 129: 623-29

Javor BJ, Castenholz RW (1981) Laminated microbial mats, Laguna Guerrero Negro, Mexico. Geomicrobiol J 3: 237-273

Javor BJ, Castenholz RW (1984) Invertebrate grazers of microbial mats. In: Cohen Y, Castenholz RW, Halvorson HO (eds) Microbial Mats: Stromatolites. Liss, New York, pp 85-94

Jørgensen BB (1988) Ecology of the sulphur cycle: oxidative pathways in sediments. In: Cole JA, Ferguson SJ (eds) The nitrogen and sulphur cycles. Cambridge University Press, Cambridge, pp 31-63

Jørgensen BB (1989) Light penetration, absorption, and action spectra in cyanobacterial mats In: Cohen Y, Rosenberg E (eds) Microbial mats: physiological ecology of benthic microbial communities. ASM Washington, pp 123-137

Jørgensen BB, Cohen Y (1977) Solar Lake (Sinai) 5. The sulfur cycle of the benthic cyanobacterial mats. Limnol Oceanogr 22: 657-666

Jørgensen BB, Des Marais DJ (1986) A simple fiberoptic microprobe for high resolution light measurements: application in marine sediments. Limnol Oceanogr 31: 1376-1383

Jørgensen BB, Revsbech NP, Cohen Y (1983) Photosynthesis and structure of benthic microbial mats: microelectrode and SEM studies of four cyanobacterial communities. Limnol Oceanogr 28: 1075-1093

Kämpf C, Pfennig N (1980) Capacity of Chromatiaceae for chemotrophic growth. Specific respiration rates of *Thiocystis violacea* and *Chromatium vinosum*. Arch Microbiol 127: 125-135

Kelly DP (1982) Biochemistry of the chemolithotrophic oxidation of inorganic sulphur. Phil Trans R Soc Lond B298: 499-529

Lassen C, Ploug H, Jørgensen BB (1992) A fibre-optic scalar irradiance microsensor: application for spectral light measurements in sediments. FEMS Microbiol Ecol 86: 247-254

Nicholson JA, Stolz JF, Pierson BK (1987) Structure of a microbial mat at Great Sippewissett Marsh, Cape Cod, Massachusetts. FEMS Microbiol Ecol 45: 343-364

Padan E (1979) Impact of facultatively anaerobic phototrophic metabolism on ecology of cyanobacteria (blue-green algae) In: Alexander M (ed) Adv Microbial Ecol, Plenum, New York, pp 1-48

Pierson BK, Oesterle A, Murphy GL (1987) Pigments, light penetration, and photosynthetic activity in the multi-layered microbial mats of Great Sippewissett Salt Marsh, Massachusetts.FEMS Microbiol Ecol 45: 365-376

Schaub BEM, Van Gemerden H (1993) Simultaneous phototrophic and chemotrophic growth in the purple sulfur bacterium *Thiocapsa roseopersicina*. FEMS Microbiol Ecol in press

Skyring GW, Lynch RM, Smith GD (1989) Quantitative relationship between carbon, hydrogen, and sulfur metabolism in cyanobacterial mats. In: Cohen Y, Rosenberg E (eds) Microbial mats: physiological ecology of benthic microbial communities. ASM Washington, pp 170-179

Skyring GW (1984) Sulfate reduction in marine sediments associated with cyanobacterial mats in Australia. In: Cohen Y, Castenholz RW, Halvorson HO (eds) Microbial mats: stromatolites. Liss, New York, pp 265-275

Stal LJ (1991) The metabolic versatility of the mat-building cyanobacteria *Microcoleus chthonoplastes* and *Oscillatoria limosa* and its ecological significance. Algol Stud 64: 453-467

Stal LJ, Heyer H, Bekker S, Villbrandt M, Krumbein WE (1989) Aerobic-anaerobic metabolism in the cyanobacterium *Oscillatoria limosa* In: Cohen Y, Rosenberg E (eds) Microbial Mats: Physiological Ecology of Benthic Microbial Communities. ASM, Washington DC, pp 255-276

Stal LJ, Van Gemerden H, Krumbein WE (1985) Structure and development of a benthic marine microbial mat. FEMS Microbiol Ecol 31: 111-125

Van den Ende FP, Van Gemerden H (1993) Sulfide oxidation under oxygen limitation by a *Thiobacillus thioparus* isolated from a marine microbial mat. FEMS Microbiol Ecol 13: 69-78

Visscher PT, Prins RA, Van Gemerden H (1992a) Rates of sulfate reduction and thiosulfate consumption in a marine microbial mat. FEMS Microbiol Ecol 86: 283-294

Visscher PT, Van den Ende FP, Schaub BEM, Van Gemerden H (1992b) Competition between anoxygenic phototrophic bacteria and colorless sulfur bacteria in a microbial mat. FEMS Microbiol Ecol 101: 51-58

Visscher PT, Nijburg JW, Van Gemerden H (1990) Polysulfide utilization by *Thiocapsa roseopersicina*. Arch Microbiol 155: 75-81

Diel and spatial fluctuations of sulfur transformations

Pieter T. Visscher[1] and Frank P. van den Ende[2]

Division of Marine and Atmospheric Chemistry, Rosenstiel School of Marine and Atmospheric Science, Miami, FL 33149-1098 USA

Introduction

Marine microbial mats often develop under conditions, where diel extremes in oxygen and sulfide are found (Fig 1), which are a result of the fluctuation of incident light. During the dark period, when photosynthesis ceases, free sulfide can reach the surface of the sediment, while during the light period super-saturating oxygen concentrations prevail at the same depth horizon. As a result of changing gradients of O_2 and sulfide concentration, the flux of these compounds displays a diel pattern as well (Fig 2).

Lamination, or stratification of microorganisms occurs (Visscher and Van Gemerden 1993; Van Gemerden 1993), because changing physico-chemical parameters set boundaries for existence of each species. When populations of sulfate reducing bacteria (SRB), sulfide oxidizers (colorless sulfur bacteria) and anoxygenic phototrophs are enumerated and compared to the total microbial population, up to 40% of the total population consists of sulfur bacteria in some microbial mats (Visscher and Van Gemerden 1993; Visscher et al. unpubl.). This is not unexpected, since sulfate reduction plays a dominant role in coastal sediments (Thode Andersen and Jørgensen 1989). Excretion and lysis products of cyanobacteria support a large population of oxygen-respiring heterotrophs, which consumption accounts for shallow oxygen penetration, as well as SRB (Van Gemerden 1993).

Since diel changes in the light condition, and accordingly oxygen and sulfide, generate the most pronounced fluctuation in microbial mats of temperate zones, its effect on different functional groups of sulfur bacteria will be discussed in this paper.

[1] Present address: Department of Marine Sciences, The University of Connecticut, 1084 Shennecossett Road, Groton, CT 06340, USA.
[2] Department of Microbiology, University of Groningen, 9751 NN Haren, The Netherlands.

NATO ASI Series, Vol. G 35
Microbial Mats
Edited by Lucas J. Stal and Pierre Caumette

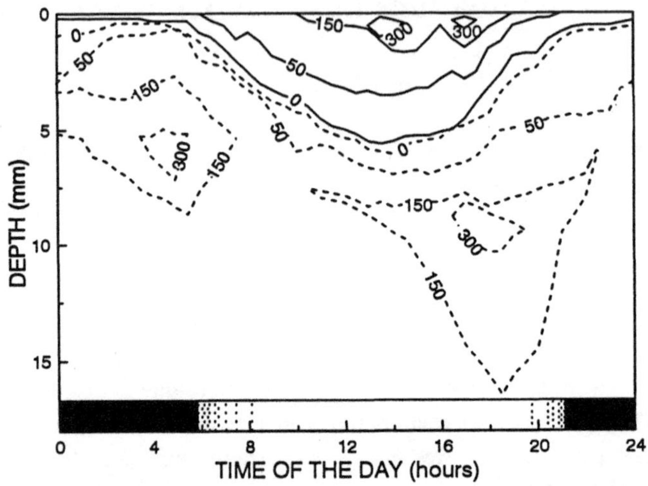

Fig. 1. Isoplethes of oxygen (solid lines) and sulfide (dashed lines) measured in a microbial mat on the Frisian island of Texel, Nl, August 1989. Numbers indicate concentration in μM; measurements taken with needle electrodes; bar represents light/dark period.

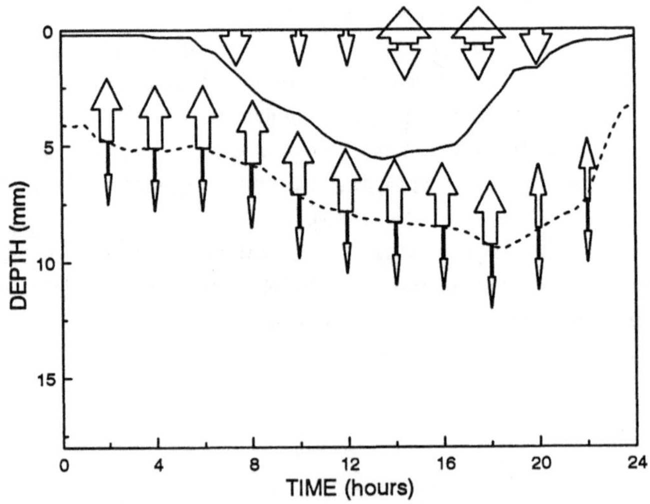

Fig. 2. Direction of oxygen (short arrows) and sulfide (long arrows) diffusion. Width of arrows indicates relative magnitude; see also legend Fig. 1.

Diel patterns of sulfide production

Sulfate reduction (SR) is not restricted to deeper permanently anoxic layers of microbial mats (Canfield and Des Marais 1991; Fründ and Cohen 1992; Visscher et al. 1992a). In a marine microbial mat on Texel (The Netherlands) highest sulfate reduction rates (SRR) were measured in the top 5 mm during the anoxic period (Fig 3). Rates during the daylight periods are underestimated, mainly due to rapid complete oxidation by a large population of thiobacilli (Visscher et al. 1991) which were found to oxidize sulfide to sulfate without intermediates under oxic conditions (Van den Ende and Van Gemerden 1993). The SRR found could be explained by isotope exchange with the large pool of reduced inorganic sulfur compounds (Fig 4). Especially polysulfides, which clearly present a sink in this microbial mat, were reported to be subject to isotope exchange (Fossing and Jørgensen 1990).

At the O_2-sulfide interface, thiosulfate and elemental sulfur can be formed when oxygen concentrations are low (Van den Ende and Van Gemerden 1993). When these sulfur species are formed, disproportionation by certain SRB can be expected.

Diel patterns of sulfide oxidation

Potential sulfide oxidation can be deduced from the depth profiles of oxygen and sulfide concentrations, SRR (Figs. 1,3) and experiments with cultures of dominant sulfide-oxidizers (see Van Gemerden 1993, and this volume for reviews). Comparison of the SRR measured and the amount of reduced sulfur in the sediment (Fig. 4) shows that a considerable part of the sulfide produced is not oxidized to sulfate , but fixed in FeS, pyrite and polysulfides. The apparent imbalance of sulfide production and oxidation is further exemplified by the enrichment of the sediment with sulfur compared to seawater, as well as a depletion in sulfate. Sulfide produced below 10 mm during the day and below 5 mm during the night diffuses downward and is likely to accumulate in the reduced sulfur pool (Fig. 2). However, most of the sulfide is produced in the surface layer (Fig. 3). All of the sulfide produced in the oxic zone is expected to be oxidized completely by CSB, as mentioned earlier. However, the maximum activity of these bacteria occurs at the O_2-sulfide interface (OSI; Fig. 5). During daylight the rate of sulfide oxidation by CSB is apparently higher than the SRR, which restricts the occurrence of free sulfide to deeper strata (Fig. 1). In the afternoon SRR again exceeds the sulfide oxidation rate, which results in a rise of the OSI towards the surface. At the OSI, sulfide will be oxidized by thiobacilli to elemental sulfur, thiosulfate or sulfate, depending on the local O_2 concentration.

In addition, the depth of the interface is also determined by PSB due to consumption of sulfide during anoxygenic photosynthesis (Fig. 6). When O_2 is present, PSB have to rely on their capacity to store sulfur compounds intracellularly (De Wit and Van Gemerden 1990; Visscher et al 1992b) and on incomplete oxidation products of thiobacilli (Van den Ende and

Van Gemerden 1993). Large pools of reduced sulfur compounds can be explained due to incomplete oxidation of sulfide (Fig 4.; Visscher and Van Gemerden 1993).

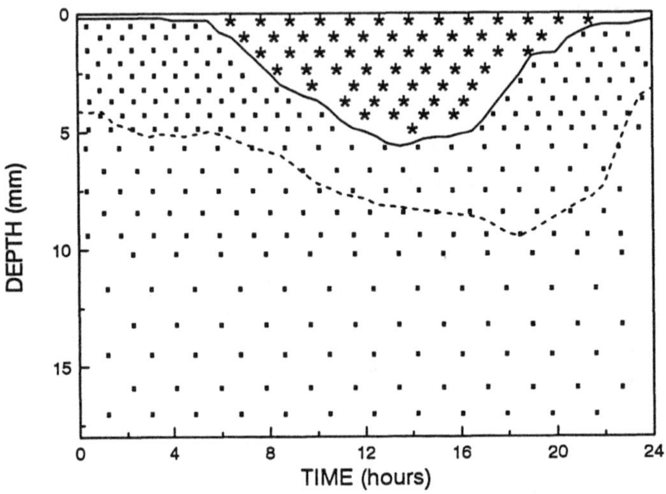

Fig. 3. Sulfate reduction rates in a microbial mat (data from Visscher et al 1992b). Dotted area: low density (150-200), medium (200-450), high (>450 nmol sulfate.cm^{-3}.d^{-1}); asterisks (150-200) represent underestimations (see text).

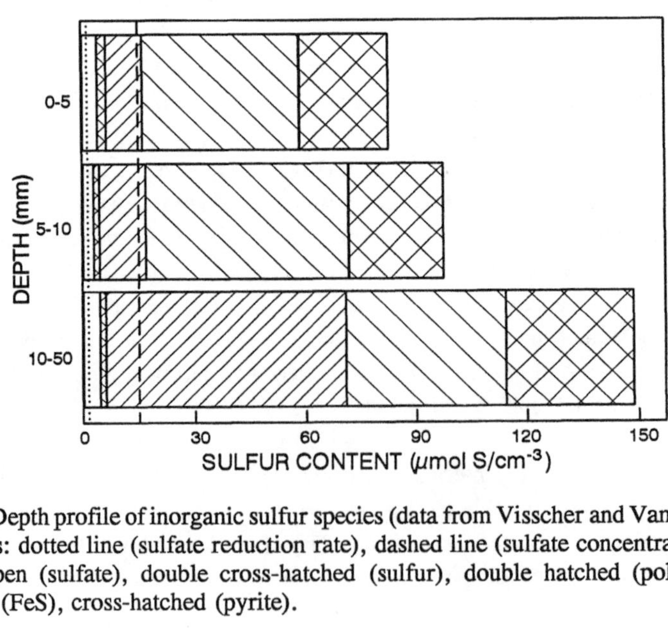

Fig. 4. Depth profile of inorganic sulfur species (data from Visscher and Van Gemerden 1993). Symbols: dotted line (sulfate reduction rate), dashed line (sulfate concentration in seawater); bars: open (sulfate), double cross-hatched (sulfur), double hatched (polysulfides), single hatched (FeS), cross-hatched (pyrite).

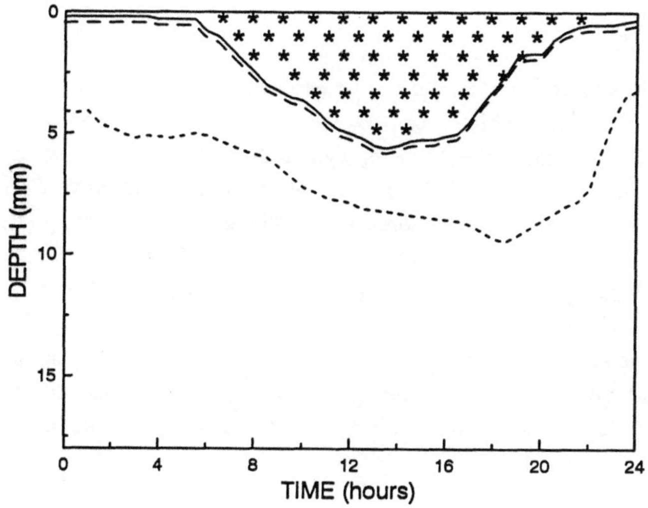

Fig. 5. Predicted spatial distribution of sulfide oxidation by colorless sulfur bacteria (long-dashed line indicates maximum activity; asterisks display additional area of potential activity). Solid line indicates depth of O_2 penetration, dashed line shows sulfide-maximum.

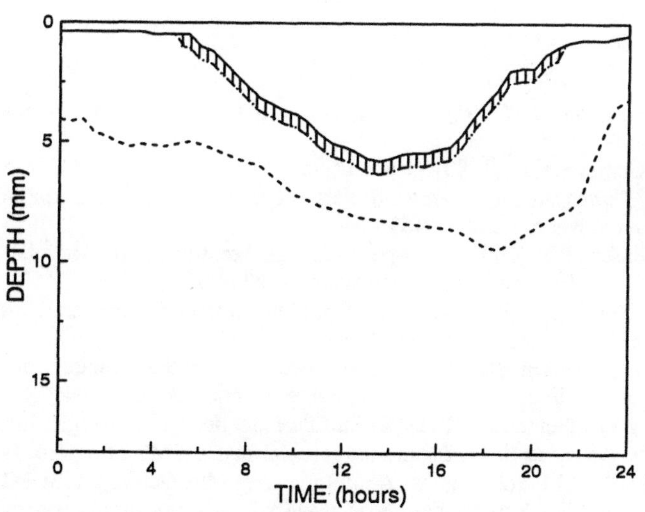

Fig. 6. Expected spatial distribution of sulfide oxidation by purple sulfur bacteria (hatched area indicates maximum activity). See also legend Fig. 5.

Conclusions

Diel fluctuation of photosynthesis is the major process which determines the depth profiles of O_2 and sulfide. This results in a high flux of O_2 into the sediment and atmosphere during the day (Fig. 2). Sulfide production takes place throughout the diel cycle, but consumption seems to be restricted during the night period. In strata where complete (biological) oxidation of sulfide is not sustained, accumulation of inorganic sulfur in insoluble fractions prevents the flux of sulfide to higher strata. Laboratory studies indicate that sulfide is the preferential substrate for CSB and PSB (Visscher et al. 1990), but also show that compounds such as polysulfides and FeS can be used as alternative substrates. However, dynamic fluctuations in O_2 and sulfide concentrations predominantly determine species composition and activity of sulfur bacteria in microbial mats. Microbial metabolism changes as a response to fluctuations, modifying physico-chemical condition in the mat, resulting in intricate biotic and abiotic interactions.

Acknowledgements

Without many stimulating discussions and farcical times in the lab and field with Hans van Gemerden, Rutger de Wit, Bart Schaub and others, this publication would not have seen daylight.

References

Canfield DE, Des Marais DJ (1991) Aerobic sulfate reduction in microbial mats. Science 251: 1471-1473

De Wit R, Van Gemerden H (1990) Growth and metabolism of the phototrophic sulfur bacterium *Thiocapsa roseopersicina* under combined light/dark and oxic/anoxic regimens. Arch Microbiol 154: 459-464

Fossing H, Jørgensen BB (1990) Isotope exchange reactions with radiolabeled sulfur compounds in anoxic seawater. Biogeochem 9: 223-245

Fründ C, Cohen Y (1992) Diurnal cycles of sulfate reduction under oxic conditions in cyanobacterial mats. Appl Environ Microbiol 58: 70-77

Thode Andersen S, Jørgensen BB (1989) Sulfate reduction and the formation of ^{35}S-labelled FeS, FeS_2 and S^o in coastal marine sediments. Limnol Oceanogr 34: 793-806

Van den Ende FP, Van Gemerden H (1993) Sulfide oxidation under oxygen limitation by a *Thiobacillus thioparus* isolated from a microbial mat. FEMS Microbiol Ecol in press

Van Gemerden H (1993) Microbial mats: A joint venture. Mar Geology 113: 3-25

Visscher PT, Nijburg JW, Van Gemerden H (1990) Polysulfide utilization by *Thiocapsa roseopersicina*. Arch Microbiol 155:75-81

Visscher PT, Quist P, Van Gemerden H (1991) Methylated sulfur compounds in microbial mats: *in situ* concentrations and metabolism by a colorless sulfur bacterium. Appl Environ Microbiol 57:1758-1763

Visscher PT, Prins RA, Van Gemerden H (1992a) Rates of sulfate reduction and thiosulfate consumption in a marine microbial mat. FEMS Microbiol Ecol 86: 283-294

Visscher PT, Van den Ende FP, Schaub BEM, Van Gemerden H (1992b) Competition between anoxygenic phototrophic bacteria and colorless bacteria in a microbial mat. FEMS Microbiol Ecol 101: 51-58

Visscher PT, Van Gemerden, H (1993) Sulfur cycling in laminated marine microbial ecosystems. In: Oremland RS (ed) Biogeochemistry of global change. Chapman & Hall, New York, pp 672 690

Panel discussion: Diel and spatial variation of physico-chemical parameters and processes in microbial mats

Marlies Villbrandt[1] (rapporteur) and Brad Bebout[2] (chairman)

Institute for Chemistry and Biology of the Sea, University of Oldenburg, Carl von Ossietzky-Straße, D-26111 Oldenburg, Germany

During the panel discussion the following 5 questions were formulated: 1) what controls primary productivity in microbial mats? is it dissolved inorganic carbon, nutrients or light? are these controling net productivity, net photosynthesis, gross photosynthesis or growth? how do we define "limitation". 2) what are the consequences of fluctuations of physico-chemical parameters for species composition of mats? 3) microzones revisted! How can we demonstrate that a cell experiences anoxic conditions or is anoxic inside in an otherwise oxic environment? What is the relevant scale, cell or environment? 4) Is fluorescence important to study in mats? 5) Are microbial mats capable of storing carbon during the day and if so, how? Are mats capable of retaining nutrients at night, and if so, how?

Comparison of laboratory culture experiments and field studies

The discussion started with a general problem microbiologists have during the enrichment of bacteria. Strains which are dominating in laboratory cultures are not necessarilly of major ecological significance *in situ*. Alternative setups, incubation conditions and isolation techniques may be applied in order to select for bacteria that are important in the natural environment. Enrichment procedures are usually determined by the metabolism of the organism that should be isolated. However, some phototrophs, for example *Thiocapsa roseopersicina*, are highly versatile and capable to live as chemoautotrophs, mixotrophs or organotrophs in the dark and as photoautotrophs or photoheterotrophs in the light. In the laboratory, growth of less important specialists is often favoured, whereas the versatile microorganisms may be outcompeted. The inability of microbiologists to obtain cultures of certain organisms with high ecological relevancet such as sulfide-oxidizing *Beggiatoa* spp. or nitrogen-fixing cyanobacteria *Trichodesmium* spp. clearly demonstrates the problem.

---

[1] Present address: Laboratory for Microbiology, University of Amsterdam, Nieuwe Achtergracht 127, 1018WS Amsterdam, The Netherlands.
[2] Max Planck Institute for Marine Microbiology, Fahrenheitstraße 1, D-28359 Bremen, Germany

NATO ASI Series, Vol. G 35
Microbial Mats
Edited by Lucas J. Stal and Pierre Caumette

The tendency to translate results obtained by competition experiments in the laboratory to the natural environment was criticized. Pure cultures are not natural and they may behave totally different in natural communities.

Competition experiments with purple sulfur bacteria showed that, although a large species diversity was expected in the natural environment, most of these were outcompeted in the laboratory. For example, *Thiocapsa* is often the dominant anoxyphotobacterium in marine mats, but species such as *Ectothiorhodospira*, *Chromatium* or *Thiopedia* may be important as well.

Sulfur cycling in marine microbial mats

Other points of discussion were (i) transformations of thiosulfate in mats of *Microcoleus chthonoplastes* and (ii) complications that arise when working with radioactive labelled sulfur compounds. This latter method is used in order to follow *in situ* the intermediate steps of disproportionation of thiosulfate, sulfite or elemental sulfur to sulfate and sulfide. It was pointed out that sulfate, but not elemental sulfur or hydrogen sulfide, may be used as radiotracer substance because the radioactive label rapidly exchanges between the different pools of sulfur compounds. Results of such experiments are difficult to interprete. It was expected that the oxidation of sulfide to thiosulfate by *Microcoleus chthonoplastes* is not important in most microbial mats because in the light sulfide rarely reaches the cyanobacterial layer. However, it should be noted that conditions in the mat may change rapidly. Fluctuations in light intensity, for instance as a result of clouds, may cause dramatic and sudden shifts of physical and chemical parameters. *In situ* measurements using microelectrodes showed that the re-oxidation of hydrogen sulfide is a fast process. Only about 10% of the sulfide is permanently bound to metals such as iron or manganese. However, polysulfides are also important compounds in mats. Iron sulfide probably does not play an important role during daytime in the layer of purple sulfur bacteria. Iron sulfide is probably continuously oxidized and reduced during a diel cycle.

Microzonations

The problem of microzonations was discussed. Although it is accepted by some authors that anoxic microzones exist in oxic environments, microelectrode measurements thus far have failed to demonstrate such sites in microbial mats. However, a microbial mat is a three-dimensional system. Therefore computer image analysis may help to characterize the micro-milieu. It was suggested that oxic microzonations in anoxic surroundings are perhaps more interesting than anoxic microzones in oxic environments. Plant roots or bioturbation may introduce oxygen into anoxic sediments. Anoxic microzones are probably not detected by microelectrodes due to the patchiness of redox gradients. The application of tetrazolium salt reduction as a tool to study microzones was discussed, but this method has some disadvantages, which makes it problematic to interprete the results obtained with it.

Another aspect relating to microzones is that the concentration of a substance may not be the same as what a single organism experiences. This is particularly the case in microbial mats or microbial aggregates. The high biomass and microbial activities will produce steep gradients of almost any substance. High rates of oxygen consumption may maintain anoxic conditions. This mechanism is known to allow the oxygen-sensitive process of nitrogen fixation in the heterotrophic *Azotobacter* spp. but also in the heterocysts of certain nitrogen-fixing cyanobacteria. The same mechanism may apply to some sulfate-reducing bacteria. Sulfate reduction is considered as an obligately anaerobic process. Yet, it has been shown in oxic environments and this was an important discovery in microbial mat research. Aerobic sulfate reduction is still a matter of dispute as long as pure cultures are lacking. The fact that high numbers of sulfate-reducing bacteria are found in the upper (oxic) layers of microbial mat can not be taken as evidence. The sulfate-reducing bacteria present in the oxic part of the mat may be inactive or perform an other type of metabolism. Several sulfate-reducing bacteria have been shown to be metabolically highly versatile.

Biodiversity

In order to cope with fluctuations of physical and chemical parameters, purple sulfur bacteria and *Chloroflexus* possess metabolic versatility. In addition, a high biodiversity (species diversity) is often found in microbial mats. There may be a vast array of different microorganisms each adapted to specific environmental conditions. This would result in a highly flexible community. Microscopical analyses often reveal the presence of different morphotypes but it is uncertain whether these represent different genotypes.

It is expected that molecular approaches will cover the main part of future microbial mat research. This may result in the rejection of the current nomenclature of cyanobacteria and replace it by one based on molecular approaches. However, traditional methods in microbiology should not be forgotten since it gives information on aspects not covered by molecular approaches.

If an organism is metabolically versatile this does not necessarily imply that this property is used in the natural environment. Changes in environmental conditions in microbial mats are generally rapid. The induction of enzymes necessary to perform another type of metabolism is usually not fast enough to respond to the rapidly fluctuating conditions. Organisms that have different metabolisms constitutively have the advantage to be reactive. On the other hand the energetic consequenses of such strategy are considerable. Organisms that are motile are not necessarlily dependent on a flexible metabolism because they can position themselves optimally along the physical chemical gradients.

Control of primary production in microbial mats

The problem of the controlling factors of primary production in microbial mats was discussed. To answer this question interdisciplinary exchange between oceanographers, geochemists, microbiologists, and others will be necessary. The microbial mat system is too complex to view it in reductionistic way.

Undoubtely, the storage of carbon and nutrients plays a role for the survival of organisms under unfavourable conditions. It was found that nitrogen and carbon may be stored in microbial mat (organisms) and may become available when required. In this context, it was stated that the rate of photosynthesis may be limited by inorganic substances.

An important question was whether a microbial mat is actually growing. Obviously, in order to be able to answer this question the limiting factor of the system must be known. Also, a microbial mat may change from actively growing, dormant or in steady state (growth equals mineralization). Examples of controling factors on biomass accumulation include the composition of autotrophic and heterotrophic organisms and the grazing pressure.

V. New physiological groups of organisms in microbial mats

Novel metabolic capacities of sulfate-reducing bacteria, and their activities in microbial mats

Heribert Cypionka

Institute for Chemistry and Biology of the Sea, University of Oldenburg, Carl von Ossietzky-Straße, D-26111 Oldenburg, Germany

A typical inhabitant of a microbial mat is able to control its metabolism or to migrate vertically in response to the diel changes of light, redox conditions, and the concentrations of O_2, sulfide, organic substrates etc. The dissimilatory sulfate-reducing bacteria, however, were considered to be strict anaerobes, active in the reduced zones at the bottom of microbial mats, only. This view has changed now. New observations concerning the metabolism of sulfate reducers are coming from activity measurements in mats and from laboratory studies. However, to this point field and pure culture results give different pictures. My aim in this review is to discuss the metabolic versatility of sulfate-reducing bacteria. Some general principles derived from pure culture studies might be worthwile as guidelines, but will not close the gap. First, I shall describe the ambivalent relations of sulfate-reducing bacteria to oxygen. Then, the consequences of varying environmental conditions for sulfur metabolism are discussed.

Electron acceptors reduced by sulfate-reducing bacteria

Most sulfate-reducing bacteria are able to reduce electron acceptors other than oxidized sulfur compounds. For example, nitrate or nitrite can be reduced to ammonia (Keith and Herbert 1983; Seitz and Cypionka 1986), or CO_2 to acetate (Klemps et al. 1985). Other inorganic electron acceptors such as selenate (Zehr and Oremland 1987), uranium (Lovley and Phillips 1992), iron (Coleman et al. 1993), and molecular oxygen are also reduced. The last one is discussed here in detail, with respect to (i) oxygen tolerance and toxicity, (ii) features of aerobic respiration, and (iii) influence of O_2 on the sulfur metabolism.

Oxyen tolerance and toxicity

High numbers of sulfate-reducing bacteria have been counted in oxic sediment zones (Laanbroek and Pfennig 1981; Jørgensen and Bak 1991; Visscher et al. 1992). In pure cultures varying degrees of O_2 tolerance are found. Many strains survive O_2 exposure for hours or even days (Hardy and Hamilton 1981; Cypionka et al. 1985; Fukui and Takii 1990 a and b). With

NATO ASI Series, Vol. G 35
Microbial Mats
Edited by Lucas J. Stal and Pierre Caumette

increasing O_2 concentrations and exposure periods, however, motility and viability decrease (Gottschal and Szewzyk 1985; Marschall et al. 1993). O_2 itself appears not to be toxic, but rather compounds that are formed by incomplete O_2 reduction. Therefore, the availability of reducing power by organic substrates and molecular hydrogen, is of importance. Especially, H_2S and thiol compounds increase the O_2 toxicity since they induce the formation of hydrogen peroxide and superoxide radicals (Carlsson et al. 1978; Cypionka et al. 1985). Catalase and superoxide dismutase have been detected in some of the sulfate reducers tested (Hewitt and Morris 1975; Hatchikian et al. 1977; Hardy and Hamilton 1981; Marschall et al. 1993). However, these activities did not reveal an unequivocal correlation with the O_2 resistance of the bacteria.

Aerobic respiration

Oxygen is not only tolerated to a limited extent, but can be metabolized by most sulfate--reducing bacteria (Abdollahi and Wimpenny 1990; Dilling and Cypionka 1990; Dannenberg et al. 1992; Marschall et al. 1993). Electron donors for aerobic respiration include H_2, formate, acetate, propionate, butyrate, lactate and pyruvate. Furthermore, sulfur compounds can be oxidized as discussed below. From the O_2 per substrate stoichiometries it can be concluded that O_2 is completely reduced to water. Depending on the strain, the substrates are completely or incompletely oxidized as with sulfate as electron acceptor. Some *Desulfovibrio* species show aerobic respiration rates comparable to those of aerobic bacteria. Sulfate reducers are microaerophilic, and reveal a very high affinity for O_2. Some show increasing respiration rates with decreasing O_2 concentrations, while at increased O_2 concentrations respiration slows down after short time. Respiration is coupled to energy conservation, as indicated by vectorial proton translocation and O_2-dependent ATP formation. Thus, sulfate reducers can carry out a true aerobic respiration (Dilling and Cypionka 1990; Dannenberg et al. 1992). In spite of this, growth with O_2 as electron acceptor is usually not observed in homogeneously aerated cultures. Not more than one doubling has been obtained with *Desulfovibrio desulfuricans* (Marschall et al. 1993). In stabilized stratified environments, however, O_2-dependent growth is possible over prolonged periods as discussed below.

The electron transport to O_2 is not clarified so far. Since respiration is insensitive to classical inhibitors such as cyanide and azide, the cells must possess terminal oxidases different from those of aerobic organisms. Cytochrome c3 has been reported to undergo autoxidation (Postgate 1984). Furthermore, an O_2-reducing rubredoxin-oxidizing flavo-hemoprotein has been described in *Desulfovibrio gigas* (Chen et al. 1993). As the capacity of aerobic respiration is constitutive, it appears probable that aerobic respiration is catalyzed by enzymes and carriers that normally are involved in sulfate reduction.

Table 1. Some reactions of the sulfur cycle catalyzed by sulfate-reducing bacteria

		$\Delta G^{\circ\prime}$ [1] kJ/mol
	Complete reduction of sulfur compounds	
1	$SO_4^{2-} + 4\ H_2 + 1\frac{1}{2}H^+ \longrightarrow \frac{1}{2}H_2S + \frac{1}{2}HS^- + 4\ H_2O$	-155
2	$\frac{1}{2}SO_3^{2-} + \frac{1}{2}HSO_3^- + 3\ H_2 + H^+ \longrightarrow \frac{1}{2}H_2S + \frac{1}{2}HS^- + 3\ H_2O$	-175
3	$S_2O_3^{2-} + 4\ H_2 + H^+ \longrightarrow H_2S + HS^- + 3\ H_2O$	-179
4	$S + H_2 \longrightarrow \frac{1}{2}H_2S + \frac{1}{2}HS^- + \frac{1}{2}H^+$	-30
	Incomplete sulfate reduction	
5	$SO_4^{2-} + 2\ H_2 + H^+ \longrightarrow \frac{1}{2}S_2O_3^{2-} + 5\ H_2O$	-65
	Disproportionation of sulfur compounds	
6	$S_2O_3^{2-} + H_2O \longrightarrow SO_4^{2-} + \frac{1}{2}H_2S + \frac{1}{2}HS^- + \frac{1}{2}H^+$	-25
7	$2\ HSO_3^- + 2\ SO_3^{2-} \longrightarrow 3\ SO_4^{2-} + \frac{1}{2}H_2S + \frac{1}{2}HS^- + \frac{1}{2}H^+$	-236
8	$4\ S + 4\ H_2O \longrightarrow 1\frac{1}{2}H_2S + 1\frac{1}{2}HS^- + SO_4^{2-} + 3\frac{1}{2}H^+$	+33 [2]
	Oxidation of sulfur compounds	
9	$\frac{1}{2}H_2S + \frac{1}{2}HS^- + 2\ O_2 \longrightarrow SO_4^{2-} + 1\frac{1}{2}H^+$	-794
10	$\frac{1}{2}H_2S + \frac{1}{2}HS^- + NO_3^- + \frac{1}{2}H^+ + H_2O \longrightarrow SO_4^{2-} + NH_4^+$	-445
11	$S_2O_3^{2-} + 2\ O_2 + H_2O \longrightarrow 2\ SO_4^{2-} + 2\ H^+$	-818
12	$\frac{1}{2}SO_3^{2-} + \frac{1}{2}HSO_3^- + \frac{1}{2}O_2 \longrightarrow SO_4^{2-} + \frac{1}{2}H^+$	-257 [3]

[1] Free energy changes calculated after Thauer et al. (1977)
[2] Promoted by precipitation of sulfide as Fe or Mn salt
[3] Incomplete O_2 reduction may form H_2O_2

Variations of sulfur metabolism

Hitherto, it was generally assumed that sulfate-reducing bacteria completely reduce sulfate or some other sulfur compounds to H_2S which is released as end product (Table 1, equations. 1 to 4). Meanwhile, it has turned out that the sulfur metabolism may change depending on the environmental conditions, especially the availability of electron donors and acceptors. Many sulfate reducers are able to (i) reduce sulfate incompletely to thiosulfate (eq. 5) or trithionate, (ii) carry out a disproportionation of partially reduced sulfur compounds (eqs. 6 to 8) or (iii) even oxidize sulfur compounds (eqs. 9 to 12).

Incomplete sulfate reduction

Sulfate reduction includes at least APS and sulfite as intermediates, with the biochemical pathway of sulfite reduction still debated. Sulfite might be reduced in a single reaction with six electrons or step by step with trithionate and thiosulfate as intermediates (Kobayashi et al. 1969). The trithionate pathway includes reactions with different redox potentials. Especially, thiosulfate reduction to sulfite and sulfide requires electrons with a strongly negative redox potential. If no powerful reductant like H_2 is present at high concentration, even reverse electron transport is necessary in order to reduce thiosulfate. In accordance with this, formation of thiosulfate (eq. 5) and of traces of trithionate has been observed with washed cells (of freshwater and marine species) or in growing cultures at conditions of electron donor limitation (Fitz and Cypionka 1990). In growing cultures up to 400 μM thiosulfate and 100 μM trithionate were formed transiently. In the presence of uncouplers, that prevent reverse electron transport, thiosulfate was the main product of sulfite reduction at low H_2 concentration (Sass et al. 1992). It should be pointed out that also in natural, and especially in marine habitats, sulfate reducers are normally under electron donor limitation.

Disproportionation of sulfur compounds

In the absence of electron donors sulfate cannot be reduced. However, many sulfate reducers are able to metabolize partially reduced sulfur compounds (Bak and Cypionka 1987). For example, thiosulfate or sulfite are disproportionated to sulfate and sulfide (eqs. 6 and 7, Fig. 1). This type of metabolism can be regarded as inorganic fermentation. It involves a partial oxidation of sulfur compounds, which proceeds via a reversal of the reactions normally involved in sulfate activation (Krämer and Cypionka 1989). The free energy change is low, and many disproportionating bacteria do not grow with thiosulfate as the only energy source. However, even the disproportionation of elemental sulfur, which is endergonic under standard conditions (eq. 8), can be utilized for growth if the sulfide formed is removed by reactions with iron or manganese (Thamdrup et al. 1993).

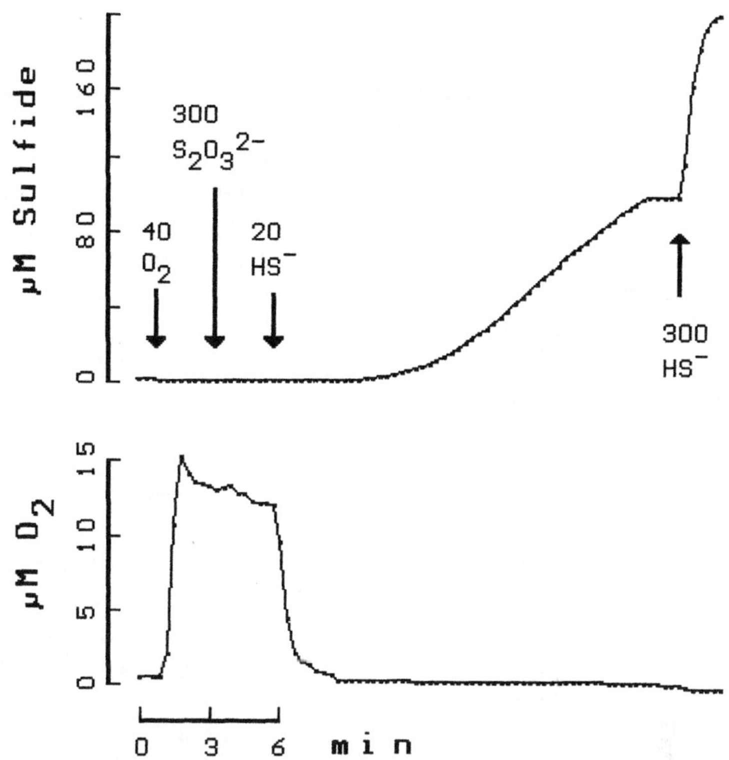

Fig. 1. Metabolism of oxygen, thiosulfate and sulfide by *Desulfobulbus propionicus*. Washed cells (1.5 mg protein/ml) were incubated at 30 °C in 3 ml N_2-saturated KCl. Sulfide and O_2 were measured simultaneously by means of electrodes. Upon O_2 addition (40 nmol), slow endogenous respiration is induced, which is not stimulated by the addition of 300 nmol thiosulfate. However, the small addition of sulfide (20 nmol) is immediately oxidized. This oxidation follows eq. 9, and is catalyzed biologically as concluded from control experiments with inactive cells. After oxygen is consumed, metabolism of thiosulfate starts. As proven by the calibration pulse at the end, one sulfide per thiosulfate is formed as expected from thiosulfate disproportionation according to eq. 6.

It has been demonstrated that the disproportionation of thiosulfate plays a major role in the sulfur cycle of marine and freshwater sediments (Jørgensen 1990 a,b; Jørgensen and Bak 1991). As indicated by tracer experiments, a significant part of sulfide reoxidation proceeds via thiosulfate, which is then disproportionated. Since the capacity of disproportionation has not been found in other sulfuroxidizing bacteria (Krämer and Cypionka 1989) it appears that sulfate-reducing bacteria take part not only in the reduction but also in the reoxidation of sulfur compounds in sediments and or microbial mats.

Oxidation of sulfur compounds

The logical step after incomplete reduction and disproportionation is the oxidation of sulfur compounds. If O_2 is available many sulfate-reducing bacteria can oxidize inorganic sulfur compounds (eqs. 9 to 12, Dannenberg et al. 1992). Sulfite, thiosulfate, polysulfide and even sulfide can be oxidized (Fig. 1). Depending on the strain, oxidation is incomplete or proceeds to sulfate. With some strains, oxidation of sulfur compounds is possible even with nitrate or nitrite as electron acceptor (Dilling and Cypionka 1990). The biochemistry of sulfur oxidation by sulfate-reducing bacteria is not completely clear. Obviously, intermediates are formed (Dannenberg et al. 1992). Although the reactions are more exergonic than sulfate reduction, aerobic growth in homogeneously agitated cultures with inorganic sulfur compounds as electron donors has not been observed so far.

Behaviour of sulfate-reducing bacteria in stratified environments

In summary, dissimilatory sulfate-reducing bacteria possess a high metabolic versatility with respect to redox reactions. They are able to catalyze all steps of the sulfur cycle in any direction. Given the capacity of respiration with various electron acceptors, allowing mediation between the cycles of oxygen, nitrogen, sulfur, iron etc., it appears unlikely that sulfate reducers should be restricted to the deepest and strictly anoxic zone in a sediment or microbial mat. Instead, their occurrence near the oxic surface is to be expected. This has been confirmed in simulation experiments with pure cultures in agar-stabilized tubes (Cypionka et al. 1985; Marschall et al. 1993). In artificial oxygen-sulfide gradients, sulfatereducing bacteria can grow directly at the oxygen-sulfide interface. They stabilize for several weeks those gradients which are oxidized within two days in the absence of bacteria. At the interface, a band of growing cells develops which is able to migrate vertically in response to changing O_2 concentrations.

The actual metabolism of the bacteria in the gradient is not easily analyzed. Under defined conditions in homogeneous cell suspensions, however, a hierarchy that follows thermodynamics is observed. As long as O_2 is available aerobic respiration dominates, while sulfate reduction is suppressed. Only at excess electron donor supply have aerobic respiration and sulfate reduction been observed concurrently. Normally, the different transformations proceeded sequentially step by step (Fig. 1). Therefore, our pure culture experiments cannot explain the field observations of sulfate reduction in microbial mats at zones with O_2 tensions up to saturation (Jørgensen and Cohen 1977; Battersby et al. 1985; Cohen 1989; Canfield and Des Marais 1991; Fründ and Cohen 1992). However, so far no isolates from microbial mats were included in the pure culture studies.

The various sulfur transformations catalyzed by sulfate-reducing bacteria can be summarized in a simplified scheme (Fig. 2). The balance from reduction to oxidation of sulfur compounds goes along with the redox gradient in a typical habitat of sulfate-reducing bacteria. The gradient of available organic substrates, however, is reversed compared to the natural conditions, since organic substrates are formed by photosynthesis within the oxic zone. Therefore, the scheme does not allow a prediction of the transformations in a sediment or microbial mat.

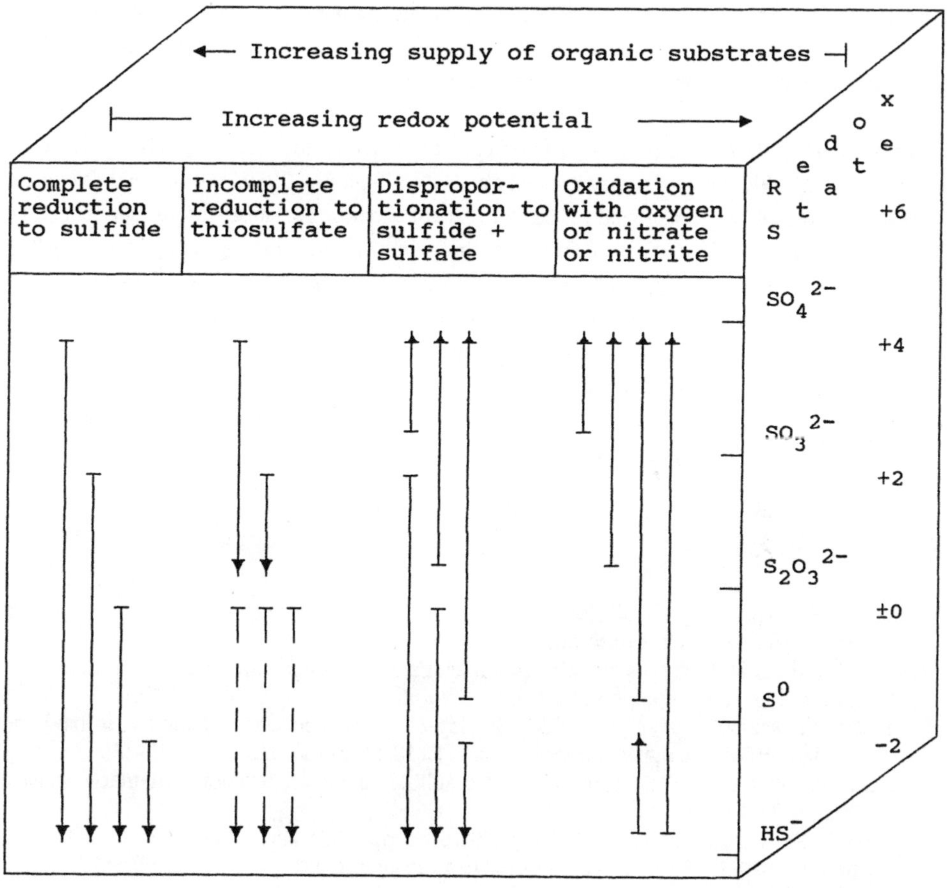

Fig. 2. Simplified scheme of the sulfur transformations by sulfate-reducing bacteria at varying environmental conditions. At sufficient supply of organic substrates, complete reduction of the sulfur compounds is to be expected. At electron donor limitation, sulfate reduction might get incomplete, and intermediate sulfur compounds might be disproportionated to sulfate and sulfide. In the presence of oxygen or nitrate, reduced sulfur compounds can function as electron donors. Note that the gradient of organic substrate availability is reversed compared to the situation in a sediment or microbial mat.

Conclusions

Two general principles can be derived from the novel observations on the activities of sulfate-reducing bacteria. First, a high degree of versatility with respect to redox reactions and sulfur metabolism must be taken into account. Almost any sulfur transformation is catalyzed by sulfate-reducing bacteria. Even reactions with positive free energy changes under standard conditions may occur when the concentrations of the reactants are kept in the appropriate range. Thermodynamics delivers a guideline, at least in pure culture studies, since it brings order in the sequence of the various possible reactions.

The second principle is that thermodynamics is not sufficient to describe biology. We have learned that the sulfate-reducing bacteria reveal a considerable metabolic diversity. While a strain like *Desulfovibrio desulfuricans* strain CSN (Cypionka 1989; Dilling and Cypionka 1990) catalyzes all reactions listed in Table 1 (except equation 9), most sulfate reducers carry out only selected reactions. For example (as demonstrated in Fig. 1), *Desulfobulbus propionicus* oxidizes sulfide completely to sulfate, and disproportionates thiosulfate to sulfide and sulfate, but does not oxidize thiosulfate. Therefore, in each investigated mat or sediment, the population of sulfate-reducing bacteria has to be analyzed carefully. The new techniques based on molecular biology are expected to be very helpful in this field. New isolates have to be characterized physiologically and identified in their natural habitats, regarding both, versatility and diversity of sulfate-reducing bacteria.

References

Abdollahi H, Wimpenny JWT (1990) Effects of oxygen on the growth of *Desulfovibrio desulfuricans*. J Gen Microbiol 136: 1025-1030

Bak F, Cypionka H (1987) A novel type of energy metabolism involving fermentation of inorganic sulphur compounds. Nature 326: 891-892

Battersby NS, Malcolm SJ, Brown CM, Stanley SO (1985) Sulphate reduction in oxic and suboxic North East Atlantic sediments. FEMS Microbiol Ecol 31: 225-228

Canfield DE, Des Marais DJ (1991) Aerobic sulfate reduction in microbial mats. Science 251: 1471-1473

Carlsson J, Nyberg G, Wrethen J (1978) Hydrogen peroxide and superoxide radical forma tion in anaerobic broth media exposed to atmospheric oxygen. Appl Environ Microbiol 36: 223-229

Chen l, Liu MY, Le Gall J, Fareleira P, Santos H, Xavier AV (1993) Rubredoxin oxidase, a new flavo-hemo-protein, is the site of oxygen reduction to water by the strict anaerobe *Desulfovibrio gigas*. Biochem Biophys Res Commun 193: 100-105

Cohen Y (1989) Photosynthesis in cyanobacterial mats and its relation to the sulfur cycle: a model for microbial sulfur interactions. In: Cohen Y, Rosenberg E (eds), Microbial Mats. Physiological ecology of benthic microbial communities. Am Soc Microbial, Washington, pp 22-36

Coleman ML, Hedrick DB, Lovley DR, White DC, Pye K (1993) Reduction of Fe (III) in sediments by sulphate-reducing bacteria. Nature 361: 436-438

Cypionka H (1989) Characterization of sulfate transport in *Desulfovibrio desulfuricans*. Arch Microbiol 152: 237-243

Cypionka H, Widdel F, Pfennig N (1985) Survival of sulfate-reducing-bacteria after oxygen stress, and growth in sulfate-free oxygen-sulfide gradients. FEMS Microbiol Ecol 31: 39-45

Dannenberg S, Kroder M, Dilling W, Cypionka H (1992) Oxidation of H_2, organic compounds and inorganic sulfur compounds coupled to reduction of O_2 or nitrate by sulfate-reducing bacteria. Arch Microbiol 158: 93-99

Dilling W, Cypionka H (1990) Aerobic respiration in sulfate-reducing bacteria. FEMS Microbiol Lett 71: 123-128

Fitz RM, Cypionka H (1990) Formation of thiosulfate and trithionate during sulfite reduction by washed cells of *Desulfovibrio desulfuricans*. Arch Microbiol 154: 400-406

Fründ C, Cohen Y (1992) Diurnal cycles of sulfate reduction under oxic conditions in cyano bacterial mats. Appl Environ Microbiol 58: 70-77

Fukui M, Takii S (1990a) Survival of sulfate-reducing bacteria in oxic surface sediment of a seawater lake. FEMS Microbiol Ecol 73: 317-322

Fukui M Takii S (1990b) Colony formation of free-living and particle-associated sulfate-reducing bacteria. FEMS Microbiol Ecol 73: 85-90

Gottschal JC, Szewzyk R (1985) Growth of a facultative anaerobe under oxygen-limiting conditions in pure culture and in co-culture with a sulfate-reducing bacterium. FEMS Microbiol Ecol 31: 159-170

Hardy J A, Hamilton W A (1981) The oxygen tolerance of sulphate-reducing bacteria isolated from North Sea waters. Current Microbiol 6: 259-262

Hatchikian CE, LeGall J, Bell GR (1977) Significance of superoxide dismutase and catalase activities in the strict anaerobes, sulphate-reducing bacteria. In: Michelson A M, McCord J M, Fridovich I (eds), Superoxide and superoxide dismutases. Academic Press, London, pp 159-172

Hewitt J, Morris JG (1975) Superoxide dismutase in some obligately anaerobic bacteria. FEBS Lett 50: 315-318

Jørgensen BB (1990a) A thiosulfate shunt in the sulfur cycle of marine sediments. Science 249: 152-154

Jørgensen BB (1990b) The sulfur cycle of fresh water sediments: Role of thiosulfate. Limnol Oceanogr 35: 1329-1342

Jørgensen BB, Cohen Y (1977) Solar Lake (Sinai). 5. The sulfur cycle of benthic cyano-bacterial mats. Limnol Oceanogr 22: 567-666

Jørgensen BB, Bak F (1991) Pathways and microbiology of thiosulfate transformations and sulfate reduction in a marine sediment (Kattegatt, Denmark). Appl Environ Microbiol 57: 847-856

Keith SM, Herbert RA (1983) Dissimilatory nitrate reduction by a strain of *Desulfovibrio desulfuricans*. FEMS Microbiol Lett 18: 55-59

Klemps R, Cypionka H, Widdel F, Pfennig N (1985) Growth with hydrogen, and further physiological characteristics of *Desulfotomaculum* species. Arch Microbiol 143: 203-208

Kobayashi K, Tachibana S, Ishimoto M (1969) Intermediary formation of trithionate in sulfite reduction by sulfate-reducing bacteria. J Biochem 65: 155-157

Krämer M, Cypionka H (1989) Sulfate formation via ATP sulfurylase in thiosulfate- and sulfite-disproportionating bacteria. Arch Microbiol 151: 232-237

Laanbroek HJ, Pfennig N (1981) Oxidation of short-chain fatty acids by sulfate-reducing bacteria in freshwater and marine sediments. Arch Microbiol 128: 330-335

Lovley DR, Phillips EJP (1992) Reduction of uranium by *Desulfovibrio desulfuricans*. Appl Environ Microbiol 58: 850-856

Marschall C, Frenzel P, Cypionka H (1993) Influence of oxygen on sulfate reduction and growth of sulfate-reducing bacteria. Arch Microbiol 159: 168-173

Postgate JR (1984) The sulphate-reducing bacteria. Cambridge University Press, 2nd edition, 208 pp

Sass H, Steuber J, Kroder M, Kroneck PMH, Cypionka H (1992) Formation of thionates by freshwater and marine strains of sulfate-reducing bacteria. Arch Microbiol 158: 418-421

Seitz H-J, Cypionka H (1986) Chemolithotrophic growth of *Desulfovibrio desulfuricans* with hydrogen coupled to ammonification of nitrate or nitrite. Arch Microbiol 146: 63-67

Thamdrup B, Finster K, Hansen JW, Bak F (1993) Bacterial disproportionation of elemental sulfur coupled to chemical reduction of iron or manganese. Appl Environ Microbiol 59: 101-108

Thauer RK, Jungermann K, Decker K (1977) Energy conservation in chemotrophic anaerobic bacteria. Bacteriol Rev 41: 100-180

Visscher PT, Prins RA, van Gemerden H (1992) Rates of sulfate reduction and thiosulfate consumption in a marine microbial mat. FEMS Microbiol Ecol 86: 283-293

Zehr JP, Oremland RS (1987) Reduction of selenate to selenide by sulfate-respiring bacteria: Experiments with cell suspensions and estuarine sediments. Appl Environ Microbiol 53: 1365-1369

Diversity of and interactions among sulphur bacteria in microbial mats

Rutger de Wit[1] **and Pierre Caumette**

Laboratoire d'Océanographie Biologique, Université de Bordeaux I,
2, Rue du Prof. Jolyet, F 33120 Arcachon. France.

Introduction

Microbial mats are found along the outflow of continental thermal springs, on marine littoral sediments, in thalassic and athalassic (inland) hypersaline ponds and lakes, and in the deep-sea along hydrothermal vents. Hydrothermal vent microbial mats consist of chemotrophic sulphur bacteria thriving in the dark on sulphide which is mainly supplied by geochemical processes. However, it has been shown that sulphate reduction in these systems does occur up to 110 °C (Jørgensen et al. 1992), thus indicating that complete sulphur cycling also takes place. Microbial mats that are exposed to sunlight often comprise dense populations of oxygenic and anoxygenic phototrophic microorganisms together with chemoorganotrophs and chemolithotrophs. In habitats where the sulphur cycling is not a dominant process, microbial mats often originate from an association of cyanobacteria with anoxygenic filamentous phototrophic bacteria. For example, in thermal mats, it was shown that the *Chloroflexus*-like filamentous bacteria incorporated glycolate which was a major excretion product of the cyanobacteria found in the same mats (Bateson and Ward 1988). This commensalistic relationship is a nice example of a positive interaction of two organisms coexisting in the same environment. In other microbial mat environments, sulphide oxidation plays a predominant role. In the absence of geochemically formed sulphide, sulphide formation results mainly from the activities of sulphur and sulphate-reducing bacteria. In this paper, we discuss the biodiversity of sulphur bacteria and their ecological interactions in microbial mats.

Along the outflow of continental sulphide-containing hot springs, microbial mats comprise: 1) sulphide-tolerant cyanobacteria; 2) *Chloroflexus* -like organisms (Giovanonni et al. 1987); 3) laminated systems of *Chloroflexus* on top of cyanobacteria (Jørgensen and Nelson 1988); 4) mixtures of *Chloroflexus* and the thermophilic purple sulphur bacterium *Chromatium tepidum* (Madigan 1984); 5) purple layers of *Chr. tepidum* (Madigan 1986), or 6) green layers of *Chlorobium tepidum* (Wahlund et al. 1991). The *Chloroflexus* strains from sulphide-containing thermal environments are dark-green in colour, contain bacteriochlorophyll *c*

[1] Centre National de Recherche Scientifique (C.N.R.S.), U.R.A. 197

NATO ASI Series, Vol. G 35
Microbial Mats
Edited by Lucas J. Stal and Pierre Caumette

(BChl*c*) and are obligately phototrophic, thus, different from *Chloroflexus aurantiacus* isolated from non-sulphidic mats.

At the sediment surface in marine and hypersaline environments, microbial mats often comprise laminated structures in which phototrophic and/or chemotrophic sulphur bacteria stratify underneath the cyanobacterial and algal layers. Most commonly, a single purple layer develops below the oxygenic phototrophs. However, more complex multilayered systems have been described for mats in Sippewissett Salt Marsh (USA) (Nicholson et al. 1987). In these systems, from top to bottom, the phototrophic sulphur bacteria constitute purple, peach and green layers corresponding to BChl*a*-, BChl*b*-containing purple sulphur bacteria and BChlc-containing green sulphur bacteria, respectively. Microbial mats are found over a high salinity range up to about 300 g/kg. Recently, several novel species of moderately halophilic purple sulphur bacteria and a sulphate-reducing bacterium have been isolated from hypersaline microbial mats (Oren et al. 1989; Caumette et al. 1988, 1991a,b).

New species and their habitats

Since 1986, many new species of sulphur bacteria have been isolated from microbial mats, which are reviewed in this chapter. These novel organisms and their habitats are listed in Table 1. However, in the mean time, many new species of sulphur bacteria have been described which were isolated from benthic environments other than microbial mats. The authors would like to emphasise that several of these species do likely occur in microbial mats as well.

Deep-sea organisms

In the deep-sea microbial mats close to hydrothermal vents in the Guaymas Basin, populations of chemotrophic sulphur bacteria have been found that consist of *Beggiatoa* filaments of hitherto unknown large size (Nelson et al. 1989). Originally, filament width was the sole criterion upon which the species within the genus *Beggiatoa* were differentiated; hence, *B. gigantea* with filament widths up to 55 μm was recognised as the largest *Beggiatoa*-species observed in field samples (Leadbetter 1974), and only one species, i.e., *B. alba* was described based on cultured type specimen (Mezzino et al. 1984). In the deep-sea mats, up to three distinct width classes of *Beggiatoa* have been observed with mean widths of 25, 42, and 120 μm, respectively. Thus, the narrowest size classes are similar in size to *B. gigantea*; whilst the broadest size represent hitherto unknown giant bacteria; several of them even exceeded 1 cm in length. The organisms contain large vacuoles devoid of cytoplasm. It was not possible to culture these giant bacteria, but ship-board experiments have been performed with mat samples. Sulphide-stimulated $^{14}CO_2$-uptake and high RuBisCo contents indicated that the organisms grow by chemolithoautotrophy (Nelson 1989). Microsensor measurements within these mats demonstrated strongly fluctuating oxygen and sulphide concentrations explained as a result of a pulsatory water-flow. This results in increased oxygen and sulphide supply to the organisms, thus, favouring the selection of giant bacteria (Gundersen et al. 1992).

Recently, a new rod-shaped chemolithotrophic sulphur bacterium was isolated from hydrothermal vents in the North Fiji Basin, and fully described by Durand et al. (1993), as new species of the genus *Thiobacillus*, i.e., *T. hydrothermalis*. The type strain was isolated from microbial mats on chimney rocks and two similar strains belonging to the same species, were isolated from the vent fluid as well. The new species is an obligate chemolithoautotroph, mesophilic (optimum temperature 35 °C), slightly halophilic and markedly halotolerant (optimum growth at 0.4 M NaCl; no growth above 2.5 M NaCl). It is barotolerant: the growth rate is maximal at a pressure of 1 atm., whilst 60 % of this value is attained at the *in situ* pressure of 220 atm. It was described as a new species on the genetic basis of the GC-content and 16-S-RNA sequence; the closestly related organism is *Thiobacillus neapolitanus*. Furthermore, five strains of sulphate-reducing bacteria have been isolated from the same environment (Elsgaard et al. 1991). Based on morphological features and lipid analyses, these strains were assigned to the genus *Desulfovibrio* and probably do not represent new species.

Thermophilic phototrophs

From the thermal environments, two novel species of phototrophic sulphur bacteria have been obtained in axenic culture, and fully described by M.T. Madigan and coworkers. *Chromatium tepidum* does occur as the predominant phototrophic bacterium in thin reddish mats (Madigan 1986) or in combination with dark-green *Chloroflexus* (Madigan 1984) along the outflow of neutral or slightly alkaline sulphidic thermal spring (Mammoth Hot Springs, Yellowstone National Park, U.S.A.). The rod-shaped, occasionally motile, BChl*a*-containing organism is a distinct *Chromatium* species due to its moderate thermophily (growth optimum 48-50 °C, no growth below 34 °C and above 57 °C), its carotenoid composition (spirilloxanthin series) characterised by an unusually high proportion of rhodovibrin (50 %), and its unique *in vivo* absorption spectrum with near infrared (NIR) peaks at 808, 858 and 920 nm. The latter peak corresponds to a novel antennae of BChl*a* (Garcia et al. 1986).

Chlorobium tepidum was isolated from acidic (pH 4.5-6) sulphidic thermal mats found in the central volcanic region in North Island (New Zealand) and fully described by Wahlund et al. (1991). It contains chlorosomes with BChlc and chlorobactene as the photosynthetic pigments. The organism is a moderate thermophile; the optimum growth temperature is 47 °C and it does not grow below 32 °C and above 53°C. Furthermore, the species grows well on thiosulphate and has excellent dinitrogen fixing capacities.

Thiorhodovibrio a new genus of the Chromatiaceae

A novel spirilloid-shaped species of purple sulphur bacterium belonging to the Chromatiaceae, was isolated simultaneously from littoral sediments of an athalassic saline lake and from marine microbial mats by Overmann et al. (1992). Similar morphological types of cells were observed in a purple layer underneath a *Beggiatoa* mat along the shore of the Bay of Århus (Denmark) (Caumette, unpublished observations). The new BChl*a*-containing spirilloid species differs, however, from the spirilloid genus *Thiospirillum*, by a higher GC-content, a smaller size and the occurrence of a monopolar flagellum rather than a flagelar tuft. Therefore, the species has been assigned to a new genus and described as *Thiorhodovibrio winogradskyi*. Sulphide, and thiosulphate (some strains) were used as electron donors. Moreover, one strain attained unprecedented high growth rates and yields with sulphur as the sole electron donor.

Filamentous phototrophic organisms

Table 1 lists several filamentous phototrophic bacteria described from mats; however, only one of them has been obtained in axenic culture. This demonstrates the technical difficulties of purifying filamentous phototrophic organisms from their heterotrophic contaminants. From hypersaline mats, D'Amelio et al. (1987) reported a filamentous purple bacterium within the sheaths of the cyanobacterium *Microcoleus chthonoplastes* that was characterised by the presence of BChl*a* and a stacked internal membrane system as found in the cells of *Ectothiorhodospira* spp. It was suggested that the organism may oxidise thiosulphate which is the product of anoxygenic photosynthesis in *M. chthonoplastes*. So far, it is not known whether it grows as a photoauto- or a photoheterotroph in its natural habitat. Green coloured *Chloroflexus* filaments have been cultured from thermal, marine and hypersaline environments. The deeply green coloured axenic isolates from sulphidic thermal springs were obligately phototrophic strains (Giovannoni et al. 1987). Several mesophilic *Chloroflexus*-like organisms have only been obtained in enrichment cultures, so far.

Halophilic purple sulphur and sulphate-reducing bacteria

From microbial mats that developed along the outflow of hypersaline sulphuric springs on the western shore of the Dead Sea (Israel), a purple sulphur bacterium was isolated after enrichment and has been described as *Ectothiorhodospira marismortui* (Oren et al. 1989).

Table 1. (overleaf). Novel species isolated from or originally observed in microbial mats including descriptions of their habitats and some characteristic morphological and physiological features (Literature data after 1986).

new species:	microbial mat habitat:	special morphological and physiological features:	Type strain[a)]	ref[b)]
phototrophic sulphur bacteria:				
Chromatium tepidum	hot spring (neutral pH; low sulphide)	moderate thermophile; antanneae BChl*a* 910 nm	ATCC 43061	1
Chlorobium tepidum	hot spring (low pH; high sulphide)	moderate thermophile	ATCC 49652	2
Thiorhodovibrio winogradskyi	marine mats and inland saline lakes	small spirilloid cells; excellent growth on S^o	DSM 6702	3
Ectothiorhodospira marismortui	hypersaline spring	moderate halophile	DSM 4180	4
Chromatium salexigens	hypersaline ponds	moderate halophile; spirilloxanthin series	DSM 4395	5
Thiocapsa halophila	hypersaline ponds	moderate halophile; major carotenoid: okenone	DSM 6210	6
filamentous sulphide-oxidizing phototrophic bacteria				
dark green *Chloroflexus* sp.	hot spring (neutral pH; low sulphide)	BChl*c*-containing obligate phototrophic strains	c)	7
Chloroflexus -like spp.	marine and hypersaline mats	contain BChl*c*	d)	8,9
filamentous purple bacterium e)	hypersaline ponds	gliding; inside bundles of *M. chthonoplastes*	f)	10
chemotrophic sulphur bacteria				
giant *Beggiatoa*-filaments	deep-sea hydrothermal vent mats	size up to 120 µm by > 1 cm; chemolithotrophes	f)	11
Thiobacillus hydrothermalis	deep-sea hydrothermal vent mats	mesophile, barotolerant; obligate chemolithotroph	DSM 7121	12
24 isolates of gram-negative rod-shaped chemolithotrophs	(hyper)-saline lakes	20 strains obligate moderate halophiles	c)	13
Thiobacillus halophilus	inland (hyper)-saline lakes	moderate halophile	DSM 6132	14
Thiobacillus sp.	marine, close to a paper mill	oxidises DMS with NO_3^- as e-acceptor	c)	15
sulphate-reducing bacteria:				
Desulfovibrio halophilus	hypersaline ponds and lakes	moderate halophile	DSM 5663	16
S.R.B.-strains (Y. Cohen)	experimental hypersaline ponds	remarkable oxygen-tolerance	c)	17

a) TYPE STRAIN deposited at a CULTURE COLLECTION; ATCC = American Type Culture Collection (Rockville, Maryland 20852, U.S.A.).; DSM= Deutsche Sammlung von Mikroorganismen (Mascheroder Weg 1b, D-38124 Braunschweig, F.R.G.)

b) REFERENCES: 1)= Madigan 1986; 2)= Wahlund et al. 1991; 3)= Overmann et al. 1992; 4)= Oren et al. 1989; 5)= Caumette et al. 1988; 6)= Caumette et al. 1991a; 7)= Giovannoni et al. 1987; 8)= Mack and Pierson 1988; 9)= Larsen et al. 1991; 10)= D'Amelio et al. 1987; 11)= Nelson et al. 1991; 12)= Durand et al. 1993; 13)= Wood et al. 1991; 14)= Wood and Kelly 1991; 15)= Visscher and Van den Ende, This Volume; 16)= Caumette et al. 1991b; 17)= Cohen pers.comm.

c) axenic cultures, but no complete descriptions and cultures have not been deposited; d) only enrichment cultures; e) presumably oxidises thiosulphate produced by the cyanobacterium *Microcoleus chthonoplastes* ; f) not cultured at all; hence, descriptions based on field material

The new species is moderately halophilic and grows at neutral pH; hence, the species is distinct from both its marine and extremely alkalohalophilic counterparts.

In coastal salterns, a stable salinity gradient, ranging from sea-water up to halite precipitation, favours a spatial separation of different biotic communities. Figure 1 gives an overview of the microbial mat communities that can be found along the salinity gradient in these coastal salterns (Thomas 1984; De Wit and Grimalt 1992; Caumette et al. in press).

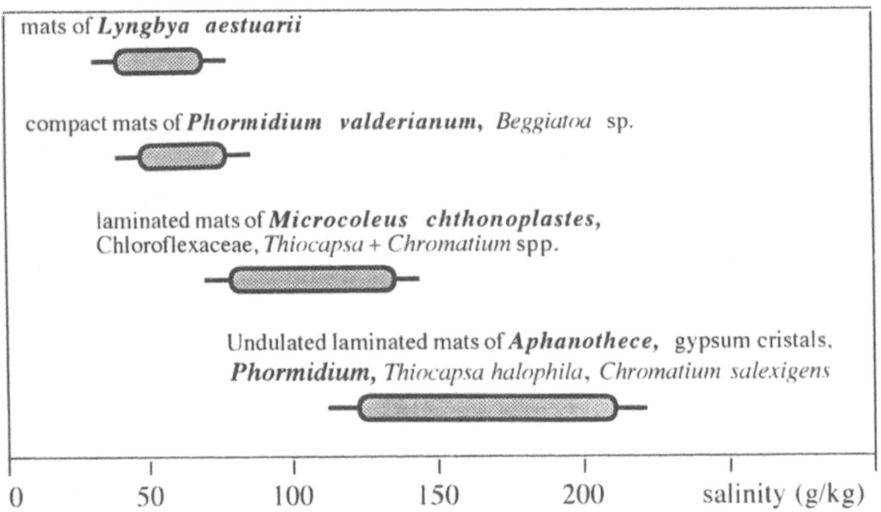

Fig.1. Microbial mat communities found along salinity gradients in coastal salterns. (cyanobacterial species in bold). After Thomas 1984; De Wit and Grimalt 1992; Caumette et al. in press.

The multilayered communities found at salinities of 130-220 g/kg (Fig. 2), comprising the cyanobacteria *Aphanothece* and *Phormidium* on top of a 3 mm thick layer of purple sulphur bacteria were studied by Caumette et al. (in press). From these communities, two novel moderately halophilic purple sulphur bacteria have been isolated and described as *Chromatium salexigens* (optimum at 100-110 g/l NaCl, no growth below 40 g/l and above 200 g/l) (Caumette et al. 1988), and *Thiocapsa halophila* (optimum from 40 to 100 g/l, no growth below 30 g/l and above 200 g/l) (Caumette et al. 1991a). Apart from morphological and nutritional differences, the species differ by carotenoid composition; *C. salexigens* and *T. halophila* contain normal spirilloxanthin series and okenone, respectively. From these mats several strains of sulphate-reducing bacteria were isolated (Caumette et al. in press) which do belong to the species *Desulfovibrio halophilus* that was originally isolated and described from

Solar Lake (Egypt) (Caumette et al. 1991b). *D. halophilus* grows optimally from 60 to 70 g/l NaCl, and in synthetic media used so far, growth does not occur below 30 and above 180 g/l.

Very recently, new strains of sulphate-reducing bacteria have been isolated from microbial mats in experimental hypersaline ponds in Eilat (Israel). The ecophysiology of these isolates has been studied in culture experiments (Cohen, personal communication). Several strains exhibited a remarkable oxygen tolerance.

Marine and halophilic chemotrophic sulphur bacteria

So far, little attention has been paid to the presence of small chemotrophic sulphur bacteria in marine and hypersaline microbial mats. Recently, *Thiobacillus* strains with novel metabolic properties have been isolated from a marine microbial mat found close to a paper mill. These strains are able to oxidise dimethylsulphide (DMS) under anoxic conditions with nitrate as the electron acceptor. Hence, DMS is demethylated, rather than being oxidised by a monooxygenase; subsequently, the sulphide liberated is oxidised (Visscher and Van den Ende, this volume).

Microbial mats are common in interdunal lakes in Western Australia in which seasonal salinity fluctuations range approximately from 20 to 220 g/kg. Twenty-four pure strains of gram-negative rod shaped chemolithotrophic sulphur bacteria were isolated from these microbial mats. Most of the strains were obligately halophilic, some of them even grew at 3 M NaCl. The strains showed considerable metabolic diversity; 13 strains were obligate chemolithotrophs, whilst 11 strains were also capable of growth on sugars, fatty acids or methylamine. From these, three strains have been described as novel isolates, but no full descriptions have been given and no cultures are deposited in culture collections (Wood et al. 1991). However, a *Thiobacillus* strain which was isolated from an inland hypersaline lake, has been fully described as *T. halophilus* (Wood and Kelly 1991). *T. halophilus* is an obligately halophilic, obligate chemolithoautotroph growing optimally at 1 M NaCl, whilst growth still does occur at 4 M NaCl. These reports by Wood and coworkers represent the first descriptions of halophilic thiobacilli after early work by Hof (1935), thus showing that thiobacilli have been neglected during a long period in studies on microbial mats.

Interactions among sulphur bacteria

Competition for light among phototrophic sulphur bacteria

On top of the sediment and within the first mm depth, phototrophic bacteria can use the lower wavelength radiation of the visible spectrum harvested by means of their carotenoids. In shallow and exposed benthic environments, these bacteria can grow deeper in the sediment, by harvesting NIR radiation, which penetrates further down than visible radiation in most sediments (c.f., Kühl et al. this Volume). NIR radiation is directly harvested by the bacteriochlorophyll antannae. The major taxonomic groups have specialised for use of different and complementary wavelength regions; thus, green sulphur bacteria with BChl*c*,*d* or *e*, purple sulphur bacteria with BChl*a*, and purple sulphur bacteria with BChl*b* may use NIR radiation of 700-800, 800-900, and 1000-1050 nm, respectively. Light measurements with fiber-optic mini-probes in the multilayered communities described by Nicholson et al. (1987) (see Introduction) , have clearly demonstrated that the irradiance of these NIR-wavelength regions is strongly attenuated in the layers of the corresponding organisms (Pierson *et al* 1990).

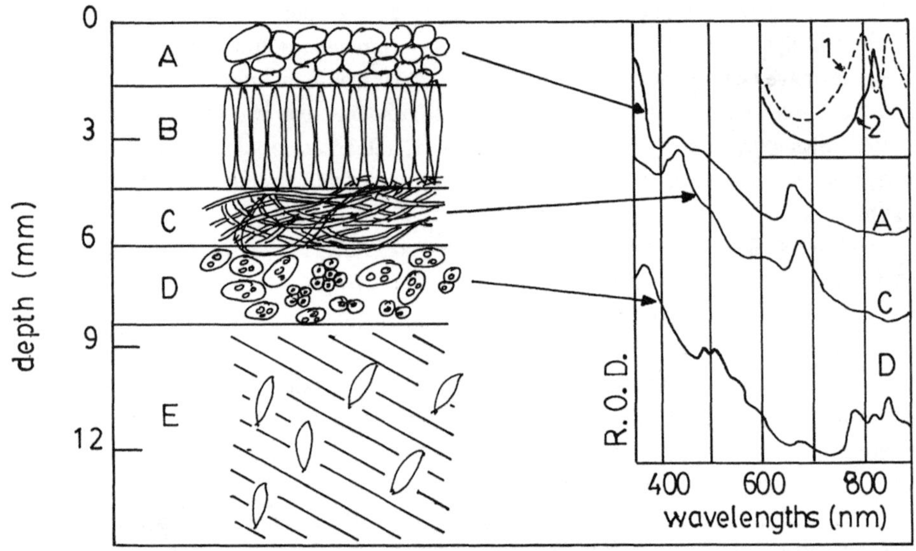

Fig. 2. Microbial mat communities which developed at 130-220 g/kg salinity in the Salin-de-Giraud (Camargue, France). Left hand panel: Stratification of the community structure (A= brown layer of the cyanobacterium *Aphanothece*; B= transparent layer composed of gypsum crystals; C= green layer of the cyanobacterium *Phormidium*; D= layer of the purple sulphur bacteria, E= black FeS-containing layer); Right hand panel: absorption spectra of the different community layers (A,C and D) and NIR absorption spectra of *Chromatium salexigens* (1) and *Thiocapsa halophila* (2). Redrawn after Caumette et al. in press.

This is an elegant example of niche differentiation evolved to prevent competition for light among the major taxonomic groups. It is surprising that the multilayered systems described by Nicholson et al. (1987) are uncommon in marine environments. Most marine microbial mats show only a single purple layer below the oxygenic phototrophs and neither the BChl*b* - containing *Thiocapsa pfennigii* nor green sulphur bacteria do occur in high densities. *Thiocapsa* *pfennigii* only develops in mats that are exposed to the atmosphere for prolonged periods, since the long wavelengths used by this organism are very strongly attenuated in the water column. The Chlorobiaceae are adapted to growth at very low light intensities and at high sulphide concentrations, whilst they appear to be unable to grow in the presence of oxygen. Likely, therefore they stratify below the purple sulphur bacteria. Most purple sulphur bacteria do tolerate oxygen and several species even grow aerobically (De Wit and Van Gemerden 1987, 1990a, 1990b; Overmann and Pfennig 1992); although no data are available for *T. pfennigii*. Perhaps, the absence of a clear layer of green sulphur bacteria in most marine microbial mats is related to frequent deep oxygen penetration. Moreover, green sulphur bacteria will have to compete for NIR radiation with *Chloroflexus*-like organisms that are more tolerant to oxygen and, consequently, may occur higher in the mat.

Specific NIR-radiation absorption differences are also found among the BChl*a*-containing purple sulphur bacteria. They correspond to different antennae constituted by different light-harvesting pigment-protein complexes in the photosynthetic membranes. It is surprising that *Chr. tepidum* is the only phototrophic organism known so far which uses light between 900 and 1000 nm (Madigan 1986; Garcia et al. 1986). Furthermore, clear absorption differences in the NIR radiation region from 800 to 900 nm are found between purple sulphur bacteria that contain okenone (a major absorption peak at 825-830 nm with a shoulder at 800 nm) and those that contain spirilloxanthin series (absorption maxima at 795-800, and 860-870 nm). In Figure 2, the NIR-absorption spectra of okenone-containing *T. halophila* and spirilloxanthin-containing *C. salexigens* are compared with the absorption spectrum of the coloured layers of the microbial mat from which both bacteria were isolated (Caumette et al. in press). Clearly, the NIR radiation absorption signal indicates a coexistence of both organisms which was confirmed by microscopy studies of the mat. This coexistence is possible due to the use of different NIR radiation. This phenomenon is another example of niche differentiation to avoid competition for NIR radiation.

Interactions between purple and colourless sulphur bacteria

Diel migrations of *Beggiatoa* and *Chromatium* species have been described by Jørgensen (1982) and by Caumette (1989). Accordingly, *Beggiatoa* located itself at the oxic-anoxic interface in the sediment during daytime. It was not able to swim in the overlying water and, consequently, had to cope with benthic anoxic conditions at night. In contrast, *Chromatium* was found to photooxidise sulphide in the illuminated anoxic layers of the sediment during daytime; whilst it was found in the overlying oxygen- and sulphide-containing water at night, thus continuing growth in a chemolithotrophic mode (Jørgensen 1982). Non-motile species like *Thiocapsa roseopersicina* require another strategy to cope with the diel shifting oxygen-sulphide gradients typically encountered in microbial mats. This species was found in depth layers that were characterised by diel oxic-anoxic alteration; i.e., it was exposed to oxic conditions during a large part of the light-period, whilst anoxic conditions occurred at night (De Wit et al. 1989). Experimental simulations of these alternating conditions showed that BChl*a*-synthesis was repressed during the oxic periods, but it did occur immediately upon the onset of anoxic conditions even in the dark. In contrast, anoxygenic photosynthesis was not directly repressed by oxygen. Hence, in experiments with anoxic periods of 3 hours and longer in a 24 hour cycle, full phototrophic growth did occur during the entire oxic light periods (De Wit and Van Gemerden 1990a,b). Thus, pigment synthesis and phototrophic growth are uncoupled in this organism and it is highly adapted to its changing environment.

The flux of oxygen towards the sulphide-containing zone results in competition between phototrophic sulphur bacteria and chemotrophic sulphur bacteria, but, generally, little is known on the actual abundance of small colourless sulphur bacteria in microbial mat communities. Nevertheless, population densities of up to 10^9 per cm^3 have been reported for a marine microbial mat (Visscher et al. 1992). Prolonged chemotrophic growth has now unequivocally been demonstrated for a large number of purple sulphur bacteria (De Wit and Van Gemerden 1987; Overmann and Pfennig 1992), but chemotrophy invariably results in lower growth yields than phototrophy per mole of electron donating substrate. Notwithstanding, purple sulphur bacteria will loose the competition from chemotrophs in constant oxic environments if sulphide is the only growth-limiting substrate (Kuenen et al. 1985). However, under oxygen limitation, it was shown that *Thiobacillus thioparus* oxidises sulphide incompletely to sulphur, thiosulphate and some minor compounds (Van den Ende and Van Gemerden 1993). These products can be used as electron donors by phototrophic bacteria, thus, allowing the coexistence of chemotrophic and phototrophic sulphur bacteria (Van den Ende and Van Gemerden, This Volume). Moreover, a mathematical model has been elaborated to predict the outcome of competition in gradient systems under diel alternations as found in microbial mats (De Wit et al. unpubl.). It was predicted that small colourless sulphur bacteria do develop in microbial mats under all circumstances, but coexistence of purple sulphur bacteria and chemotrophs must be expected, provided that NIR radiation sufficiently penetrates the sediment and both benthic sulphate-reduction rates and illumination period exceed threshold values. In several cases, phototrophs will exceed chemotrophs in biomass by a factor up to 10.

In summary, it must be anticipated that purple and colourless sulphur bacteria coexist in microbial mats. However, the purple sulphur bacteria may be extremely successful, since they have special adaptations to cope with the fluctuating conditions. In this respect, the high phototrophic yield, their versatility with respect to reduced sulphur compounds and the independence of oxygen-fluxes are a competitive advantage for the purple sulphur bacteria. In contrast to the conspicuous phototrophs, the presence of the small colourless sulphur bacteria in microbial mats has received too little attention so far, since they escape detection if no special detection techniques are used. More information will be required in the future on their species composition and role in sulphur transformations in microbial mat communities.

Interactions for organic substrates and growth-factors

Cyanobacteria do excrete organic compounds in microbial mats, both during day (e.g., Bateson and Ward 1988) and night periods (Heyer and Krumbein 1991). Moreover, several species of phototrophic and colourless sulphur bacteria use or excrete organic compounds or do both. Hence, these sulphur bacteria establish competitive, commensalistic and mutualistic interactions

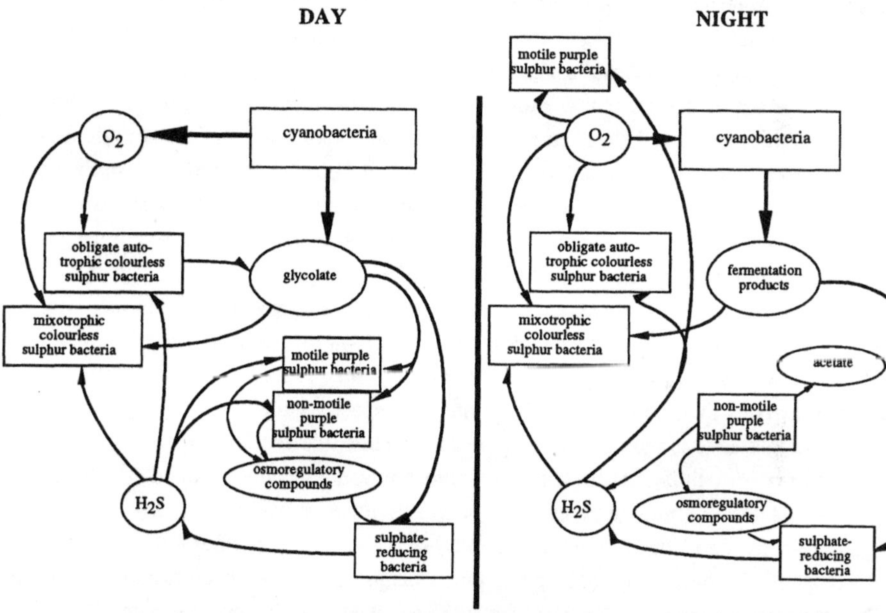

Fig. 3. Scheme of the major flows of organic compounds, oxygen and sulphide among the cyanobacteria and different sulphur bacteria during day and night, in saline and hypersaline environments.

with the sulphate-reducing bacteria. A scheme of the major flows of organic substrates and of O_2 and sulphide among cyanobacteria and different sulphur bacteria during day and night times is depicted in Figure 3. Obligately autotrophic thiobacilli excrete glycolate (Cohen et al. 1979), which is a product of oxygenase activities of RuBisCo, as in cyanobacteria and algae. These excreted compounds may directly favour the sulphate-reducing bacteria even under oxic conditions (Cohen, personal communication.). These compounds could also be incorporated by mixotrophic colourless sulphur bacteria (Gottschal et al. 1979) and by purple sulphur bacteria (Caumette, unpublished results). The importance of these different fluxes is not known so far. However, an indication that purple sulphur bacteria directly use organic compounds from cyanobacteria was encountered in marine microbial mats (De Wit et al. 1989). It was found that the *T. roseopersicina*-like cells consisted of two subpopulations; one closely adhered to *M. chthonoplastes* -bundles, and one deeper in the sediment consisting of microcolonies and single cells that occurred on sand grains. The cell size, which is often positively correlated with growth rate, was largest in the population that strongly adhered to the cyanobacterial bundles. Since this subpopulation was not in a very favourable position with respect to sulphide-supply during day-time, it appears that growth occurred on organic excretion products.

During the night, anaerobic degradation of glycogen in cyanobacteria may give rise to a whole spectrum of fermentation products (Heyer and Krumbein 1991) which can be directly metabolised by the sulphate-reducing bacteria. Whilst the motile purple sulphur bacteria may move to oxygen-sulphide coexistence zones (see above), the non-motile purple sulphur bacteria are trapped in the anoxic zones. It was shown that *T. roseopersicina* is capable of degrading glycogen to acetate coupled to the reduction of intracellular sulphur to sulphide (De Wit 1989). Thus, in this species, the dark metabolism of glycogen is different from the one described for *Chromatium vinosum* where sulphur reduction is coupled to poly-ß-hydroxybutyrate formation (Van Gemerden 1968). This intracellularly stored compound is metabolised during the following light period. In comparison, in the dark, acetate production probably results in higher ATP-production per mole of glycogen monomer, but the excreted acetate is lost for *T. roseopersicina* and used by sulphate-reducing bacteria. Upon diffusion to the oxic zone, the excreted sulphide may be favourable to colourless and other purple sulphur bacteria. This example demonstrates that mutualistic interactions are especially important during the night.

Positive interactions may not only involve substrates that are further metabolised, but also the transfer of growth factors. In the hypersaline environments it was found that the novel purple sulphur bacteria *C. salexigens* and *T. halophila* synthesise osmoregulatory compounds mainly glycine-betaine and in addition sucrose and N-acetyl-glutaminylglutamine amide (Welsh and Herbert 1993). In contrast, the salt-loving mechanisms have not yet been elucidated for the novel *Desulfovibrio halophilus*, although it may accumulate salts in its cytoplasm (Galinski, personal communication). Preliminary experiments have demonstrated that the uptake of exogeneous glycine-betaine results in increased salt-tolerance in some sulphate-reducers. Probably, in the microbial mat, these bacteria obtain compatible solutes arising from lyses of or from active excretion by purple sulphur bacteria. It is still somewhat puzzling that so few halophilic sulphate reducers have been isolated from hypersaline environments so far. Perhaps, several halophilic sulphate-reducing bacteria rely on uptake of external compatible solutes and will show no or very poor growth in synthetic hypersaline culture media unless these compounds are supplemented.

Conclusions

The last seven years have shown the isolation, characterisation and full descriptions of many new sulphur bacteria from microbial mats. Especially, new species have been isolated from extreme environments. Thus, hitherto unknown moderately halophilic purple sulphur bacteria, sulphate-reducing bacteria and thiobacilli-like colourless sulphur bacteria have been isolated from hypersaline environments. A new mesophilic barotolerant *Thiobacillus* species has been obtained from mats along hydrothermal vents in the deep-sea and hitherto undescribed thermophilic phototrophs, i.e., *Chromatium* and *Chlorobium* species, have been fully characterised. However, isolation has not usually been successful for filamentous organisms, mainly due to difficulties in purifying the filaments from their heterotrophic contaminants. These bacteria merit more attention, since they appear to be important sulphur oxidisers in several microbial mat habitats.

Considerable increase has been obtained in the understanding of interactions for light among the anoxygenic phototrophs. The use of small fiber-optic sensors has been paramount for this improvement. Niche differentiation for NIR radiation does not only occur among the major taxonomic groups, but, also within the purple sulphur bacteria. Moreover, more light has been thrown on the interactions between colourless and purple sulphur bacteria. Many purple sulphur bacteria are extremely well adapted to the fluctuating conditions in microbial mats and do cope in different ways to frequent exposure to oxygen. However, it is to be anticipated that small colourless sulphur bacteria will prove to be much more important in microbial mats than previously thought.

Many questions remain concerning the flow of organic compounds in microbial mats. In this respect, it is important to study to what extent the cyanobacterial excretion products are used directly for photoheterotrophic growth of anoxygenic phototrophic bacteria, used by the sulphate-reducing bacteria, or by fermentative organisms. Moreover, in sulphidic springs both on the continent and in the deep-sea, more attention should be paid to the presence and activities of sulphate-reducing bacteria to quantify the role of biogenic sulphate-reduction in comparison to the geochemical sulphide input.

References

Bateson MM, Ward DM (1988) Photoexcretion and fate of glycolate in a hot spring cyanobacterial mat. Appl Environ Microbiol 54: 1738-1743

Caumette P, Baulaigue R, Matheron R (1988) Characterization of *Chromatium salexigens* sp nov, a halophilic Chromatiaceae isolated from Mediterranean Salinas. System Appl Microbiol 10: 284-292

Caumette P (1989) Ecology and general physiology of anoxygenic phototrophic bacteria in benthic environments In: Cohen Y, Rosenberg E (eds) Microbial Mats; Physiological Ecology of Benthic Microbial Communities ASM, Washington DC, pp 283-304

Caumette P, Baulaigue R, Matheron R (1991a) *Thiocapsa halophila* sp nov, a new halophilic phototrophic purple sulfur bacterium. Arch Microbiol 155: 170-176

Caumette P, Cohen Y, Matheron R (1991b) *Desulfovibrio halophilus* sp nov, a halophilic sulfate-reducing bacterium isolated from Solar Lake (Sinai). Syst Appl Microbiol 13: 33-38

Caumette P, Matheron R, Raymond N, Relexans, J-C (in press) Microbial mats in the hypersaline ponds of Mediterranean salterns (Salins-de-Giraud, France). FEMS Microbiol Ecol

Cohen Y, De Jonge I, Kuenen JG (1979) Excretion of glycolate by *Thiobacillus neapolitanus* grown in continuous culture. Arch Microbiol 122: 189-194

D'Amelio ED, Cohen Y, Des Marais DJ (1987) Association of a new type of gliding, filamentous, purple phototrophic bacterium inside bundles of *Microcoleus chthonoplastes* in hypersaline microbial mats. Arch Microbiol 147: 213-220

De Wit R (1989) Interactions between phototrophic bacteria in marine sediments. PhD-thesis, University of Groningen

De Wit R, Van Gemerden H (1987) Chemolithotrophic growth of the phototrophic sulfur bacterium *Thiocapsa roseopersicina*. FEMS Microbiol Ecol 45, 117-126

De Wit R, Jonkers HM, Van den Ende FP, Van Gemerden H (1989) *In situ* fluctuations of oxygen and sulphide in marine microbial sediment ecosystems. Neth J of Sea Res 23: 271 - 281

De Wit R, Van Gemerden H (1990a) Growth of the phototrophic purple sulfur bacterium *Thiocapsa roseopersicina* under oxic/anoxic regimens in the light. FEMS Microbiol Ecol 73: 69-76

De Wit R, Van Gemerden H (1990b) Growth and metabolism of the purple sulfur bacterium *Thiocapsa roseopersicina* under combined light/dark and oxic/anoxic regimens. Arch Microbiol 154: 459-464

De Wit R, Grimalt JO (1992) Microbial ecosystems in Spanish coastal Salinas; an ecological and geochemical study of biomarkers. Limnetica 8: 205-212

Durand P, Reysenbach A-L, Prieur D, Pace N (1993) Isolation and characterization of *Thiobacillus hydrothermalis* sp nov, a meesophilic obligately chemolithotrophic bacterium isolated from a deep-sea hydrothermal vent in Fiji Basin. Arch Microbiol 159: 39-44

Elsgaard L, Guezennec J, Benbouzid-Rollet N, Prieur D (1991) Fatty acid composition of sulfate-reducing bacteria isolated from deep-sea hydrothermal vents (13°N, East Pacific Rise) Kieler Meeresforsch, Sonderh 8, 182-187

Garcia D, Parot P, Vermeglio A, Madigan MT (1986) The light-harvesting complexes of a thermophilic purple sulfur bacterium *Chromatium tepidum*. Biochim Biophys Acta 850: 390-395

Giovannoni SJ, Revsbech NP, Ward DM, Castenholz RW (1987) Obligate phototrophic *Chloroflexus*: Primary production in anaerobic hot spring microbial mats. Arch Microbiol 147: 80-87

Gottschal JC, De Vries S, Kuenen JG (1979) Competition between the facultatively chemolithotrophic *Thiobacillus* A2, an obligately chemolithotrophic *Thiobacillus* and a heterotrophic *Spirillum* for inorganic and organic substrates. Arch Microbiol 121: 241-249

Gundersen JK, Jørgensen BB, Larsen E, Jannasch HW (1992) Mats of giant sulphur bacteria on deep-sea sediments due to fluctuating hydrothermal flow. Nature 360: 454-455

Heyer H, Krumbein WE (1991) Excretion and fermentation products in dark and anaerobically incubated cyanobacteria. Arch Microbiol 155: 284-287

Hof T (1935) Investigations concerning bacterial life in strong brines. Extrait du Recueil des Traveaux Botaniques Néerlandais 32: 92-173

Jørgensen BB (1982) Ecology of the bacteria of the sulphur cycle with special reference to the anoxic-oxic interface. Phil Trans R Soc London B298: 543-561

Jørgensen BB, Nelson DC (1988) Bacterial zonation, photosynthesis, and spectral light distribution in hot spring microbial mats of Iceland. Microbiol Ecol 16: 133-147

Jørgensen BB, Isaksen MF, Jannasch HW (1992) Bacterial sulphate reduction above 100 °C in deep-sea hydrothermal vent sediments. Science 258: 1756-1757

Kuenen JG, Robertson LA, Van Gemerden H (1985) Microbial interactions among aerobic and anaerobic sulfur-oxidizing bacteria. In: Marshall KC (ed) Advances in Microbial Ecol Vol 8 Plenum Press, New York, pp 1-59

Larsen M, Mack EE, Pierson BK (1991) Mesophilic *Chloroflexus* -like organisms from marine and hypersaline environments. Abstr VII Int Symp Photosynthetic Prokaryotes, Amherst, USA, pp 169

Leadbetter ER (1974) Beggiatoaceae. In: Buchanan RE and Gibbons NE (eds) Bergey's Manual of determinative bacteriology, 8th ed The Williams & Wilkins Co, Baltimore, pp 112-116

Mack EE, Pierson BK (1988) Preliminary characterization of a temperate marine member of the Chloroflexaceae. In: Olsen JM, Ormerod JG, Amesz J, Stackebrandt E, Trüper HG (eds) Green Photosynthetic Bacteria. Plenum Publ Corp, New York, pp 237-241

Madigan MT (1984) A novel photosynthetic purple bacterium isolated from a Yellowstone hot spring. Science 225: 313-315

Madigan MT (1986) *Chromatium tepidum* sp nov, a thermophilic photosynthetic bacterium of the family of the *Chromatiaceae.* Int J of Syst Bact 36: 222-227

Mezzino MJ, Strohl WR, Larkin JM (1984) Characterization of *Beggiatoa alba.* Arch Microbiol 137: 139-144

Nelson DC, Wirson CO, Jannasch HW (1989) Characterization of large, autotrophic *Beggiatoa* spp abundant at hydrothermal vents in the Guaymass basin. Appl Environ Microbiol 55: 2909-2917

Nicholson JAM, Stolz JF, Pierson BK (1987) Structure of a microbial mat at Great Sippewissett Marsh, Cape Cod, Massachusetts. FEMS Microbiol Ecol 45: 343-364

Oren A, Kessel M, Stackebrandt E (1989) *Ectothiorhodospira marismortui*, sp nov, an obligately anaerobic, moderately halophilic purple sulphur bacterium from a hypersaline spring on the shore of the Dead Sea. Arch Microbiol 151: 524-529

Overman J, Pfennig N (1992) Continuous chemotrophic growth and respiration of Chromatiaceae species at low oxygen concentrations. Arch Microbiol 158: 59-67

Overmann J, Fischer U, Pfennig N (1992) A new purple sulfur bacterium from saline littoral sediments, *Thiorhodovibrio winogradskyi* gen nov and sp nov. Arch Microbiol 157: 329-335

Pierson BK, Sands VM, Frederick JL (1990) Spectral irradiance and distribution of pigments in a highly layered microbial mat. Appl Environ Microbiol 56: 2327-2340

Thomas JC (1984) Formations benthiques à Cyanobacteries des salins de Santa Pola (Espagne): composition spécifique, morphologie et caractéristiques biologiques des principaux peuplements. Rev Invest Geol 38-39: 139-158

Van den Ende F, Van Gemerden H (1993) Sulfide oxidation under oxygen limitation by a *Thiobacillus thioparus* isolated from a marine microbial mat. FEMS Microbiol Ecol 13: 69-78

Van Gemerden H (1968) On the ATP generation of *Chromatium* in darkness. Arch Mikrobiol 64: 118-124

Visscher PT, Prins RA, Van Gemerden H (1992) Rates of sulfate reduction and thiosulfate consumption in a marine microbial mat. FEMS Microbiol Ecol 86: 283-294

Wahlund TM, Woese CR, Castenholz RW, Madigan MT (1991) A thermophilic green sulfur bacterium from New Zealand hot springs, *Chlorobium tepidum* sp nov. Arch Microbiol 156: 81-90

Welsh DT, Herbert RA (1993) Identification of organic solutes accumulated by purple and green sulphur bacteria during osmotic stress using natural abundance ^{13}C nuclear magnetic resonance spectroscopy. FEMS Microbiol Ecol 13: 145-150

Wood AP, Kelly DP (1991) Isolation and characterisation of *Thiobacillus halophilus* sp nov, a sulphur-oxidising autotrophic eubacterium from a Western Australian hypersaline lake. Arch Microbiol 156: 277-280

Wood AP, Burke CB, Knott B, Kelly DP (1991) Chemolithotrophic sulfur bacteria in sediments, mats, and stromatiles of Western Australian saline lakes. Geomicrobiol J 9: 41-49

Phototrophic oxidation of ferrous minerals - a new aspect in the redox microbiology of iron

Armin Ehrenreich and Friedrich Widdel

Max-Planck-Institut für marine Mikrobiologie, Fahrenheitstraße 1, D-28359 Bremen, Germany

Among the metals involved in biological processes, iron stands forth by a multiplicity of physiological and biochemical functions. It appears that organisms make use of iron in so many ways because this element not only exhibits a versatile chemistry, but is also wide-spread and in principle available in almost every type of environment.

Indeed, iron is the most common heavy metal in the Earth's crust and presumably the predominating metal in the cosmos. The ^{56}Fe nucleus is the energetically most stable of all elements, and energy-yielding nuclear fusion sequences in stars must come to a halt at this state (Greenwood and Earnshaw 1984).

The biological significance of iron lies mainly in its redox behaviour, Fe(II) and Fe(III) being the most common oxidation states. The redox potential of the Fe(III)/Fe(II) pair is largely influenced by the type of ligands. The E^0 of the free Fe^{3+}/Fe^{2+} pair, with only water as ligand, is +0.77 V. Stability constants of complexes of the two oxidation states can differ by several orders of magnitudes. E.g., oxygen-containing ligands, especially if charged (e.g., OH^-, $R{\cdot}COO^-$), have a higher affinity for Fe^{3+} than for Fe^{2+} and thus increase the reducing power (Greenwood and Earnshaw 1984; Luther et al. 1992). As a result, various Fe(III)/Fe(II) pairs are scattered over a wide range of the redox scale.

One may distinguish between three major functional modes (1-3) of iron in biology.

(1) Iron proteins have catalytic functions, mostly by acting as electron carriers, in almost every living organism. The anabolic reactions ranging from uptake of iron to final incorporation of Fe^{2+} into biomolecules (e.g., Loehr 1989) may be summarized as *assimilatory* iron metabolism.

(2) Fe(III) or Fe(II) can serve as the principal electron acceptor or electron donor, respectively, in a respiratory process. This *dissimilatory* iron metabolism is found in specialized bacteria only, usually comprised as iron-reducers and lithoautotrophic iron oxidizers, respectively. Fe(III) reduction as a type of bacterial metabolism, after its first demonstration in a pure culture using H_2 (Balashova and Zavarzin 1980), has been recognized as a means of completely oxidizing organic compounds under anoxic conditions (Lovley 1991). The end product is siderite ($FeCO_3$) or magnetite ($Fe_3O_4 = Fe_2O_3{\cdot}FeO$). The reverse reaction, the oxidation of Fe(II), is carried out by bacteria in acidic and neutral environments. Under acidic conditions, the autotrophic *Thiobacillus ferrooxidans* oxidizes Fe^{2+} to Fe^{3+}; this is a key reaction during leaching of pyrite (Ehrlich et al. 1991; Luther et al. 1992). The best studied

NATO ASI Series, Vol. G 35
Microbial Mats
Edited by Lucas J. Stal and Pierre Caumette

neutrophilic Fe(II) oxidizer is *Gallionella ferruginea*, which has also been shown to grow autotrophically (Hallbeck and Pedersen 1991). It has to compete with the rapid chemical oxidation of Fe(II) as it occurs in neutral solutions. The amounts of iron transformed per amount of cell mass in these dissimilatory processes are in principle by orders of magnitudes higher and presumably biogeochemically more significant than the amounts transformed by assimilation.

(3) A third mode of iron transformation is the controlled formation of intracellular magnetite particles of a physically defined size, the magnetosomes, in magnetotactic bacteria (e.g., DeLong et al. 1993; Sakaguchi et al. 1993).

Several authors have discussed the possibility of a naturally occurring oxidation of ferrous iron without free oxygen and the implications which such a process would have for our understanding of the origin of precambrian banded iron formations. On the one hand, a UV light-driven reaction of ferrous iron with H_2O to yield ferric iron and H_2 has been demonstrated experimentally, and the significance of this reaction on the archaean Earth without UV-absorption in the anoxic atmosphere has been discussed (e.g., Braterman et al. 1983; François 1986). Former claims that CO_2 can be reduced to a significant extent to the formaldehyde level during photochemical iron oxidation have been doubted more recently (Mauzerall et al. 1993). On the other hand, there have been speculations about a biological oxidation of ferrous iron by photosynthetic organisms without the appearance of free oxygen (Cohen 1983; Hartmann 1983; Walker 1983). Cohen (1983) could measure an Fe^{2+} -dependent photoassimilation of CO_2 by cyanobacteria from mats while inhibiting the oxygenic photosystem. In this paper we describe recently isolated (Widdel et al. 1993) purple bacteria which grow by using ferrous iron as electron donor for anoxygenic photosynthesis, i.e. via one photosystem.

For understanding dissimilatory transformations of iron, the chemistry mainly of inorganic iron is relevant. The decisive chemical factor that influences the redox potential of inorganic Fe(III)/Fe(II) couples is the pH (Fig. 1), because of the high affinity of OH^- ions for Fe^{3+} (Garrels and Christ 1965; Luther et al. 1992; Stumm and Morgan 1981). At pH values higher than 2-3, ferric iron forms hydroxy and oxyhydroxy compounds of which neutral polymers are the stable forms at pH 7. The concentration of Fe^{3+} which at pH 7 is in equilibrium with ferric iron of an assumed formula $Fe(OH)_3$ is 10^{-17} M or less. Ferrous iron can also form precipitates around pH 7. However, with the exception of FeS, their solubility is by several orders of magnitudes higher than the solubility of ferric hydroxy or oxyhydroxy complexes.

Materials and methods

Mud samples for enrichments were taken from ditches in Schaumburger Wald (District of Hannover) and near Lübeck, and from the intertidal mud flat of Jadebusen, North Sea.

Anoxic defined mineral media buffered with bicarbonate (20 mM) were prepared and kept under a mixture of N_2 and CO_2 (80/20 to 90/10, vol/vol) as described (Widdel and Bak 1992; Widdel et al. 1993). Analytical procedures for the determination of iron species and bacterial cell mass were recently described (Widdel et al. 1993).

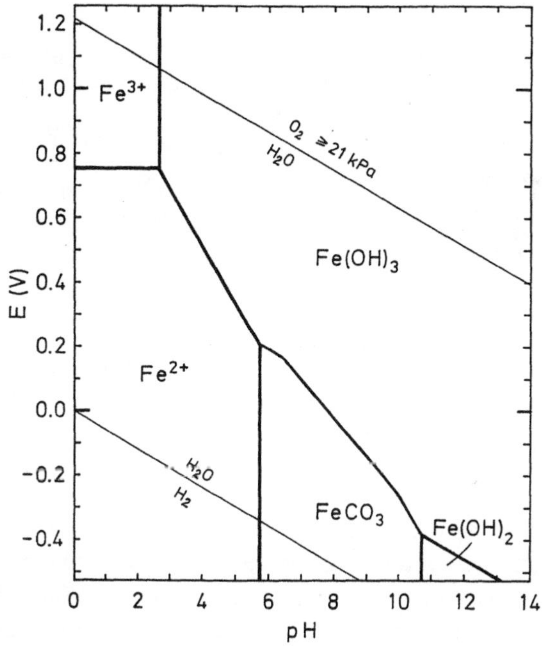

Fig. 1. Simplified Eh-pH-diagram of iron species and O_2 /H_2O at 25 °C. Calculated for $1 \cdot 10^{-3}$ mol l^{-1} total iron and $2 \cdot 10^{-2}$ mol l^{-1} dissolved inorganic carbon (CO_2 + HCO_3^- + CO_3^{2-}) at an ionic strength of 0.05. (For further information see Garrels and Christ 1965; Stumm and Morgan 1981; Widdel et al. 1993)

Results

Enrichment and isolation of Fe(II)-oxidizing phototrophic bacteria

Anaerobic, black mud samples from ditches, ponds and tidal areas were exposed to light. After 2-4 weeks, some showed development of intensely brown, rusty patches on the glass wall. If brown, iron-rich sediment samples were incubated anaerobically, they first turned black, probably due to formation of FeS as a result of sulfate reduction with organic matter. However, after 1-3 weeks, rusty crusts reappeared in the light. Microscopic examination of subsamples from rusty spots did not reveal autofluorescent, chlorophyll *a*-containing cyanobacteria or algae. To study the phenomenon in more detail, anoxic, bicarbonate-buffered mineral media with 5 mmol FeS or 10 mmol $FeSO_4$ per l were inoculated with subsamples. $FeSO_4$ formed a white precipitate, most of which was presumably $FeCO_3$. All media turned brown in the light after 1 to 2 weeks.

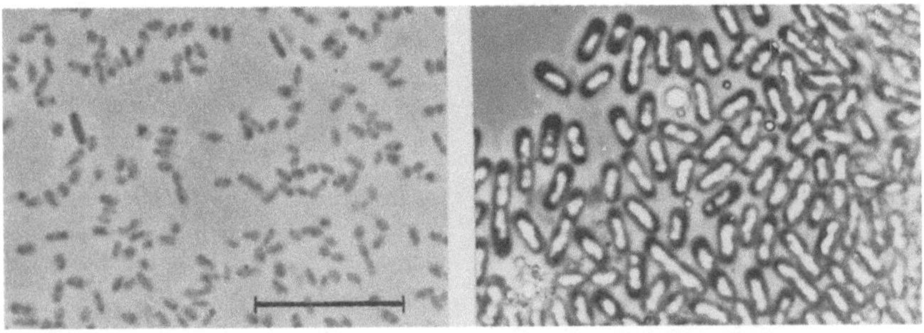

Fig. 2. Viable cells of phototrophic purple bacteria able to oxidize ferrous iron. Phase contrast, bar = 10 μm. Left: strain SW2 grown on acetate. Right: strain L7 grown autotrophically; precipitated ferric iron is visible

After inoculation of agar dilution series with added $FeSO_4$, coloured colonies of phototrophic bacteria developed many of which appeared opaque due to the deposition of greyish to brownish minerals. Two principal types of bacteria were isolated from freshwater enrichments (Fig. 2). One type, represented by strain SW2, had small oval to irregular, sometimes motile cells which were difficult to detect in the orange-brown precipitate. The other type, represented by strain L7, had relatively large, oval, nonmotile cells that often occurred in clumps or chains. Gas vesicles were formed towards the end of growth. Strain L7 exhibited essentially the same morphology as *Thiodictyon elegans* (Pfennig 1989). Cells were clearly separate from the formed yellowish-brown precipitate. The precipitate deposited by this strain was more compact and appeared more crystalline than the ferric mineral formed by strain SW2. From another enrichment with ferrous iron, a strain of *Rhodomicrobium vannielii* was isolated by Schink and coworkers (for description see Widdel et al. 1993). Attempts to isolate pure cultures of iron-oxidizing phototrophs from marine enrichments have thus far not been successful.

Physiology

Growth experiments revealed a production of 7.6 mg cell dry mass per mmol of Fe(III) formed (Widdel et al. 1993). This result is in good agreement with the following growth equation, assuming CH_2O as the approximate formula of cell mass (nitrogen neglected):

$$4\,FeCO_3 + 7\,H_2O \rightarrow (CH_2O) + 4\,Fe(OH)_3 + 3\,CO_2 \qquad (1)$$

Phosphate scavenging from the medium (as ferric phosphate) is not considered here. As long as the exact nature of the ferric iron formed is unknown, the product is simply expressed as ferric hydroxide.

In the dark, neither oxidation of Fe(II) to Fe(III) nor growth of cells was observed. In sterile controls incubated in the light, the content of Fe(II) remained unchanged. Growth was never observed in the absence of ferrous iron, indicating that neither ammonia nor water could be used as electron donors.

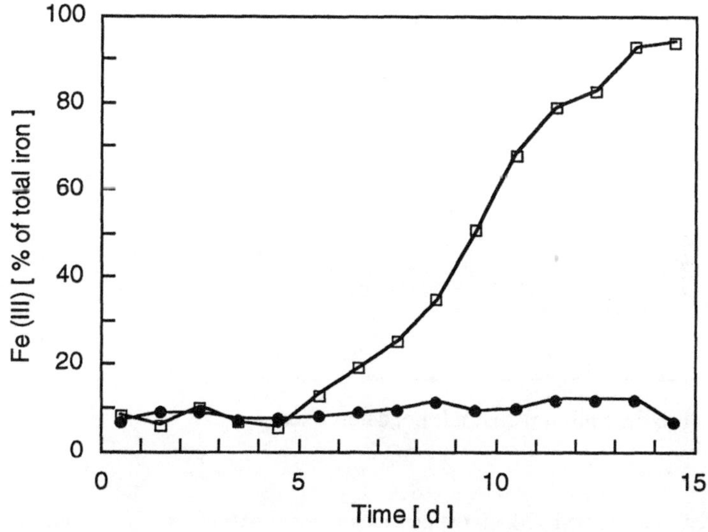

Fig. 3. Ferric iron production by strain L7 growing autotrophically in a medium with initially 10 mmol l^{-1} ferrous iron as the sole source of electrons. □, ferric iron in the illuminated culture; ●, ferric iron in the dark control

Under autotrophic conditions on $FeCO_3$ with an inoculum of 5 %, iron was fully oxidized by strain L7 after approximately 15 days (Fig.3) and by strain SW2 after about 30 days. There were striking initial lag phases within these periods, the real growth phases being approximately 10 and 16 days, respectively.

Autotrophic growth of strain SW2 on FeS (4 mmol l^{-1}) was somewhat faster than on

$FeCO_3$, in contrast to strain L7 which needed almost two months to oxidize FeS. The intensely black color of FeS disappeared completely. FeS remained unchanged in dark or sterile controls. In a culture of strain SW2 on FeS, 60-70 % of the iron was recovered as Fe(III) and 30 % of the sulfur as sulfate. Obviously, a portion of the FeS was converted to acid-insoluble products or intermediate oxidation states of sulfur which await identification.

Table 1. Physiological properties of iron-oxidizing phototrophs [a].

	Strain SW2	Strain L7
pH range for Fe(II) oxidation	5.5-7.0	6.0-7.5
pH optimum for Fe(II) oxidaton	6.5	7.0
Electron donors and carbon sources		
$H_2 + CO_2$	-	+
Formate	+	-
Acetate	+	+
Propionate	+	-
Lactate	+	-
Succinate	-	-
Glucose	+	-
Alanine	+	nd
Aspartate	+	nd
Glutamate	+	-

[a] Symbols: +, utilized; -, not utilized; nd, not determined

In contrast to described *Thiodictyon* species (Pfennig 1989), strain L7 did not oxidize free sulfide, even though it tolerated concentrations of 0.5-1.0 mM in the presence of organic substrates.

Both types of isolates were able to use organic electron donors and carbon sources (Table 1); especially strain SW2 exhibited a striking nutritional versatility.

If organic electron donors were added together with iron(II), the strains exhibited different growth responses. Strain L7 utilized ferrous iron and acetate or glucose simultaneously (mixotrophic growth), whereas oxidation of ferrous iron by strain SW2 started only after the organic substrate had been consumed completely. After exhaustion of acetate (1 mM), strain SW2 oxidized ferrous iron (10 mmol/l) within 6 days.

The isolates did not grow on ferrous iron if the incubation temperature exceeded 20 - 22 °C.

Photoheterotrophically grown cells of strains L7 and SW2 were purple and yellowish-brown, respectively. Absorption spectra of both strains exhibited peaks characteristic of Bacteriochlorophyll *a*, which again is proof of the anoxygenic type of photometabolism in these organisms.

Sequencing of 16S rRNA from strain L7 revealed a clear affiliation with the γ-group of the proteobacteria and hence a relationship to Chromatiaceae. More detailed analyses will show if it belongs to the genus *Thiodictyon*. The phylogenetic position of strain SW2 is still unknown.

Rhodopseudomonas and *Rhodospirillum* species which had formerly been isolated on organic substrates were also tested in ferrous iron media. They were either unable to oxidize ferrous iron, or they oxidized only a small portion; in the latter case, a product with a greenish appearance, as common for a mixture of the two oxidation states, remained. *Rhodobacter capsulatus* carried out a slow, almost complete oxidation of ferrous iron, but only in the presence of an additional electron donor; the process did not continue in subcultures. Authenticated *Thiodictyon* species and *Chlorobium limicola* did not oxidize ferrous iron.

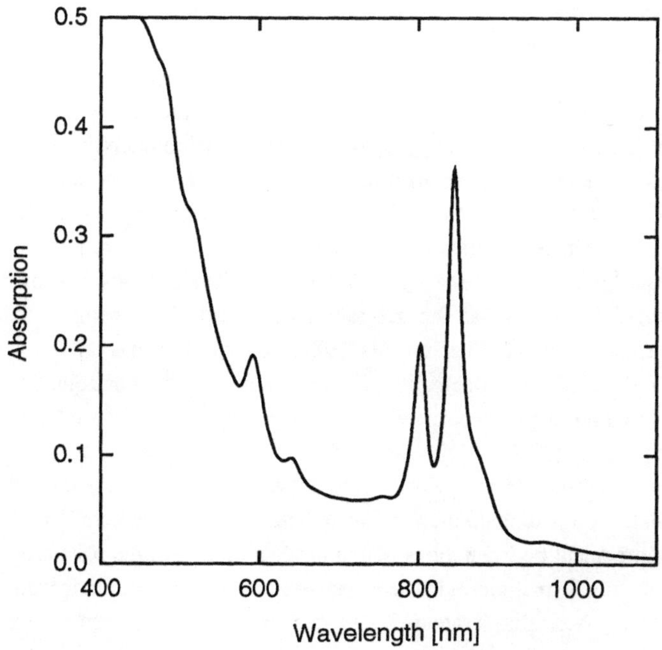

Fig. 4. Absorption spectrum of strain SW2 showing characteristic peaks of Bacteriochlorophyll *a* in the long wavelength range (800-850 nm)

Discussion

The use of ferrous iron as an electron donor by phototrophic bacteria which possess only one photosystem is comprehensible from a bioenergetic point of view. The redox potential of the reaction center in its non-excited state is approximately +0.45 V (Norris and Katz 1978). The $Fe(OH)_3 + HCO_3^- / FeCO_3$ couple at a concentration of inorganic carbon (CO_2 + HCO_3^-) of around 20 mM has a redox potential under neutral conditions of $E' \approx +0.1$ V. The topography of the iron-oxidizing system is of special interest. Oxidation in the periplasm or inside the cell would destroy the cell structure by deposition of immense quantities of insoluble ferric iron (around 14 g ferric hydroxide per g cell dry mass). The cells never showed any sign of intracellular accumulations. Export of ferric iron by means of chelators would be wasteful for the metabolism. Hence, one has to postulate that the oxidation of iron(II) takes place on the cell surface, i.e. in contact with the outer membrane in which a protruding redox-active factor may be embedded. One would further have to assume an electron carrier which mediates between the outer primary electron acceptor and the periplasmic site of the photosystem. A similar periplasmic electron shuttle system has been suggested to function in the acidophilic iron oxidizer, *Thiobacillus ferrooxidans* (Ehrlich et al. 1991).

If one assumes that a phototrophic metabolism with only one photosystem evolved before the two step oxygenic photosynthesis (Pierson and Olson 1989), the existence of iron-oxidizing purple bacteria should have implications for our understanding of the formation of banded iron formations. Many models are based on a chemical reaction of dissolved ferrous iron with photosynthetically produced oxygen. According to such models, oxygen came into contact with ferrous iron either by transport through the atmosphere and surface waters, or in the immediate vicinity of oxygenic phototrophs. With the isolation of iron-oxidizing purple bacteria, it appears in principle possible that such organisms contributed to banded iron formations during the Archaean before oxygenic phototrophs developed and became abundant. Such an idea appears as attractive as the previously discussed photochemical oxidation of ferrous iron (Braterman et al. 1983; François 1986; Mauzerall et al. 1993). During the optimum growth period of autotrophically growing cells in flat, horizontally incubated cultivation bottles (medium depth, 4 cm; cell dry mass per area, 0.4-0.5 g m^{-2}) with modest light from a tungsten lamp (32 W m^{-2}), the production rate was 390 mg ferric iron (as Fe) m^{-2} h^{-1} or 1.7 kg ferric iron m^{-2} h^{-1} (with 12 h light per day). Even if the conditions in the experiment may not be comparable to those at the archaean ocean and lagoons, there seems to be no principal disagreement between the potential production rate of the phototrophs and the rates at which banded iron formations must have been deposited. E.g., for ferric iron deposition in the Hamersley Basin, Australia, a rate of 0.2 kg m^{-2} yr^{-1} has been estimated (for literature see Kump 1993).

Still, the assumption of anoxygenic phototrophs involved in the deposition of banded iron formations poses new questions (Kump 1993). It is assumed that the oxidation of ferrous iron occurred mainly at deep oceanic sites (Beukes and Klein 1992) where light had no access. Also, banded iron formations are almost deficient in organic carbon. One may therefore argue that cells of phototrophs associated and deposited together with formed ferric iron should have been preserved as organic carbon; if not, they should have been degraded and utilized by chemotrophs for a re-reduction of the ferric to the ferrous state. However, there are also viewpoints in favour of a deposition of ferric iron by anoxygenic phototrophs. Exept for the reactivity towards sulfide, the phototrophically produced ferric iron seemed to be chemically rather sluggish. The enrichment cultures (before purification) which in addition to the phototrophs contained chemotrophic commensals did not reduce the rusty precipitate if incubated in the dark for several weeks. Furthermore, cells of the *Thiodictyon*-like strain did not adhere to the formed ferric iron. Since they tended to form cell aggregates and become buyoant by means of gas vesicles, an easy hydromechanic separation from the relatively dense, finely grained ferric iron seems possible. In case of adherent cells, such as those of strain SW2, gradual lysis and anaerobic degradation of cells with formation of soluble end products may have caused a loss of the bulk of organic carbon from the ferric mineral during sedimentation.

Acknowledgement

We wish to thank Ralph S. Wolfe for encouraging discussions about bioenergetics of iron.

Referenes

Balashova VV, Zavarzin GA (1980) Anaerobic reduction of ferric iron by hydrogen bacteria. Microbiology (transl. from Russian) 48: 635-639

Beukes NJ, Klein C (1992) Models for iron-formation deposition. In: Schopf JW, Klein C (eds) The proterozoic biosphere. Cambridge University Press, Cambridge, pp 147-151

Braterman PS, Cairns-Smith AG, Sloper RW (1983) Photo-oxidation of hydrated Fe^{2+} — significance for banded iron formations. Nature 303: 163-164

Cohen Y (1983) Oxygenic photosynthesis, anoxygenic photosynthesis, and sulfate reduction in cyanobacterial mats. In: Klug MJ, Reddy CA (eds) Current perspectives in microbial ecology. American Society for Microbiology, Washington DC, pp 435-441

DeLong EF, Frankel RB, Bazylinski DA (1993) Multiple evolutionary origins of magnetotaxis in bacteria. Science 259: 803-806

Ehrlich HL, Ingledew WJ, Salerno JC (1991) Iron- and manganese-oxidizing bacteria. In: Shively JM, Barton LL (eds) Variations in autotrophic life. Academic Press, London, San Diego, New York, pp 147-170

François LM (1986) Extensive deposition of banded iron formations was possible without photosynthesis. Nature 320: 352-354

Garrels RM, Christ CL (1965) Solutions, minerals and equilibria. Harper & Row, New York

Greenwood NN, Earnshaw A (1984) Chemistry of the elements. Pergamon Press, Oxford, New York

Hartmann H (1983) The evolution of photosynthesis and microbial mats; a speculation on the banded iron formations. In: Cohen Y, Castenholz RW, Halvorsen HO (eds) Microbial mats: stromatolites. Alan R. Liss, New York, pp 441-454

Hallbeck L, Pedersen K (1991) Autotrophic and mixotrophic growth of *Gallionella ferruginea*. J Gen Microbiol 137: 2657-2661

Kump L (1993) Bacteria forge a new link. Nature 362: 790-791

Loehr TM (1989) Iron carriers and iron proteins. VCH, New York, Weinheim

Lovley DR (1991) Dissimilatory Fe(III) and Mn(IV) reduction. Microbiol Rev 55: 259-287

Luther GW, Kostka JE, Church TM, Sulzberger B, Stumm W (1992) Seasonal iron cycling in the salt-marsh sedimentary environment: the importance of ligand complexes with Fe(II) and Fe(III) in the dissolution of Fe(III) minerals and pyrite, respectively. Marine Chem 40: 81-103.

Mauzerall D, Borowska Z, Zielinski I (1993) Photo and thermal reactions of ferrous hydroxide. Orig Life Evol Biosphere 23:105-114

Norris JR, Katz JJ (1978) Oxidized bacteriochlorophylls as photoproduct. In: Clayton RK, Sistrom WR (eds) The photosynthetic bacteria. Plenum Press, New York, pp 397-418

Pfennig N. (1989) Genus VII. *Thiodictyon*. In: Staley JT, Bryant MP, Pfennig N, Holt JG (eds) Bergey's manual of systematic bacteriology, vol 3. Williams & Wilkins, Baltimore, pp 1650-1651

Pierson BK, Olson JM (1989) Evolution of photosynthesis in anoxygenic photosynthetic procaryotes. In: Cohen Y, Rosenberg E (eds) Microbial mats. Physiological ecology of benthic microbial communities. American Society for Microbiology, Washington DC, pp 402-427

Sakaguchi T, Burgess JG, Matsunaga T (1993) Magnetite formation by a sulphate-reducing bacterium. Nature 365: 47-49.

Stumm W, Morgan JJ (1981) Aquatic chemistry, 2nd edn. John Wiley & Sons, New York.

Walker JCG (1983) Was the Archaean biosphere upside down? Nature 329: 710-711.

Widdel F, Bak F (1992) Gram-negative mesophilic sulfate-reducing bacteria. In: Balows A, Trüper HG, Dworkin M, Harder W, Schleifer KH (eds) The prokaryotes, 2nd edn, vol IV. Springer-Verlag, New York, Berlin, pp 3352-3378

Widdel F, Schnell S, Heising S, Ehrenreich A, Assmus B, Schink B (1993) Ferrous iron oxidation by anoxygenic phototrophic bacteria. Nature 362: 834-835.

Heterocystous versus non-heterocystous cyanobacteria in microbial mats

Lucas J. Stal, Hans W. Paerl[1], Brad Bebout[2] and Marlies Villbrandt[3]

Laboratory for Microbiology, University of Amsterdam, Nieuwe Achtergracht 127, 1018 WS Amsterdam, The Netherlands

Introduction

Relative to biological demands, marine waters are often depleted in combined forms of nitrogen (Dugdale 1967). As such, the development of microbial mats in coastal marine environments is dependent on the fixation of atmospheric dinitrogen as the dominant source of biologically available N.

Nitrogen fixation is known as an anaerobic process, confined to prokaryotic organisms. It is widespread among different groups of bacteria. N_2 fixation has relatively high energy demands (Jensen 1983). Among phototrophic cyanobacteria, energy demands can be readily met by photosynthetically generated reducing equivalents. Photosynthesis transforms light into chemical energy (ATP); during this conversion electrons are liberated from water and used to reduce ferredoxin. It is therefore not surprising that a large number of cyanobacteria possess the capacity to fix nitrogen. On the other hand, cyanobacteria are oxygenic phototrophs and it is well-known that nitrogenase is irreversibly inactivated *in vitro* by oxygen (Gallon 1992). Cyanobacteria accordingly should have a suite of physiological, morphological and ecological strategies in order to prevent nitrogenase from inactivation by oxygen. (i) A large number of cyanobacteria are capable of nitrogen fixation only under anoxic conditions. (ii) Heterocystous cyanobacteria separate nitrogenase spatially from oxygenic photosynthesis. These organisms differentiate special cells, heterocysts, which have lost the capacity for oxygenic photosynthesis, but by using the anoxygenic photosystem I, are nevertheless capable of converting light energy into chemical energy. Heterocysts have thickened cell walls which serve as diffusion barrier. This, and a respiratory oxygen uptake provide an anoxic micro-environment within the heterocyst. (iii) A third group of unicellular and filamentous nitrogen-fixing cyanobacteria does not show cell differentiation, but are nevertheless capable of fixing nitrogen aerobically. Although a considerable number of papers have addressed means by which these organisms may protect

[1] Institute of Marine Sciences, University of North Carolina, 3407 Arendell Street, Morehead City, NC 28557 USA,
[2] Max-Planck Institute for Marine Microbiology, Fahrenheitstraße 1, D-2800 Bremen, Germany,
[3] ICBM, University of Oldenburg, Schleusenstraße 16, D-2940 Wilhelmshaven, Germany

NATO ASI Series, Vol. G 35
Microbial Mats
Edited by Lucas J. Stal and Pierre Caumette

nitrogenase against oxygen, underlying mechanisms are not well understood. As a functional equivalent to spatial separation of nitrogen fixation and photosynthesis in heterocystous species, it has been suggested that non-heterocystous species separate these incompatible processes temporally. Indeed, a majority of non-heterocystous species fixes nitrogen during the dark when they are grown under alternating light-dark cycles. The same day-night pattern has been observed in microbial mats.

An obvious and fundamentally-important question is, given what appears to be an indispensible and persistent need for N_2 fixation as a source of "new" nitrogen for supporting mat formation, combined with the strong environmental constraint that O_2 represents, why don't heterocystous cyanobacteria play a more dominant role as diazotrophs? Curiously, not all mats are devoid of heterocystous cyanobacteria. In tropical reef and lagoonal (including hypersaline) systems, certain heterocystous genera (*Scytonema*, *Calothrix*) can dominate mat formation. Typically, these mats exhibit highly irregular, nodular morphologies, distinct from the smooth, laminated mats composed primarily of non-heterocystous genera.

Non-heterocystous cyanobacterial microbial mats

A majority of microbial mats are dominated by non-heterocystous cyanobacteria, yet these systems depend on N_2 fixation as the source of "new" N. It could be argued that nitrogen fixation in such mats is not solely mediated by the cyanobacteria but also by other bacteria present in the mat (Paerl et al. 1989b). Particularly, the purple sulfur bacteria and sulfate reducing bacteria have to be considered as potential nitrogen fixers. Although we cannot exclude these organisms as potential contributors to nitrogen fixation, there are a number of arguments which favor cyanobacteria as the major nitrogen fixers.

Cyanobacteria that fix nitrogen in pure culture have been isolated from most of the non-heterocystous nitrogen-fixing microbial mats. The majority of these isolates belong to the *Oscillatoria* or *Lyngbya* type (Paerl et al. 1991). One argument that the cyanobacteria are responsible for a bulk of mat nitrogen fixation is that nitrogenase activity (per unit chlorophyll *a*) in a mat corresponded with that measured in a culture of *Oscillatoria limosa* isolated from the same mat (Villbrandt et al. 1990). Moreover, *O. limosa* often forms a mixed community with the cyanobacterium *Microcoleus chthonoplastes* and nitrogenase activity in such systems corresponds with the frequency of *O. limosa* (Stal et al. 1984). The pattern of nitrogen fixation in cultures of *O. limosa* grown under alternating light-dark cycles was similar to diel nitrogenase variations in the natural mat (Stal and Heyer 1987). The inhibitor of oxygenic photosynthesis (PS II), 3-(3,4-dichlorophenyl)1,1-dimethylurea (DCMU), stimulates nitrogenase activity during short-term incubations while longer treatments result in decreased activity (Bebout et al. 1993, Stal et al. 1984) (Fig. 1). The latter can be explained by the depletion of reducing equivalents while the short-term effect is dominated by the cessation of oxygen evolution a concomittant decrease in pO_2 in the medium and hence reduced O_2 inhibition of nitrogenase.

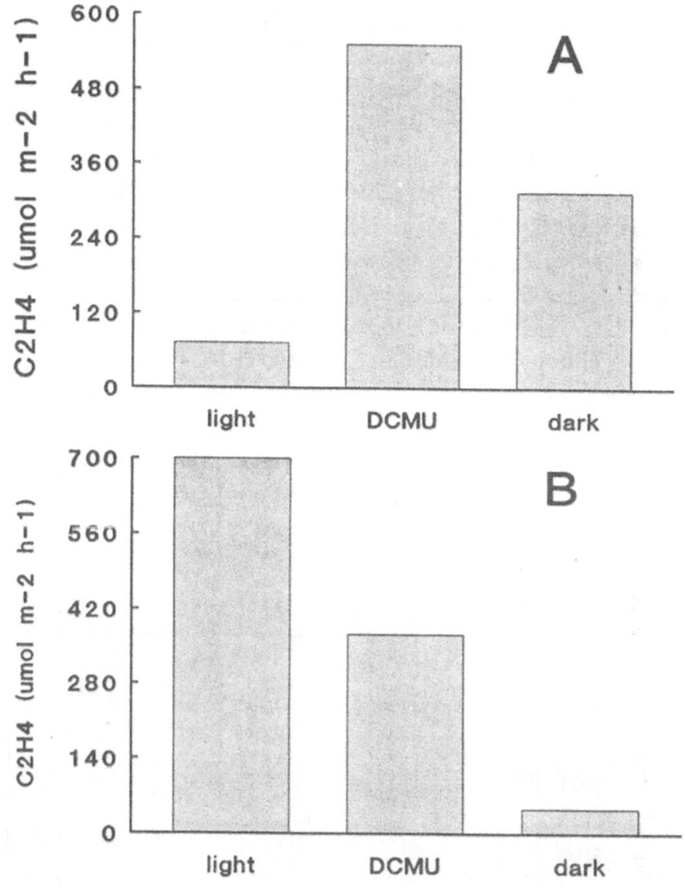

Fig. 1. Effects of light and oxygenic photosynthesis on nitrogenase activity (acetylene reduction) on mats of (A) the non-heterocystous cyanobacterium *Lyngbya aestuarii* and (B) of the heterocystous *Calothrix* sp. of an intertidal lagoon (Guerrero Negro, Baja California Sur, Mexico). Oxygenic photosynthesis was inhibited by adding DCMU (10^{-5} M). The samples were incubated under natural light and ambient temperature from 7-10 am.

The results depicted in Fig. 1 also show the stimulation of nitrogenase activity by light. Another observation in favor of cyanobacteria as nitrogen fixers, is that nitrogen fixation expressed on the basis of the biomass of the cyanobacteria is highest in spring, when the microbial mat develops (Stal et al. 1984). Such mats do not contain a large amount of organic matter and therefore heterotrophic diazotrophic bacteria, including sulfate-reducing bacteria can be expected to be low in number. As a result, sulfide is virtually absent, which can be seen by the absence of black iron sulfide layers. These mats also do not have a visible layer of purple sulfur bacteria. It seems unlikely that these organisms would contribute significantly to nitrogen fixation at this time of the year.

The mechanism(s) by which non-heterocystous cyanobacteria protect nitrogenase from inactivation by oxygen are not well known (Gallon and Stal 1992). It is well known that nitrogenase is a highly-conserved enzyme complex among all diazotrophs and that nitrogenase of non-heterocystous cyanobacteria is as sensitive to oxygen as other nitrogenases. This means that, as is the case in heterocysts, nitrogenase in non-heterocystous cyanobacteria must function in an anoxic micro-environment.

It has been suggested that nitrogenase in cyanobacteria can be present in active and non-active, modified forms. The latter appears to be oxygen insensitive (Ernst et al. 1990). This however, does not alter the need for an anoxic micro-environment for nitrogenase to be active. The advantage of modification could be that nitrogenase is not irreversibly inactivated during periods where an anoxic micro-environment can not be provided. This mechanism is perhaps only relevant for short term (minutes) exposures to oxygen, since diel variations of nitrogenase activity appear dependend on *de novo* protein synthesis (Stal and Krumbein 1985a, Capone et al. 1990, Villbrandt et al. 1992).

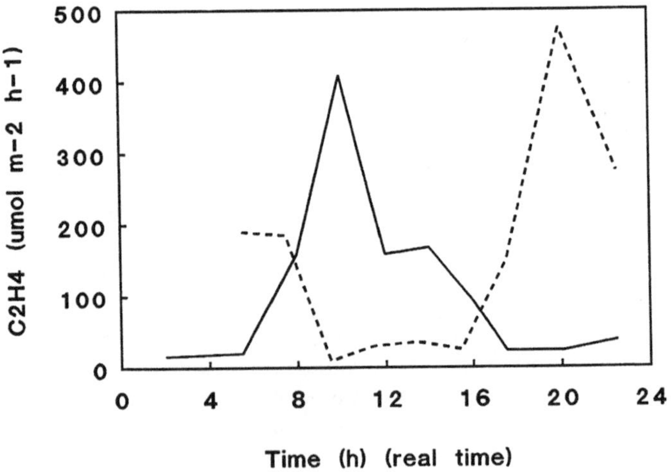

Fig. 2. Day-night variations of nitrogenase in mats of the heterocystous cyanobacterium *Calothrix* sp. (solid line) and of the non-heterocystous cyanobacterium *Lyngbya aestuarii* (dashed line). The mats were sampled from the same location as Fig. 1. The mats were incubated under natural light and ambient temperature.

Unlike heterocystous species, non-heterocystous cyanobacteria have to cope with photosynthetically-produced oxygen presumably in the same cells. Compared to the heterocystous cyanobacteria, which have separated nitrogen fixation in the heterocyst spatially from oxygenic photosynthesis in the vegetative cells, it has been suggested that non-heterocystous cyanobacteria separate these processes temporally (Gallon and Stal 1992). In nature this can be achieved by fixing nitrogen during the dark (Fig. 2). If the temporal separation of nitrogen fixation and photosynthesis will prove indeed to be the mechanism by which these organisms are capable of fixing nitrogen, it is not likely to universally apply to all non-heterocystous genera. One notable exception is the non-heterocystous marine planktonic species *Trichodesmium* spp. which fixes nitrogen during daylight, revealing a similar diel N_2 fixation pattern as heterocystous species (Capone et al. 1990, Ohki et al. 1992). Paerl et al. (1991) presented some evidence for the potential of spatial separation of N_2 fixation and CO_2 fixation in *Lyngbya* sp. An additional example is the unicellular cyanobacterium *Gloeothece*, which has been shown to fix nitrogen during the dark when grown in batch culture, but is capable of fixing nitrogen at high rates during the light period when grown in continuous culture (Ortega-Calvo and Stal 1991). Also, some non-heterocystous cyanobacteria (*Lyngbya*) can fix N_2 during daylight, but at reduced rates (especially under low light levels). Moreover, non-heterocystous cyanobacteria can also be cultured diazotrophically under continuous light concomitantly producing oxygen and fixing nitrogen. Although it may be possible that individual cells can alternate between oxygenic photosynthesis and nitrogen fixation, there is evidence that indeed a cell may evolve oxygen and fix nitrogen simultaneously (Ortega-Calvo and Stal 1991). This is also in agreement with the stimulation by DCMU of nitrogenase activity in non-heterocystous cyanobacteria (Stal and Krumbein 1981). Considering the fact that nitrogenase can only function in the absence of oxygen, spatial separation of photosynthesis and nitrogen fixation at the subcellular level must be invoked. Hitherto, no evidence for such spatial separation has been generated. Rather, immunological studies hint to fairly homogeneous distributions of the enzyme (Paerl et al. 1989a; Stal and Bergman 1990). The concept of autoprotection of nitrogenase may perhaps offer a solution to this dilemma (Thorneley and Ashby 1989). In this concept it is proposed that O_2 itself is a substrate for nitrogenase which reduces it to H_2O_2. Like all other reactions catalized by nitrogenase this reduction is at the expense of reduced ferredoxin and ATP. If this mechanism would apply to non-heterocystous cyanobacteria an interesting situation would arise that for an effective protection of nitrogenase, photosynthetically-derived energy and reducing equivalents are necessary. Paradoxically, this would result in the production of even more oxygen. Contrary, in darkness either energy and reducing equivalent supplies are not sufficient to prevent oxygen from inactivating nitrogenase, or the autoprotection reaction by nitrogenase competes with the reduction of nitrogen. In both cases, the result would be reduced nitrogenase activity (acetylene reduction) under oxic conditions in the dark.

Microbial mats of heterocystous cyanobacteria

The cell wall of the heterocyst is quite different from that of a vegetative cell and its thickness can be altered by the ambient concentration of oxygen (Kangatharalingam et al. 1992). The thick cell wall presents an efficient diffusion barrier for gases (Walsby 1985). It was shown that due to this diffusional barrier and to the fact that the heterocyst can consume oxygen, the interior of the heterocyst is completely anoxic (Walsby 1985). Nitrogenase in the heterocyst is fueled by PS I mediated photophosphorylation. Reductant is derived from neighbouring vegetative cells, catabolized and used for the reduction of ferredoxin in a PS I mediated uphill electron transport or respired. Nitrogen fixation in heterocysts is therefore strongly dependent on light. Heterocystous cyanobacteria generally show very low nitrogenase activity in the dark (Fig. 1). In contrast to non-heterocystous cyanobacteria, DCMU is not expected to be stimulatory to nitrogenase because PS II is absent from the heterocysts. In fact, it may inhibit nitrogenase activity (Fig. 1). This can be explained by the cessation of generation of reducing equivalents by the vegetative cells or by other toxic effects of DCMU on cell metabolism (Stal and Krumbein 1985b). Also, natural populations of heterocystous cyanobacteria show typical day-night variations, with virtually all the nitrogenase activity is confined to daytime (Fig. 2).

Interestingly, when heterocystous cyanobacteria dominate microbial mats, they often form the mat surface with filaments firmly oriented in a uniform manner. *Calothrix* spp., for example, has tapered filaments with terminal heterocysts located at the innermost (away from the surface) position. These filaments do not glide freely, as do many non-heterocystous taxa. Another heterocystous mat builder, *Scytonema* spp., likewise reveals distinct vertical orientation, with heterocysts (while not terminal) located towards the center of spherical aggregates. Again, no motility is evident in this genus. Microbial mats often develop in environments characterized by high rates of sediment deposition. Cyanobacteria that built microbial mats are usually motile (by gliding), facilitating optimal vertical positioning. Gliding also allows the cyanobacteria to compensate for rapidly shifting physico-chemical gradients (Castenholz 1982). However, it is well known that in cultures of freshwater planktonic heterocystous genera such as *Anabaena* and *Nostoc*, filaments tend to break at the heterocyst-vegetative cell junctions when exposed to turbulence-induced stress. The unique "weak link" and morphological features associated with heterocysts may result in filament breakage due to the shear forces produced during gliding in highly compressed laminated mats.

Microbial mats built by heterocystous cyanobacteria are known from relatively few environments. In an intertidal lagoon near Guerrero Negro, Baja California Sur, Mexico, 2 types of microbial mats occur. The smooth mats of the non-heterocystous *Lyngbya aestuarii* occur next to irregular nodular mats of the heterocystous *Calothrix* sp. (Javor and Castenholz 1981). Both mats show high rates of nitrogen fixation. The *Calothrix* mats are situated at higher places in the intertidal flat and for that reason they are less often and for shorter periods inundated. The *Calothrix* mats are rather porous in structure in contrast to the dense mats of *Lyngbya*. Similar mats of heterocystous cyanobacteria are known from stromatolites such as those built by *Scytonema* in San Salvador Island, Bahamas or by *Calothrix* from Walker Lake, Nevada (Bebout 1992). The lithification of these stromatolites as a result of calcium carbonate precipitation results in a rather porous structure in which relatively little organic matter is retained. This results in fairly uniform O_2 concentrations.

Oxygen dynamics and diffusion

Microbial mats often show very steep gradients of oxygen that shift up and down in the mat as a result of diel changes and short term fluctuations of light (Revsbech et al. 1983). During daylight, cyanobacteria produce oxygen as a result of their photosynthetic activity. Oxygen is removed from the mat by the respiratory activities and by diffusion. The mat is a diffusion-limited system (Decho, this volume; Jørgensen, this volume) which results in the accumulation of oxygen in the mat which may reach levels of several-fold air saturation. During the night oxygen is rapidly consumed and due to the same constraints on diffusional oxygen input from the atmosphere or from the overlying water which is too slow to compensate for its removal from the system. As a result, the mat will turn anoxic during the night.

During oxygen supersaturation, non-heterocystous cyanobacteria may not be able to fix nitrogen. For instance, in *O. limosa*, nitrogenase activity decreases sharply at oxygen levels of air saturation and higher (Stal and Krumbein 1985b). If a mechanism such as autoprotection is present in non-heterocystous cyanobacteria, it may have limited capacity. Nitrogenase may not be capable of removing all of the oxygen, leaving it irreversibly damaged, or the elevated level of oxygen results in the diverting of a larger amount of the available energy and reducing equivalents towards its reduction. The latter hypothesis would help explain the switch-off/on of nitrogenase due to short term exposure to oxygen (Stal and Krumbein 1985b).

During night when anoxic conditions prevail, nitrogen fixation is not susceptible to oxygen inactivation. At that time non-heterocystous mat-forming cyanobacteria may utilize a fermentative mode of energy generation. Nitrogen fixation under anoxic conditions in the dark is possible (Stal and Heyer 1987), and is indeed observed in the field (Bebout et al. 1993). However, the amount of energy that can be obtained from the fermentative degradation of storage glucan may be low, and given high energy requirements of nitrogenase, only low activities are expected. Some microbial mats fix the bulk of nitrogen under anoxic conditions in the dark (Bebout et al. 1993). Other mats tend to fix nitrogen at sunrise when the mat is still anoxic but light overcomes the energy limitation of nitrogenase (Villbrandt et al. 1990). The latter diel pattern of nitrogenase activity has also been shown in the laboratory (Stal and Heyer 1987).

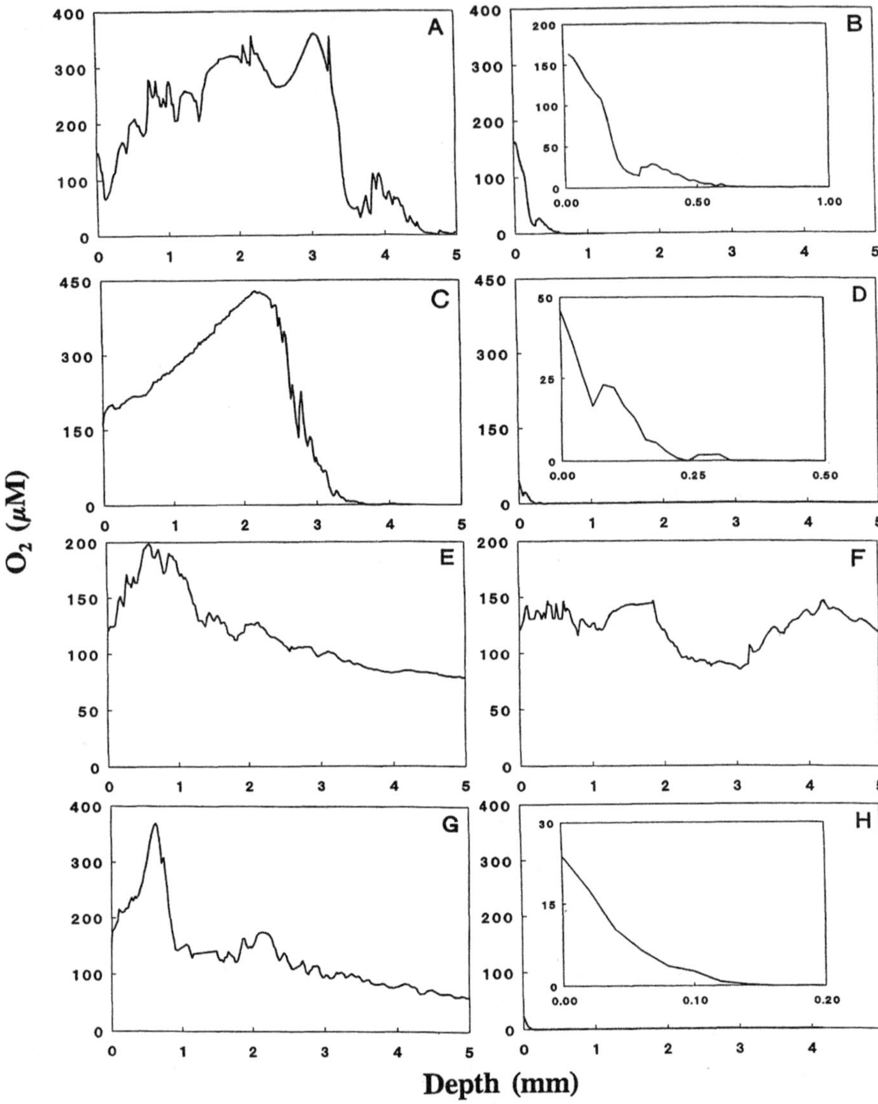

Fig. 3. Oxygen profiles measured with microelectrodes in mats of *Lyngbya aestuarii* and *Calothrix* sp. Profiles were measured either while illuminated under natural light or after 2-3 h of darkness. Both mats were investigated while inundated (simulating high tide) or exposed to air (simulating low tide). A-D mat of *Lyngbya aestuarii* E-H mat of *Calothrix* sp. A, C, E, G illuminated mat; B, D, F, H dark. A, B, E, F exposed to air (low tide) and C, D, G, H inundated (high tide).

The *Calothrix* sp. mats in Baja California are less frequently and more intermittently inundated than the *Lyngbya* mats at the same location. A comparison was made between the oxygen dynamics in *Lyngbya* and *Calothrix* mats during inundation or exposed to the atmosphere. It is clear that *Lyngbya* mats show strong spatial and temporal shifts in pO_2 (Fig. 3). In the light, oxygen accumulated, resulting in a several fold supersaturation, whereas during the dark most of the mat wasanoxic. In this mat there was no difference between low or high tide (inundation or exposure of the matsurface to the atmosphere). The *Calothrix* mat resembled the *Lyngbya* mat during conditions of inundation (which is rare for this mat), while oxygen was more uniformely distributed throughout the diel cycle when the mat was exposed to the atmosphere.

Vertical distribution of nitrogen fixation

In some non-heterocystous mats nitrogenase activity is present throughout the day regardless of oxygen supersaturation (Bebout et al. 1987). Under these conditions there may be spatial separation of nitrogen fixation and photosynthesis. Oxygenic photosynthesis most likely is localized in the upper layers of the mat while nitrogen fixation is confined to the deeper layers (Stal et al. 1984). Light attenuation in microbial mats differs according to wavelength. Far red light penetrates better than shorter wavelengths (Kühl and Jørgensen 1992). This means that in the deeper layers oxygenic photosynthesis is not possible, while anoxygenic photosynthesis may continue. The cyanobacteria in the lower part of the mat may therefore fix nitrogen mediated by PS I driven energy production. By moving up and down in the microbial mat, cyanobacteria may be able to switch between oxygenic photosynthesis and nitrogen fixation. The nitrogen fixing cyanobacteria in the deeper layer may even be capable of phototrophic growth if sulfide is present. Some mat-forming cyanobacteria are capable of sulfide-mediated anoxygenic photosynthesis and CO_2 fixation (Cohen et al. 1986).

Effects of sulfide and anoxic dark conditions on heterocystous cyanobacteria

Not much is known about the effects of dark anoxic conditions on heterocystous cyanobacteria. None of the heterocystous species that were isolated from a microbial mat shows nitrogenase activity under dark anoxic conditions (Stal and Krumbein 1985c). To our best knowledge, no information exists concerning fermentation in heterocystous cyanobacteria. Fermentation of endogenous storage glucan may be an important physiological feature of non-heterocystous mat-forming cyanobacteria which allows them to generate energy for maintenance purposes in order to survive dark conditions (Moezelaar and Stal, this volume). Moreover, there are indications that heterocystous cyanobacteria are more sensitive to sulfide then non-heterocystous species (Howsley and Pearson 1979). These aspects may be additional factors for the exclusion of heterocystous cyanobacteria from microbial mats.

Conclusions

Heterocystous cyanobacteria are well-suited for contemporaneous nitrogen fixation and oxygenic photosynthesis and would therefore be ideal organisms for colonizing nitrogen-depleted coastal sediments. However, morphological and, perhaps, physiological constraints exclude them from such environments. Mat-forming organisms must be motile in order to position themselves optimally in the fluctuating physico-chemical gradients and also because of the high rates of sedimentation in such environments. Motility demands of heterocystous cyanobacteria in densely-packed microbial mats may lead to detachment of the heterocyst from the vegetative cell. The absence of a fermentative mode of energy generation and a high degree of sensitivity to sulfide may also be factors responsible for the exclusion of heterocystous cyanobacteria from potential habitats. Consequently, microbial mats in environments having relatively low sedimentation rates and which do not turn anoxic during the night or accumulate sulfide, are often built by heterocystous cyanobacteria. However, such mats are relatively rare, leaving non-heterocystous cyanobacteria as the dominant group. Non-heterocystous cyanobacteria are versatile organisms that can cope well with the fluctuating physical-chemical conditions in the mat. They are highly motile by gliding movement, which allows them to position themselves optimally in the mat. They are also capable of fermentation and anoxygenic photosynthesis or tolerate high levels of sulfide. Overall, nitrogen-fixing non-heterocystous cyanobacteria are capable of diazotrophic growth under rapidly shifting, environmentally-diverse conditions. They thus exhibit reasonably effective mechanisms for protecting nitrogenase from oxygen inactivation in the highly dynamic microenvironment which mats represent.

Acknowledgements

We would like to acknowledge financial support from the European Union (MAST grant 0027C), the Fulbright Program to LJS and from NSF grants OCE 9012496 and OCE 9115706 to HWP. We thank Dr. Dave DesMarais for inviting us to Baja California, Mexico.

References

Bebout BM (1992) Interactions of nitrogen and carbon cycling in microbial mats and stromatolites. PhD Dissertation, University of North Carolina at Chapel Hill

Bebout BM, Fitzpatrick MW, Paerl HW (1993) Identification of the sources of energy for nitrogen fixation and physiological characterizations of a marine microbial mat community. Appl Environ Microbiol 59:1495-1503.

Bebout BM, Paerl HW, Crocker KM, Prufert LE (1987) Diel interactions of oxygenic photosynthesis and N_2 fixation (acetylene reduction) in a marine microbial mat community. Appl Environ Microbiol 53:2353-2362

Capone DG, O'Neil JM, Zehr J, Carpenter EJ (1990) Basis for diel variation in nitrogenase activity in the marine planktonic cyanobacterium *Trichodesmium thiebautii*. Appl Environ Microbiol 56:3532-3536

Castenholz RW (1982) Motility and taxes. In: Carr NG, Whitton BA (eds) The biology of blue-green algae. Blackwell Scientific Publishers, Oxford, pp 413-439

Cohen Y, Jørgensen BB, Revsbech NP, Poplawski R (1986) Adaptation to hydrogen sulfide of oxygenic and anoxygenic photosynthesis among cyanobacteria. Appl Env Microbiol 51:398-407

Dugdale RC (1967) Nutrient limitation in the sea: Dynamics, identification and significance. Limnol Oceanogr 12:685-695

Ernst A, Reich S, Boger P (1990) Modification of dinitrogenase reductase in the cyanobacterium *Anabaena variabilis* due to C starvation and ammonia. J Bacteriol 172:748-755

Gallon JR (1992) Reconciling the incompatible: N_2 fixation and O_2. New Phytol 22:571-609

Gallon JR, Stal LJ (1992) N_2 fixation in non-heterocystous cyanobacteria: an overview. In: Carpenter EJ, Capone DG, Rueter JG (eds) Marine pelagic cyanobacteria: *Trichodesmium* and other diazotrophs. NATO ASI Series C, Vol. 362, Kluwer Academic Publishers, Dordrecht, pp 115-139

Howsley R, Pearson HW (1979) pH dependent sulphide toxicity to oxygenic photosynthesis in cyanobacteria. FEMS Microbiol Lett 6:287-292

Javor BJ, Castenholz RW (1981) Laminated microbial mats, Laguna Guerrero Negro, Mexico. Geomicrobiol J 2:237-273

Jensen BB (1983) Energy requirement for diazotrophic growth of the cyanobacterium *Anabaena variabilis* determined from growth yields in the dark. J Gen Microbiol 129:1633-1640

Jørgensen BB, Cohen Y, Des Marais DJ (1987) Photosynthetic action spectra and adaptation to spectral light distribution in a benthic cyanobacterial mat. Appl Env Microbiol 53:879-886

Kangatharalingam N, Priscu JC, Paerl HW (1992) Heterocyst envelope thickness, heterocyst frequency and nitrogenase activity in *Anabaena flos-aquae*: influence of exogenous oxygen tension. J Gen Microbiol 138:2673-2678

Kühl M, Jørgensen BB (1992) Spectral light measurements in microbenthic phototrophic communities with a fiber-optic microprobe coupled to a sensitive diode array detector. Limnol Oceanogr 37:1813-1823

Ohki K, Zehr JP, Fujita Y (1992) Regulation of nitrogenase activity in relation to the light--dark regime in the filamentous non-heterocystous cyanobacterium *Trichodesmium* sp. NIBB 1067. J Gen Microbiol 138:2679-2685

Ortega-Calvo JJ, Stal LJ (1991) Diazotrophic growth of the unicellular cyanobacterium *Gloeothece* sp. PCC 6909 in continuous culture. J Gen Microbiol 137:1789-1797

Paerl HW, Bebout BM, Prufert LE (1989) Naturally occurring patterns of oxygenic photosynthesis and N_2 fixation in a marine microbial mat: Physiological and ecological ramifications. In: Cohen Y, Rosenberg E (eds) Microbial Mats. Physiological ecology of benthic microbial communities. ASM, Washington, pp 326-341

Paerl HW, Priscu JC, Brawner DL (1989) Immunochemical localization of nitrogenase in marine *Trichodesmium* aggregates: relationship to N_2 fixation potential. Appl Environ Microbiol 55:2965-2975

Paerl HW, Prufert LE, Ambrose WW (1991) Contemporaneous N_2 fixation and oxygenic photosynthesis in the non-heterocystous mat-forming cyanobacterium *Lyngbya aestuarii*. Appl Environ Microbiol 57:3086-3092

Revsbech NP, Jørgensen BB, Blackburn TH, Cohen Y (1983) Microelectrode studies of the photosynthesis and O_2, H_2S and pH profiles of a microbial mat. Limnol Oceanogr 28:1062-1074

Stal LJ, Bergman B (1990) Immunological characterization of nitrogenase in the filamentous non-heterocystous cyanobacterium *Oscillatoria limosa*. Planta 182:287-291

Stal LJ, Groβberger S, Krumbein WE (1984) Nitrogen fixation associated with the cyanobacterial mat of a marine laminated microbial ecosystem. Mar Biol 82:217-224

Stal LJ, Heyer H (1987) Dark anaerobic nitrogen fixation (acetylene reduction) in the cyanobacterium *Oscillatoria* sp. FEMS Microbiol Ecol 45:227-232

Stal LJ, Krumbein WE (1981) Aerobic nitrogen fixation in pure cultures of a benthic marine *Oscillatoria* (cyanobacteria). FEMS Microbiol Lett 11:295-298

Stal LJ, Krumbein WE (1985a) Nitrogenase activity in the non-heterocystous cyanobacterium *Oscillatoria* sp. grown under alternating light-dark cycles. Arch Microbiol 143:67-71

Stal LJ, Krumbein WE (1985b) Oxygen protection of nitrogenase in the aerobically nitrogen-fixing, non-heterocystous cyanobacterium *Oscillatoria* sp. Arch Microbiol 143:72-76

Stal LJ, Krumbein WE (1985c) Isolation and characterization of cyanobacteria from a marine microbial mat. Bot Mar 28:351-365

Thorneley RNF, Ashby GA (1989) Oxidation of nitrogenase iron protein by dioxygen without inactivation could contribute to high respiration rates of *Azotobacter* species and facilitate nitrogen fixation in other aerobic environments. Biochem J 261:181-187

Villbrandt M, Stal LJ, Bergman B, Krumbein WE (1992) Immunolocalization and western blot analysis of nitrogenase in *Oscillatoria limosa* during a light-dark cycle. Bot Acta 105:90-96

Villbrandt M, Stal LJ, Krumbein WE (1990) Interactions between nitrogen fixation and oxygenic photosynthesis in a marine cyanobacterial mat. FEMS Microbiol Ecol 74:59-72

Walsby AE (1985) The permeability of heterocysts to the gases nitrogen and oxygen. Proc R Soc London B 226:345-366

Development of versicolored microbial mats: Succession of microbial communities

Isabel Esteve, Dolores Ceballos, Maira Martínez-Alonso, Núria Gaju and Ricardo Guerrero[1]

Department of Genetics and Microbiology and Institute for Fundamental Biology. Autonomous University of Barcelona, 08193 Bellaterra (Barcelona), Spain.

Today, microbial mats are confined to a restricted range of habitats including hypersaline and coastal marine environments, hot springs and alkaline lakes. In the last few years, these systems have been considered a major subject of study and many articles (Cohen et al. 1984; Cohen and Rosenberg 1989) have been published, which extensively revise the structure, physiology, genetics and evolution of microbial mats.

A preliminary investigation of these microbial ecosystems distributed along the west Mediterranean coast, from Almería (SE corner of the Iberian Peninsula) to the Ebro Delta has been done (Esteve et al. 1992b). More exhaustive investigations have been done about microbial mats of La Banya spit, the southern peninsula of the Ebro Delta (Mir et al. 1991), which thrive in several areas that may be flooded during the occasional rises in the sea level due to tides, waves and storms.

Here we describe the structure and the development of communities in different microbial mats in la Banya spit, the ultrastructure of individual cells, as well as the spatial and temporal variations of these communities.

Distribution and typology of the Ebro deltamicrobial mats

The studied microbial mats are located at the Ebro Delta (40°35' N, 0°40' E), 50 km south from Tarragona. They occur all along the coast in the narrow ephemeral ponds of the backshore, on the flanks of the storm inlets and, most commonly, on the sand flats and channels of La Banya spit.

In the last few years, we have studied ten sites along La Punta de la Banya, which were selected in order to study spatial variations. All of them lay 1 to 7 cm below the water surface during the flooded periods. The temperature of water covering these microbial mats ranged from 12 to 30°C; conductivity from 59 to 105 $mS.cm^{-1}$; salinity from 40 to 75 ‰, and pH from 7.5 to 9.0.

[1] Department of Microbiology, University of Barcelona, Diagonal 645, 08028 Barcelona, Spain

NATO ASI Series, Vol. G 35
Microbial Mats
Edited by Lucas J. Stal and Pierre Caumette

The same genera were usually present in all the sites, but at different ratios. The filamentous cyanobacteria *Microcoleus* sp. and *Lyngbya* sp. predominated over the coccoid forms in most of the mats. *Spirulina* sp., another filamentous cyanobacteria, was also identified, but it was only predominant in one of the selected sites. A well-developed purple layer was seen at different sites, which were submerged when sampled, whilst the presence of a black layer of sulfate reducing bacteria was ubiquitous and it ranged from 3 mm to several cm (Rampone et al. in press).

To study temporal variations, three sampling sites (P1, P2 and P3) that represented different stages of development were chosen. The structure and the vertical distribution of the undisrupted communities of these three sampling sites were studied by the combined use of light microscopy, scanning electron microscopy and pigment analysis (Mir et al.1991).

Ultrastructure of a well developed microbial mat community

Light microscopy offers a first direct approach to the study of microbial communities. However, in most cases a higher resolution is required due to the small size of the organisms. Transmission electron microscopy (TEM) has been frequently used to identify microorganisms in natural habitats (Stolz 1984; Guerrero et al. 1987) and to study the interactions among them (Esteve et al.1992a; Gaju et al. 1992).

We have studied the microorganisms living in the photic zone of site 3 (P3) by using TEM. This mat was vertically stratified into three pigmented layers. In the uppermost yellow-brown layer, diatoms and coccoid cyanobacteria dominated. The second layer was composed mostly by the cyanobacterium *Microcoleus* sp. (Fig. 1a). This bacterium presents bundles of trichomes in a common sheath. Cylindrical trichomes have a diameter of 2.7 μm. The ultrathin sections of individual cells show the typical radial arrangement of thylakoids. Phycobilisomes are located in the interthylakoidal space next to the photosynthetic membrane. Other species of filamentous cyanobacteria could also be found. Figure 1b shows cells of *Lyngbya* sp. which present a thick sheathed filament of the granulated trichomes. Moreover, different coccoid cyanobacteria forming a tetrade of cells surrounded by a thick sheath (Fig 1c) could also be observed.

In situ ultrathin sections of the deepest pigmented layer, purple in color, revealed: cells of *Chromatium gracile*, which have been identified from axenic cultures as a motile rod shaped bacterium (1 μm x 2.5 μm), with an intracytoplasmic membrane system (ICM) of vesicular type (Fig. 1d); spherical non-motile cells surrounded by a capsule of *Thiocapsa* sp. showing inside vesicles (Fig. 1e) and two unidentified purple phototrophic bacteria. The first one is a filamentous segmented bacteria with a lamellar photosynthetic membrane system (Fig. 1f). The second one is ovoidal and exhibits phenotypic characteristics similar to *Chromatium* cells (Fig. 1g). The major difference between *Chromatium* and this bacterium lies in the nature of the intracellular membrane system (vesicles and lamellae respectively) (Fig. 1h).

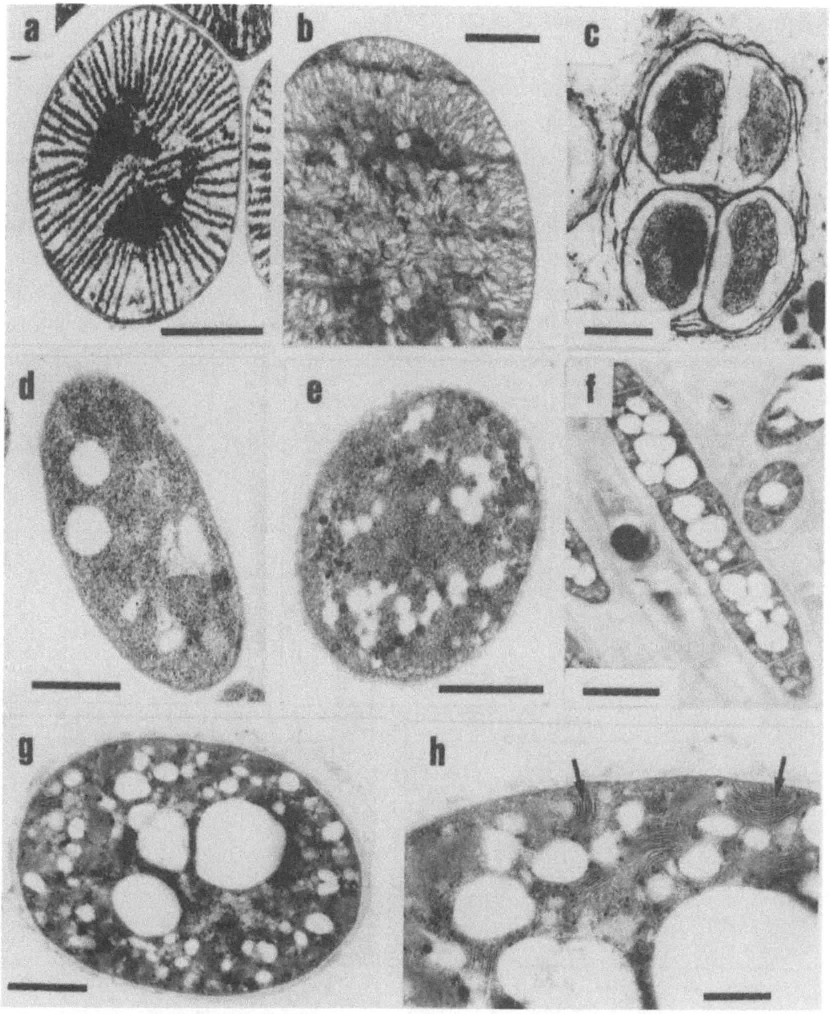

Fig. 1. **(a)** *Microcoleus chthonoplastes* (bar= 1 μm), **(b)** *Lyngbya* sp. (bar= 1.5 μm), **(c)** tetrade of coccoid cyanobacteria (bar= 0.5 μm), **(d)** *Chromatium gracile* (bar= 0.5 μm), **(e)** *Thiocapsa* sp. (bar= 1 μm), **(f)** unidentified phototrophic filamentous microorganisms (bar= 1.5 μm), **(g)** unidentified purple phototrophic bacteria with a lamellar membrane system (bar= 1.5 μm) and **(h)** detail of ICM indicated by arrows (bar= 0.5 μm).

Also at the Ebro Delta microbial mats (sites P1 and P3), the largest free-living spirochete known has been identified and described for the first time. It has been named *Spirosymplokos deltaeiberi*, it can be up 100 μm long and 0.4 μm to 3.0 μm wide, and it presents a composite structure (Margulis et al. 1993; Guerrero et al. 1993).

Succession of Ebro delta microbial mats

Microbial mats have been differentiated in the literature into several types (Jørgensen et al. 1983; Stal et al. 1985).The microbial mats at the Ebro Delta have been classified by Rampone et al. (in press) considering the different ratios of cyanobacteria, as well as the presence or absence of anoxygenic sulfur phototrophic bacterial layers.

We have studied the dynamics of predominant pigments in two different microbial mats (P2 as an example of initial colonization and P3 as an example of a well-developed microbial mat) belonging to main phototrophic populations in relation to both the mat structure and environmental parameters (Martínez-Alonso et al. In prep). Figure 2 shows the vertical distribution of chlorophyll *a* and bacteriochlorophyll *a* from sites P2 and P3 during the summer season in 1989 and 1991.

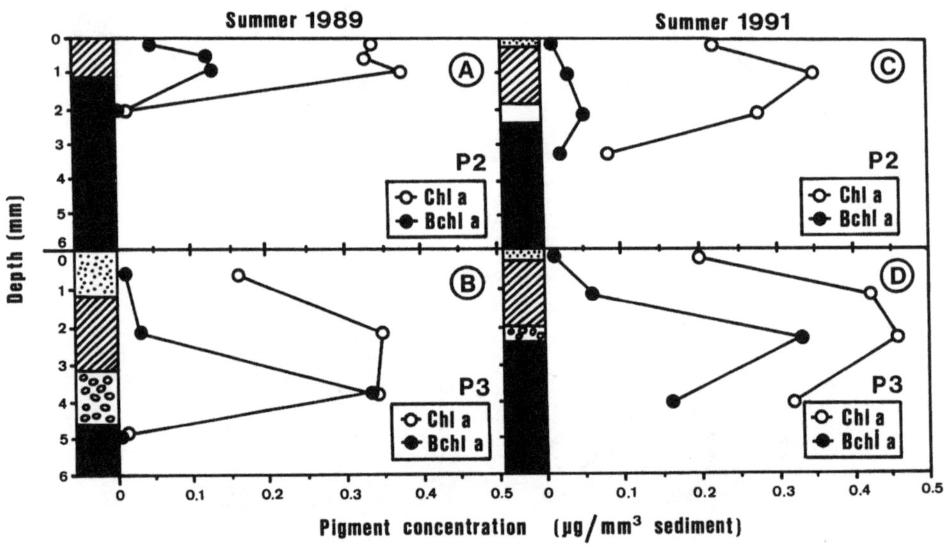

Fig. 2. Pigment distribution with depth in sites 2 (P2) and 3 (P3) (methanol extracts). The pigment concentration from each layer is represented in the middle of the layer. The distinct colored layers are designated by different symbols in the core diagram at the left:

Yellow-brown layer, green layer, white layer, purple layer, and black sediment.

The thickness of the colored zone at site P2 increased from summer 1989 to summer 1991, and thus structure and pigment concentration of the mat changed. In 1989 only a dark-green layer was present at which *Spirulina* sp. was the dominant filamentous cyanobacterium. The highest concentrations of chlorophyll *a* and bacteriochlorophyll *a* were 3.72×10^{-1} $\mu g.mm^{-3}$ of sediment and 1.27×10^{-1} $\mu g.mm^{-3}$ of sediment respectively, both of them located at the deepest part of this dark green layer (Fig.2 A). In 1991 three pigmented layers could be distinguished; a yellowish-brown layer at the top, a second layer, green, with the highest chlorophyll *a* concentration (3.5×10^{-1} $\mu g.mm^{-3}$ sediment) and, finally, a white layer with purple spots, in which the highest concentration of bacteriochlorophyll *a* was detected (5.3×10^{-2} $\mu g.mm^{-3}$ sediment) (Fig. 2C).

In site 3, the thickness of the pigmented zone decreased during the same period. Nevertheless, the species composition (Fig. 1), the number of pigmented layers, and the pigment concentration (Fig. 2B and D) did not change. The highest concentrations of chlorophyll *a* and bacteriochlorophyll *a* were very similar during the summer season both in 1989 and 1991.

Changes in oxygen microgradients as well as, temperature, salinity and water covering were the most important factors controlling microbial mat communities development.

The main objectives of our current studies are: the characterization of phototrophic bacteria, and the assessment of the nutritional state of the microorganisms from Ebro Delta microbial mats

Acknowledgements

We gratefully acknowledge M. Piqueras, J. Mir, and J. Urmeneta for their advice and collaboration and J.M. Cuartero and J. Checa for technical assistance. This work was supported by grants CICYT PB-90-0697 to I.E. and MAR91-0874 to R.G. D.C. research was supported by a scholarship of the Autonomous Government of Catalonia. We also thank the Autonomous University of Barcelona for a research grant to M. M-A.

References

Cohen Y, Castenholz R W and Halvorson HO (eds) (1984) Microbial mats: Stromatolites. Alan R Liss Inc. New York

Cohen Y , Rosenberg E (eds) (1989) Microbial mats. Physiological ecology of benthic microbial communities. American Society for Microbiology, Washington, D.C.

Esteve I, Gaju N, Mir J, Guerrero R (1992a) Comparison of techniques to determinate the abundance of predatory bacteria attacking Chromatiaceae. FEMS Microb Ecol 86: 205-211

Esteve I, Martínez-Alonso M, Mir J, Guerrero R (1992b) Distribution, typology and structure of microbial mat communities in Spain. A preliminary study. Limnetica 8: 185-195

Gaju N, Esteve I, Guerrero R (1992) Distribution of predatory bacteria that attack Chromatiaceae in a sulfurous lake. Microb Ecol 24: 171-179

Guerrero R, Esteve I, Pedrós-Alió C, Gaju N (1987) Predatory bacteria in prokaryotic communities. The earliest trophic relationship. Ann NY Acad Sci 503: 238-250

Guerrero R, Ashen JB, Solé M, Margulis L (1993) *Spirosymplokos deltaeiberi* nov. gen., nov. sp.: Variable diameter composite spirochete from microbial mats. Arch Microbiol 160: 461-471

Jørgensen BB, Revsbech NP, Cohen Y(1983) Photosynthesis and structure of benthic microbial mats: Microelectrode and SEM studies of four cyanobacterial communities. Limnol Oceanogr 28: 1075-1093

Margulis L, Ashen JB, Solé M, Guerrero R (1993) Composite, large spirochetes from microbial mats: Spirochete structure review. Proc Natl Acad Sci USA 90: 6966-6970

Mir J, Martínez-Alonso M, Esteve I, Guerrero R (1991) Vertical stratification and microbial assemblage of a microbial mat in the Ebro Delta (Spain). FEMS Microbiol Ecol 86: 59-68

Rampone G, Urmeneta J, Puigdefàbregas C, Guerrero R (in press) Geographical distribution of microbial mats in the Ebro Delta: Structural diversity and role in the stabilization of the sediments. Verh Internat Verein Limnol

Stal LJ, van Gemerden H, Krumbein WE (1985) Structure and development of a benthic marine microbial mat. FEMS Microbiol Ecol 31: 111-125

Stolz JF (1984) Fine structure of the stratified microbial community at laguna Figueroa, Baja California, Mexico: II. Transmission electron microscopy as a diagnostic tool in studying microbial communities *in situ*. In: Cohen Y, Castenholz RW, Halvorson HO (eds) Microbial Mats: Stromatolites. Alan R. Liss Inc, New York, pp 23-38.

Light-driven sulfate reduction and methane emission in hypersaline cyanobacterial mats

Yehuda Cohen, Yael Helman and Pavel Sigalevich

The Moshe Shilo Center for Marine Biogeochemistry, Institute of Life Sciences, The Hebrew University of Jerusalem, Jerusalem 91904, Israel

Introduction

Sulfate-reducing bacteria and moreover methanogenic bacteria have long been considered to be strictly anaerobic organisms (Morris 1979; Skyring 1987; Widdel 1988). No light harvesting system has ever been described for these bacteria, on the contrary, several methanogens were found to be inhibited by exposure to light. It is therefore surprising to find accumulating data indicating that elevated levels of sulfate reduction and methane emission are found to be associated with photosynthetically active cyanobacterial mats when exposed to high intensity of solar irradiation in close proximity to supersaturation concentrations of oxygen (Cohen 1984; Giani et al. 1984; Canfield and Des Marais 1991; Fründ and Cohen 1992). Oxygen, which was found to be toxic at much lower concentrations to many isolates of both sulfate reducing and methanogenic bacteria does not seem to inhibit both processes (Hastings and Emerson 1988). Moreover, there are some indications that these processes are enhanced in cyanobacterial mats during noontime when supersaturation of oxygen is present in the photic zone (Fründ and Cohen 1992).

The careful examination of possibly oxygen-tolerant sulfate reduction and methane emission in cyanobacterial mats requires a great degree of caution given the finely laminated nature of the benthic microbial community of cyanobacterial mats. Different layers of distinctly different microbial communities are found one below the other along steep attenuation of light spectra (Kühl and Jørgensen 1994) and chemical microprofiles (Jørgensen et al. 1983; Revsbech et al. 1983) which are established to a large extent as a function of oxygenic photosynthesis of cyanobacteria. The entire photic zone may be less than 1mm as it is the case in the shallow mat of the Solar Lake (D'Amelio et al. 1989). Hence a series of novel microtechniques are required to allow resolution on microscales relevant to the zonation of the microbial community.

A series of specific microelectrodes are presently available for the measurement of O_2, H_2S, NO_3^- and N_2O (Revsbech and Jørgensen 1986) along with measurement of light spectra by the use of fiber-optic coupled to a diode array (Kühl and Jørgensen 1994). The combination of these tools allows the microscale measurement of oxygenic photosynthesis, aerobic respiration and denitrification. Novel techniques are required for the measurement of sulfate reduction, methanogenesis and methane oxidation at submillimetric scale. A micro-technique

NATO ASI Series, Vol. G 35
Microbial Mats
Edited by Lucas J. Stal and Pierre Caumette

for the *in situ* measurement of sulfate reduction is presently available and will be discussed later, while novel techniques for the microscale determination of methane formation and oxidation still need to be developed.

The microscopic identification of the vertical distribution of sulfate reducing and methane producing and oxidizing bacteria are methodologically problematic. It requires the construction of specific fluorescent probes either by developing specific antibodies or oligonucleotide probes. Both techniques require the isolation and growth in pure culture of reference microorganisms of the ecologically important strains. This is a challenging task for innovative microbiologists requiring a great deal of in depth understanding of both the ecology of the natural community and the physiology of axenic cultures of what may be ecologically important microorganisms. Furthermore, the microbial mat community is largely embedded in the copious amount of excreted polysaccharides, it may well be that some diel migration of microorganisms may occur along with the diurnal shifts of the chemical microgradients.

Diurnal cycles of sulfate reduction in microbial mats

High rates of sulfate reduction of up to 3 μmol $SO_4^{=}$ reduced cm^{-2} h^{-1} were found to occur at noontime in the upper zone of laminated, hypersaline cyanobacterial mats which exhibit diurnal fluctuations of over 1mM oxygen during daytime to 200μM H_2S at night (Fründ and Cohen 1992). Much higher sulfate reduction activities of up to 30 μmol $SO_4^{=}$ reduced cm^{-2} h^{-1} were found by the autoradiogramming of $Ag_2{}^{35}S$ trapped on silver wires coated with $Na_2{}^{35}SO_4$ which were introduced to the cyanobacterial mats (Helman and Cohen unpublished). Using kinetic models for concomitant sulfide oxidation and fast isotopic exchange of $^{32}S^{=}$ and $^{35}S^{=}$ (Fossing and Jørgensen 1990) occurring on the silver wires we could quantify the above mentioned rates of sulfate reduction activity within the oxic zone of the cyanobacterial mats. The sulfate reduction activity was then compared to microprofiles of oxygen and sulfide determined by specific microelectrodes.

High rates of sulfate reduction activity were found to occur for the first time at close proximity to oxygen and well in oxygen supersaturated zones within laminated hypersaline cyanobacterial mats when using the silver wire technique. This yet not fully explained activity is of prime importance in understanding the microbial mechanisms controlling the recycling of organic matter at the sediment-water interfaces.

Three distinct zones of sulfate reduction occur along a vertical profile within the cyanobacterial mat: an upper peak in the oxic layer which is preferentially stimulated by the addition of glycolate; a major peak at the chemocline within the mat which is primarily stimulated by the addition of acetate and a third peak in the anoxic zone where lactate and ethanol show to serve as the preferred carbon source.

We presently postulate that the high rate of sulfate reduction in the oxic zone is due to the concomitant production of O_2 and H_2 as a result of photosynthetic activity of the major mat-building cyanobacterium *Microcoleus chthonoplastes* when exposed to high light intensities under temporal CO_2 limitation. This hypothesis will be examined by the examination of recycling of O_2, H_2, H_2S and CH_4 using a micro-diffusion probe presently developed at the Max-Planck Institute for Terrestrial Microbiology in Marburg using gas chromatography and later a quadropole gas analyzer.

Diurnal cycles of methane production in microbial mats

Rates of methane emission from the hypersaline cyanobacterial mats were found to be as high as from the classical methane generating ecosystem-rice paddies. Yet, the methane reservoir is considerably lower (Conrad and Cohen unpublished). Using specific inhibitors for methane oxidation (dimethylether and methyl bromine) we could not detect any activity of methane oxidation in this ecosystem. This may be because of sensitivity of this group of microorganisms to elevated salinity or because methane concentrations are too low for the K_m for methane by methylotrophs in these hypersaline environments where methane saturation occurs at considerably lower concentrations compared to fresh water environments. The lack of *in situ* methane oxidation requires a different explanation other than methane oxidation for the isotopically heavy CH_4 of $\delta^{13}C$ -35 ‰ to -42 ‰ typically found in this habitat (Gerling, Faber and Cohen unpublished). Dimethylation of compounds such as methylamines and mcthylsulfides may serve as an important source for methane in this environment which may result in isotopically heavy methane.

Like sulfate reduction, also methane emission shows higher rates at noon time under elevated oxygen concentrations. Oxygen-tolerant methane producing organisms are presently being enriched and isolated and hence the physiology of this interesting group of microorganisms is not known as yet.

Physiology of oxygen-tolerant sulfate reducing bacteria

Novel sulfate reducing bacteria were isolated using fluctuating O_2/H_2S gradients established in shake agar tubes using lactate or acetate as carbon source, provided by a 2mM Na_2S at the bottom and a top layer of an axenic culture of hypersaline cyanobacterium *Synechococcus* sp. The shake agar tubes were incubated in a 12 hours light/dark cycle resulting in oxygen-sulfide fluctuating gradients determined by specific microelectrodes for oxygen and sulfide. After 3 days incubation the agar containing the colonies developing under exposure to oxygen was sliced to 2mm sections and incubated on anaerobic Petri dishes containing medium for SRB and Fe^{++}. Other isolates were obtained in fluctuating chemostat experiments under alternating N_2/air atmosphere using lactate or hydrogen as electron donors. Other oxygen-tolerant sulfate reducing bacteria growing on glycolate and acetate are presently being isolated and characterized in a different set of continuous culture experiments.

Several *Desulfovibrio* type sulfate reducing bacteria were shown to be able to reduce sulfate either autotrophically or in the presence of lactate; when exposed to 5% O_2 these isolates carried out a *Thiobacillus*-like mode of growth whereby sulfide was oxidized to sulfate determined in an oxygen-limited continuous culture experiments. In the absence of reduced sulfur compounds, aerobic growth could be observed along with a drastic increase in cytochrome *c* contents in the culture. These strains could also grow at higher oxygen concentrations of 20% provided that 5% H_2 was added to the gas mixture.

Discussion

Shallow water cyanobacterial mats (Krumbein et al. 1977) are exposed to marked light-dependent diurnal cycles mainly as a function of diel variation of oxygenic photosynthesis. During nighttime the microbial mats are usually fully reduced, except for the uppermost layer which is in direct contact with the overlaying oxygenated water. Oxygenic photosynthesis commences with sunrise and after a lag time of 1-3 hours afterwards excess photosynthetic oxygen starts to build up. The rate of oxygen production increases gradually with increasing solar irradiation towards noontime. By this time of the day oxygen production rates are extremely high and result in the ebullition of oxygen bubbles during the duration of exposure to high solar irradiation. The intense production of oxygen bubbles is obvious when observing coastal microbial mats throughout the globe. At maximal solar irradiation of above 1500 $\mu E\ m^{-2}\ sec^{-1}$, photoinhibition processes can than be observed resulting in lower rates of oxygen production. Simultaneously, excretion of glycolate is observed (Cohen unpublished). pH increases to maximal values of 10-10.5 as a result of high rates of CO_2 photoassimilation. Oxygen concentrations of up to 1200 μM have been found in coastal hypersaline cyanobacterial mats together with maximal pH of 9.8 (Jørgensen et al. 1983) or even up to 2400 μM and maximal pH of 10.5 were recorded in cyanobacterial mats from Christmas Island, Central Pacific Ocean (Cohen and Ideses 1989) and in coastal lagoons on Turks and Caicos in the Bahamas (Cohen unpublished).

These conditions prevail usually during 4 hours at noontime and the early afternoon hours. At about 3 pm with lowering of light intensities a temporal increase in the maximal oxygenic photosynthesis rate is observed which lasts about an hour before this activity becomes light limited and gradually decreases with decreasing light intensities towards sundown. We may therefore divide the diurnal cycle of oxygenic photosynthesis into the following types of activities:

A. 6-10am: Light limitation, low oxygen concentration, pH 8-8.5.
B. 10-11am: Optimal light, oxygen saturation, pH 9-9.5.
C. 11am-3pm: Photoinhibition, oxygen supersaturation, pH above 9.5.
D. 3-4pm: Optimal light, oxygen supersaturation, pH above 9.5.
E. 4-6pm: Light limitation, oxygen supersaturation to saturation, pH above 9.5.

Oxygenic photosynthesis is carried out in the cyanobacterial mats under oxygen saturation or oxygen supersaturation at elevated pH and for 4 hours around noon time under photoinhibiting light intensities. Under these conditions significant amounts of glycolate are expected to be excreted via photorespiration. Attempts to measure *in situ* levels of free glycolate in the interstitial water showed only low levels of up to 10 μM to accumulate in the photic zone at noon time. No glycolate could be found in samples collected during other times of the day.

The stable carbon isotopic compositions of the produced organic matter of coastal cyanobacterial mats were measured in the Solar Lake, Sinai (Aizenshtat et al. 1984) in the Laguna Ojo de Liebre in Baja California, Mexico (Des Marais et al. 1989) and in the Salt Pond in Eilat, Israel have all $\delta^{13}C$ values which are exceptionally heavy, ranging from -6 ‰ to -13 ‰. These findings strongly indicate photosynthetic activity under temporal CO_2 limitations during most of the daytime. Furthermore, under these conditions hydrogen was found to be produced as a result of photoreduction of water by the cyanobacteria at high light intensity combined with CO_2 limitation. Indications for CO_2 limitations for cyanobacterial photosynthesis come from both the isotopic composition of the produced organic matter and from hydrogen production, both of these findings require CO_2 limitations. Yet it is not easy to consider CO_2 limitation in the light of the possible operation of a CO_2 concentrating mechanism in cyanobacteria (Schwarz et al. this volume) and in an environment which precipitates $CaCO_3$.

What are the environmental limitations for oxygenic photosynthesis? Why is there a 2-fold increase in oxygen buildup in the Christmas Island mat? It may be a function of limitations of available photosynthetic action spectra or CO_2 limitations or availability of nutrients such as nitrogen and phosphate (Rothschild and Mancinelli 1990). The microbial mats there are built over a fossilized coral reef composed primarily of aragonites. Alkalinity of 20mM was measured in the interstitial water compared to less than 1mM measured in the Solar Lake. At Christmas Island supply of CO_2 from dissolution of the aragonite prevents the temporal CO_2 limitation observed in other coastal cyanobacterial mats. The isotopic signature of the produced organic matter in the Christmas Island mat is substantially higher compared to that found in the Solar Lake: $\delta^{13}C$ -18 ‰ compared to $\delta^{13}C$ -8 ‰ for the Solar Lake (Aizenshtat et al. 1984; Des Marais et al. 1989).

Aerobic sulfate reduction is probably an important process in environments where H_2 is produced within oxic conditions, as it is the case in the cyanobacterial mats. The exact physiology of the involved microorganisms awaits further physiological characterization which may result in the widening of the spectrum of metabolism recently described for classical *Desulfovibrio* species (Dilling and Cypionka 1990; Dannenberg et al. 1992; Marschall et al. 1993; Cypionka this volume). These new isolates probably resemble *Desulfovibrio halophilus* isolated by Caumette from the Solar Lake (Caumette et al. 1991).

Acknowledgements

This research is sponsored by a grant of the Israel Ministry for Science and Technology and GSF, München, Germany, a grant of the German-Israel Science Foundation and the Moshe Shilo Center for Marine Biogeochemistry.

References

Aizenshtat Z, Lipiner G, Cohen Y (1984) Biogeochemistry of carbon and sulfur in the microbial mats of the Solar Lake (Sinai). In: Cohen Y, Castenholz RW, Halvorson H (eds) Microbial mats: stromatolites. Alan R Liss, New York, pp 281-312

Canfield DE, Des Marais DJ (1991) Aerobic sulfate reduction in microbial mats. Science 251: 1471-1473

Caumette P, Cohen Y, Matheron R (1991) Isolation and characterization of *Desulfovibrio halophilus* sp. nov., a halophilic sulfate-reducing bacterium isolate from Solar Lake (Sinai). Systematic and Appl Microbiol 14: 33-38

Cohen Y (1984) Oxygenic photosynthesis anoxygenic photosynthesis and sulfate reduction in cyanocterial mats In: Klug MJ, Reddy CA (eds) Current perspectives in microbial ecology. Amer Soc Microbiol, Washington DC, pp435-441

Cohen Y, Ideses R (1989) Hypersaline cyanobacterial mats: environmental manipulations for the production of specific carotenoids and specific polysaccharides. Proc First International Symp Mar Biotechnol, Tokyo, pp 135-138

D'Amelio ED, Cohen Y, Des Marais DJ (1989) Comparative functional ultrastructure of two hypersaline submerged cyanobacterial mats: Guerrero Negro Baja California Mexico and Solar Lake Sinai Egypt, pp 97-113

Dannenberg S, Kroder M, Dilling W, Cypionka H (1992) Oxidation of H_2, organic compounds and inorganic sulfur compounds coupled to the reduction of O_2 or nitrate by sulfate-reducing bacteria. Arch Microbiol 158: 93-99

Des Marais DJ, Cohen Y, Nguyen H, Cheatham M, Cheatham T, Manuez E (1989) Carbon isotopic trends in hypersaline ponds and microbial mats at Guerrero Negro, Baja California Sur, Mexico: Implications for Precambrian stromatolites. In: Cohen Y, Rosenberg E (eds) Microbial mats: Physiological ecology of benthic microbial communities. Amer Soc Microbiol, Washington DC, pp 191-206

Dilling W, Cypionka H (1990) Aerobic respiration in sulfate reducing bacteria. FEMS Microbiol Lett 71: 123-128

Fossing H, Jørgensen BB (1990) Isotopic exchange reactions with radiolabeled sulfur compounds in anoxic seawater. Biogeochemistry 9: 223-245

Fründ C, Cohen Y (1992) Diurnal cycles of sulfate reduction under oxic conditions in cyanobacterial mats. Appl Environ Microbiol 58: 70-77

Giani D, Giani L, Cohen Y, Krumbein WE (1984) Methanogenesis in the hypersaline Solar Lake (Sinai). FEMS Microbiol Lett 25: 219-224

Hastings D, Emerson S (1988) Sulfate reduction in the presence of low levels of oxygen in water column of the Cariaco Trench. Limnol Oceanogr 33: 391-396

Jørgensen BB, Revsbech, Cohen Y (1983) Photosynthesis and structure of benthic microbial mats: Microelectrode and SEM studies of four cyanobacterial communities. Limnol Oceanogr 28: 1075-1093

Krumbein WE, Cohen Y, Shilo M (1977) Solar Lake (Sinai) 4. Stromatolitic cyanobacterial mats. Limnol Oceanogr 22: 635-654

Kühl M, Jørgensen BB (1994) Spectral light measurements in microbenthic phototrophic communities with a fiber-optic microprobe coupled to a sensitive diode array detector. Limnol Oceanogr (in press)

Marschall C, Frenzel P, Cypionka H (1993) Influence of oxygen on sulfate reduction and growth of sulfate-reducing bacteria. Arch Microbiol 159: 168-173

Morris JG (1979) Nature of oxygen toxicity in anaerobic microorganisms In: Shilo M (ed) Strategies of microbial life in extreme environments. Dahlem Konferezen Berlin, pp 149-162

Revsbech NP, Jørgensen BB (1986) Microelectrodes: their use in microbial ecology. Adv Microbiol Ecol 9:293-352

Revsbech NP, Jørgensen BB, Blackburn TH, Cohen Y (1983) Microelectrode studies of the photosynthesis and O_2, H_2S and pH profiles of a microbial mat. Limnol Oceanogr 28: 1062-1074

Rothschild LJ, Mancinelli RL (1990) Model for carbon fixation in microbial mats from 3500 Myr ago to the present. Nature (London) 345: 710-712

Skyring GW (1987) Sulfate reduction in coastal ecosystems. Geomicrobiol J 5: 295-374

Widdel F (1988) Microbiology and ecology of sulfate- and sulfur-reducing bacteria. In: Zehnder AJB (ed) Biology of anaerobic microorganisms. Wiley, New York, pp 469-585

Panel discussion: New physiological groups of bacteria

Pieter T. Visscher[1] (rapporteur) and Hans van Gemerden[2] (chairman)

Division of Marine and Atmospheric Chemistry, Rosenstiel School of Marine and Atmospheric Science, 4600 Rickenbacker Causeway, Miami, FL 33149-1098 USA

The discussion on new physiological groups of bacteria focused on iron and sulfur metabolism emphasizing on versatility of both known and newly isolated organisms involved in iron and sulfur transformations. The question arises what the advantage is of (energy) costly diverse metabolic capabilities for an organism that on one hand might benefit from these because of fluctuations of physicochemical parameters but on the other hand might loose the competition for substrates with microorganisms that are specialists. Panel Chairman was Hans van Gemerden, members included Pierre Caumette, Heribert Cypionka, John F. Stolz, Bruno Risatti and Fritz Widdel.

Since the previous meeting in Eilat (Cohen and Rosenberg 1989), specific molecular probes have been developed to detect "assemblages", or functional groups of bacteria, especially sulfate-reducing bacteria (see Chapters by Stahl, Muyzer, Ramsing in this volume). According to Stahl (pers. comm.) 2-100% of the community composition can be determined by probes not relying on cultured organisms. Data obtained with general probes, such as applied in Baja California mats (see Risatti, this volume) need to be interpreted with caution (i.e. specific RNA content; Risatti, pers. comm.): they represent a static situation (Stolz, pers. comm.) and yield numbers of bacteria, and do not reflect their activity. Nevertheless, molecular techniques contribute significantly to the understanding of community structure of microbial mats.

Extremely high sulfate reduction rates under oxic conditions ($< 18\ \mu mol.cm^{-2}.d^{-1}$) have been reported in a mat in Baja California (Canfield and Des Marais 1991, 1993). These rates require either specific activities or population sizes 3-4 orders of magnitude higher than reported in the literature (Risatti, this volume; Jørgensen 1978; Jørgensen and Bak 1991; Parkes et al. 1993). Others have reported sulfate reduction in the presence of oxygen as well (Cohen, pers. comm.; Fründ and Cohen 1992; Visscher et al. 1992; Jørgensen, in press). Interestingly, Ohmoto et al. (1993) found evidence from isotopic studies that pyrite, which was formed 3.4 billion years ago, was of biogenic origin and deduced that sulfate reduction rates must have been similar to those reported for microbial mats. If sulfate reduction only occurs

[1] Present address: U.S. Geological Survey, 345 Middlefield Road, Menlo Park, CA 94025, USA.
[2] Department of Microbiology, University of Groningen. Kerklaan 30, 9751 NN Haren, The Netherlands.

NATO ASI Series, Vol. G 35
Microbial Mats
Edited by Lucas J. Stal and Pierre Caumette

under anoxic or microaerophilic conditions, the existence of microzones, which sustain sulfate reduction in marine snow (Shanks and Reeder 1993), becomes tangible. Clearly, pure cultures are necessary to study ecophysiological aspects of sulfate respiration. The metabolic versatility of some sulfate-reducing bacteria range from sulfate- and nitrate reduction to use of O_2 as electron acceptor as well as fermentation of organic and inorganic compounds (see Cypionka, this volume). New isolates from environments where sulfate reduction was measured in the presence of oxygen have been obtained (Cohen, pers. comm.) and will be used in physiological laboratory studies.

As a result of this metabolic versatility, definition of "physiological group" needs revision, i.e. are certain sulfate reducers indeed sulfate reducers or rather nitrate reducers (Stolz, pers. comm.), or, to pose this question differently: which metabolic pathway is used at a certain time and location by certain microorganisms that seems to be so diverse, and why do certain organisms possess several metabolic pathways (Oren, pers. comm.). Under rapidly changing environmental conditions specialists are believed to do better than generalists, and organisms with constitutive enzyme systems are better prepared when the appropriate conditions prevail. However, different functional groups of bacteria within the mat community might compete for the same substrate: for example, thiosulfate is utilized by sulfate reducers, thiobacilli and anoxygenic phototrophs (Van den Ende and van Gemerden, this volume), as well as by cyanobacteria and heterotrophs. Study of growth kinetics for certain substrates and *in situ* activity measurements will answer questions of substrate turn-over through different pathways.

De Wit and Caumette (this volume) list the most important novel isolates of phototrophic and chemolithotrophic sulfide oxidizers. Most of the species described recently can be classified as "sulfur" bacteria, either producers of sulfide (predominantly sulfate reducers) or consumers which gain their energy from oxidation of sulfur compounds. Therefore, microbial mats are a classical example of "sulphureta" (Baas Becking 1925). As a result, the main emphasis is on bacteria involved with sulfur metabolism, while other functional groups of microorganisms, such as aerobic heterotrophs, are neglected. The use of iron and manganese as terminal electron acceptor in the absence of oxygen has been reported in the literature (Lovley 1987), but has received little attention in mat research (Stal, pers. comm. and this volume; see also Ghiorse 1989). Lovley et al. (1987) pointed out that instead of magnetotactic (aerobic) bacteria, iron reduction by ferric iron reducing (anaerobic) bacteria might play a dominant role in magnetite (Fe_3O_4) formation. Ferrous iron formed during (chemical and/or biological) iron reduction reacts with ferric iron to form Fe_2O_3 (Stolz, pers. comm.). In addition, greigite is produced by magnetotactic bacteria under anoxic conditions (Stolz, pers. comm.).

A combination of the use of reduced sulfur and iron species was reported by Ehrenreich and Widdel (this volume), who showed the capacity of anoxygenic phototrophic bacteria to utilize ferrous iron as electron donor for CO_2 fixation. Such organisms could have played an important role in depositing oxidized iron in early Precambrian life. Provided light was available, these anoxygenic phototrophs could have contributed to banded iron formation (Cloud and Licari 1968). However, the hypothesis that banded iron formation occurred in the anoxic Archaean biosphere at locations with low primary production (Walker 1987) seems to contradict this, unless export of organic matter took place. Nevertheless, the use of reduced iron as electron donor for CO_2 fixation will spark the search for this type of phototrophic metabolism, which might occur at very low light intensities (Des Marais, pers. comm.) in mats around iron-containing hot springs (Cohen, pers. comm.).

In conclusion, traditionally, the main focus in attempts to isolate new (physiological groups of) bacteria has been on extreme environments, such as hot springs, hydrothermal vents, and hypersaline environments. Recent observations of *in situ* population dynamics raises the question whether an approach of functional groups of bacteria rather than pure cultures of a single species is more meaningful to understand microbial mat communities. Clearly, a combination of both with the aid of phylogenetic probes and functional probes is necessary to unravel the community structure (Pearl, pers. comm.).

Acknowledgement

Valuable comments by John F. Stolz and appetite for sucrose by Fifi contributed to the manuscript in its present state.

References

Baas Becking LGM (1925) Studies on the sulphur bacteria. Anal Bot 39: 613-651

Canfield DE Des Marais DJ (1991) Aerobic sulfate reduction in microbial mats. Science 251: 1471-1473

Canfield DE, Des Marais DJ (1993) Biogeochemical cycles of carbon sulfur and free oxygen in a microbial mat. Geochim Cosmochim Acta 57: 3971-3984

Cloud Jr PE, Licari GR (1968) Microbiotas of the banded iron formations. Proc Nat Ac Sci USA 61: 779-786

Cohen Y, Rosenberg E (1989) Microbial mats. Physiological ecology of benthic microbial communities. ASM Washington DC

Fründ C, Cohen Y (1992) Diurnal cycles of sulfate reduction under oxic conditions in cyanobacterial mats. Appl Environ Microbiol 58:70-77

Ghiorse WC (1989) Manganese and iron as physiological electron donors and acceptors in aerobic-anaerobic transition zones. In: Cohen Y, Rosenberg E (eds) Microbial mats. Physiological ecology of benthic microbial communities. ASM Washington DC, pp 163-169

Jørgensen BB (1978) A comparison of methods for the quantification of bacterial sulfate reduction in coastal marine sediments. III Estimation from chemical and bacteriological field data. Biogeochem J 1: 49-64

Jørgensen BB Sulfate reduction and thiosulfate transformations in a cyanobacterial mat during a diel oxygen cycle. FEMS Microbiol Ecol in press

Jørgensen BB, Bak F (1991) Pathways and microbiology of thiosulfate transformations and sulfate reduction in a marine sediment (Kattegat Denmark). Appl Environ Microbiol 57: 847-856

Lovley DR (1987) Organic matter mineralization with reduction of ferric iron: a review. Geomicrobiol J 5: 375-399

Lovley DR, Stolz JF, Nord Jr GL, Phillips EJP (1987) Anaerobic production of magnetite by a dissimilatory iron-reducing microorganism. Nature 330: 252-254

Ohmoto H, Kakegawa T, Lowe DR (1993) 34-billion-year-old biogenic pyrites from Barbeton South Africa: sulfur isotope evidence. Science 262: 555-557

Parkes RJ, Dowling NJE, White DC, Herbert RA, Gibson GR (1993) Characterization of sulphate-reducing bacterial populations within marine and estuarine sediments with different rates of sulphate reduction. FEMS Microbiol Ecol 102: 235-250

Shanks AL, Reeder ML (1993) Reducing microzones and sulfide production in marine snow. Mar Ecol Prog Ser 96: 43-47

Visscher PT, Prins RA, Van Gemerden H (1992) Rates of sulfate reduction and thiosulfate consumption in a marine microbial mat. FEMS Microbiol Ecol 86: 283-294

Walker JCG (1987) Was the Archaean biosphere upside down? Nature 329: 710-712

VI. Bioremediability and Biogeochemical cycles (Panel discussion)

Bioremediability and biological value of microbial mats

Rutger de Wit (chairman) and Maria-Carmen Guerrero[1] (rapporteur)

Laboratoire d'Océnographie Biologique, Université de Bordeaux I, 2, Rue du Prof. Jolyet, F 33120 Arcachon, France.

The round table discussion on "bioremediability and biological value of microbial mats" was unique within the context of the workshop for two reasons. Firstly, it was the only round table session that was not preceeded by a series of lectures. Secondly, this session was clearly focused on the interface between society and science. It thus draw attention to the role and the specific responsabilities of the scientist, and it posed the question how to develop applied research proposals. The session was split up in two parts focusing on the following subjects: 1) the use of microbial mats for bioremediation and in environmental technology; 2), biodiversity and conservation of microbial mats.

The use of microbial mats in environmental technology

The discussion was structured along two clear and simple questions: 1), can microbial mats be applied technically for human purposes ? ; 2), do you actually have applied research proposals in this respect ? In general, it was recognised that microbial mats do spontaneously develop in several human-made environments where they often perform a beneficial role. Often these environments represent so-called extreme environments characterised by the virtual absence of algal and cyanobacterial grazers. Thus, microbial mats do develop on oil-spills, but it is still not documented whether this really results in an effective spontaneous biodegradation of noxious compounds. Strikingly, in the Gulf area, *Microcoleus* sp. dominated mats developed on the extensive oil-spills resulting from the Gulf war, interpreted as a first sign of self-cleaning (Sorkhoh et al. [1992] Nature 359: 109). Nevertheless, the beneficial effects of microbial mats in waste-water treatment plants where they develop on trickling filters are well documented. Microbial mats are a rich source of natural products, they have been shown to be beneficial in solar saltern ponds, they are used in mining and their ability to accumulate heavy metals suggests a possible role in detoxification of water masses. Furthermore, microbial mats provide an important resistance to erosion in coastal areas. This feature is used in so-called "soft-option" coastal engineering, which aims to make use of natural ecosystems for coastal protection rather than applying expensive and ecologically harmful civil works. However, it is still difficult to stimulate microbial mat development on demand in the coastal areas. Thus,

[1] Laboratory of Limnology, Dept. of Ecology, Universidad Autónoma de Madrid, 28049 Madrid, Spain.

NATO ASI Series, Vol. G 35
Microbial Mats
Edited by Lucas J. Stal and Pierre Caumette

"soft-option" engineering in practice often results in "doing nothing" and just monitoring where microbial mats do spontaneously develop. In general, a similar feeling was expressed for the other areas in which one makes use of spontaneously established microbial mats. Often, one is convinced of the usefulness of these microbial ecosystems, but detailed information on the processes and their potential functions is still scarce; hence, so far, we are hardly able to engineer microbial mat habitats for performing certain processes.

We discussed the possibilities of introducing genetically engineered microorganisms, but scepsis was expressed to the possible dangers of introducing such novel organisms in the environment. Doubt was also expressed concerning the usefulness of microbial mats for production of natural products and degradation of noxious compounds, rather than using genetically engineered organisms in controlled fermentors. However, it was noticed that the high population densities and the close interactions between microorganisms in microbial mats may favour the evolution of organisms with novel capacities useful for human-beings.

In conclusion, although our knowledge and skills are still limited, microbial mats are potentially useful objects in future biotechnology, thus justifying more research. Basic research is needed today, whilst, applied research proposals are awaited for the future.

Biodiversity and conservation of microbial mats

The discussion in the second part centered around two questions. Firstly, does the biodiversity in microbial mats or the diversity of microbial mats habitats decrease due to anthropogenic effects as it has now been thoroughly documented for animals and plants? Secondly, what are the responsabilities and the possibilities of the scientists to act in favour of the conservation of microbial mat environments?

A clear answer on the first question is not possible, since we do not possess a clear idea of biodiversity at the microbial level, and in microbial mats, the first biodiversity studies are just arising (e.g. Ward et al. this volume). However, we agreed that efforts should be made to protect certain microbial mat habitats. In this respect, it is important that the protection of microbial diversity is recognised in international treaties like e.g., the biodiversity convention of Rio de Janeiro. Secondly, nature reserves could be created and managed with the aim to protect the microbial mat habitats. These habitats are often very susceptible to quantitative and qualitative changes of the ground water and, therefore, their protection is strongly linked with the general problematics on wetland conservation. The scientists have a responsability to explain society on values of microbial mat habitats and help to ensure their conservation. Nevertheless, this is often problematic in practice, due to the fact that the microbial world is often conceived by the general public as something devoid of any aesthetic values. To change this conception it is recommended that the scientists explain their observations to the general public, preferably nicely illustrated with coloured drawings and photographs. Furthermore, emphasis should be paid to microbial mats as Precambrian relicts and their role in geochemical processes, especially concerning their N_2- and CO_2-fixation capacities.

New concepts in biogeochemical cycling and ecology

John F. Stolz (rapporteur) and David J. Des Marais[1] (chairman)

Department of Biological Sciences, Duquesne University,
Pittsburgh, PA. 15282-1502 USA.

The discussion on new concepts in biogeochemical cycling and geophysical implications was chaired by D. Des Marais with H. Pearl, B. B. Jørgensen, and B. Bebout as panelists. Five major topics were addressed, controls on productivity, physical architecture of mats, nutrient recycling and conservation, the basis of depth zonation, and diversity and versatility in mat communities. A general model for productivity was proposed :

```
              CO2                    N?                       P?
               ↘                      ↘                        ↘
Gross photosynth.-----> Net photosynth. -----> Net diel product. -----> Net annual product.
               ↘                      ↘                        ↘
              Photo                  Mat                      Undermat
              resp.                  resp.                    resp. etc.
```

A major factor which may affect gross productivity is the interaction of light with the microbial mat. B.B. Jørgensen noted that while planktonic systems have been well characterized, the little information which is know for benthic systems show they have properties which differ substantially. For example, because light is scattered more strongly and absorption effects are different in the benthos, the potential exists for much higher productivity than that observed for plankton. However, the interactions between light and the solid mat are still poorly understood. Also, it is not known whether residing on a sediment surface or being distributed down into the sediment is more advantageous. The scalar irradiance increases in a light scattering environment because the photons are reflected multiple times in the photic zone, increasing their probability of absorbance by the biota. This effect has not been studied quantitatively. As light conditions change, mats can modify their inventories of pigments more rapidly than their generation times. This is perhaps one of several mechanisms for maintaining high rates of productivity.

Nutrient limitation of productivity is much better understood in planktonic than in benthic environments. H. Pearl interjected that the habitats needed to study the effects of nutrient perturbation are not available for benthic systems over a range of timescales. For example, small scale closed systems collapse after two to three days; large open systems are experimentally impractical. D. Canfield commented that the observational database for

[1] NASA Ames Research Center, Mail Stop 239-4, Moffet Field, CA 94035-1000, USA

NATO ASI Series, Vol. G 35
Microbial Mats
Edited by Lucas J. Stal and Pierre Caumette

phosphorus is very small and no in-depth studies have been done with mats. Can pore water analysis be done effectively and at the spatial resolution necessary? Again H. Pearl stressed the need for novel approaches to understand the actual fluxes of nutrients which occur in such systems.

Infauna may also affect productivity. It would be useful to take inventories of macrofauna such as nematodes. They ingest the flora and regenerate nutrients. Little is known of their distribution, or specialization and impact upon productivity. D. DesMarais suggested that small animals might create microenvironments which support key populations of bacteria such as methanogens, both in their guts and in their excretion products. Animals might affect the diversity of flora by selective grazing. In Guerrero Negro *Microcoleus* mats, J. Farmer has observed between 1600 and 2300 nematodes per cm^3. Grazing of diatoms probably occurs. In thermal springs, R. Castenholz observed that the distribution of *Phormidium* is influenced by grazers. B.B. Jørgensen noted that some nematodes and ciliates harbor consortia of bacteria such as methanogens. As the animals migrate, the bacteria go with them, and therefore the bacteria might occur in unlikely places. Examples might be anaerobic organisms situated in the oxygenated zone. R. Conrad mentioned that such consortia involving protozoa can be important also in lake sediments, making key contributions to the methane budget. The high populations of nematodes in mats argue for their importance in biogeochemical cycles. Microbial mats may have witnessed the evolution of metazoans. Many of the exotic groups observed in mats diverged early in the metazoan evolutionary tree.

The role of the mat superstructure or architecture merits more attention and several questions were raised by the chairman. Depth zonation is the aspect which has been most commonly studied. However, do the exopolymers act as channels or barriers to the movement of organisms or chemical species within the mat? Does the solid architecture of the mat contribute to the high gross productivities of the mats? D. Ward asked whether the laminations cause any discontinuities. B.B. Jørgensen responded that gradients of oxygen do not reveal any discontinuities between observable layers in the mat. Mats can be perfused with indicator dyes such as tetrazolium salts, so they must be uniformly permeable to at least some chemical species. However, the diffusivities of mat exopolymers have not been studied with respect to various ions or organic compounds. Does the exopolymer assist in the retention or exclusion of certain species? For example, are the toxic effects of heavy metals mitigated by exopolymers? How can diffusivities and chemical effects of exopolymers be studied? It can be challenging to distinguish experimentally between a mat's biological and physical/chemical properties. For example, killing the biota in order to study the mat's physical properties is complicated by the fact that the sterilization procedure usually also alters the structure of the exopolymer. The polymeric matrix contains charged groups and thus enrich concentrations of charged species from the environment. The polymers might be gelled or dispersed by changes in the chemical environment. Binding of various species might play an important role by buffering changes in the concentrations of those species. H. Pearl added that some polymers appear to attract specific bacteria, which themselves can act either as filters or regenerators of nutrients.

Most microbial mats display heterogeneity, not only with depth in the mat, but also laterally. One expression of this is the topographic relief of the mat surface. This surface is important because it controls the flux of species in and out of the mat. It is also important because it defines the fabric which is observed in stromatolites. The laminated texture of a stromatolite basically is a set of nested former-surfaces of the mat community which built the stromatolite. The relative extent to which this surface is controlled by biota versus abiotic environmental factors is not well known.

A. Kaplan stressed the importance of considering three dimensional structures in the movement of an important species such as CO_2 out of a cyanobacterial carboxysome. Concentration gradients require unusually low resistance of the carboxysome shell when only two dimensions are assumed. Normal diffusivity through the shell can account for the observed gradients if three dimensional diffusion is assumed. Still, mat textures are not uniform. What supports the heterogeneity's observed in the populations of organisms? N. Ramsing mentioned studies on 5mm thick biofilms which showed great heterogeneity in populations based on in situ labeling techniques. Do they reflect heterogeneities in the concentrations of key substrates such as organic compounds? However, B.B. Jørgensen said that although microelectrodes should be able to detect microenvironments the data did not support this. D. Canfield pointed out the need to detect fermentation products. New techniques must be developed to explore three-dimensionally in mats. How can the surface be mapped more efficiently, along with it diffusion boundary layer? Can flow patterns over mats be efficiently documented? Can oxygen microgradients be mapped? Regarding mat interiors, D. Castenholz suggested that perhaps magnetic resonance imaging could be used.

Deposition of minerals such as silica and carbonates in mats has received little attention by biogeochemists. In the daytime, high pH and low CO_2 concentrations should favor rapid carbonate precipitation in marine hypersaline mats, yet little precipitation occurs in many environments. L. Stal queried if mineral nuclei were armored by polysaccharides? If so, why do carbonates form anywhere in environments dominated by prokaryotes? Perhaps the calcium is bound to some extent by the polymers, thus inhibiting calcification. G. Zavarzin suggested that the epitaxial constraints on crystal growth seen in coccoliths is the same in mats. In some environments, if another metal is present which can be bound more strongly, it could displace this calcium and make it available for carbonate precipitation. In many hypersaline mats, carbonate precipitation around cyanobacterial sheaths is frequently observed at depth. J. Farmer noted that three different types of carbonate are found in the lower mat in Guerrero Negro.

The next topic of discussion were the factors controlling the depth zonation of organisms other than light, O_2 and sulfide. To some extent such zonation might be controlled by the diffusion gradients of key species such as sulfate, but D. Canfield suggested that dissolved organic carbon distributions could also be important. As depths increase, diffusive or convective transport from above weakens, relative to the production of these species by the decomposition of solid organic matter at depth. B.B. Jørgensen stressed the need to consider time and spatial aspects. It also depends on the substrate, R. Conrad noted.

Environmental change appears to be the norm for a microbial mat, and it is important to understand how the mat adapts. In hot springs, D. Ward noted that variations in temperature are accommodated by populations (labeled "guilds" by D. Ward) of bacteria, which includes several closely-related but genetically distinct strains each of which is best suited to a different range of temperatures. In contrast, some bacteria can substantially change their physiology to adapt to environmental change. How common is the first versus the second strategy in nature? How do populations respond to changes in salinity, for example? *Chloroflexus* species display a wide range of morphotypes, are these examples of genetically distinct organisms, or do they represent just a few versatile bacteria? It would be useful to chart the distribution of guilds across natural variations in temperature, salinity, light intensity, etc. It is highly challenging yet rewarding to demonstrate the full versatility of microorganisms.

General considerations

Cyanobacterial mats in general biology

George A. Zavarzin

*Institute of Microbiology, Russian Academy of Sciences,
117811 Moscow, Russia*

General considerations

What have we learned from studying cyanobacterial mats? What is the influence of this study on microbiology and general biology?

Cyanobacterial mats and microbial biofilms have received considerable attention as subjects of reviews and symposia. Research into biofilms has undergone several changes: 1) A shift from species to community physiology and biochemistry; 2) A switch in focus from single, tightly controlled populations to complex communities having diverse population-level growth rates. Cyanobacterial community structure and function are perhaps best understood in mats. The reason for this is that cyanobacteria represent a structural and functional "key species". A better term was proposed for plant communities by V.Sukachev who designated the leading prime producer as "edificator". Cyanobacteria are well known to be involved in shaping mat structure by, among other things, mucilage, sheath and glycocalix formation in response to environmental stimuli. But it is the prokaryotic community which is subject of study and this makes a great difference with the dominating ideas in microbiology and biology of the past, which were concentrated on the species evolution and cell biology. I propose to differentiate cyanobacterial and cyano-bacterial communities. The sign "-" indicates that there are other bacteria in addition to oxygenic phototrophs. In this regard, a relevant question is if the cyanobacterial community could exist autonomously i.e. can the cyano-bacterial community be self-sustaining?

Evolution of the species remains a general theme in biology, and in some instances there is an attempt to explain everything in terms of Darwinian evolution. In studying cyano-bacterial communities, can we purely invoke Darwinian thinking, or should we investigate other postulates? The search for alternative paradigms is not anti-Darwinian, but rather non-Darwinian. As an indication for existence of non-Darwinian field in Natural Sciences, that is Earth and Life Sciences, it should be noted that the most ambitious Global Change Program has found no place for evolution based on natural selection.

The alternative idea is what might best be designated "vernadskian", after the Russian scientist V.I. Vernadsky (1926) who concentrated his studies on the interaction of biota with the Earth system and stressed that there was no possibility for a single species to survive outside from the biogeochemical system built by functionally diverse species. This means that the community is shaped by the environment and that species diversity is dictated by the

NATO ASI Series, Vol. G 35
Microbial Mats
Edited by Lucas J. Stal and Pierre Caumette

community. The community is not constructed by divergence from a single species, e.g. a cyanobacterium, in spite of its versatile metabolism. The community is often composed of phylogenetically unrelated organisms. This type of diversity evolution is entirely different from individual mutation and competitive survival of progeny, but could find its parallel in symbiogenetic constructions. In the bacterial community there is some kind of "planned economy" based on the combination of interlinked functional niches determined by the flow of matter and energy. Predetermination of these niches by the system structure makes an apparent goal for its fulfilment either by divergence or by recruitment from outside. In the vernadskian paradigm, evolution of the community is caused by imbalance in the biogeochemical cycling. The highest priority should be given to the trophic structure of the cyano-bacterial community.

Autotrophic cyanobacteria are limited by nutrients, among which one is particularly important: phosphorus. Cycling of matter within the community seems to be not so important for cyanobacteria. On a mega-annum time scale, production of oxygen is not counterbalanced. In the Earth system oxygen in the atmosphere is stoichiometrically equal to organic carbon. At ecosystem level destruction of organic matter is restricted which makes it a global source of O_2 and sink of CO_2. Aerobic oxidation of organic carbon by microbes is the main sink for excessive oxygen in a balanced biosphere. Oxygen evolved by cyanobacteria into the atmosphere is counterbalanced by kerogen originating from thylakoid membranes that do not decompose in the anaerobic environment. The Vernadskian approach is incompatible with Oparin's scheme of evolution by sequential development from anaerobic to aerobic mode of life. In reality, from the very beginning of oxygenic metabolism in the benthic community there is a danger of trapping oxygen within the fabric and the possibility of oversaturation with oxygen. Cyanobacterial mats are the most primitive community recorded on Earth as shown from the fossil record. However, there is no factual possibility to study the origins of a subject which had no predecessors. The real meaning of phylogenetic constructions, if not supported by geochronology, is the measured similarity between elements deductively ordered by arbitrary rules. It is possibly a domain. However there is a problem of correlation between the deduced phylogenetic and evolutionary order which can be consistent with ecological theory.

In fact, students of cyano-bacterial communities have the reverse problem to evolutionists - the question is not to look into phylogeny of the species but to understand why they are persistent, why was there no evolution? There is no word of denying Darwinian evolution but the field has undergone considerable modification.

Hierarchical approach

In any study there is a set of objects which are in relation $A \subseteq B \subseteq C$ where A are elements (in this case: microbial species). All A are components of B (in this case: communities). All B are components of larger system C, e.g. ecosystem, landscape or biome. Considering B the focus of our study, we will not concern ourselves about the origin of A's, taking their diversity as precondition for constructing B. In the study of cyanobacterial mats the focus is on B as a system of interlinked elements which fits to its existence in conditions determined by C (Zavarzin 1992).

The hierarchical approach implies that objects might be organized in such an order that each combination represents a subset of a set of a higher order and signals are submitted to the set of higher order with lower frequency as smoothed integrated messages (O'Neil et al. 1986). This means that a signal from A might reach C in an entirely transformed manner. The sophomoric mistake of fitness A to C should be avoided, since as it is best exemplified by the mat community, within this community a local environment might be quite different from that in C, which resulted from interaction of many B's.

With B represented by cyano-bacterial mat as prokaryotic community structured both in physical and interactive way, we are in the non-darwinian domain since our aim is to elucidate fitness of B to C and not the origins of A in B. However, this does not mean that there is no selective pressure on cyanobacterial community and its competition with other benthic communities which resulted in its sustainability over the time in changing C or sequential states of biosphere. It also does not mean that there is no selective pressure for A to exist within the community. Sustainability of B is determined by its structure and it predetermines a functional niche which constitutes a trap for random variability. The problem is that the community cannot exist for a long period if the essential functions are not realized.

A number of rules which were studied in the so called "organismal" manner of thinking, introduced first by Winogradsky (1897) which was a starting point for modern microbial ecology concept. Winogradsky considered a diversity of microbes ("ferments") as components of an "organism" or as we now recognize as "system". This way of thinking was rejected by continualists who studied evolution of plant communities.

Microbial communities are differentiated from plant or animal communities by their modes of interaction. Representative types of interactions in prokaryotic community include a food chain such as $a_0 \rightarrow A_1 \rightarrow a_1 \rightarrow A_2 \ldots\ldots \rightarrow A_n \rightarrow a_n$ where a is a substance consumed or produced by microbial species. That type of trophic interaction might be represented by a graph with branches and loops making a transport network. The main rule for a microbial community is material balance over all transformations carried out by microbes. A community is autonomous and sustainable if $a_0 \approx a_n$, that is: if its trophic structure is cyclic (Zavarzin 1992). If some functional A_i is absent the chain is either interrupted (the community is not stable because of product accumulation) or bypassed. If A_i is not known but a_{i-1} does not accumulate, there is a chance to use an elective procedure as a main approach for elucidating trophic structure of community from A_i to A_n and describe new A_i. These functional A's represent seeming (teleonomic) goals of species evolution described as a trap for mutant or for a recipient in lateral gene transfer. It was found that two trophically linked A in a prokaryotic food web often belonged to phylogenetically remote branches (Zavarzin 1989). There are many examples of such interactions. One of the best is H_2- or acetate- consuming methanogenic *Archaea* and their bacterial predecessors including syntrophic ones; the mutual interdependence is predetermined.

According to Darwinian theory, evolution is mutation-limited and can be treated as a sequence of single gene substitutions; the rate of evolution is determined primarily by the random occurrence of favorable mutations which are selected if they fall into a stable niche which can be regarded as a trap for an evolutionary random walk (Metz et al. 1992). In the community the general tendency is to fill the trap by genetic "minimal walk" regardless of phylogeny.

The position in a community depends on functional "essence", related to "Wuchsformen", and was reflected in old days bacterial classification of morpho-physiological genera. In his last paper criticizing C.B. van Niel's approach, S.Winogradsky (1952) insisted that bacterial classification can not go along the same phylogenetic lines as for animals. However the next step is radiation within a certain physiological group which leads to adaptation, to fitness to specific conditions. That is, *aliquid* or accidental which remains in Darwinian domain of the origin of the species. This type of diversification might best be illustrated within the cyanobacteria which occupy different ecological niches but have essentially the same type of metabolic requirements for growth.

Traditionally, the physiology of cyanobacteria has been studied under permanent light in liquid pure cultures (Rippka et al. 1979). Innovative changes were brought about by students of M.Shilo, who studied metabolism of cyanobacteria in the night and/or anaerobic conditions. This gives a unique feature of the cyanobacterial mat as a community which is permanently in transition state. Bacteria which accompany cyanobacteria in mats are attracted to them (i.e. chemotaxis) and should be accommodated to diurnal changes from hyperbaric oxygen to excess of sulfide. This type of community physiology has been demonstrated within mats by measuring oxygen and sulfide gradients and their dynamics by microelectrode techniques (Revsbech and Jørgensen 1986). However, oxygen and sulfide dynamics are poorly understood as factors that induce stress and adaptation mechanisms in bacterial cells.

Bacteria in cyano-bacterial community

"When interest in laboratory studies with mixed cultures arose, it became more and more clear that the sum total of the results obtained with pure cultures does not really provide an accurate picture of naturally occurring processes, where interactions between microbes appear to have a profound influence on the behavior of individual species" - H.Veldkamp (1985) followed S. Winogradsky. Metabolism of mats as an integral system was studied by L. Stal who has demonstrated the influence of ecological factors. Regulation in the community as an entity follows from kinetics of chemical reactions in a network performed by the members of the community in the way analogous to that within the cell. However, the point is that until now diversity of bacteria in cyano-bacterial mats is hardly touched. Many bacteria observed in mats are not cultivated; the number of species in this category may be hundreds. Does a significant part of those species cover all fundamental types of bacterial diversity determined by phylogenetic branches thus giving a pluralistic starting point to the whole microbial evolution? How many types of bacteria co-exist in cyano-bacterial mats? Which lineages are missing from mats? Which are found in other relict communities? The first step has already been taken for the simplest thermophilic community (Ward et al. 1992).

In general, the cyano-bacterial community within mats might be regarded as a complex of cyanobacteria living together with anaerobic decomposers. The reason is that the pool of dissolved oxygen in mats, in contradistinction to planktonic communities, is quite limited. Living at the highest possible density, bacteria are limited by minimal substrate concentration except for transitional periods in the diurnal cycle. The anaerobic decomposition pathway in mats at first glance is organized along the same lines as in the sulfidogenic or methanogenic community. However, the cycles seem to be closed within the anaerobic part of the community with various sulfur shunts. The bacterial filter of aerobic dissipotrophs which intercept volatile substances from an anaerobic zone, depends on the activity of cyanobacteria.

Functional groups in the cyano-bacterial community include bacteria which could be descendants of common ancestors. A good example might be given by radiation of *Proteobacteria* as members of the cyano-bacterial community. The phylogenetic α-, β-, γ-, δ-branches harbor representatives of oligotrophic, gazotrophic, copiotrophic, sulfur-cycle bacteria which might have had their origin mainly in aerobic trophic pathways as consorts of cyanobacteria - edificators of community (Zavarzin et al. 1991). For aerobic microbes which completely oxidize organic matter to CO_2, trophic diversity is based on their spectrum of available substrates. Bacteria which utilize simple monomeric compounds belong to the ecological group of dissipotrophs - bacteria which utilize substances dissipating from the site of their origin. Most dissipotrophs are oligotrophic: they possess high affinity towards the narrow spectrum of substrates and effective transport systems. Substrates might dissipate either from the sites of hydrolysis of an insoluble substrate which is characteristic for prostheco-bacteria in α-branch or from the anaerobic zone as for "Knallgasbakterien" or "*Athiorhodaceae*" in β-branch. Purple bacteria of γ-branch and sulfate reducers of δ-branch are obviously trophically interlinked and very characteristic for mats. Such branches as *Chloroflexaceae* and green bacteria are usual inhabitants of mats. A more detailed example of diversification caused by specific substrate availability is presented by the members of phylogenetically independent branch of *Haloanaerobiaceae* adapted to degradation of osmolytes in the halophilic cyanobacterial community (Zhilina and Zavarzin 1991). Bacteria such as *Halobacteroides* and *Haloincola* are specified for disaccharides and *Acetohalobium* - for betaine. Following this line of argumentation, we can trace origins of many phylogenetic branches in the ribosomal RNA classification of bacteria as suggested by C. Woese for trophic groups in cyano-bacterial communities. However, cyanobacterial mats appear to be the most ancient communities in fossil record on the Earth. The trophic structure of these relict communities is organized along much the same lines as contemporary anaerobic communities and this may help explain why bacteria are preadapted to such artificial sites as e.g. landfills (Zavarzin and Nozhevnikova 1993).

Relict communities

The knowledge of early communities is summarized by W. Schopf (1983). Various oxygen-producing photosynthesizers may have been extant as early as the Early Archean 3.465 Ga ago (Schopf 1993). Trichomic oscillatoriaceans were identified by their similarity with younger progenitors. The fossil record is more or less continuous and relatively well known from about 2.1 Ga to present. The oldest microfossils are part of an evolutionary continuum. Stromatolitic structures of similar age were reported from Archean rocks indicating that they might be of cyano-bacterial origin. Arguments for this include the sedimentary organic matter of appropriate isotopic composition in kerogen and the presence of banded iron formations (BIF's) which indicate a geological sink for O_2. These facts do not entirely exclude the involvement of anoxygenic photosynthesis. But there are no clear objections against the idea that the ancient community was similar to that of nowadays. If this is true, this would give us the unique possibility to study prokaryotic communities from the Archean to the Middle Proterozoic and fundamental processes caused by their activity. However, there are problems in drawing conclusions by analogy as is the case in any actualistic approach.

There was a unidirectional evolution of the Earth under actualistic factors such as weathering-sedimentation pattern accumulation of sedimentary rocks. Changes driven by biota are well known from biogenic rocks accumulation. However, much more impressive are the biologically-induced changes in the composition of the atmosphere, first formulated by V.I. Vernadsky. Half a century later, this statement was reformulated as "Gaia-hypothesis". The main innovation was that biota generate their own favorable environment, particularly including climate changes through S.Ahrrenius greenhouse effect. Bacteria may be capable to determine all biological transformations of components of the atmosphere (Zavarzin 1984).

In relict communities we have "natural experiments" depicting atmospheric changes. Exhalations of volcanoes model the degasation of the Earth and its primary atmosphere. Thermophilic cyanobacterial communities develop in hydrotherms which might transform exhalations. Studying transformations caused by thermophilic cyanobacterial communities from Uzon in Kamchatka, we found that *Mastigocladus laminosus* and *Phormidium laminosum* dominated communities transform exhalations into simulated air mixture containing approximately 20% O_2. Reduced gas species like H_2 were eliminated when passing through mat layers but there was no oxidation of CH_4 in spite of its production and presence of thermophilic H_2-consuming methanogens and new aceticlastic *Methanothrix* discovered there. The purple bacteria were present only in the "tail" of mat in stream. The cyanobacterial community proliferates in gas stream relevant to primary atmosphere of the Earth and dominated by CO_2 at moderately elevated temperature. Cyanobacteria seem to have the best niche in this ancient environment.

Another example of a relict community is represented by stromatolite forming cyanobacterial mats (Walter 1983). Similarity between the laminated structure of cyanobacterial mats and microfossils in stromatolites proves that things are very close to one another, but until now the process of lithification is not understood: something may be missing, since recent mats do not lithify, in terms of connecting to modern-day conditions. Another problem which needs clarification is the extinction of stromatolites towards the end of Proterozoic. It is not less enigmatic than extinction of dinosaurs but much more important for the biosphere. The cyanobacterial community remained unchanged and there are no indications that significant change on species level occurred through the Earth's history, then the reason is in chemical changes. These might be changes in lithification pattern which cause the ceasing of stromatolite formation. The rare stromatolites that can be found nowadays occur mainly in seepages of underground waters which resemble an ancient environment. Is it an indication to find stromatolites as ecotones in the mixing zone of sea and underground terrestrial waters? Of course there are other biogerm forming communities like calcareous algae but in essence the function of biogenic carbonate precipitation was taken by skeletal fauna and flora about Vendian time.

Another impressive example of unidirectional changes are banded iron formations (BIF's), which do not form nowadays; this magnificent process appears to have ceased in the Precambrian. P. Cloud suggested that iron served as a sink for O_2 produced by cyanobacteria and that this sink is responsible for more than 1/3 of the oxygen ever produced by photosynthesis. There was local saturation of oxygen from the very beginning in the benthic oxygenic cyanobacterial (or prochlorophytic) community since O_2 is trapped by the fabric of mat. But by the same reason of transport limitations, such a community cannot be the site of BIF laminate ferrous/ferric and silica deposits in spite of the fact that there are ancient iron stromatolites.

There is evidence from microelements that iron in BIF's had a volcanic origin. But that needs an indication of indigenous rocks depleted from iron. In oligotrophic oceans biogenic sulfide formation might be limited and not intervene into iron precipitation. These restrictions caused K.M.Towe (1990) to support P.Cloud's hypothesis of BIF's origin, with the addition that O_2 appeared earlier than any known stromatolite or microfossils of Pilbara in Australia. There is a need to explain the BIF's of Greenland, which are more ancient. Cyanobacteria - with few exceptions like *Trichodesmium*- are in general not indigenous to the ocean. Stromatolite- forming cyanobacterial communities are inhabitants of ecotones with the front towards the sea and the rear in a terrestrial environment. However, organisms needed for saturation of the oceanic water with the oxidant might be represented by the type of prokaryotic picoplanktonic *Prochlorococcus* (Chisholm et al. 1988).

There is a multitude of aerobic ferrihydrite producing bacteria. Some facultative anaerobic bacteria reduce ferric iron when oxygen is not available. Recently, iron oxidation has been found in anaerobic environments. That is specific for magnetite producing *Geobacter* discovered by D. Lovley. There is also a non-specific magnetite formation in alkaline medium by primary anaerobes which support E_h-pH field of magnetite stability. There is anaerobic photooxidation of ferrous iron by *Rhodomicrobium*: phototrophic counterpart of the well known aerobic iron accumulating *Pedomicrobium* (Widdel et al. 1993; Ehrenreich and Widdel this volume). But all these findings indicate on A's which are sufficient for diagenetic transformations of iron-bearing sediments, but not on B - the community which accumulates these deposits.

There are a number of questions which should be answered concerning such B relevant to C-system. The question is: what is the deposition site which allows to form thin layered sediments of such a scale? It was most probably undisturbed deep water. But deep water excludes benthic photosynthesis because of light penetration and turbidity. There is a lot of silica which indicates alkalinity consistent to magnetite stability. But there is no indication of massive sulfur species deposition which should occur in sulfate bearing oceanic water in the presence of organic matter.

All these indicate on large epicontinental water bodies with alkaline water in the vicinity of volcanic activity up-the-stream as a possibility. Something similar was suggested by V.I. Lazurenko (1989) for Ukrainian shield and deposits when he insisted on bacterially induced diagenesis.

"Hot-spots" of biodiversity origin

Relict communities in extreme environments by definition are the blind-alleys for evolution since they may have persisted largely unchanged. By the same reasoning, sites of biodiversity origin should be occupied now by plants and designated as "normal" conditions for life. As an indication of original cyanobacterial communities we can observe ephemeral cyanobacterial mats on ground in the humid belt or cyanobacterial communities in the arid zone of Sahara-Gobi botanical province (Novichkova-Ivanova 1980).

But there is one predominantly cyanobacterial community in extreme environment which to a certain extent might be relevant to centers of prokaryotic origin. These would be continental lacustrine environments in rift zone. In the African Rift there is a variety of lakes, with some of them being quite alkaline. Such lakes existed more than 2.3 Ga ago. Arguments for extensive studies of such environments were suggested (Zavarzin 1993). In contrast to zoologists who claim that "life comes from the sea" there is a possibility to investigate the statement that "life comes from continents". To support this thesis one should demonstrate that in alkaline soda lakes characteristic for terrestrial origin, exists an autonomous prokaryotic community represented by most of prokaryotic lineages. There is active cyanobacterial production and no accumulation of organic matter indicating complete destruction: local consumers such as flamingo's could not have eaten all the biomass. However, in spite of the fact that some alkaliphilic prokaryotes like *Natronococcus* or *Ectothiorhodospira* were extensively studied, there is no general picture of the alkaliphilic prokaryotic community. A survey of cyanobacterial life in African lakes was made by French researchers. But the first comprehensive study of microbial communities in continental lakes was done by B.I. Issatschenko in the 1920's for the present Kazakhstan lakes.

We have been able to demonstrate representatives of many functional groups of the anaerobic destructive pathway at pH > 10. There is experimental evidence that the alkaliphilic community constitutes a complete food web. However, by now we can not talk about phylogenetic affiliage of new alkaliphilic homoacetogens, sulfate reducers, cellulose decomposers, primary anaerobes.

Conclusion

The cyano-bacterial mat is the first prokaryotic community available for studying the post-biochemical period of development in microbiology. It is also the best indicator of the persistence and continuity of microbial life during Earth's history and is relevant to considerations of central problems in general biology - including the changing paradigms associated with the non-Darwinian domain in evolution.

References

Chisholm SW, Olson RJ, Zettler ER, Goericke R, Waterbury JB, Welschmeyer NA (1988) A novel free-living prochlorophyte abundant in the oceanic euphotic zone. Nature 334:340

Lasurenko VI (1989) Geological activity of iron bacteria Naukova Dumka Kiev

Metz JAJ, Nisbet RM, Geritz SAH (1992) How should we define "fitness" for general ecological scenarios? Trends Ecol Evol 7:198-202

Novichkova-Ivanova LN (1980) Soil algae in phytocoenoses of Saharo-Gobi desert region. Nauka Leningrad

O'Neil RV, DeAngelis DL, Waide JB, Allen TFH (1986) A hierarchical concept of ecosystems. Princeton, New Jersey

Revsbech NP, Jørgensen BB (1986) Microelectrodes: their use in microbial ecology. Adv Microbial Ecol 9:293-352

Rippka R, Deruelles J, Waterbury JB, Herdman M, Stanier RY (1979) Generic assigments, strain histories and properties of pure cultures of cyanobacteria. J Gen Microbiol 111:1-61

Schopf JW (ed) (1983) Earth earliest biosphere. Princeton Univ Press New Jarsey

Schopf JW (1993) Microfossils of the Early Archean Apex chert: new evidence of the antiquity of life. Science 260:640-646

Towe KM (1990) Aerobic respiration in the Archaean. Nature 348:54-56

Veldkamp H (1985) Microbial interactions in nature and in the laboratory. Anthonie van Leeuwenhoek 51:457-472

Vernadsky VI (1926) La biosphere. Alcan Press, Paris 234 pp.

Ward DM, Bateson MM, Weller R, Ruff-Roberts AL (1992) Ribosomal RNA analysis as they occur in nature. Adv Microbial Ecol 12:219-286

Walter MR (1983) Archean stromatolites: evidence of the Earth's earliest benthos. In: Schopf JW (ed) The Earth's earliest biosphere. Princeton Univ Press, New Jersey

Widdel F, Schnell S, Heising S, Ehrenreich A, Assmus B, Schink B (1993) Ferrous iron oxidation by anoxygenic phototrophic bacteria. Nature 362:834-836

Winogradsky SN (1952) Sur la classification des bacteries. Ann Inst Pasteur 82:125-202
Winogradsky SN (1897) On the role of microbes in the general turnover of life. Separate issue reprinted in: Ivanov MV (ed) (1989) Chemosynthesis Nauka Moscow
Zavarzin GA (1984) Bacteria and the composition of atmosphere. Nauka Moscow
Zavarzin GA (1989) Microbial community in the past and the present. Microbiol Zhurn 51:3-13
Zavarzin GA (1992) Biodiversity and sustainability of microbial community. Zhurn Obsch Biol 53:92-106
Zavarzin GA (1993) Epicontinental water bodies as relict biotopes for the development of terrestrial biota. Microbiol 62:789-800
Zavarzin GA, Nozhevnikova AN (1993) Landfills and waste deposits producing greenhouse gases. Current Opinion Biotechnol 4:356-362
Zavarzin GA, Stackebrandt E, Murray RGE (1991) A correlation of phylogenetic diversity in the Proteobacteria with the influence of ecological forces. Can J Microbiol 37:1-6
Zhilina TN, Zavarzin GA (1991) Anaerobic bacteria participating in organic matter destruction in halophilic cyanobacterial community. Zhurn Obsch Biol 52:302-318

Subject index

NATO ASI Series G

Vol. 1: **Numerical Taxonomy.** Edited by J. Felsenstein. 644 pages. 1983. (out of print)

Vol. 2: **Immunotoxicology.** Edited by P. W. Mullen. 161 pages. 1984.

Vol. 3: **In Vitro Effects of Mineral Dusts.** Edited by E. G. Beck and J. Bignon. 548 pages. 1985.

Vol. 4: **Environmental Impact Assessment, Technology Assessment, and Risk Analysis.** Edited by V. T. Covello, J. L. Mumpower, P. J. M. Stallen, and V. R. R. Uppuluri. 1068 pages.1985.

Vol. 5: **GeneticDifferentiationand Dispersal in Plants.** Edited by P. Jacquard, G. Heim, and J. Antonovics. 452 pages. 1985.

Vol. 6: **Chemistry of Multiphase Atmospheric Systems.** Edited by W. Jaeschke. 773 pages. 1986.

Vol. 7: **The Role of Freshwater Outflow in Coastal Marine Ecosystems.** Edited by S. Skreslet. 453 pages. 1986.

Vol. 8: **Stratospheric Ozone Reduction, Solar Ultraviolet Radiation and Plant Life.** Edited by R. C. Worrest and M. M. Caldwell. 374 pages. 1986.

Vol. 9: **Strategies and Advanced Techniques for Marine Pollution Studies: Mediterranean Sea.** Edited by C. S. Giam and H. J.-M. Dou. 475 pages. 1986.

Vol. 10. **Urban Runoff Pollution.** Edited by H. C. Torno, J. Marsalek, and M. Desbordes. 893 pages. 1986.

Vol. 11: **Pest Control: Operations and Systems Analysis in Fruit Fly Management.** Edited by M. Mangel, J. R. Carey, and R. E. Plant. 465 pages. 1986.

Vol. 12: **Mediterranean Marine Avifauna: Population Studies and Conservation.** Edited by MEDMARAVIS and X. Monbailliu. 535 pages. 1986.

Vol. 13: **Taxonomy of Porifera from the N. E. Atlantic and Mediterranean Sea.** Edited by J. Vacelet and N. Boury-Esnault. 332 pages. 1987.

Vol. 14: **Developments in Numerical Ecology.** Edited by P. Legendre and L. Legendre. 585 pages. 1987.

Vol. 15: **Plant Response to Stress. Functional Analysis in Mediterranean Ecosystems.** Edited by J. D. Tenhunen, F. M. Catarino, O. L. Lange, and W. C. Oechel. 668 pages. 1987.

Vol. 16: **Effects of Atmospheric Pollutants on Forests, Wetlands and Agricultural Ecosystems.** Edited by T. C. Hutchinson and K. M. Meema. 652 pages. 1987.

Vol. 17: **Intelligence and Evolutionary Biology.** Edited by H. J. Jerison and I. Jerison. 481 pages. 1988.

Vol. 18: **Safety Assurance for Environmental Introductions of Genetically-Engineered Organisms.** Edited by J. Fiksel and V.T. Covello. 282 pages. 1988.

Vol. 19: **Environmental Stress in Plants. Biochemical and Physiological Mechanisms.** Edited by J. H. Cherry. 369 pages. 1989.

Vol. 20: **Behavioural Mechanisms of Food Selection.** Edited by R. N. Hughes. 886 pages. 1990.

NATO ASI Series G

Vol. 21: **Health Related Effects of Phyllosilicates.**
Edited by J. Bignon. 462 pages.1990.

Vol. 22: **Evolutionary Biogeography of the Marine Algae of the North Atlantic.**
Edited by D. J. Garbary and G. R. South. 439 pages. 1990.

Vol. 23: **Metal Speciation in the Environment.**
Edited by J. A. C . Broekaert, Ş. Güçer, and F. Adams. 655 pages. 1990.

Vol. 24: **Population Biology of Passerine Birds. An Integrated Approach.**
Edited by J. Blondel, A. Gosler, J.-D. Lebreton, and R . McCleery.
513 pages. 1990.

Vol. 25: **Protozoa and Their Role in Marine Processes.**
Edited by P. C. Reid, C. M . Turley, and P. H. Burkill. 516 pages. 1991.

Vol. 26: **Decision Support Systems.**
Edited by D. P Loucks and J. R. da Costa. 592 pages. 1991.

Vol. 27: **Particle Analysis in Oceanography.** Edited by S. Demers. 428 pages. 1991.

Vol. 28: **Seasonal Snowpacks. Processes of Compositional Change.**
Edited by T. D. Davies, M . Tranter, and H . G. Jones. 484 pages. 1991.

Vol. 29: **Water Resources Engineering Risk Assessment.**
Edited by. J. Ganoulis. 551 pages. 1991.

Vol. 30: **Nitrate Contamination. Exposure, Consequence, and Control.**
Edited by I. Bogárdi and R. D. Kuzelka. 532 pages. 1991.

Vol. 31: **Industrial Air Pollution. Assessment and Control.**
Edited by A. Müezzinoğlu and M. L. Williams. 245 pages. 1992.

Vol. 32: **Migration and Fate of Pollutants in Soils and Subsoils.**
Edited by D. Petruzzelli and F. G. Helfferich. 527 pages. 1993.

Vol. 33: **Bivalve Filter Feeders in Estuarine and Coastal Ecosystem Processes.**
Edited by R. F. Dame. 584 pages. 1993.

Vol. 34: **Non-Thermal Plasma Techniques for Pollution Control.**
Edited by B. M. Penetrante and S. E. Schultheis.
Part A: Overview, Fundamentals and Supporting Technologies. 429 pages. 1993.
Part B: Electron Beam and Electrical Discharge Processing. 433 pages. 1993.

Vol. 35: **Microbial Mats. Structure, Development and Environmental Significance.**
Edited by L. J. Stal and P. Caumette. 481 pages. 1994.

FSC
www.fsc.org
MIX
Papier | Fördert
gute Waldnutzung
FSC® C083411

Kolibri 360 GmbH
Ferdinand-Jühlke-Straße 7
99095 Erfurt, Deutschland
produktsicherheit@kolibri360.de